THE CONSTRUCTION MANAGER
IN THE 80'S

King Royer, P.E., has been a carpenter, job superintendent, project manager, or chief engineer for construction firms and public agencies in several states and five countries. He has been a general contractor, is a professional engineer and land surveyor, and has earned several graduate degrees. His previous books are *Desk Book for Construction Superintendents* (Prentice-Hall, Inc.) and *Applied Field Surveying* (John Wiley & Sons, Inc.).

THE CONSTRUCTION MANAGER IN THE 80'S

KING ROYER

Construction Engineer

Prentice-Hall, Inc., *Englewood Cliffs, N.J. 07632*

658.9
R89c

Library of Congress Cataloging in Publication Data

Royer, King.
   The construction manager in the 80's.

   Bibliography: p.
   Includes index.
   1.   Construction industry—United States—Management.
I.   Title.
HD9715.U52R68 1981   624'.068   80-19977
ISBN 0-13-168690-9

Editorial/production supervision
   and interior design by Lori Opre
Manufacturing buyer: Anthony Caruso

Previously published by Prentice-Hall, Inc. as THE CONSTRUCTION MANAGER, © 1974

Printed in the United States of America

10 9 8 7 6 5 4 3 2 1

PRENTICE-HALL INTERNATIONAL, INC., *LONDON*
PRENTICE-HALL OF AUSTRALIA PTY. LIMITED, *SYDNEY*
PRENTICE-HALL OF CANADA, LTD., *TORONTO*
PRENTICE-HALL OF INDIA PRIVATE LIMITED, *NEW DELHI*
PRENTICE-HALL OF JAPAN, INC., *TOKYO*
PRENTICE-HALL OF SOUTHEAST ASIA PTE. LTD., *SINGAPORE*
WHITEHALL BOOKS LIMITED, *WELLINGTON, NEW ZEALAND*

I tell this tale, which is strictly true,
Just by way of convincing you
How very little, since things were made,
Things have altered in the building trade.

From *A Truthful Song*
Rudyard Kipling 1865–1936

CONTENTS

Chapter 5
PURCHASING AND SUBCONTRACTING   168

Chapter 6
ACCOUNTING   208

LIST OF FIGURES AND TABLES

CHAPTER 7

CHAPTER 8

PREFACE

My earlier *Desk Book for Construction Superintendents* is directed to new superintendents. It explains procedures followed by construction firms, and is expected to be read by trade foremen who have become or will be superintendents. It sells to the trade and to junior colleges. This book, *The Construction Manager in the 80's* is directed to new project managers, who may previously have been superintendents or estimators. Since college graduates in engineering and construction often become project managers with little experience, this book is appropriate for an undergraduate course in this area. At last count, thirty-four schools have used the previous edition as a text. It is descriptive of the responsibility of the person initiating estimates, purchases, and cost control procedures. Small contractors do this personally; in medium-size firms a project manager is designated, and in large companies, the project manager is assisted by accounting, estimating, scheduling, or other departments. This description may serve as a definition of small, medium, and large firms.

The term *construction manager* has come into use in recent years to mean a consultant firm that performs a variety of management or consulting functions for the owner, and some have considered the term confined to that use. A construction manager to anyone in the business is the person who ultimately makes management decisions, and it is used in this sense here. The title antedates the oil price increases and consequent importance of construction management firms in the Near East.

It is intended to present the options popular in the industry(and by this is meant the industry of building, not of design), but my own preferences naturally take precedence. Contractors do not exchange information on their methods freely, nor are there standardized details of operation. For example, some firms with a gross of several million dollars a year do not use written purchase orders, and other similar firms prepare a dozen copies. Some contractors consider methods shown here as prohibitively expensive because of the detail work involved; others consider them too simple to provide full control. Idle speculations and theoretical preferences have usually been avoided; even the most peculiar methods are or have been in use by a firm of my acquaintance.

This book is for a construction management course in engineering, architectural or business school (although the only architectural school using it, to my knowledge, has a course taught by an active project manager in construction), and is useful for any person, including construction superintendents, engineers, or architects, who intend to manage or invest in a construction business. It is an aid to accountants for construction firms. The necessity for such management practices as contract and purchase order negotiations, bid review, cost reports, and scheduling, is explained in detail. Construction methods and technical review of estimating are not included, to make room for extended explanation of topics more directly related to the receipt and spending of money. A certain amount of general knowledge is assumed of the reader, although no great depth of knowledge in any academic area is required. Elementary courses in accounting, business law, or construction estimating and methods are valuable prerequisites or corequisites to this text.

Topics included here but not previously treated in the literature, are detailed explanations of contracts and subcontracts, cash requirement, union practices, and arbitration. Cost controls are presented as the heart of the construction business. The annotated bibliography is the most detailed now in print, of books pertinent to construction management.

This text may be combined with my *Desk Book for Construction Superintendents* for a practical course, with the *AGC Cost Control and CPM in Construction* for planning orientation, or with William E. Coombs' *Construction Account and Financial Management* for a construction business course.

The questions for discussion at the end of each chapter lead far beyond the content of the text.

Readers' corrections and suggestions are invited.

Many construction terms are rarely written and use of them varies. A bricklayer may *twig* a line, although the dictionary says trig. *Off-balance* may indicate a bidding method in heavy construction, or payment accounting to a building contractor. The jackhammer drill has been corrupted in popular speech to apply *jackhammer* to a paving breaker, and has returned to be used this way by many construction workers. *Bulldozer* is used to refer to any tracked vehicle, regardless of whether a dozer blade is installed. The vocabulary used varies with the area, type of construction, and size of the firm. Any such terminology in this text may be but one of several common usages.

King Royer

1237 S.W. 9th Road
Gainesville, Florida 32601

CHAPTER *1*

THE CONSTRUCTION BUSINESS

The construction business furnishes capital improvements for the nation. *Capital improvements* include nearly all of what an economist calls *investment,* as opposed to *expenses,* which provide only immediate, not future, benefit. Investments finance structures representing money spent for future production (factories) or future use (housing). About 10 percent of the gross national product is spent on construction, with the same percentage of the working force being engaged in it.

Since the construction industry primarily represents investment, construction drops more than other industries during recessions. Construction almost vanished during the depression of the 1930s, and again during the war period of 1944–1945; it has dropped severely in a number of other instances. A large number of construction workers change to other types of jobs and back again as work load fluctuates. As a consequence, there is less security in construction than in many other industries, but wages and salaries must be higher in order to regain workers previously lost to other industries. Also, the individual nature of the work done, in which the output of one person can be identified, makes skill more immediately rewarded and those with lack of skill more readily dismissed to go into other work. Those of us in the industry therefore feel that we are individually more important and collectively more efficient and skilled than persons in any other field of endeavor.

1.1 ELEMENTS OF THE CONSTRUCTION INDUSTRY

Construction requires designers, such as architects and engineers, who prepare plans for construction; engineering surveyors, who make surveys of elevations and drainage of sites; and contractors to do the field work. There are manufacturing firms to furnish materials, and fabricating plants to furnish specially fabricated parts. A considerable construction equipment industry furnishes and maintains equipment for the industry.

Of these elements, this text focuses on the contractor, who comes in contact with all the others. Designers are employed by independent architect–engineer firms, by contractors who both design and build construction work, by public agencies for their own construction, and by private owners, usually manufacturing firms, who build highly specialized plants.

Of those people who work for the contractor, the project manager deals with the greatest variety of firms and people. The project manager, who receives his authority as *all he can take,* has the authority of a contractor and the detail work required of the superintendent. This text deals with his duties and presents an introduction to the business aspects of the industry.

1.2   THE CONSTRUCTION INDIVIDUAL

On a national or even a local scale, the construction industry is not organized. Since business organization presupposes a limited market, the result of organization has usually been higher prices. In home building in 1977, nearly all aspects had become organized except site labor; consequently, site labor had dropped under 20 percent of nominal selling price. In terms of the homebuyer's eventual dollar paid, including interest, the construction worker received less than 7 cents!

The average firm in the construction industry employs about three workers. By all modern industrial management logic, such firms cannot exist, much less be efficient; yet these firms are the backbone of the industry. It is easy to become a businessman, and equally easy to fail. One must apply construction reasoning to construction firms; this reasoning has long been that if workers have adequate incentive (which they call money), other considerations are of little importance. Of course, these workers are more technically qualified, as the work requires it. The contractor is neither ignorant nor stupid; it is no accident that Frank Gilbreth, an early methods study consultant for industries, was first a contractor.

1.3   FORMS OF BUSINESS OWNERSHIP

The *individual proprietorship* is the form most construction contractors use, but it is limited to small firms. In this form, the owner and his business are not separate. The owner is directly responsible for all business debts, and all income is personal. This arrangement is simple—no papers are needed, and only a single income tax return is necessary.

Partnerships are combined individual proprietorships. This form enables two owners to combine their capital and talents, with no papers except a contract between themselves, any way they want to do it.

Disadvantages of a partnership are:

1. As each partner can obligate the partnership, the partners must have full confidence in each other.
2. If one partner dies, the partnership no longer exists.
3. A properly made partnership agreement must be prepared by an attorney, is complicated, and as compared with a corporate charter, is more difficult to change.

Partnerships are most successful when the partners are closely related, or man and wife. The man-and-wife arrangement must contend with the possibility of divorce, but this contingency exists for any business enterprise owned by a married couple. The advantages of a partnership can be obtained by a corporation, plus the benefits of incorporation.

In *limited partnerships,* some of the partners may avoid liability for business debts while maintaining the partnership form.

Corporations are almost universally used by every firm with over three or four employees. There is a cost of incorporation, although this is quite modest—from $75 to $600, depending on whether an attorney is used, and if he is obligated to a minimum-fee agreement with the local bar (attorney's) association. Such minimum-fee agreements are illegal, but this does not prevent their operation by lawyers.

A corporation may be owned by one person, or state laws may require initial incorporation by three persons, and may require more than one director and corporate officer.

The advantages of incorporation include the following:

1. Income tax avoidance. There are many ways to handle income taxes in the form of options by the owners. Small corporations may be free of income tax by passing all income to the owner, who pays taxes as if he were a proprietor. Other options may reduce tax rate paid on the income. Like all tax avoidance methods, tax is eventually paid, although it may be years later.

2. Owners of a corporation are not responsible for business debts of the corporation if it should fail. This protection is not important for small corporations who do public work; the surety who must guarantee payment will require that the owner also be responsible for debt. However, investors can more easily be persuaded to invest if they are not responsible for debts beyond their investment, and if their shares are salable.

3. Certain benefits, some tax-free, are available to the owners of a corporation if they are also furnished to other employees. These include health and life insurance, unemployment insurance, and sometimes pension plans.

4. Owners of the corporation may remain anonymous.

5. By using a separate corporation for each development, some tax advantages may be obtained.

6. The death of the owner need not terminate the operation of the firm. Sureties (bonding companies) who guarantee a contract try to avoid a situation where a death can cause them serious losses.

Since Delaware has a more liberal law on incorporation than do most other states, a disproportionate number of corporations are incorporated there. Among larger corporations, 38 percent of corporations listed on the New York Stock Exchange and 33 percent of those on the American Stock Exchange are Delaware-based. Information is available from Enterprise Publishers Company, 1000 Oakfield Lane, Wilmington, Delaware 13810, on low-cost incorporation in that state.

1.4 MANAGEMENT INFORMATION SYSTEMS

During the past 30 years, the former accounting departments of large firms are realizing the possibilities of electronic computers. They are growing from a service department, which determines if operations are profitable, to a separate division which directs the firm through the information it furnishes or fails to furnish. This division may be called ADP (automatic data processing), MIS (management information systems), or a variety of other names, sometimes rather derisively the "number crunchers" or "bean counters." The manager of this department, because of the complexity of the system he operates, must be a computer specialist. He usually is responsible to ("reports to" is the more common term) the manager (president) of the firm. Many company presidents are former MIS directors, to the disadvantage of their knowledge of the actual operation of the business, as compared to the operation as reported by the data processor.

As applied to construction, this author has found this system unreliable in its reporting and too late in its report lag time. The person receiving such reports is usually unable to use them for action, as he is too far from the work, and lower-level supervisors who could take action are not given this information in a manner they can use it.

Theoretically, there is no reason why this should be so, and there are many people—usually not working managers—who can explain its proper use. At present, construction organizations must become decentralized before they become large enough to profit from complicated systems, and the reliable operation of such smaller units requires additional managers with long experience, much of it in the field. Attempts to make managers of people without field experience (at least in job offices) has generally failed. Any student who plans to be a general manager should work in the field as long as possible, both before and after he obtains a college degree. An educated competent field man has to resist office assignments rather than seek them, if he is to remain in the field.

1.5   MANAGEMENT FUNCTIONS

The functions of a manager (or in more general terms a member of "management," a rather vague term) are considered to be

1. Organizing
2. Staffing
3. Directing
4. Planning
5. Controlling

These functions are partly well established in construction and each supervisory level is responsible, within fairly well established limits, for the lower levels.

The job superintendent is responsible for these functions at the job level. Organizing his subordinates is his responsibility. His staffing may be restricted by being obliged to first use of available men from other jobs. His directing is limited by subcontract requirements. His planning does not usually include estimating or purchasing. His controlling is subject to operations he can see each day, and dependent on his own judgment; cost controls are seldom provided him in sufficient detail and in time to be useful to him.

The other functions are split up among project managers and corporate officers in a variety of ways; only the top 5,000 or so contractors have enough office people to provide an option in their organization.

Contractors are often criticized for their failure to use modern business methods, and to place primary reliability on specific technical experience for executives. Some of this is justified. However, the Levitt & Sons story illustrates what inexperienced managers can do to a construction firm. In the days following World War II, Levitt was one of the firms with the greatest application of new ideas in the construction business. It was sold to a conglomerate, International Telephone & Telegraph. Its 1974 loss was $20 million on revenues of $98 million; the trustee noted the fixed cost (overhead) rate to be 25 percent, more than twice appropriate levels and noted an almost *total lack of management experienced in the home-building industry.* Fewer than half of the 12 regional managers had direct home-building experience before taking these jobs.[1] This firm was under the control of one of the largest international firms in the world. They had emphasized "managing" almost to the exclusion of "doing."

Management specialists, like other specialists, have a proper place in large construction firms, but they must not be allowed to fill all management positions with persons like themselves. If technical nepotism of managers is unavoidable, one must choose a manager for his ability to choose subordinates with experience least like his own.

Larger firms follow some modification of *functional, matrix,* or *project* organizations.

The *functional* organization is illustrated in Figure 1-1. This type is most similar to that of a manufacturing company, and since the "number crunchers" or MIS people exercise the greatest power without responsibility in this arrangement, they actively promote it. It is based on the assumption that each employee can operate most efficiently if he is doing the type of work he can most efficiently handle. This is not easy, since a person's capabilities are constant and the work to be done varies both in type and volume. But there is no doubt that similar workers grouped together can operate more efficiently, by transfer of work load from one to another, than the same workers can do if divided into small groups, each supervised separately and located in separate locations.

Let us follow a job through this type of organization. The first contact is the estimating (sales) department, which prepares an estimate and proposal. If the proposal is accepted and a contract signed, the estimating department prepares budgets for MIS, and Design prepares plans and perhaps material lists. The lists go to Purchasing, who buys the materials, Personnel hires people, and Operations (the construction department) starts to work. All are feeding in-

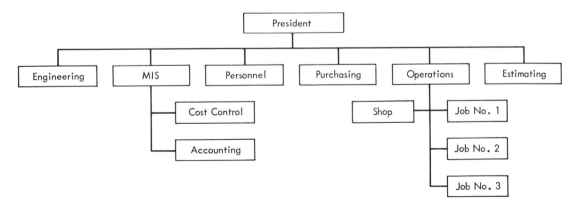

Figure 1-1
FUNCTIONAL ORGANIZATION

[1] *Wall Street Journal,* April 16, 1975, p. 12.

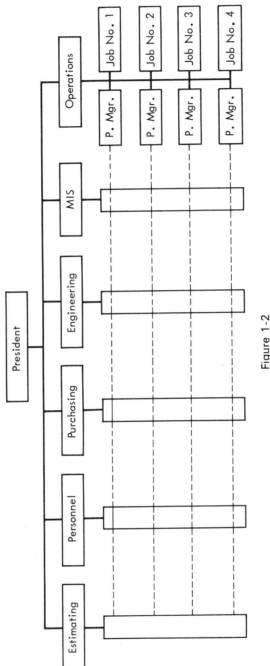

Figure 1-2
MATRIX ORGANIZATION

formation on expenses, estimates, money committed (undelivered purchases), operations schedules, and purchased items to MIS, who furnishes reports to the managers. Accounting pays bills and prepares paychecks. Each such functional department—responsible for a certain function, rather than a certain site—processes papers and passes them to the next department. Each department works on each job each day.

What is wrong with this? It does not take priorities into account. How do people in each department decide what to do first, so that important items do not get lost among the routine papers? No matter how it is done, there will be delays which represent job delays and waste. So a project manager is added.

This type of operation requires that each unit, which may be one person, has two supervisors; his functional or technical supervisor, who tells him how to do things, and the project manager, who tells him when to do it. Operation is the same as before, but now each worker can consult a project manager to determine what is important, and the project manager follows his work through the network so that important matters will get prompt action. To eliminate conflicts between project managers, there may be a supervisory project manager (sometimes called the contracts manager) to set priorities. In some ambitious organizations, the MIS department reduces all actions to paper and monitors their status, reducing the responsibilities and conflict of the project managers.

This *matrix* organization, as shown in Figure 1-2, has been well established for decades in the U.S. Navy. The "project managers" are military line officers—a captain of a ship, for example. The "functional managers" are technical bureaus in Washington. Each man is assigned two chains of command. A captain of a ship may not install another gun on deck, by technical instructions of the Bureau of Ordinance and Bureau of Ships. But if he does so, he must report the reasons for it. A gunner has technical instructions on how to fire a gun, but operational instructions on who to shoot at and when. If the captain tells him to remove the guard which prevents him from hitting his own conning tower, although technical instructions forbid its removal, he does as the captain says.

Most A/E (architectural–engineering) firms operate on a matrix organization, and the contractor calls the project manager, who contacts the people involved, about any technical point.

However, when a job gets large enough, following the work through many departments daily becomes an impossible task. The comparative simplicity of the support (office) functions and complexity of the field work requires close support even at the expense of office worker efficiency. An organization is then formed that looks like the one shown in Figure 1-3.

In Figure 1-3, the MIS department has been changed to Accounting, to indicate the reduced importance of this type of work in project management. It is apparent that the functional organization of Figure 1-1 has reappeared in triplicate, which is the case. The project type of organization is therefore called *decentralized.* If the "project manager" is a "regional manager," this diagram represents a national firm with regional offices. If the projects grow to separate corporations, the diagram represents a conglomerate (firm made up of many corporations or separate business) and its affiliate parts. As a matter of practice, few firms decentralize their management reporting systems or their accounting.

The large project has the same faults as the functional organization, so some form of the matrix organization becomes efficient. The most efficient operation is usually by a mixture of organizations each properly suited to its task. An executive is seldom qualified or disposed to set up such an organization, but it is done by economic conditions and has become the modern construction industry, which is shaped by wholesale failure of firms who do not operate efficiently at what they try to do. In large firms, one hears "we can't do small jobs" or "we can't

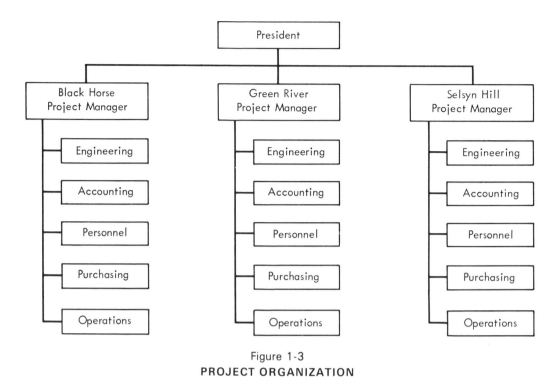

Figure 1-3
PROJECT ORGANIZATION

do electrical work economically." This means that in a small firm management is lacking for this type of work, or in a large one the managers are not experienced in a particular type of work and are not inclined to delegate work for that reason. Many firms "aren't set up" to do work of one type or another, not because they do not have the people, but because it would require a relinquishment of authority by some of the managers. The managers prefer to give authority to someone who guarantees his own efficiency—that is, a subcontractor.

1.6   THE OWNER

The smallest owner is the homeowner who wants a new furnace in his house; the largest, national governments and international corporations. Owners have in common that they are buying an improvement about which they know only the results to be obtained. Modern business organizations standardize their people and their methods for a particular product. Firms who build water pumps cannot efficiently design pump factories, and factory designers cannot build the factories they design.

These design and construction functions are sometimes combined, but when this is done the department which gains control usually emphasizes its own operations until the business does little work in other departments. Few firms do several functions well.

Practically every business firm is an owner of construction work. Some developers of apartment buildings build, own, and operate the property after construction; practically no other firms do so. Government often combines ownership and design, and may also engage designers. They very rarely engage in construction.

1.7   THE CONSTRUCTION PROCESS

Typically, a person or firm, the *owner*, decides that a capital investment is justified on the basis of future demand for a product the owner provides. The firm must have a portion of the cost, termed the *down payment*, *equity*, or *front money*. Additional money or *financing* may be obtained from a variety of sources, such as *commercial banks* (banks with checking deposits), *savings and loan associations, mortgage brokers,* or *insurance companies.*

Plans and specifications are prepared by *architects* and several specialties of *engineers,* either employed by the owner in his own organization, or by a separate consulting firm who prepares plans and supervises construction for a fee, under contract. For large jobs (usually above $20 million) a separate *construction management* firm may be employed. The duties of the construction manager are not well established by custom or law; he may be contracted merely to advise the owner, or his duties may be extended to act as an agent for design and construction contracts.

The construction work, consisting of purchasing materials, hiring workers, and installing the materials is done by a *general contractor,* who may *subcontract* parts of the work to others. The relationship is generally illustrated in Figure 1-4.

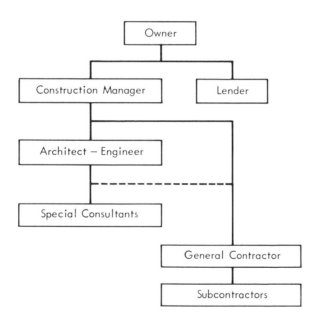

Figure 1-4
ELEMENTS OF THE CONSTRUCTION INDUSTRY

1.8   CONSTRUCTION MANAGEMENT FIRMS

A corporate owner often does not have a person to represent the firm in construction transactions, either because of the owner's lack of qualifications or of adequate time to qualify a person. If he does, the owner may employ a person in his own organization, designated *project manager, construction manager,* or *construction engineer.*

Architects have professed to represent the owner in such matters. Engineers have done the same, but firms employing consulting engineers are usually well qualified technically and have no problem in explaining their requirements. However, firms offering both design and construction services can build while design is under way, offering substantial time advantages as compared with separate design and construction contracts. Until recently, architect-directed firms were prevented from also doing construction work by their own code of ethics.

Construction management firms furnish the service which was previously provided by the owner's own people, by the architect-engineer (A/E), or by the design-build contractor. The construction manager may act as the owner's agent in letting contracts and inspecting the work, or perform only advisory work, approving documents for the owner's acceptance.

One of the largest construction managers was the U.S. government itself, when the U.S. Corps of Engineers assumed a construction management role in the 1970s for military construction in Saudi Arabia. Several other construction management firms staffed by retired Corps of Engineers officers also supervised considerable work in the Middle East in this capacity.

1.9   ARCHITECT–ENGINEERS

Architects are licensed professionals educated and experienced in design of artistic elements, layout, finish materials, and structural design of buildings. Architects normally employ structural, electrical, and mechanical engineers, either on salary or by contract for these features of buildings. *Engineers* are licensed professionals who design engineering elements of buildings, as well as highways, bridges, and a wide variety of specialized construction. Architectural and engineering firms are licensed as such, separately from each other. Each may subcontract design work from the other, although architects have a strong disinclination to act as sub-contract designers.

Design firms operated by both architects and engineers are *architect–engineers,* abbreviated A/E. In this text, A/E refers to architects, engineers, or a combination of the two.

1.10   FINANCIAL INSTITUTIONS

Permanent financing—that is, loans designated to be paid back out of the income of the property—is often furnished by firms who will make the loan only when the project is finished. The owner must therefore obtain *construction financing* from a firm specializing in short-term loans, to pay bills during the construction period. Commercial banks are limited by state and national laws in the type and length of loans they can make; they usually have short-term loans available, but can make few long-term mortgages.

A number of institutions, including savings and loan institutions and the Farmers Home Administration, pay out the permanent loan as construction progresses, so that construction loans are not needed.

1.11 THE DESIGNER

Architects often are interested in proper workmanship to the extent that they neglect the importance of competition. Designation of a brand name increases the cost by notifying the distributor that he has no competition; and stating that substitutes may be acceptable does not undo the damage. Sometimes the architect will allow substitutions only if the contractor notifies him during bid time, so all bidders may bid the substitution! This is shortsighted, since no bidder has any incentive for an economical substitute if his competitors will also get the benefit of the substitution. It is unrealistic to consider the contractor as interested in reducing the cost of the job; his interest is in reducing his cost below that of his competitors.

By inducing an out-of-town bidder to bid a job, the architect obtains a lower price even though the out-of-town bidder is not low. The outside man increases the distribution of the plans to subcontractors, many of whom also give their quotations to the in-town contractors. The low bid is therefore lower because it represents lower subcontract bidders. Architects show little interest in distributing plans to subcontractors, usually allowing them no refund for return of plans; one of the functions of the general contractor is distribution of plans. The overall price of building is established as much or more by the subs as by the general construction estimate.

1.12 THE CONTRACTOR

Construction firms are classified according to the kind of projects they construct. The AGC (Associated General Contractors) classifies contractors as building, highway, heavy, industrial, municipal utilities, railroad, and foreign. Many contractors, including nearly all of the larger ones, engage in more than one of these classifications of work.

Building contractors need less working capital and equipment than other contractors with the same gross business, as their work is predominantly hand labor, and most of this is sublet to specialty contractors. Some contractors sublet all or nearly all of their work, and are referred to as *brokers,* not really contractors. Federal, and sometimes state, regulations, set a minimum percentage of work which must be done by forces of the prime contractor—usually about 30 percent. The strength of building contractors lies in their management and estimating ability, particularly in judging the capability of subcontractors.

Highway contractors do excavation and paving. They may do culverts, curbs, and small bridges, or subcontract them to others. Some contractors work only on publicly owned jobs, refusing subdivision and driveway work, so they may have only one customer—their state highway department.

Heavy contractors build bridges, piers, foundations, marine work, and other specialties characterized by high risk, high capital requirements, and high equipment costs. The managers are heavily engineering- and equipment-oriented.

Industrial contractors specialize in manufacturing plants. They sublet a small proportion of their work, and managers are heavily experienced in technical specialties such as paper mills or refineries. The work force contains a large proportion of millwrights, pipefitters, and welders; some industrial contractors do their own electrical work.

Municipal utility contractors specialize in water and sewer pipelines, and treatment plants.

Like heavy contractors, their equipment cost is high. They usually will do private work when opportunity offers.

Railroad contractors' work now consists primarily of renovation or roadway improvement. There are many small contractors in this field who depend on subcontract work for spurs incidental to industrial construction.

Foreign contractors, in the United States, refers to U.S. contractors who work in other countries. They may specialize as do domestic contractors, but do not have the subcontractors available and cannot economically construct small jobs. Foreign work by U.S. firms is almost entirely in undeveloped countries, as the industrial countries have their own very competent firms. Consequently, foreign construction requires managers who can adequately staff the jobs and purchase all tools and materials. The lack of tool and material sources, as well as the lack of skilled labor, requires management skills not necessary for domestic work. With these skills, the international firm undertakes nearly any kind of work with small changes in its own organization.

1.13 CONSTRUCTION METHODS

A *construction method* may be:

1. The *method of accomplishing the objective* of the users, for example, an architect, may consider a brick wall as compared with a wood-frame wall. In this sense the wall is used by an architect as the method of obtaining the required result—to enclose a space. This is the way in which architects speak of construction methods in design.

2. The *method of doing the work,* as used by contractors. A contractor considers building by shoved brick (putting mortar on the wall) as compared with buttered brick (putting the mortar on the brick). Digging a trench by hand as compared with digging by machine are alternative construction methods to a contractor.

3. The work *methods,* as used by time-and-motion-study engineers. This choice of methods refers to the way men are used to do the work, as whether a bricklayer picks up one brick at a time or two. A choice is made between assigning workers to pull screeds (guides for leveling the concrete surface) continuously, or using workers occasionally, taking them from other work. There is little formal analysis, that is, by industrial engineers, of construction methods. This is done by the foremen and superintendents on the basis of their own experience. To the extent that the work is within past experience, this is efficient but requires highly experienced people. New kinds of work require formal analysis. Knowledge of construction methods in the contractor's sense is the primary attribute of construction managers, from foremen to project managers, although this knowledge may not be essential for positions in which the principal skill is to pick subordinates.

Construction management methods are the methods of collecting information and distributing directions, and as such are clerical procedures. A construction manager makes decisions based on information obtained by management information methods, and enforces his decision by a management method. The management method is the way a construction work method is determined and brought about, theoretically; in practice, management methods are not organized or dependable, in general, in the construction industry.

1.14   THE COLLEGE GRADUATE IN CONSTRUCTION

The student with no industrial experience wants to know: *How do I get a job, and what kind of job might I get?*

The engineering, construction, or architectural graduate each has a different skill useful to the contractor, such as estimating or surveying. Some contractors require draftsmen, but there are not many such openings. Project managers spend a great deal of time in obtaining approvals and submitting materials which comply with specifications. In this work they deal with architects, and the knowledge that an architect requires of materials is essential. In college, however, the comparison of actual available products with specifications is taught very little if at all, so the recent graduate has no advantage in this regard. Most contractors expect only that college graduates have a facility to learn, and many doubt the usefulness of a college education to a contractor at all.

The first job you are likely to get may be as a quantity surveyor, who determines quantities of materials in a job. In this country, this is a "takeoff man" and is done by a less-experienced estimator, or by an estimator who also makes cost estimates. In British countries, the quantity surveyor is a profession in itself, and is part of the A/E organization. In an estimating position, you will learn procedures, interpretation of plans, and will gain a great deal of experience with different kinds of plans and projects. "*Plan reading*" includes wide differences in skill; you will never stop learning new types of drawings. The estimator often goes into a project manager type of position, but it will be many years before he is the kind of project manager in complete charge of a job, unless he also acquires field experience.

Field work *is* construction; any other work for a contractor is at the edges of the construction management profession, not in it. The recent graduate is best suited for a position as engineering assistant (in the older sense of engineers as a surveyor), for layout, drawings, interpretation of plans, cost reporting, progress reporting, and anything else he can do for the superintendent. On larger jobs, this work may require a jobsite department, and the office engineer in charge has a responsible position. On small jobs, the new worker acquires valuable experience both on drawings and office procedures, as well as on construction methods.

Only a few of the larger contractors send interviewers to universities, and these will be firms with most specialized positions. Some contractors send requests to your department head or university placement office; apply for any in which you may have a chance. Most small and medium-size contractors (a medium-size contractor may gross $10 million a year) have no definite idea of what a new man will do either immediately or in the future. Consider this an advantage rather than a drawback; if you want to show what you can do, you can accomplish this sooner in a flexible organization.

If you are interested in particular area, make up a mailing list of contractors who were awarded recent contracts, from back issues of the *Engineering News-Record*. This publication is the Bible of the construction industry, and is where you will find the greatest number of help wanted ads by firms throughout the country. Another good source of prospective employers is the list of AGC (Associated General Contractors of America) members in the area, from the annual directory issue of the *Constructor*. You need only one employer, but may have to send hundreds of resumes to find him. If you want to stay in one city, get your mailing list from the local chapter of the AGC and the members' list and lists of contractors bidding local work at your local Builder's Exchange or Dodge room. See "Construction Reports," "Contractors," and "Builders" listings from the telephone directory.

Give your telephone number in all cases. If a contractor responds by sending you an application blank, it is a weak gesture and indicates little interest. A letter response is a little better, but if a contractor really wants to hire you, he will tell his secretary "Ask this man to come see me" and give her your letter. No one but the manager usually knows what he has in mind. Often he will interview you for a project which has not yet begun; perhaps he has bid it and wants to cover himself on future needs. This is particularly true of contractors interviewing superintendents and project managers. If a message requires travel but does not say so, it usually means that the contractor will pay travel expenses, but it is well to ask. Travel expenses include mileage or taxi to the airport at both ends, parking, fares, and meals. Many contractors will pay expenses in cash at the time of interview if you have the information and receipts with you, but most of them ask you to submit an invoice. Even new college graduates may travel thousands of miles for an interview. New graduates are seldom considered for foreign work. Managers interviewed for foreign work often go to the work site for an interview, and some firms make a policy of inviting—even requiring—that the spouse also see living conditions.

You will usually be interviewed by the person who hires you and sets your salary. If he introduces you to his superior, it may be only a walk-in, which indicates that you are being actively considered. If he makes no offer, it does little good to push him. By the apparent stature of the person in the firm and the amount of time he gives you, you can get some idea of how many people he is interviewing. A personnel manager may interview a large number of applicants; a line manager will often interview only the best applicant presented to him by the personnel manager, or will hire young people without interview. It is popular among large corporations for a prospective employee to interview several people who may be interested in his work.

Very little is accomplished by interviews; in construction probably only one-fifth of the interviews affect whether the applicant is hired or not. People who are compulsive talkers often interview well. Managers realize that little can be learned in a few minutes, especially of an experienced applicant, but a few people can be eliminated. Most interviewers talk about the job and their own firm.

You do not know what the interviewer may be looking for, so it is usually safe to dress conventionally for applicants, that is, a suit and tie, or at least a tie. If you are being interviewed on a construction job for a field position, clothes adapted for work are indicated; you do not want to give the impression that you are afraid of mud.

A small company will usually look for an exceptional man, as it will not have many college-trained men. A large company is more concerned with your ability to follow than to lead, and may stress academic grades to the exclusion of work experience.

Fringe benefits are fairly well standardized in all industries, for salaried employees. Non-salaried, or daily, employees are usually not paid for days they do not work, regardless of the reason. Fringe benefits include two weeks' vacation a year, group medical and life insurance of which the employer pays a part or all, and a pension plan which is of no value unless you remain for five or ten years—for young men, it is more likely to be ten. "Profit-sharing" plans are usually a type of pension plan designed to make a long career with the employer financially attractive. The usual profit-sharing plan is a pension fund of which employees and employers pay a portion. The employee's payment is refundable if he quits or is discharged, but he acquires no pension rights for a considerable period. The details of these plans are usually in language that does not say a great deal. Many employees therefore acquire no pension rights at all during their lifetime, as they change employers too often. Some concerns have annual bonuses based on profits—the prospective employee should inquire what type of bonus payments the firm has, and how much was paid out during the previous year. Firms working in Latin America are usually oriented to sizable Christmas bonuses, as they are required by law in many countries.

1.15   CONSTRUCTION MANAGEMENT

Construction management as a profession rather than a business—that is, for salaried people rather than for firms—is the organization and direction of men, materials, and equipment to accomplish the purpose of the designer. This is undoubtedly an essential part of the contractor's business, but the sale of the product is also essential. The contractor must price his work before he starts. Estimating is critical; determining his efficiency before the work is done may be as important as doing it efficiently. The management cycle is an endless chain of estimate–spend–estimate–spend, in which the weakest link determines the profit of the firm.

The project manager who does not estimate work is still deeply concerned with this cycle, as he uses estimates for production cost control.

Management, as further explained in this text, is direction of construction work requiring:

1. *Planning* before work is done.
2. *Direction* of the work while in progress.
3. *Comparison* of results and use of them in the next planning cycle.

Management responsibility does not fall solely on a class of people who are managers. Management responsibility is exercised by each employee to the extent of his authority. In factory work, each individual's task can be planned with considerable accuracy. Construction work, which is done at widely separated places, under varying and unpredictable conditions, and with changing labor forces, requires the greatest self-reliance, experience, and intelligence of the managers at every level, in order to accomplish a reasonable part of the efficiency that is possible under controlled conditions. That is, one can calculate that a worker can do a certain amount of work in a day, by motion study; it may be difficult to obtain 25 percent of this in actual practice.

Management is distinguished from direction in that people may be directed, sometimes very capably, by persons who have a very limited planning ability. A foreman or a journeyman must plan the number of workers and tools assigned to a job; but if he does not think, plan, and observe in terms of *time* as well as excellence and efficiency, he is not a manager. *The time sense is crucial.*

Technical supervision, which may be engineering or architectural, requires an appreciation of what is necessary in the completed product and how it is affected by the method of construction. It is not management, although a high degree of skill is required.

1.16   TIME SENSE

A manager must have a sense of time, as expressed in job progress, possibly subsequent delays, and cost. This is the basic skill for estimating and scheduling, which are different uses of the same process; estimating for cost, scheduling for time. Time can usually be calculated in terms of cost, and academic planners have quite sophisticated techniques in which they attempt to do this in complicated situations.

This time sense must be consciously learned by trial and error. It is not necessary for a journeyman or foreman to have it. A superintendent on a small job may have a limited requirement for timing. These people may have a very wide experience but may not be able to remember the time they have taken, even for jobs they have done a great many times. This is a common reason that contractors fail; they underestimate labor required, underestimate job time required; and the effect is an underestimation of their direct costs, overhead costs, and money requirements.

An estimator learns the *value of labor* items largely by rote, and from value he can *calculate time.* For this reason, and because an estimator is part of the management cycle due to his need for new unit costs, he is often made a project manager. Since advancement on this road to management is more rapid than the path through field supervision, the project manager may be paid less than is the field superintendent.

This is because people with short training periods, who learn faster and have greater exposure to new experiences, can be replaced more easily than can people with long training periods. There are more of the former. Neither office or field men are usually satisfied with the relative pay scales. College training represents a short training period as compared with the time necessary to produce a line manager. Salaries are determined by supply and demand, and particularly by the fact that an exceptional field man can demonstrate his ability directly in dollars and cents. No amount of education or knowledge can replace the ability to direct, plan, and complete one's plan.

In summary, the kinds of direction required for construction are:

1. *Business direction,* typified on a project by the project manager.
2. *Production direction,* or superintendents, by the job superintendent and his subordinates.
3. *Technical direction,* or architectural–engineering supervision, by an A/E representative or inspector. The inspector, in most cases, has no technical knowledge beyond that of the superintendent. He is the policeman of the technical process.

Architects and engineers perform substantially the same design and supervisory services separately or in combined firms. *Design professional* is often used to avoid the cumbersome term architect/engineer. Throughout this book, A/E refers to the person or firm with design or technical supervisory duties. The learned professions employ *professional* to indicate one of their own group. Among the public it indicates a person qualified and experienced for the work he is doing.

The project manager usually is the contractor's representative and the superior of the superintendent; the inspector is an employee of the A/E and of governing bodies. The A/E often designates a project manager in his own office, perhaps a senior designer supervising work on a particular project rather than designing in his own specialty for a large number of projects.

1.17   CONTRACTOR'S ORGANIZATION

A construction contractor typically is an independent businessman who contracts to construct a structure for which he is given detailed plans and instructions (specifications), for a fixed price. The contract price may be negotiated—that is, agreed between the owner and contractor without competition—or it may be bid by several contractors competitively. Instead of a fixed price, the contractor may be paid his costs plus a fixed fee, or a fee which varies with cost. If the fee varies with cost, it may be greater with increased cost, or less with increased cost.

Many contractors become *turn-key* operators, who design as well as construct projects, under a single contract. Turn-key indicates that the owner need do no more than turn a key to occupy the completed project; no planning, payments (until occupation), perhaps no necessity to purchase property.

A construction firm may be described in terms of numbers of office and general supervisory employees, and the responsibility they exercise in three sizes or phases of development. Of course, a firm may go up in this series by growth, downward by reduction in operations, or may stay in one phase indefinitely. A firm may also split into two others of less complicated organization.

Phase I—one person. This is the smallest operation—a contractor who does everything himself, with a part-time secretary when needed (Figure 1-5). He assumes all duties except manual labor, but may do this as well. An accountant audits his records, monthly or annually. Such a firm may gross $100,000 to $700,000 per year, and the firm's profits correspond to the owner's earnings for his time.

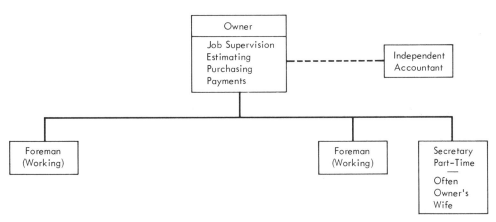

Figure 1-5
COMPANY ORGANIZATION—PHASE I
Minimum-Size Contractor

Phase II—four persons. At this stage (Figure 1-6), an office is maintained with a secretary–bookkeeper for clerical chores. An estimator–quantity surveyor makes the owner's estimating easier, and a general superintendent is responsible for the field work. The contractor now has free time to effectively supervise the work. It is still a one-man company as far as direction is concerned, since all daily decisions regarding purchases and estimates are made by the owner. The firm may gross $700,000 to $5 million, or even more, with such an arrangement. The work would usually be competitively bid commercial or public work in the $100,000 to $2 million class.

The transition to the next stage is critical, as authority must be delegated. As the small organization takes on more work, detailed decisions by the contractor require longer hours

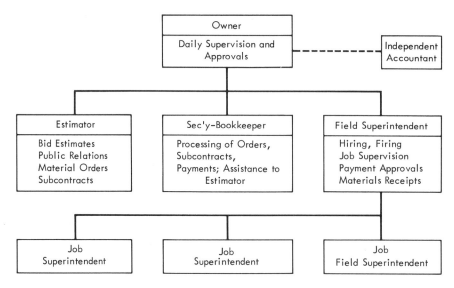

Figure 1-6
COMPANY ORGANIZATION—PHASE II
Medium-Size Contractor

and higher nervous strain. If the contractor in Phase I or II dies or becomes ill, the business may have to be liquidated.

Most people think of the Type II or III organization as typical; among college people, it is Type III. In relative numbers of firms, however, it is probable that of 100 firms, over 90 remain as Type I, one-man operations, indefinitely. Of the remaining 10, only one or two become a real Type III organization. The salaried man is interested primarily in the large firms, but the person interested in independent employment considers the first two smaller types.

Phase III—seven persons or more. With a competent estimator-manager and assistant, an accountant and a clerk, a secretary, and a field superintendent (Figure 1-7), the contractor now leaves daily routine to others and can spend his own time on approval of bids, hiring and firing supervisory personnel, and negotiation of larger purchases. He is now able to manage by *exception*—that is, spending his time where trouble is indicated rather than attempting to review *every* decision to be made in the firm. He is able to investigate projects outside his routine work, such as turn-key or contractor-owned projects, and to investigate other use of his capital (if he is weak in management) or of his management (if he is strong in management). He has the framework into which he can fit a design section or fit people engaged to look for new investment opportunities or negotiated contracts.

The difference between Phases II and III in the diagrams (Figures 1-6 and 1-7) has been minimized to emphasize that the difference in growth is primarily one of decision rather than of organization. It may be impossible to tell from the number of people in the office or the gross of the company whether the firm can be classified as in Phase II or Phase III. Phase III exists when the head of the company may check a job he believes to be of importance in the morning rather than coming to the office when a bid proposal is submitted. He depends on the estimator for the bid. Phase III exists if the field superintendent, without speaking to the president, tells the job superintendent to buy a truck or has the estimator order a couple of carloads of form material. Phase III exists if the accountant rather than the president calls the bank for a loan.

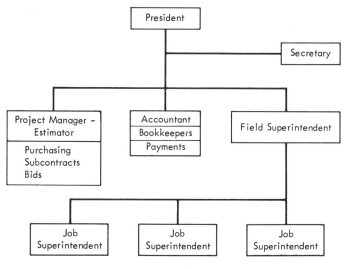

Figure 1-7
COMPANY ORGANIZATION—PHASE III
Large Contractor

After this delegation of authority has been made, any number of variations of organization may be adopted. The field superintendent position is usually the first to be eliminated, as there are so many jobs and so much time is spent traveling that he cannot get around the jobs often enough to be effective. Project manager positions may be set up on the job or in the field, and they may or may not include estimating and purchasing duties. Once the department heads or other subordinates can deal directly with each other rather than through the president, a specialized organization becomes workable. This is not necessarily profitable for the company, which may be headed either toward bankruptcy or profitable expansion, depending on the direction of the new type of organization and the manner in which the president handles it.

1.18 TYPES OF CONSTRUCTION ORGANIZATION BY FUNCTION

General is freely used in the industry as a short term for *general contractor*, a firm that accepts general responsibility for the work and usually does the work of the *basic* (structural) trades. A *prime contractor* contracts directly with an owner, and the *subcontractor* has a contract for a part of the work from the prime contractor. Usually generals are prime contractors, and *specialty contractors*, who do certain parts, are subcontractors. However, there is no fixed custom or law. An electrical contractor, who is a specialty contractor, may take a contract for a power plant, and let the subcontract for structural work to a general. The electrical contractor has become a general in this case, but does not so label himself.

General contractors usually contract the entire project directly with an owner. They may be *building contractors*, who build architect-designed works, or *heavy* or *engineering contractors*, who specialize in works designed by engineers rather than architects, as highways, bridges, water treatment plants, and pipelines.

Generals are also termed *constructors,* preferred by the Associated General Contractors of America, but less widely used. In earlier times they were *builders.* There are now builders in the old sense, mostly in rural sections, who do the work of all trades as well as design structures. The term builder is now limited to the definition below. A firm that does work on its own property and owns both the land and the structure is a *developer* or *builder*.

A *developer* supplies the initial capital and organization to obtain financing, technical help, and construction facilities to create a new subdivision, apartment complex, industrial park, or even a complete city. A developer may be a designer or contractor, or he may hire this work done. A developer may be a single person or an organization of hundreds. Since he may engage firms to do any of his functions, the size of the developer's own organization has no relationship to the size or number of projects he develops.

A *builder* is a special kind of developer, who constructs buildings—often by contracting them entirely to others. The typical builder buys lots and sells homes directly to the public. If he buys the property as acreage and installs streets, sewers, and other improvements, he is also a developer.

Subcontracting—or rather, letting a number of prime contracts to specialty firms—is characteristic of home builders, and most small contractors do their work entirely in homebuilding. Home builders speak of *subcontracting* their work, but actually write direct contracts for the entire construction. In this they differ from nearly all contractors in other areas, who do at least the basic trades with their own forces. Home builders are primarily sellers and designers of homes, and the building portion is not the best organized (in the general contractor's sense) part of their business.

1.19   INTERNAL RELATIONSHIPS IN THE FIRM

Delegation of authority in a firm must be accompanied by a clear understanding among the various specialists. "We just work together—we don't need divisions" is a common attitude of an owner-manager, and the result is lost time and the necessity of the manager's decision in almost everything. Some managers do it in order to retain detailed control. Supervisors do their work at the last possible time because they expected someone else to do it. The aggressive employee is eliminated, as he is never sure when he is doing what he should. The less competent hang on longer, and there is no spare time to take care of unanticipated emergencies and increased work load, which causes loss of interest by everyone.

1.20   LINE AND STAFF RELATIONSHIPS

A decision may be given to the person who takes action on it in one of two ways: as an *order* or as *advice*. For example, a contractor gives a superintendent an instruction, and the superintendent passes it on, as an order, to a foreman, who passes it to a journeyman; this is a line relationship between the parties concerned. If an accountant advises the contractor on how costs should be kept, he is in a staff relationship. Of course, anyone may advise his superior, but the accountant has a *duty* to give advice to his superior; this gives him a staff position.

To be effective, a staff member's advice must, at some level, be made a line order. Some organizations operate without this step at all; for example, a small engineering design office may have an accountant who specifies, directly to the draftsman or designer, how time is to be reported. The necessity is recognized and no order is required. Managers of this type of organization often do not understand the resistance they meet when they try to use the same method on a different type of operation, with different persons.

The characteristics of an organization are largely determined by the following factors:

1. The existence and responsibility of staff supervisors.
2. The level of line management at which the staff advice becomes a line order.
3. The level of management at which the staff advice is given.

1.21 AUTHORITY AND RESPONSIBILITY IN THE FIRM

In a construction organization of any size, the first staff adviser of importance is the controller or accountant. There is no doubt that he must have the responsibility for tax returns, and he handles the money. He may have the duty of making cost reports of work in progress. His work, as that of most staff employees, includes separate duties and responsibilities. He must have the authority to enforce regulations relating to his *responsibilities*; he need not have an authority relating to his *duties*. For example, although he handles the cash, his responsibility is very limited. He is responsible for dispensing money in accordance with *proper authorization*; but since he does not determine the proper authority, he cannot be responsible for wasteful, or sometimes for duplicate, payments. He cannot be responsible for embezzlement if it occurs as a result of an authorization he is bound to respect. He may review payments and question those which appear improper, but he is not in a position to certify them as properly spent if this cannot be determined from the information given him.

His duties may include service functions, such as recording and arranging cost reports, for example, or calculating payrolls. He may keep personnel records of employees. He is not responsible for initiating measures, except as they are necessary for his quarterly and annual government reports and for profit and loss statements. He is responsible for cash forecasts, but here also he must rely on reports furnished to him from other departments.

Since the accountant may have assistants and even a full-fledged department, he is also a line supervisor insofar as his own people are concerned.

1.22 LEVEL AT WHICH AN ORDER IS GIVEN

The organization would be most definite if all staff advice were given to the highest line manager, and most effective if all advice were given at the lowest possible level. In practice, a staff department acts at many levels, depending on the matter at hand. The accountant would deal directly with the top manager in regard to corporation reports, profit and loss statements, and tax reports. But he would deal with the project manager in regard to questions on invoices, cost reports, and cash flow. He is advising the project manager, who may accept the advice and direct his subordinates to follow it (for matters of this sort, the job superintendent is nearly always the project manager's subordinate, even when the superintendent's independence in other respects is absolute; that is, he is independent of everyone else in the company). If relations are smooth, any office employee should be able to advise any person in the project manager's organization. However, there should be no doubt that this is being done with the project manager's permission; chaos can result when people on the job accept instructions from various people in accounting. The project manager must have full control of his people, and they must be well enough informed that the minor employee (usually the timekeeper) on the job knows if new instructions are of importance to the project manager.

For example, a great deal of paperwork is saved if the payroll clerk in the office can call the timekeeper for missing data on time cards, about rearrangements to fit new forms or methods of accounting, for requesting different or earlier methods of transmittal of data, or the like. On the other hand, if an employee setting up a data processing system gives instructions to the timekeeper which change the items of reporting costs or the method of calculating paychecks, this person can wreck the cost reporting system on the job and cause serious labor problems. Where trouble results from subordinates taking action without proper authorization, contact between subordinates may have to be suspended entirely. The job superintendent is entitled to receive *all* his direction from the project manager; it is not part of his duties to decide matters of cost control and cost responsibility.

The same restrictions apply to other staff departments, such as personnel, engineering, or estimating. Estimating itself is a staff function—that is, the estimator advises the manager of the cost of a new project and the bid price—but the estimator may also be a line manager or subordinate in his department.

1.23 LEVEL AT WHICH ADVICE IS GIVEN

Advice may be given at any level, so long as it is recognized as advice. Since many matters, if not accepted as orders, are sent to the accountant's superior and eventually to the project manager or general manager, an attitude may develop that the accounting employee is going "over the head" of the timekeeper or project manager.

Such advice should therefore be given as a suggestion, so the line employee feels free to reject it for the reason that he lacks authority to change a system which has basically been given him by the project manager or superintendent. Also, the general manager should not order changes requested by one department and resisted by another without discussion with the objecting department. Neglect of this simple rule causes more irritation and resentment than may a dozen really important decisions.

As a contracting organization grows, it experiences "growing pains," which can bring expansion to a stop without the manager realizing what has happened. In the early stages of organizational growth, construction managers are predominantly ex-tradesmen not familiar with accounting and engineering functions. The staff departments are set up to give them instructions in these matters, and when managers with more extensive experience are hired, friction develops. The general manager looks on the newcomer as uncooperative and the established department looks on him as stealing its responsibilities and reducing the staff's importance. So the new construction manager, in order to keep his job, must either do the work a job superintendent has been doing, or must get the general manager's duties—neither of which improves the organization as the general manager had hoped.

Staff departments are much less important to a construction organization than they are to similar-size firms in other businesses, because staff departments of the industry are separate firms. Design is by several different firms, with details such as electrical and plumbing design often executed by subcontractors. A contractor who assumes the entire responsibility for conception and execution—and sometimes for operation—has a number of specialized departments for these functions.

1.24 SPAN OF CONTROL

Span of control refers to the degree of supervision that may be assumed by one individual. A construction industry study on this topic, the Educational Survey of General Building Contractors, 1961, conducted by Penn State University Continuing Education Services, reported that in Philadelphia, superintendents supervised an average of 37.5 persons, but the report also questioned the validity of the survey. This number is quite large; it undoubtedly included subcontractors over which the superintendent has operational rather than technical control; one person can supervise almost any number of such foremen. The Penn State survey observed that superintendents are doing too many things to handle them all properly. The survey empasized the development of subordinates who can work with the minimum of supervision. Supervision is less detailed and immediate in the construction industry than in manufacture; this is shown by larger numbers of employees per supervisor. The problems in construction are more difficult, and the superintendent or project manager must be well qualified and experienced for his post; but he can expect to have responsible employees.

Line and staff supervision by a single individual is also characteristic of construction. The superintendent has a staff relationship (that is, he does not hire and fire men or specify how the work is to be done) with a large number of subcontractors, as do the project manager and other managers in the company. Frederick Taylor[2] advocated a number of supervisors for each individual, each to be responsible for a different characteristic of the work. This was never attempted in construction, although some foremen may have a number of supervisors. The insulation foreman, for example, is responsible to his manager for production and waste and for hiring and firing, to the job superintendent for when work is to be done, and to a fitter foreman (who may not be on the job all the time) for whom his employer is a *sub-subcontractor*.[3]

1.25 SUBORDINATES AND REPRIMANDS

Occasionally, a subordinate does something he shouldn't, or fails to do something he should. The two situations are entirely different. There are few men who do things without being told, so any correction must be in such terms that the subordinate will not lose his initiative. Foremen and superintendents in the construction business rarely need to be told that their jobs depend on their performance; they are quite aware of it. Consequently, one danger of giving estimate amounts to the superintendent is that he may be unable to meet them; if he can't, he may leave, assuming that his manager is dissatisfied with him. Field men have a tendency to believe that office estimators can make accurate estimates, or at least make estimates that the office man *thinks are accurate.*

Unless a manager takes the attitude, "I'm *never* satisfied," any statement he makes is taken seriously by the superintendent. The most severe reprimand I ever received was when I approved an invoice on the strength of the superintendent's previous approval. The general manager, going over the bills, asked, "So-o, you're a good fellow, eh?" in a soft but obviously threatening voice. The validity of the invoice should have been investigated.

A subordinate who oversteps his duties may purchase with no authority to do so, may kick a subcontractor off the job when he is not authorized to do so, or may occasionally tangle with

[2]Taylor was an engineer of the late 1800s, known for his work in time study. Most time studies in construction of the last 20 years start where he left off, as little work was done in this area during the years 1910–1960.

[3]A subcontractor to a subcontractor—a clumsy term, but no one has come up with a better one.

persons not under his control who report back to his superior. It is usually enough for the manager to state without any particular emotion that such and such is not the subordinate's responsibility, and to clear it with the manager first. This may well be tempered with permission to do certain things in an emergency, so it does not have to appear as a reprimand; and particularly it should not appear that his authority has been reduced. In any argument, always leave the other party "a horse and saddle to ride away"; that is, your opponent should always be granted a partial justification of his point of view.

A supervisor's failure to act is more serious. The common faults are failing to check plans properly, not carrying out minor instructions involving changes or priorities, or forgetting to inform the manager when an item is delayed. Here you can promise to help the superintendent; he may be loafing, but if so, he will think your offer to help is sarcasm. If he is not loafing—and it is very difficult to tell—he will take the offer seriously. Don't take his failure to carry out a special order from you as a personal affront. He may well think the work is unnecessary and that you'll forget it in a few days anyway. Come back in an offhand way in a day or two, or an hour or two, depending on the circumstances, with "How_____coming?" He will be able to furnish a plausible excuse and still get the work done in a reasonable time.

New college graduates, who have learned how businesses operate from the viewpoint of lower-paid employees, often have an unrealistic attitude and may require more direct criticism. Not having the experience of age, they will usually accept such correction; if not, you haven't lost much. A person from some other industry who asks, "Do you have a regular salary review schedule in the company?" and gets for an answer, "Yes, each afternoon at five," will realize the short-term nature of construction work.

Drunkenness is more of a problem. The man who drinks on the job, or who, after return from lunch misuses machinery or makes some obvious mistake, must be discharged with no delay. But many men drink heavily the previous night or into the morning hours and have alcohol on their breath the following morning. Some men can absorb large quantities of alcohol without apparent effects. Just tell the man what the facts of his appearance or actions appear to be and let him decide what to do about it.

You should never find fault with a subordinate who you have decided should be discharged. Likewise, there is little point in telling a man why he is being discharged unless the reason is lack of work for him. Regardless of the reason, you will not be on better terms with him later by being friendly and then discharging him. A man being discharged will very rarely agree that his work or conduct is poor; and, if he does agree, he will resent even more that you know it, too. If there is obviously a need for his skills, a statement such as, "You do not fit the present needs of the project," is enough. Never discharge a person in such a way that you cannot hire him again if you need him, or in a way that he won't hire you if he needs you. The discharged person spreads the word about you to other firms and people. Also, do not pass the blame on to your superior, unless the superior not only has ordered the discharge but has directed it over your protest. Your superior should not direct the discharge of any person by name, although he may do so by an order to cut labor forces or overhead, and if he does so, your loyalty need not go so far as to cover up his action.

1.26   CHARACTERISTICS OF SUPERVISORS

You have undoubtedly recognized that good people are hard to get. It follows that people must be hired for their best characteristics and used in this way. A young project manager often expects the capable man to conform to certain preconceived ideas, chief among which is a willingness to comply with instructions and a blameless personal life. Compliance with instructions is considerably different from a *willingness* to do so. The "Yes, sir" superintendent is a poor subordinate for a young man. A young manager needs the ability to reason out his instructions, and to prove them to his own satisfaction and that of others. He needs to learn what is wrong with the ways he wants to do things. He should therefore realize that characteristics of men are not good or bad; they just are. A good characteristic has a bad side in many instances, and it is necessary to know the overall result before making a judgment.

For example, aggressiveness is necessary in a superintendent. He must take action when necessary, give orders, and, in general, feel confident. This is the way the work is done, and the superintendent's personality shows in the finished job. A superintendent who is fully informed about what is to be done, but unable to enforce his ideas on his crew and subcontractors, is ineffective.

But aggressiveness has a bad side also. The superintendent may not allow his subordinate to present ideas; by his dominant personality, he submerges others. If one of his foremen is also aggressive, there may be a clash. The "super" will be unable to tolerate competition on the job and will demand dismissal of the other foreman. This is a common situation. If the foreman is fired, a valuable backup man for the superintendent's job is lost; if he is not, the job suffers, morale of the foreman's men is low, and production falls. A superintendent may also clash with an equally agressive project manager. *The project manager, as the executive and presumably more broadly trained man, should make the adjustment in such a case.* Also, the project manager's contacts with the superintendent are not so important to him as they are to the superintendent. The aggressive supervisor often undercuts his own foremen and gives instructions to the men, thus failing to develop foremen or to build an organization which will continue to function in his absence. He fails to delegate authority, thus requiring considerable work of himself. These tendencies can be modified and shaped, but to eliminate them is to eliminate the desirable traits as well. A manager, to a greater degree than a superintendent, must accept inefficiency in some places in order to concentrate his time where inefficiency is greater, and must be lazy enough to delegate authority, in order to build an organization rather than merely to run a job. In addition, the manager's work in progress should be continuously documented so that if he quits or is incapacitated, another can pick up the work immediately. Otherwise, the work is not only delayed while material orders are checked, but also considerable trouble with subs and vendors occurs because of arrangements never committed to paper.

1.27 DEPARTMENTAL ORGANIZATION

A contractor has, within his organization, personnel for the basic *functions;* there may be other functions, either separate or included in the basic organization. The basic functions are:

1. *Sales,* or obtaining new business. This is usually handled by the estimator.

2. *Production,* or supervising personnel engaged by the contractor at the site of the work. These people may be direct employees or employees of subcontractors.

3. *Accounting and finance,* usually combined. The accountant keeps accounts required by law and other records required by the manager. Finance is the planning of cash available to pay bills, and of borrowing as necessary.

4. *Purchasing,* including purchasing materials, and subcontracting work that is not to be done with the contractor's own forces.

When operations become so extensive that several persons are required for one function, a *departmental organization* is created to efficiently use specialists of varying degrees of ability. This organization facilitates interchange of information between these specialists, designates what this information represents (such as orders, recommendations, or facts), and defines the authority of the persons involved.

1.28 SALES

Most contractors keep informed of projects through the Dodge reports or builders' exchange reports. The Dodge reports (see Figure 1-8) are issued by the F.W. Dodge Division of McGraw-Hill Information Systems Company, several times weekly, and daily. Subscriptions for these reports are sold by size and type of jobs and by area; for example, you may subscribe for building construction reports only, or for projects over a million dollars, in specified counties or cities. Reports are issued several times for each project—when it is first reported as planned, when drawings are begun, when an approximate date for bid is set, when the contract is awarded, when subcontracts are wanted, and when subcontracts are awarded. Contractors often furnish information to the Dodge service to get free advertising. Most of the information

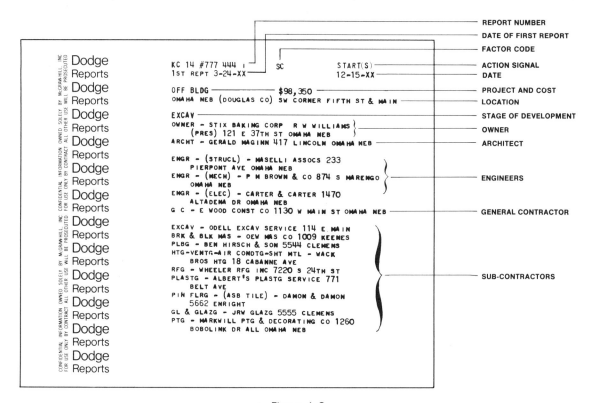

Figure 1-8
A DODGE CONSTRUCTION REPORT

on Dodge reports originates with architects and engineers and can never be complete; some A/E's do not cooperate because they want "closed" bidding; that is, they accept bids only from selected contractors whom they believe to be reliable. Dodge reports are not as important to A/E's as to contractors, since the selection of the designer is often made before the project is first reported to the public. Dodge reports also offer the sale of microfilmed plans and specifications for current projects. Local builders exchanges, which are associations made up of construction firms in an area, put out construction reports of their own. Although builders' exchange reports are valuable in some areas, the reports are virtually nonexistent in many areas.

Since estimating is part of the sales function, its organization is basic to both costs and sales. A building contractor who estimates 30 percent of the construction he bids and uses subcontractors' estimates on the rest may spend on estimating from $1/10$ percent of the bid amount on jobs of $1 million each, to ½ percent or more on smaller jobs. If he gets 10 percent of his bids, a not unlikely figure with an average of seven bidders (since many jobs are not built at all), his cost of estimating may be 1 percent of gross—a considerable item where net profit[4] often averages less than 3 percent in successful firms.

1.29 MANAGEMENT BY ESTIMATOR

Both estimators and estimator–managers are former job superintendents or engineering graduates in the specialty of the contractor—civil, mechanical, electrical, or construction.

After a contractor hires his first estimator, he decides if the estimator is also to be a manager. As the firm grows, estimators become either segregated in a single department or become independent estimator–project managers. Both methods are in common use for firms of all sizes.

The separate estimating department has the following advantages:

1. The full-time estimator becomes more proficient in quantity surveying and acquires a wider knowledge of prices and materials.
2. With no interruption for other work, the estimator makes fewer errors and can plan his time more efficiently.
3. All estimators may be more easily required to use the same procedures, standardizing costs and units and reducing labor by computer use.

Some disadvantages of a separate estimating department are:

1. The estimator and the project manager, each confined to one part of the work, do not improve their management ability. The office as a whole does improve, but the manager's constant attention is required to assure operation of the management cycle, returning actual costs to correct new estimates.

---

[4]Net profit is gross profit less all office overhead costs, manager's salary, and interest available on capital invested. Contractors often use gross profit in discussions when they want to appear more prosperous. The gross profit may be 10 percent or more, depending on the accounting method followed. Net profit in the industry usually is figured without deducting interest available on capital, if it were loaned, and often without deducting the owner-manager's salary or possible salary if he were employed elsewhere.

2. The estimator and the project manager are not general managers. A tremendous investment is therefore being made in one man—the general manager—who is not readily replaceable by a subordinate.

3. The estimator does not become familiar with construction methods, so he cannot price labor items himself the first time a method new to him is used.

4. Many potentially good people, who would tolerate estimating as part of the more interesting project manager job, would not continue in a job devoted only to estimating.

5. The necessary feedback of costs to the estimator, an essential feature of the cost control system, is made imprecise as information passes through several people, and is quite likely to be neglected since only the general manager can define the items and can discover and correct errors.

In summary, the separate estimating department organization is best suited for firms with repeated similar work. More building contractors than heavy contractors are in this category. Heavy contractors are intimately influenced by variations in natural conditions, requiring experienced judgment for each job, and it is important that their estimators have field experience.

1.30 PRODUCTION

The production manager on a construction job is the job superintendent. Through the trade general foremen and foremen, he directly supervises the hiring of men, the methods to be used, and the organization of crews. When several foremen of one trade are required on a job, the general foremen supervise foremen. Although labor unions have considerable power regarding work methods, they do not run the job. Only in a few cases, particularly among the bricklayers, is there direct dictation of work methods. In some areas there is resistance to prefabricated work, but the objection to using material fabricated by nonunion workers is often confused with objection to any prefabricated work at all.

The superintendent generally has the responsibility for planning and specifying methods, but some large companies have a separate group, usually of young engineers, for planning work methods. These people are presumed to be free of preconceived ideas of how work is to be done, and they therefore will more probably come up with new ideas. They must question the tradesmen to find out possible methods. Since they start with the best the tradesmen have, and since their ideas are reviewed by a competent manager, they will undoubtedly come up with improvements.

The job superintendent is responsible for the work subcontracted as well as that done by direct labor. He is the immediate inspector of the subcontractor's work, although he is rarely qualified to determine if electrical work or a mechanical job (heating, air conditioning, and piping) is properly done. He may also schedule the work and notify subs in advance when they will be needed.

In a typical job organization (shown in Figure 1-9), the division of responsibilities between the job superintendent and the project manager will vary with the company and with different jobs in the same company, depending on the size and type of job and the abilities of the individuals. If the project manager is a former superintendent, he may make decisions regarding

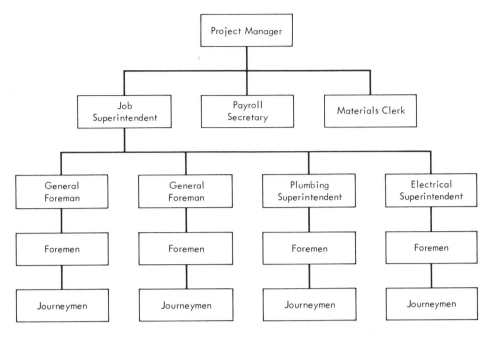

Figure 1-9
JOB ORGANIZATION CHART

scheduling and subcontractors without consulting the superintendent, but these decisions are usually made after joint discussion.

The job superintendent is a permanent employee of the contractor, so far as there are permanent employees in this industry. To keep their superintendents, established contractors will use them as estimators, will carry them on the payroll with no work, or will put them on other jobs as foremen when they are short of work. They are the key people in obtaining production. They may go back to working with their tools (as journeymen) in slack times, at superintendent pay rate. When not needed, they may be paid half wages and not be required to work at all.

Job superintendents are superintendents for general contracts and trade superintendents are superintendents for subcontractors. *Superintendents* are usually former craftsmen, for general contractors on buildings; they are nearly all former carpenters. They are often union members, but need not be, on union jobs. A superintendent on open-shop work may be a union member, but he *can* be fined by the union if discovered. The union to which a superintendent belongs expects him to comply with union rules. Since the superintendent can always quit the union, the union will rarely take action against him. However, if he is acting as a general foreman by supervising several foremen of one trade, he may not, by some union rules, be a superintendent; that is, he may not supervise other trades as well. Unions rarely enforce this rule, as it is seldom to their advantage to do so.

Foremen are former journeymen selected as supervisors. They *lay out* the work; that is, they specify who is to do what and how, within the limits of the instructions they receive. One study showed that the average carpenter who served an apprenticeship could expect to be a foreman within 4 years after completing his apprenticeship. This advancement occurs because most carpenters have not served an apprenticeship and do not have the more general training it provides. The better foremen are kept by a contractor, even though the company often must move them from job to job. Because of both expense and union rules, most foremen are hired locally by out-of-town contractors. Some specialty contractors, such as Gunite (sprayed concrete) contractors, move all their foremen from city to city for each job, and sometimes move their laborers as well.

A company with several small jobs will have in the office a project manager who supervises the superintendents or a project manager who may be responsible only for purchasing. In the latter case, a *general field superintendent* may be responsible for the field work as a whole. The general field superintendent is not effective on widely separated jobs, since he spends much of his time traveling from one job to another. When he is on the job, he does not know if what he observes is accidental or typical; to learn why work is being done in the manner it is, he must take up the job superintendent's time, which reduces efficiency. Adequate records of purchases and subcontracts are seldom maintained by the job superintendent, so the field superintendent must return to the office for information. Too often, on small jobs, he tries to deliver material and payroll checks, reducing his ability to be where he is most needed at the time. Many construction firms have eliminated the position of general or field superintendent, supervising job superintendents through spot checks by a general executive, often a vice president.

If the ability of the job superintendent and the project manager is such that cost controls cannot be used as the primary tool of supervision, the *field superintendent* is most effective by:

1. Not attempting to make deliveries or carry payroll, unless it can fit *his* schedule for the particular day.
2. Acting as an inspector making spot checks where trouble occurs, or better, where it is most likely to occur.
3. Making each visit as a completed task, not leaving work held up pending his decision or pending confirmation or instructions.
4. Exercising the deciding voice in assignment of superintendents and in hiring and firing them.
5. Advising superintendents, not giving orders, but approving actions superintendents believe appropriate.

The foregoing assumes that the jobs are of such size that only two visits can be made weekly. If a field superintendent can spend as much as 4 hours every 2 days on a job, he may function more nearly as a job superintendent himself.

If an adequate and trustworthy cost control system is in operation, a superintendent may have greater authority and independence: the costs show immediately if he is efficient. This requires that the management system of pricing and comparing actual costs is essential, and that the firm has confidence in its pricing. If the estimate is off, the superintendent is considered to be inefficient. Of course, this means the general executive need only give attention to those jobs which are running over the estimate, to see if the superintendent is to blame. There is little more the executive can do, without replacing the superintendent. Since in most cases the superintendent makes the estimates of work done which is the basis of cost control, a check is needed on cost status preparation in the field. Some firms have an independent branch which does nothing but periodically check the status of the work in the field, both to check accuracy of the superintendents' reports and to detect fraud.

The relationship between a job superintendent and his superior is such that the executive advises rather than directs the superintendent. Each has a definite area in which to make decisions. A superintendent is in intimate contact with the workers and the qualify of work, and decisions based on observation on the job must necessarily be left to the superintendent's judgment.

1.31   ACCOUNTING AND FINANCE

Accounting refers to the process of making payments and keeping a record of receipts and expenditures. *Finance* refers more specifically to planning the availability of cash, by borrowing or investing as the cash situation dictates. In most construction firms, the same person is responsible for both functions, although the finance duties may be personally executed by the chief accountant or financial officer. Even though financial duties may be light, the title is more prestigious than "accountant."

Accounts are required by the Internal Revenue Service to substantiate costs for tax purposes. The minimum requirement is that the contractor must be able to substantiate his payments as being business-connected, and must distinguish between material and labor expenses and purchases of equipment which is an asset after the work for which it was purchased is completed. Receipts, if deposited in a bank account, are recorded and the IRS (Internal Revenue Service) will consider all receipts as business income unless one can prove otherwise. The receipts one might want to prove are not income may be money borrowed or sale of assets.

Most contractors, both for cost control and for tax purposes, are able to prove their costs for jobs completed to date, by jobs; that is, the simplest stage of construction accounting requires that costs be segregated not only by income and borrowings, but also by jobs. This may be done by keeping paid invoices in boxes—one box for each job—and then adding them up at the end of the year. If an invoice is missing, the tax auditor may balk at accepting an entry or check as evidence that the payment was for business, not personal, expenses. The one-person construction business may also file invoices with copies of checks in order, without separating them by job; the description on the check includes the job designations. A bookkeeper may then summarize the checks by jobs at any interval requested, even annually. And if the contractor is paying taxes on a cash received basis, he need not separate his payments by jobs.

Payroll accounting is more detailed. For each person in each payroll period, the records must show a check or cash payment, payroll deductions, and an entry on the worker's individual card. The accounting and payroll work take relatively little time of the one-person firm and, except for payroll, need not be done so often that it interferes with daily job supervision. The growing contractor first finds accounting very time-consuming when he writes checks and substantiates deliveries, that is, determines if a bill is really due when it is presented. As long as the contractor is personally receiving shipments and inspecting subcontract work, he dispenses with paperwork for orders and receipts. When his business expands enough that others must order and receive goods, he must save these delivery slips. The small contractor can often be recognized by the sheaf of delivery slips and invoices he carries in his pickup truck, to work on at night. The one-person contractor must do his sales and office work at night and on weekends.

The accounting department may also keep a *current record of job costs*. If these records are in sufficient detail, they show when corrective action is needed, such as replacement of the superintendent or other supervisors, correction of errors in the reporting system, or revision of errors in estimating. The one-person contractor hasn't the time for cost control accounting practices, although he can keep a surprisingly complete record in his memory. His lack of written records is a serious disadvantage in bidding, as he may use approximations which simplify his estimating but tend to give him the bad jobs, that is, the ones for which the low bid is unprofitable.

The accounting department also, in its finance function, plans *cash flow* for the firm. The term cash flow in many usages means income, even though it is not profit. For the contractor, it means keeping at least a zero position, in which income equals outgo. The small contractor has no time for such calculations; he tries to keep a month's living expenses and a month's payroll ahead, and he pays his suppliers out of receipts when the money comes in. If he runs short, he borrows short-term money (90 days or less) from a bank.

As the business grows, the contractor's first expansion into Phase II is to hire a bookkeeper. From then on, expansion depends on how completely he wants to do certain work, such as cost control and personnel records, in the accounting department. He may also use outside services, particularly computer services and outside auditors, to reduce manpower requirements. In addition, he has options in approval of invoices, such as "Is receipt of materials to be verified by the project manager, or by the accountant?" A contractor may have as extensive a business as several multimillion dollar contracts under way at once, with only two or three people in the accounting department, necessary manual operations at the job site, and mechanized operations contracted with a computer service.

1.32   PURCHASING AND SUBCONTRACTING

Paper work can be eliminated on the part of the one-man contractor by using the sub's proposal form and an oral acceptance. If the matter is definitely agreed and the parties are honest, an oral contract is as good as any other. A written contract is sometimes defined as a "confession of mutual distrust," but it is better for both parties that a complete written record (contract) be made. If the supplier or subcontractor is larger than a one-man firm, he may make a proposal that, when accepted, constitutes an adequate contract. It is a tribute to the honesty of business people that verbal agreements work as well as they do; trouble usually arises not because of dispute over what is said but over matters which were not discussed. When each party compromises by giving one point and getting another, the final agreement is easily forgotten if not reduced to writing. A contractor who uses oral agreements in doing business with new firms is more likely to have trouble than is the contractor doing repeated business with the same subs.

Few contractors have employees functioning as purchasing agents with no other duties. The project manager or an estimator-project manager usually handles all purchases for his project. A large firm may have a purchasing agent, but his function is usually more clerical than managerial in that he makes purchases as designated by project managers. He supervises the clerical details, sees that the orders are complete, and may combine orders given him to get a lower price. He acts as an exchange between project managers, so that low prices and notice of good or poor performance of vendors is known to all managers. He may buy standard materials, but the variation of materials and locations limits the number of combined purchases. When equipment purchases and subcontracts are to be made, he must consult the superintendents or project managers about both the technical and ethical questions that may arise. The ethical question is usually to determine if the firm has become obligated to another firm during the preparation of the bid.

For very large projects, overseas projects, and for builders with repetitive work, the purchasing agent assumes full responsibility. He receives a list of materials and required delivery, then obtains quotations and places orders.

1.33 PERSONNEL RECORDS AND HIRING

Many large contractors have personnel managers, but these managers rarely have the power of their counterparts in industrial firms. The smaller size of construction firms and their treatment of nearly all field employees as temporary reduces the area of responsibility for a personnel manager, and his functions are absorbed by other managers in the firm. For example, the Hunt Building Corporation, with 1971 revenues of $36 million and a construction *backlog* (work contracted and not completed) of $34 million, had only 149 permanent employees (of which 59 were engineers and architects, as this firm does design and development work as well as construction). In construction, there were about 90 regular employees; yet the gross receipts indicate that about 2,500 persons, including subcontractors' employees, were employed on the work sites. The personnel manager has a staff (advisory) position, but some direct powers may also be delegated to him. The general manager considers these powers to be explanatory of the policy approved by the general manager, but the result is often that the personnel manager makes rules intended to be applications of the manager's policy, but that are a new policy. This is a failure of most large organizations, but less so in construction than in other industries. Responsibilities of the personnel manager may include:

1. Hiring new employees.
2. Maintaining individual records of employee pay and deductions (although these are more often maintained by machine in the payroll department).
3. Administering policies regarding overtime, vacations, pension plans, and other fringe benefits.
4. Recommending salary scales.
5. Maintaining vacation and sickness time records.
6. Publishing a company newspaper or bulletin.
7. Negotiating with labor unions.

The industrial relations manager, who in construction negotiates with labor unions, is often a separate position from personnel manager. Union negotiations are primarily daily discussions on pay, benefits, and jurisdiction, rather than wage negotiations.

1.34 SUMMARY

Management, as opposed to *direction,* is a trial-and-error process whereby time and money are estimated, the work is done, and, as the work proceeds, the accuracy of the estimate is continually checked against actual costs. The new costs are then used both to change the supervision of the work and to revise estimates for new work.

Direction is included in construction management, by the contractor, but *technical direction* is usually independent of the contractor.

The contractor's organization may be divided into three phases by size and extent of development. The first phase is a one-man organization, with every decision made by the contractor personally. In the second phase, clerical and routine work is taken by assistants, but daily decisions and, therefore, the business operations depend on the contractor's full-time availability. In the third phase, daily decisions are delegated and the contractor becomes a manager *by exception;* he spends his time on places where trouble occurs rather than being obliged to initiate routine work.

The principal functions of the contractor's organization are sales, production, accounting, and purchasing. Organizations differ mainly in whether they use one person or several persons to handle the sales and purchasing functions for a single construction project. Both methods are widely followed, but the writer favors the combination of these functions in the estimator–project manager for long-range benefits, particularly for executive development and up-to-date accurate pricing.

1.35   QUESTIONS FOR DISCUSSION

1. Under what conditions do you believe a contractor in Phase I or II of his firm's development should not expand to the next stage?

2. Under what conditions might a contractor be forced to expand to the larger state in order to stay in business?

3. What circumstances might cause the success or failure of a recent college graduate entering the construction business? Of a construction superintendent? Of an experienced estimator?

4. As a new college graduate, would you rather enter a company with estimating and purchasing functions combined or separate? Why?

5. If management of a firm does *not* follow the estimate–buy–estimate cycle, what may cause losses? How do many contractors avoid these results?

6. If you were employed as a general field superintendent, how would you avoid hauling and delivery duties?

7. Who handles, in detail, the duties listed for the personnel manager when no such position exists?

8. While in a contractor's office for interview, you overhear a conversation between the contractor and a project manager, as follows: "Boss, did you agree to buy this lumber from Botts Lumber Company?" "Well, sort of, but just verbally." What would your reaction be?

CHAPTER 2

THE CONSTRUCTION PROJECT

A construction site, if the work is large and complex and particularly if it includes several separate contracts or different kinds of construction, is a *project*. The project manager may be in charge of only one such project and is usually located at the site of the work. *Project cost* and *project personnel* refer not just to field costs and personnel but to all costs and all persons associated with the work, including office personnel and sometimes designers or other persons in the organization. Often, "project" designates the contractor's office people, working on the job, but located off the site.

A construction *job* is a somewhat smaller work, and *job costs* and *job personnel* refer more strictly to field personnel. A project manager may manage several small jobs from the central office, or one project from an office on the project. *Jobs* are terms used by contractors, suppliers, and subcontractors. *Project* is more often used by designers; a designer's project manager is the person in his office charged with coordinating the plans from various branches of the office.

In this book "job" and "project" are often used indiscriminately, as many contractors use them.

2.1 THE PROJECT MANAGER

The organization of a construction project follows, in general, the organization of a construction firm. The more extensive office organization usually requires a smaller job organization, and vice versa. The supervisory work, particularly purchasing and subcontracting, is much greater proportionately for smaller than for larger jobs; a $1 million housing project may have no greater volume of purchases, and often less, than two $250,000 projects. A small job, where one man is responsible for all supervisory and purchasing functions, is comparable to the one-man contractor, and job organizations can vary through the same phases as do contractor organizations.

The contractor with a four-man or Phase II organization (as defined in section 1.17) may, for example, contract a larger job—as $5 million building with kitchens, grading, and utilities, as a barracks or hotel—and may find that with his existing organization he has more office work than he can handle. If he attacks the new job with proper vigor, he will occupy his entire staff, will acquire no new work, and then be virtually out of business when the large job is completed. His other options are to expand his existing organization or to hire a new staff for the new job. Unless he intends the expansion to be permanent, he will suffer less from expanded office, added responsibility, and changes in organization by manning the new job as a separate organization.

To do this, he requires a project manager—a position the Phase II company does not have. Let us suppose that he hires such a person, and let us follow this project manager as he proceeds with his work. The new executive would be supervised by the chief executive officer (usually president) of the firm, or by another senior officer. Such a person should not have to be told what to do, but he needs to ask several questions to establish the limits of his authority and the limitations on his methods, such as the following:

1. What is the extent of my authority?
2. To what companies or procedures are we committed?
3. What support can I expect from the existing organization?

The first question is to define the size of purchases or subcontracts which the contractor may personally want to approve before the firm is committed. Decisions are needed as to whether a separate bank account is to be used and who will sign checks; about any limitations on salaries or limitations on purchase of construction equipment; about whether the contractor wants to receive copies of purchase orders, correspondence, or contracts; about whether work reports or cost reports are required, and, if so, in how much detail and at what intervals?

The commitments referred to in question 2 may be internal, such as the promise of a job on the project to someone, or a restriction by company policies regarding salaries or fringe benefits.

The commitments may be external, in that during the preparation of the bid the firm has become committed to firms who submitted low subcontract bids. Other commitments may have arisen from bids given on other jobs, design work done by a subcontractor, financial aid furnished by vendors or banks, or from interlocking ownerships. Some contractors have a friendly relationship with certain subs and give them preference. The contractor may have greater confidence in the judgment of subcontractors with whom he has had a long relationship than he does in a newly hired project manager; if the project manager does not give preference to such a favored firm, he may be accused of bad management by such a sub, and even be discharged. This is a hazard encountered by any new manager entering an operating firm. Such a contractor may deny that he is being partial to any subcontractor and really believe it. A sub who was refused cost-plus work by a project manager was even able to get the manager discharged.

Other limitations, such as a policy to subcontract or not to subcontract certain trades, may reflect an unfortunate experience by the contractor in years past. The new project manager does well not to resist these prejudices until he has the employer's confidence and can attack them one by one.

Question 3 refers primarily to foreman and construction equipment. If these are not forthcoming, the contractor may say, "No support at all." Still the home office may support him in many other ways, such as these:

1. Use of the corporate name, with its operating license, liability and compensation insurance, and its withholding tax and social security registration. Contractors use a single corporation for many jobs, but builders quite often use a new corporation for each project.
2. Preparation of social security, tax, insurance, and other periodic reports based on payroll.
3. Sharing contracts for services the home office has already arranged, such as auditing and payroll preparation.
4. Information files on suppliers, and previous prices.
5. Membership in associations that give credit reports to members. Credit reports may be obtained by membership in Dun & Bradstreet or other associations, or the contractor's bank may provide such a service as courtesy. The manager needs reports on all subcontractors, both on their net worth and on their general business situation and background.
6. Use of standard company stationery and supplies.

In some cases, as with services of the company accounting department, the line between "support" and "restriction" is rather fine, with the distinction that "support" is requested, but "restrictions" are placed on the project manager if requested or not.

Having established the framework of his responsibilities and restrictions, the project manager then proceeds with the work and planning.

Stationery and office supplies, such as letterheads, envelopes, purchase orders, business cards, foremen's time reports and time books, and equipment report forms are ordered. These are custom-printed and are needed at once; lack of such a small item as business cards can cause the firm to be forgotten by a vendor or subcontractor. Letterheads from the home office may be used, with the job or local office added under "Please reply to," but two addresses on a letterhead often cause misdirected mail.

The manager needs all estimates made of the job, the plans actually used for bidding, and all quotations or other correspondence.

The *estimate* is often found to be disorganized. The most common faults are that the original calculation sheets are unreadable, unexplained changes or corrections have been made in totals, items are not clearly identified, the dates of estimates or of drawings are not identified, or figures from several pages have been added or even multiplied and then transferred to another sheet, with no notation about how it was done. The estimator may also have had unrecorded supplementary instructions from the architect or contractor.

The first step is to obtain *all* papers relative to the job and to review them sheet by sheet with the original estimator, marking corrections and inserting omissions. Unfortunately, many estimators resent this, as they know it may result in exposure of their errors (for all of us do make errors). The manager makes his own estimate from the beginning, if he has the time. The contractor will more readily accept estimate errors at this stage than later, after estimating error has made costs run high (*overruns*). Overrun, to a contractor, means cost in excess of his bid estimate, or in excess of budget made from the bid estimate.

Some companies make less careful estimates than others, designating small items by "allowances" (not to be confused with architects' "allowances," where the owner is responsible for overruns). Some estimators omit items but compensate by a larger "profit and overhead" markup than is actually expected. They say they "always run a little over the estimate." If a more thorough estimator makes an estimate for budget purposes, he will pick up these items. When the indicated profit margin on a low bid is over 3 percent on larger buildings, 5 percent on houses, or 15 percent on repairs and remodeling, it may indicate that some of the markup represents omitted items. Where possible, overhead should not be figured as a percentage; overhead costs consist of direct cost pertaining to a particular job (as estimating *successfully,* purchasing negotiations, and supervising) and dead costs, which are unsuccessful attempts to attain business. These dead costs are a deduction from profit; the separation of the two is significant when different types or sizes of jobs are being obtained. Overhead is a small percentage of a large job but a large percentage of a smaller one. If in calculating the profit a project manager makes for the firm, office overhead is calculated as a percentage for such services,derived from smaller jobs than this, the contractor should change the percentage for the larger job.

Many contractors hire a project manager first as an estimator, then let him manage the job he gets. This eliminates any possible estimator–project manager disagreement on overruns.

Once a corrected estimate is complete, the next step is to prepare the *budget.* The budget is an estimate of the work, divided into parts on each of which the actual after-completion cost can later be determined, with letters or numbers suitable for the cost reporting system (a clerical part of the cost control system) to be used. Along with this budget, the manager designs and orders various forms to be used, if the firm does not already have them. A firm accustomed to $100,000 to $300,000 jobs very likely will not keep in operation a system that is suitable for a $3 million job.

After the estimate and budget are complete, the plan, schedule, and cash flow are prepared. A *payment schedule* is prepared for the architect as an advance notice of how much money will be requested at different stages of the work, that is, when foundations are complete, masonry finished, roofing laid, and so on. On a job of this size (the $3 million example), payment requests will normally be based on percentage completion of each large component of the work.

The amount requested is usually more than the actual cost of work for the items first completed, so the contractor may put as little of his money as possible into the work; that is, so that it requires as little investment as possible to do the work. The architect—who makes all decisions, acts as the agent of the owner, and authorizes the owner on payments—usually holds out 10 percent of the payments due until the end of the job, as an incentive to finish without delay and as insurance that the contractor is not overpaid. Since very few jobs yield 10 percent profit to the general contractor, he cannot avoid an investment in the job at the end; nor can he avoid it at the beginning, before he has made a billing.[1] But he can get far enough ahead at the other times that he may have more paid to him than he has paid out. He tries to collect enough so that he has no money in uncollected bills in all his work combined. This practice is not dishonest, but contractors with adequate reserves who do not engage in it are just more solid and straightforward than others. Overbilling is used by firms becoming insolvent, but its use does not in-

---

[1] Unless a "mobilization" item is allowed in the breakdown, as an initial pay item which may be drawn before the work is done. This is becoming more common, and the contractor has nothing to lose by asking for it. In British practice, such items can sometimes be inserted by the contractor in his proposal.

dicate even that the capital volume of the firm is low.[2] A firm may have a low net worth or liquid assets and prefer to conceal such a condition because, if exposed, its *bonding capacity* (amount of work their bonding company will guarantee) will be reduced. The *cash flow,* calculating the cash requirements of the job, is explained in Chapter 6, on *accounting,* and the *planning and scheduling* process is explained in Chapter 7.

The general contractor holds back a 10 percent retainage on his subcontractors. If the general contractor has sufficient capital, he may get a lower price by holding the 10 percent only until the completion of each sub's work, not until the entire job is finished. The retainage difference is significant at times, for subs who finish their work early in the job are then paid off even though the architect is holding back part of the payment to the general contractor. If the work is 70 percent sublet, then the retention on the general construction work done directly by the general contractor will be 10 percent times 30 percent, or 3 percent of the value of the whole amount, at the end of the job. This will often be sufficient also for cash required before the first payment is received, and it is about the minimum cash needed by the contractor—5 percent is more commonly accepted. Bonding agencies desire 10 percent working capital,[3] for the new firms; and more may be required.

For our $3 million job, we can plan that about $150,000 in cash will be needed; this depends on a proper analysis of scheduled receipts and payments from the budget, and an assurance from the architect that he will pay for materials delivered but not yet installed—an important item often not mentioned in the specifications. If the cash needed or which may be needed is not on hand, a *line of credit* should be arranged at the contractor's bank. Under this arrangement, additional money may be drawn when needed to a limit established at the bank for short periods—usually 90 days, renewable. The cash required may be optimistically planned in regard to money on hand, but there should nevertheless be borrowable reserves, or an unexpected cash shortage could be a disaster for the firm, ranging from loss of credit standing to bankruptcy.

As many material invoices are due on the tenth of the month following delivery, they can be paid out of current billings if the owner pays promptly. This considerably reduces necessary cash. Labor, also, is paid normally two or three days after the end of the work week, which reduces payroll cash requirements. Bills for some materials—notable concrete and steel—may be payable in 10 or 15 days. Arrangements can be made for later payments, but the cost will then be greater, either by credit charges, or higher quoted prices to the contractor.

Next, the project manager obtains the *bond,* if required. The bonding company will already have the data and will have furnished a bid bond, if so stated on the invitation. A bid bond is a guarantee that the contractor will furnish a performance bond but specifies a comparatively small penalty—usually 5 percent of the bid price. The bid bond is either free or is given by the bonding company for a small fee. The performance bond is for the amount of the contract, and the cost is about 1 percent of the contract price. If a contractor refuses the contract or cannot obtain a performance bond, he may forfeit the amount stated in the bid bond. After the contract is made, the contractor cannot avoid his responsibility by any size of forfeit (payment).

Permits must be obtained. There may be a permit required for the company to do business in a new state; business and contractor's licenses (they are separate) for the state, county, township, and city; and construction permits for the particular project from the same agencies (but usually not from the state). If the work includes wrecking or blasting a separate permit may be

---

[2] "Low" is relative. By overbilling, any firm can show a larger net worth than otherwise, and therefore can bond more work. See Chapter 6.

[3] The bonding agency, or *surety,* is a firm that guarantees the solvency of a contractor on particular jobs.

required. If the project involves a city street, permission is needed to cross the curb with entrances to the work, and to obstruct traffic if that is required. In connection with these permits, the manager lists the names of all inspectors for utility companies which have lines across or adjacent to the work; telephone, electricity, gas, sewers, water supply, and railroads. He obtains or verifies location of underground utilities; the A/E's plans should not be used for construction until underground work is verified. Sanitary and electrical permits are usually obtained and paid for by the plumbing and electrical contractors, but the general contractor may pay the permits for temporary services.

Next, the temporary facilities, such as temporary utilities, storage, and parking, are planned, and work is started.

Although a rough progress schedule was made for forecasting money and temporary facilities, a *final schedule* requires participation by the mechanical and electrical subcontractors. These contracts and others are let as rapidly as possible. In order to plan their work, the mechanical contractors must have firm delivery dates on their materials and equipment. They can get these dated only by placing an order, so they need firm contracts themselves. At one time, contractors delayed contract awards as long as possible to get better prices, a practice discontinued since rising prices began. Contracts must be let as soon as possible for the following reasons:

1. The subcontractor is not available for scheduling early in the job.
2. The job will be delayed, if not for physical material and labor (which often occurs), then by the apparent disregard of time by the contractor.
3. Forming a firm budget is delayed.
4. In recent years, failure to place orders will result in higher prices later.

The writer favors bid dates and publicly read bids for subcontractors, but subs are not greatly interested in this procedure. Although the subs are needed to schedule the job, scheduled dates must be included in the subs' contracts. The project manager may insert his own idea of the progress schedule into the contract, making dates late enough so there should be no problem in meeting them. He may schedule the work in the *critical path* (critical work, or work items that hold up progress of another trade, and determines overall construction time) and leave less important dates for others. He is usually able to do this from experience. The manager, if he does not have specialized help, then makes up a final schedule himself. This may be by any of the methods later described—CPM, PERT, Gantt chart, trade-lag, or manpower. Each of these has its purpose, and several types may be required. If the manager has assistance of people experienced in the various trades, planning is better and the first schedule may not have to be redone. Assistance of specialists in the CPM procedure itself, rather than in the work, is useful for an inexperienced manager.

At this point, the manager has completed the planning part of his work. He has placed orders and laid out an office and organization; he now has time, money, and manpower estimates. He will start field work almost simultaneously with planning, and the items of work will not be entirely completed at one time or done in the order he would like. If the job is complex, he will have acquired an assistant long before he completes all work to this point and may have started field work before the formal award of the contract. Some specifications require the contractor to start work by owner's acceptance of proposal, but this is unusual.

The project manager by this time will have encountered discrepancies and omissions in the plans and specifications. He thoroughly studies them and lists such items for the A/E's clarification, or for approval of the manager's suggestions. In extreme cases, the project manager will have to have the drawings redrawn; if such an expense is necessary, he should accept it rather than attempt to build with incomplete information. Shop drawings are required for fabrication of specialties, reinforcing steel, and structural steel; preparation of these drawings must be scheduled and expedited as carefully as the field work.

Before completion of the drawings, the A/E has checked them for compliance with local codes. The manager provides the city or county with a copy for the building permit, and their inspectors also check them. In many cases, particularly large jobs, the plans are submitted to the city or county for approval in advance of applying for the permit, to avoid later delays when the permit is requested. Permits may be obtained by either the contractor or by the owner, in most jurisdictions; but since the fee is significant and the permit expires if work is not started, the application is delayed until the contractor is ready to start work. There is often a requirement, not in standard AIA (American Institute of Architects) general conditions, that the contractor comply with local codes. The AIA conditions provide only that the contractor is responsible if he *knowingly* follows erroneous plans. If the superintendent proceeds with plans after he has told someone that they do not comply with the building code, he can cause a serious damage to the contractor—even if the superintendent did not *know* that the plans were faulty. A superintendent is chosen, among other things, for his ability to keep quiet when he has no actual knowledge of a condition.

The manager should check all drawings and specifications he understands, although he is forced to rely on specialists for some sections, for errors, inconsistencies, and departures from local codes, as well as for apparently faulty structural design, omissions, safety, or unsuitability, such as water leakage. This should be done early in the job, although many contractors do not do it. Many points on drawings need clarifications, and additional detail drawings may be required of the architect. Most managers do not do this completely, leaving details to the field superintendent and the inspector. Each such indefinite detail causes a delay, waiting for materials or for the A/E decision, and much of the time of both job and general field superintendents is spent on such details.

Although the architect is responsible for added details of some types, it is often to the contractor's advantage to submit a low-cost suggestion or interpretation rather than to ask the A/E for it. The A/E is more receptive to a suggested detail which saves him drafting cost than to a lower-cost substitution after the A/E has made a detail drawing. During estimating and budget preparation, the drawings are examined very thoroughly and questionable items should be listed or marked on the drawings. Contractors are required to submit requests for clarification during the bid period, but many details do not materially affect cost, and since the estimator is always pressed for time, he often does not make such requests.

From the drawings and specifications, the manager prepares a list of samples, test reports certificates, and any other submissions required. Although most of these are the subcontractor's responsibility, they may cause delays in the entire work and are too important to trust to the sub without a follow-up procedure. *It must always be assumed that every such requirement will be enforced by the architect.*

With paperwork out of the way, the manager can give his undivided attention to the field work.

2.2 JOB PERSONNEL

The first people to be hired will be a *secretary–bookkeeper* and a *job superintendent.* If the project manager is on the job, he can handle much work usually done by the job superintendent; however, he needs a man who can work full time in the field. The duties of the superintendent are listed here, in general order of importance:

1. To hire workers and foremen as needed.
2. To direct the men employed by the contractor in the most efficient way; that is, to set up the order of work and determine the size of crews to be used for each job.
3. To observe the men at work, to correct and to direct the foremen, and to discharge the foremen when necessary. The project manager, if he works on the job, might be consulted before discharging a foreman. The superintendent should never discharge a worker directly, and the workers should not know that anyone other than the foreman is exercising authority over them.
4. To check the drawings to determine that the work is being done properly, and to make spot checks of layout dimensions.
5. To review foremen's time reports for the number of men reported, to avoid "padding the payroll," and to see that the costs of the work are charged to the proper work or cost item. *Padding the payroll* is a form of fraud by which the foreman reports time for workers not actually on the job, in order to keep all or part of the resulting wage payments for himself.
6. To inspect incoming materials for quality and to supervise the clerk's receipt of such materials, assuring that counts are being taken where needed, and that satisfactory estimates are being made of material which was not counted or weighed.
7. To schedule the subcontractors, within the general framework given the subcontractor by the project manager, and to determine where the subcontractor's men are, that is, if they are conforming daily and hourly to the requested schedule.
8. To inspect the subcontractors' work for quality and compliance with the plans and specifications.
9. To enforce all safety rules, both of the subcontractors' men and of his own.
10. To make regular reports of quantity and percentage of work completed in each cost control category, or to direct the foremen in doing so.
11. To assure that materials are ordered as needed and that delivery to the job is requested in time to keep the workers working without interruption.

These duties are somewhat less than would be expected of a superintendent on a large job, but more than those on a small one. The project manager's field duties would be the same, but in reverse order of priority: the project manager would start with material orders and work backward through the list. At various stages of work, or even on different days, some material expediting and reports would be done by the project manager, at another time, by the superintendent. The details would be worked out according to the demands on the project manager's time and the number of subordinates the superintendent had. On days or weeks that the manager was not on the job, the superintendent would have to assume full responsibility, or vice versa—the manager might have to assume field supervision when the superintendent was absent.

However, the project manager is responsible for meetings, which would have to be off-site, and the superintendent would normally be the person always on the job. The project manager would be the person usually available at the office and the person usually available at the office and the telephone, so he would be the contact man for the company. Both of these people may wear radios so that they will be constantly available. Radios are inconsistent in their operation under conditions where either distance or density of structure interferes with them; on a job covering several miles, the weight of personal radios required would be too heavy to carry, and communications are confined to automobile radios and telephones. Radios should not be overused; that is, there should be a person in the office able to handle office business, and the superintendent should get around the job often enough and should foresee problems so that radio traffic is necessary only for emergencies or circumstances which could not be readily foreseen.

Hiring of new foremen, as opposed to foremen who are transferred from other branches of the company, is a duty of the superintendent. If the superintendent is a local person, he will be acquainted with the people he wants to hire, both foremen and journeymen, and will know if they are union. If he is in a new area, he may work through union sources for foremen when these sources are dependable. In most cases, the union business agents' recommendations are useful. The union will naturally recommend foremen who they know comply with union rules, but this still leaves the contractor a wide latitude in selecting from them. The business agent is as cagey when recommending a foreman as the American Medical Association is when recommending a doctor; he will give several names, usually of men either unemployed or working as journeymen at the time. It is desirable to know the foremen working for other contractors; once the superintendent has started to hire people, he can get some information from them about foremen on other jobs. Also, newspaper ads may be useful. In most locations, the contractor may hire foremen who are not members of the local union but of a local in another city. Through relatives, usually the spouse's, classified ads can reach local men who are working out of town. If you are offering more than the union scale for foremen, you may so advertise. The business agents like to see their foremen get more pay (although they are not anxious that you pay the journeymen more than the scale) but like also to maintain good relations with their other contractors. Consequently, they may not be cooperative in hiring away another company's foreman.

The BA (*business agent of the union*) does know when men will be laid off on other jobs and can sometimes bring you an entire crew at a time. This leads to a high level of efficiency if the crew you are hiring has been culled out over a period of time by another contractor. Such opportunities are rare, because layoffs usually are made by laying off the poorest men first, and the poorer men will be among the unemployed. You can sometimes find competent foremen through the local state employment service, but they seldom register men unless they are drawing unemployment compensation. Most public employment agencies require so much time from the working day of the registrants that only an idle man feels that he can afford to seek employment through a state service. Local private employment agencies can be utilized if the local agency is familiar with construction practices and people. The person you want as foreman will probably be registered as a superintendent. Employer-paid agencies are best, as they are much more active in locating employees and are not confined to looking among those who will pay for their services. These agencies charge about 15 to 25 percent of the employees annual earnings.

Classified ads, when used, are more efficient if they are run as display—that is, in quite large type, for which space is paid for by the inch of column rather than by the word. You can expect blind ads (listing only a box number) to receive fewer responses than do those showing the company and person to be contacted. Many firms do not give their names, perhaps because they do not want to be bothered with calls from people who are not qualified. However, at the foreman level and up, it is justified to handle a large number of telephone calls. Employed foremen often will not answer blind ads because they cannot talk directly to the employer, just wait several days or more without knowing if they are even being considered, and run the risk of their application going to their own employer or someone who will tell their employer of the application.

When recruiting supervisory people, remember that there are highly qualified people with a very restricted experience in job seeking. They may have worked 10 or 15 years with one employer, and have received such successive promotions that they have never felt the need to look for work. These people may never have filled out an application blank or answered a position wanted advertisement in their lives. The idea of a job interview may be very difficult for them, one which they think is embarrassing. Such people will not be reached except by advertisements which:

1. Give a clear idea of the position, location, conditions, and salary.
2. Explain who to call, by name.
3. State who may be contacted without the necessity of first writing a letter or taking time off from work.

The idea that there are many people seeking supervisory positions, and that one may choose from the applications submitted, leads to poor supervisory help when jobs are plentiful. The employer must search out the superior employee. The writer, in a strange city for a job interview, was once greeted at the receptionist's desk by a sign, "Welcome King Royer." Such treatment is long remembered and costs nothing.

2.3 WAGES

The manager should not be satisfied with average and less competent people. A dominant feature of American contractors, as opposed to those of other countries, particularly in less industrialized areas, is that the American firm recognizes the importance of paying higher wages; we do not confuse wages and costs. Consequently, American firms do best at those jobs most technically advanced, which require the cream of the local market; and the people they hire remain proud of having been selected as superior workers as much as for the high pay.

A construction company needs superior employees to survive in competition, and there is no reason to believe that it can get such people by paying less than average. Therefore, it must pay more than average. How much more requires close attention to the employees' attitudes. If a person says he is leaving for higher pay elsewhere, it means someone who does not know him will pay more than someone (yourself) who does. If he is that kind of worker, your own organization should long ago have given him a raise or discharged him. A leading writer of construction management contends there is no personal contact between management and workers in construction. If this is true, it indicates a very poor organization—one that dispatches information downward, but receives none in return.

The union wage or union contract is no reason to ignore individual differences; it indicates only that the manager has competition; he must have better contacts with his subordinates than the union has. Many firms reason that they can get better workers by assuring them of regular work. This is true if the people they hire happen to be out of work at the time they are first hired, or can be retained during times when others are unemployed; but it is inadequate to get people to change from one employer to another. It may take a long time to build up a good work force by assuring them continuous work. The contractor in one city may be able to do this; the outside contractor entering a new area cannot wait until people of his choice are unemployed.

It is a valid argument against paying no more than the union scale or local acceptable wage that if every contractor does so, wages rise and become the scale (union rate); but the writer does not know of an instance where it happened. It is doubtful that the contractors who pay only the scale would adequately recognize better workers well enough to pay them higher wages. It is futile to hire better people if you cannot recognize them once they are hired. This requires competent foremen who are in a position to recognize superior journeymen.

The variation between individual workers and between supervisors is recognized, but seldom is it given the importance it deserves. Our accounting methods are adequate for discerning which are better superintendents, but inadequate for discrimination between foremen.

On one job when a group of laborers had nothing to do, their foreman told them to sit down behind a pile of lumber. One man, over six feet tall, stood up when the superintendent came by, and was discharged for loafing. The unseen, sitting men and the foreman were retained by the superintendent. The individual, rather than the foreman who had failed to plan work for them, was discharged. It is not surprising that the contractor soon became bankrupt.

When workers are hard to get through the normal union source, you may guarantee a designated amount of overtime in newspaper advertisements. As a supplier of labor, the BA normally does not want to take members from a job, since this will be resented by the employer losing them. In many trades the BA avoids bringing in members from other areas. A manager can apply pressure on the BA by newspaper advertising, which brings in both local members from other jobs and members from other cities. The BA is then under pressure from members of his own union for the higher wages for local, rather than outside, members. He cannot, even if he wants to, do anything about members in his local who want to change jobs; but when they change without his knowledge or permission, he cannot be blamed by the contractor for losing the workers.

Higher wages are often paid by guaranteed overtime. It must be for a specified period of time, or it means no more than a week's overtime to the prospective worker. This guarantee raises the average hourly wage for the week, but overtime can be terminated later with less resentment than lowering the basic wage would have done. Because of this, many contractor's organizations agree to double-time provisions for overtime in union agreements. The employer does not intend to work overtime when there are enough workers anyway. The double-time provision is a way to get workers from outside the area.

The manager must not accept the attitude that the labor union restrictions make improvement in efficiency impossible. It is quite possible to "turn the job over to the union," that is, to accept the union representative's requests and even suggestions on manning, designation of foremen, and assignment of men. The results are almost always bad. Although union people are often quite competent to run a construction project—in fact, many BA's are ex-superintendents—they are working with a different objective than efficiency. Where union people are competent, you may get rid of an overzealous job steward (who represents the union on the job) by promoting him to foreman. He is then disqualified as steward, and may be just as zealous in your interest as he was for the union. You may also have the reverse situation; you would like to use the steward for a foreman, but are afraid that a friendly steward would be appointed in his place. The unions realize they have a stake in their contractors' success, and are partners in his business—often with beneficial results for all.

2.4 WORK THROUGH SUPERVISORS

Routing of complaints through *channels*—that is, through workers' superiors—is sometimes slow and inefficient; but the alternative of direct instructions and complaints is much worse. The positions of foremen and superintendents must be protected; they must preserve their authority in the eyes of their subordinates. This is not just a matter or pride or "pecking order"; if the contractor is to expect his workers to be loyal and efficient, he must show himself to be reasonably efficient himself. If an employer is seen to knowingly accept a poor supervisor, the worker will have less belief that his own efforts will be rewarded. An employer may have a poor supervisor, and this is accepted as something that must occasionally happen; but when the employer shows, by giving direct orders to the men, that his own superintendent is *known* to be incompetent, it is expected that the employer will discharge the superintendent. Construction workers know who is competent and who is not; and they have no respect for an employer who does not develop the same knowledge. This is part of *morale*.

The superintendent may decide or approve whether or not a man may be discharged, praised, or given a raise, but the action itself should be carried out by the foreman, who has the option of telling—or not telling—the employee who it was who originated the order. Direct orders to a supervisor, in matters involving his men, should be kept to a minimum. The superintendent should know the capabilities of his people; if he does not, a new superintendent is needed. On a union job, the men will complain immediately if a foreman is bypassed, and a man laid off by anyone other than his foreman may have a valid claim of payment.

Field supervisors, and executives who were formerly field supervisors, are very jealous of their prerogatives to supervise their men exclusively. In the field, unlike in the office, workers are not constantly supervised and may make costly errors by misunderstandings resulting from receiving direction from two sources.

2.5 THEFT

Theft from construction sites represents a considerable expense, as much for time lost in the work and cost in reordering as for the cost of materials stolen. Nearly all such theft is by the workers on the job, or by persons assisted by site workers; most thefts occur during working hours. Sometimes workers hide material during working hours where they can come back for it later.

On small jobs the cost of a guard, day or night, may be greater than the loss prevented. On large jobs, the presence of a guard requires thieves to limit themselves to what they can carry without a vehicle; but since people learn where materials are and where the guard is, the guard cannot entirely protect the work even at night. The magazine *Construction Methods and Equipment* estimated that a contractor loses 1 to 6 percent of his equipment each year. For 1970 the Subcontractors Trade Association of New York put the average subcontractor firm's annual loss at nearly $16,000 and their members' total theft losses at $50 million—which the Association considered a low estimate. Some theft was reported as *intended* to cause delays and overtime pay, and the overtime pay due to theft was estimated at $6 million in the same year. In one instance, $50,000 worth of cable was installed on a weekend but was stolen by Monday morning.[4]

To reduce theft, the work should be isolated as much as possible from surrounding public areas. Workers should be given a building for their tools and required to use it, so they do not carry a box back and forth each day—which box may carry stolen materials or tools—or the box itself may be stolen. The contractor must then assume responsibility for workers' tools left in the shack—which is required by some union contracts anyway. Losses of workers' tools, both real and claimed, will be a fraction of that which might otherwise be carried away in toolboxes. Automobiles should be parked at a distance from the work, and the building lot fenced so that workers may be observed entering and leaving the work. Trucks leaving the job should be inspected, with no covered loads allowed to leave without passing inspection. Trucks with small shipments, also carrying shipments for others, should unload at a receiving point where they are not required to go through the site. The cleanup crew, who have one of the few loaded trucks leaving the job, must be closely supervised, with occasional changes of personnel. Persons working overtime must be checked in and out. If the work can be adequately fenced, dogs are useful at night; some firms rent dogs who will not allow anyone other than their trainer on the property. But dogs have also been stolen. Adequate tight storage should be provided for such valuable and useful items as transits, power tools, and door hardware. People will steal material they cannot possibly use, such as high-voltage light bulbs, out of ignorance or curiosity. Theft is as likely in a high-class residential area as in a slum; what would a person without money to buy material, who owns no house, do with a power saw? He can sell it, but this requires a second—and more risky—operation. A professional man, however, may stop to put a two-by-four or a power saw in his Cadillac.

The job superintendent will usually decide or recommend if he considers guards necessary, on the basis of theft losses on his jobs or on similar jobs similarly located. The contractor may be responsible to neighbors not only for children injured outside of working hours, for lack of a guard, but for fire damage as well. In a recent case, a building, while being wrecked, burned, damaging adjacent buildings; the contractor was obliged to pay for damage to the buildings, as he had not provided a guard on the work. The cause of the fire was unknown, and no one had seen anyone enter the site, but the lack of a guard was held to be in itself negligent.[5]

Vehicles without steering wheel locks, with ignition locks only, can be started in a few minutes by even an amateur mechanic. They should be guarded as if unlocked, and parked at night in a locked area. Small equipment such as vibrators, pumps, and saws, with factory serial numbers, should be marked with the company name and the serial number recorded.

2.6 COST CONTROL

Control, as applied to costs, reflects some forgotten optimist who hoped, by calling cost accounting *control,* that he could make it so. Literally, to control means that one can cause an operation to come out one way or another, as desired. But the system provides control as a steering system provides a car; you can guide it to some extent but you cannot stop it. The reporting system is a tool the manager uses more to find out where he is going than to directly

[4]Allan J. Mayer in the *Wall Street Journal* of June 19, 1972. There are no reliable or overall figures available.
[5]*Aetna Insurance Company* v. *Three Oaks Wrecking and Lumber Company,* 382 N.E. 2d 283 (Ill. App. 1978), reported in *Engineering News-Record,* January 25, 1979, Vol. 202, No. 4, p. 43.

stop the journey. In the construction sense, *cost control* means that costs are planned, may be determined as the work proceeds, and if not in accordance with the plan, the reasons can be determined. *Cost control* may also refer, where used by a firm in another business, as the entire management process, in that work or expense without authorization is not started, but the term is not so used in this text.

Some reasons for cost *overruns,* or in excess of planned costs, include poor management at any level. *Out of control* refers to costs which cannot be forecast, and which may increase without authorization or subsequent explanation. It may also refer to increasing costs of labor or materials, which cannot be forecast at the time. If no cost control system exists, the conditions out of control cannot be determined until some kind of cost report is made. This cost report may be the final determination of cost at the end of the project. As used in this text *cost report* and *cost analysis* are identical, although cost analysis implies a certain amount of detail which may be omitted from the cost control report, and a rationalization or explanation of variances.

Control as applied to time and scheduling is more definite. It is the ability of the manager to keep construction on a previously determined schedule, except as affected by changed circumstances. If a job can be kept on schedule, there is an implication that it could be made to go faster. There is always a variation of cost with speed of construction, although increased cost may result from either more rapid or less rapid progress. Progress scheduling, to be effective, must have available additional resources—material, money, people, and management—to make it effective. If the cost variation due to changes in rate of progress is known and can be applied to increase progress, the project is fully under time control. The popular concept that planned time is always exceeded is due to the restrictions of money—if additional money cannot be spent to increase progress, there is no control of progress. The job stays on time as long as costs stay in line and cutting the work force is a way to cut unit labor costs in many situations, but causes a later completion date.

The popular conception is that a project should be completed as rapidly as possible, as long as an increased construction cost does not result. This is not true in many cases; if construction is planned for completion when needed, early completion results in increased interest and maintenance costs. This is apparent only on large jobs—when housing is being completed, for example, the rate of completion should not exceed rate of sale or leasing. If one part of a project depends on another—as a power transmission line does on a generating plant—it is more costly to complete one earlier than the other. On the other hand, such an order as "Don't order any material that will arrive before it is needed" is utter nonsense; material *must* arrive before it is needed.

2.7   FUNCTION OF A COST CONTROL SYSTEM

A young accountant once defined cost control as "accounting for expenditures the way the home office wants them." Such confusion of the clerical system with what it is intended to accomplish is only too common. Each department sees its part of the whole; only the manager has an overall view. The manager, therefore, must be as close to the work as possible. If the foreman can use the system, it can be summarized for the general manager's use; but if made only for the manager, the foreman cannot use it, nor can anyone below the level of manager. People at lower levels will not take an interest in it, and will give whatever figures to the manager that may make him happy at the time. Hence the saying, "the first 90 percent of the work takes 90 percent of the money, and the last 10 percent of the work takes 90 percent of the money." That is, reports will follow what is expected, until at the end this is impossible.

If reports are made for the manager, even he can tell what is wrong only in large units—which means that he knows costs are overrunning but can do nothing effectively about it. In the construction industry, *cost control* and *job cost control* are synonymous.

To be effective, a cost control system must furnish as many as possible of the following results.

1. A constant and timely measure of the individual ability of as many employees as possible, which may serve as a basis for improvement of efficiency by promotion, bonuses, and discharge.
2. Accurate data, clearly defined in detail, for future estimating.
3. A currently corrected estimate of total cost, and consequently cash requirements and profits.
4. The cost of extra work, for justifying invoices. Practically all firms are forced to keep this aspect of cost records, even if they do not keep other cost accounts.

It may also furnish current invoicing, progress, and profit accounting information. The progress information may be used to update the final completion date through CPM or other scheduling methods, as explained in Chapter 7.

Some authorities differ on whether the cost data can actually be obtained in detail, with manual methods. There is no practical problem if the effort is well organized, as any other part of the work. The AGC book *Cost Control and CPM in Construction* is rather vague about how to obtain cost control data, apparently leaving open a less detailed option in order to make the manual as widely acceptable as possible. Data processing people insist that the time-consuming detail for weekly labor cost reports must be mechanized and, being unfamiliar with the difficulty in obtaining reliable data, often overrate the reliability of the results.

2.8   COST CONTROL AND ACCOUNTING

The accounting department, in general, is not able to establish and maintain adequate labor cost control. There is no reason why accountants cannot learn this skill; but the writer's experience is that accounting-department-centered systems are not often satisfactory. When an accountant understands what is required, he often prefers it to be done by a project clerk or cost engineer, on the site.

The chief difficulties in keeping these records in the accounting department are the following.

1. *Time.* Labor cost reports should be completed weekly. Accounting reports are made at longer intervals, and people may be diverted from weekly reports to handle the accounting reports considered more important. Labor cost reports are needed in a few days, while accountants expect to take a much longer time to prepare them. In general, reaction time for correcting a bad situation will take three report periods—one to show up, one to confirm, and one to take action. Such action is important and definite—like discharging a superintendent or subletting a part of the work—and reports are individually unreliable. So several are needed.

2. *Complexity.* Accountants deal with basically simple situations, within the framework of purchasing and approval authority adopted by the organization, but cost reports are needed promptly on many items. The data are used for only a week before a new report is received. To keep the clerical labor required in reasonable bounds, a number of shortcuts are used which are not suitable for profit accounting and are therefore not used by accountants. Among these are checking some costs and not others and ignoring failure to balance; or in some cases, by single-entry bookkeeping. The distribution of premium overtime may not be consistent, and percentages rather than actual computation are often used, as for payroll fringe benefits and freight.

3. *Technical decisions.* These are woven throughout the fabric of cost control. The decision of how to separate items, what figures are currently useful, and the estimated percentage of completion of labor items all require technical knowledge of the particular job, and considerable field experience.

4. *Transmittal time losses.* Most of the required information must come from the job, and results must be returned to the job site. If this must pass through the mail, prompt use of the data requires that they either be calculated on the job site or transmitted by telephone. Telephone transmittal is possible by teletype, which transmits sound, but its application on construction jobs is not common. Large jobs may have quite detailed on-site systems. On foreign work, conventional private-line teletype is often easier to obtain than reliable telephone service, as it bypasses politically operated telephone systems, so this teletype is used for cost reporting.

5. *Orientation.* Accounting is intended basically to provide information to the general manager. Accounting responsibility must be reoriented to furnish information useful at lower levels; this will be done only if the manager takes an active interest in seeing that it does. This is especially true in a firm of which only one branch is devoted to construction. The accounting system is then tailored for the other departments or branches. Manufacturing and engineering firms seldom have adequate cost control on their construction. For this reason they usually prefer to have their construction done by outside firms—oddly to the writer, even on cost-plus contracts. Profit accounting is by and for merchants; cost accounting is by and for engineers.

6. *Lack of uniformity.* A characteristic of construction is its lack of uniformity. An accountant prefers to standardize the construction cost system with systems with which he is familiar; and if a company has other interests, he likes to make a system uniform throughout the company. To the extent that the construction reports resemble those from farming, entertainment attractions, or assembly-line operations, to that extent they are inadequate for construction control.

Cost accounting is so different from profit accounting that a duplication of entries is usually necessary; a cost accountant keeps separate records that he hopes to reconcile with the general accounts. In such a case, it is immaterial whether the accounts are kept under the supervision of the accountant or of the project manager, as the cost accounting will be free of profit accounting technical and time restraints. Theoretically, machine data processing systems can handle both types of accounts with but a single entry, but the data processing people and the construction managers are usually too limited each in his knowledge of the others' methods to design a combined system. Also, the time constraints of the cost reports may be impractical to build into a system for different end products. Two data systems are usually necessary; and if CPM scheduling is used, three separate systems may be used.

2.9   OPERATION OF A COST CONTROL SYSTEM

A cost control system includes the following basic elements.

1. Designation of a piece of work for which costs are to be kept separate from any other work, and the kind of costs; for example, labor forming exterior walls.

2. Identification of the cost with the designated item. There must be an identification in the mind of the estimator, project manager, and foreman marking the time card that a certain number means certain labor operations, identifiable with hours of individual workers. This identification is very often not clearly available to the foreman in the same manner as it is to the estimator and manager.

3. Accumulation of the costs according to the item. This is a clerical procedure of calculating and adding the day's time costs in each cost item. To a bookkeeper, this process is posting costs to a cost ledger account. These reports need not be part of the permanent books of account, and may be discarded when no longer needed.

4. Summary of the costs by items in a useful form. Any detailed system will include, at any time, considerable amounts of data not immediately useful. Computer runs disgorge hundreds of pages of such data. Some methods of summary are: to report only on items with considerable variance, items with large current payroll, or items which have overrun several periods in succession.

5. Comparison of costs with work done. The value of work done must be calculated by units or percentages. This is the most difficult part of cost reporting, as this figure is largely, and sometimes entirely, a personal estimate based on experience. This is usually done by the superintendent or other executive; the only way to cultivate the talent is to start doing it and to compare results with estimates later. Accurate results should not be expected until the individual is well experienced; usually this takes several years.

The data described above must be evaluated for action by the manager. However, most of the value of cost reporting is attained by the same superintendent whose activities are reported on. This is a self-administered system, used by the persons who create it; it is the basic document of construction management.

The description in this chapter is confined to *labor and equipment cost.* Material costs are fairly well known when the work begins, change slowly, and are not under the control of the project manager, as are labor costs. Costs of materials cannot normally be blamed on any person in the organization, so no action can be taken. The contractor is in the position of separating the cost elements he can do something about, and disregarding the rest. Material costs may be handled as part of the normal profit and loss accounting system, with overruns and losses discovered through accurate ordering and overrun purchase orders for shortages.

Equipment cost, for job costing, may be kept on an hourly basis, as are labor costs, with a standard hourly cost. The relationship between the standard hourly cost and any of the other equipment costs computed may be very great, but the writer prefers this method for heavy equipment as being most indicative of efficiency of management. Some firms ignore equipment costs in job costs entirely, assigning a lump sum or monthly figure.

Designation of a piece of work as a cost item is done by the estimator or project manager and is an important part of preparing the job budget. A numbering system is often used for identifying items, and many contractors use the same numbering system when they estimate items. For example, if 101.16 is used for ladder reinforcing in blockwork in one job, it will be used for this item in all jobs. See the *Uniform Construction Index* in the Bibliography for one such system, which has the advantage of the same number headings or divisions as architects usually use in building specifications. The advantages of using the same numbering system are as follows.

1. The same checkoff list is used for all estimates.
2. The particular item can be readily located in the estimate, from a master index.
3. The information can be filed for estimating information according to the master index.
4. The use of decimals, although it is not characteristic of the system, makes summarizing by trades simpler, as all items pertaining to a larger category have the same first number.

These advantages are greatest when jobs are small or have little variation from one job to another. However, when uniform cost codes are applied to large or complex jobs, or jobs differ radically, as from a hospital to an apartment house, or a water treatment plant to a bridge, there are disadvantages, as follows.

1. Many items in the index become so small as to be negligible on some jobs, and when this happens each item is still shown separately, wasting time on small amounts of work without giving useful information. Of course, an item that does apply at all is not shown.
2. Checking for possibly omitted items in the case of the decimal system requires time to check an index. With a consecutive whole-number system, omitted items show directly as missing numbers. If a decimal system is made consecutive to eliminate checking an index, the number to the right of the decimal point must be reassigned for each job, again wasting time.
3. With a system with standardized numbers for *all* items, not just major designations, the numbering must be changed for any item so large that it needs to be broken apart into several smaller ones.
4. A decimal system is subject to more errors in reporting, as it requires more digits in each number and, most important, requires a period. A period is difficult to recognize when written under field conditions, as to be noticeable it must either be a line or a circle—each of which can be interpreted as a number. A hyphen is preferred for this reason.

When machine data processing is used and numbers for the same type of work vary from one job to another, the difficulty of filing estimating data can be readily overcome by cross-indexing numbers for a particular job with the master index in the machine itself. Coombs presents the argument that estimate items should *not* have a uniform numbering system but should have a system that will force estimators to obtain *details* of the previous unit cost before they use it. The availability of microfilmed plans as permanent records, to accompany job files, increases the availability of complete cost information. Similar microfilmed plans may be retained for jobs previously bid but not built; this information is useful in following bids of the same type of construction.

Many contractors use a work designation such as columns, fireplaces, or slabs, rather than numbers. These may be standardized and extended to any size operation, and a word index is necessary. The disadvantage of such an index is that the name has a specialized meaning nearly as abstract as a number, but people using the system may think they know what it is without referring to the index.

The numbering system will have to be modified or enlarged for each new unit price for an item, or a complete new system may be used for each job. Three methods of coding (numbering) cost items are compared in Table 2-1. There are an unlimited number of variations. The first column, shown as names, may be abbreviated still further, with letters used for words, or a combination of letters and words used. The fixed code shown could as easily have been one using whole numbers, provided that the total number of digits used is uniform. The Uniform Construction Index numbers shown can vary also; the first two digits, 03, refer to concrete. Numbers are allotted 03100 to 03149 for concrete formwork for example; all may be used, or only one of them. The fifth digit and any further digits are assigned by the contractor's own method; the Uniform Index does not extend that far. In this entire column, another contractor using the uniform Index might have all numbers different other than the initial 03.

Table 2-1. Variation in Cost Account Numbers

| Abbreviated Name Code | | Fixed Number Codes | | Variable Number Codes | |
		Contractor	Uniform Construction Index	Decimals	Whole Numbers
Columns	1st floor forms labor	12.1	03101	1.10	2 (w.o. #2)*
"	material	12.2	03102	1.11	0310
"	1st floor concrete labor	13.1	03303	1.13	3 (w.o. #3)
"	material	13.2	03114	1.14	0310
"	1st floor bars labor	14.1	03205	1.15	4 (w.o. #4)
"	material	14.2	03206	1.16	0310
Slab on grade	Forms labor	15.1	03107	2.10	1 (w.o. #1)
	Concrete material	15.2	03308	2.11	0310
Slab	2d floor forms labor	16.1	03109	3.10	5 (w.o. #5)
	Concrete labor	16.2	03311	3.11	0310

*Work order.

In the variable decimal system, the first digit represents a class, such as first-floor columns; the decimals are items within the class. The writer favors numbering as in the last column, where labor items are carried entirely separately, as they will be on a separate cost report, and all similar materials are combined into a single number, here a convenient one from the Uniform Index. Since 0310 is the code for concrete formwork, its use here is a variation of the Uniform Index. If a contractor says he uses the Uniform Index, this does not give you a great deal of specific information, as illustrated here.

In the last column, labor items are numbered in the order that the work will be done, and they are reported in a separate budget. The cost code number shown is the same as the work order number, but it does not have to be. The job people can use the work order number, which is then coverted to a more complicated account number on the page on which cost data are recorded. With mechanical data processing, this conversion may be programmed into the machine. An accountant is not concerned about a work order No. 5, which he is going to post as 1963.218Cjk; both numbers may be shown on the ledger page or on the work order, and work orders may be combined as is convenient for other records. Only on the labor cost report, which is used primarily on the job and need not be sent to the home office at all, and on the time sheet, is the work number order number used.

This description appears to be a lot of concern about a minor clerical point, which is true; but many writers and accountants attach a great deal of importance to the particular system in use. As the manager can often revise the system completely to his own needs and leave the accountants unconcerned as long as their numbering system is untouched, the manager must be willing to work with any system.

2.10 IDENTIFICATION AND CLASSIFICATION OF COST ITEMS

Identification, as mentioned, is the description of the cost item. There is often confusion because the estimator sees it as a physical entity, and the foreman must record the time spent on it. However, a considerable amount of time is spent on activities which include physical work other than that which can be identified with a particular portion of the construction. Quantity identification—that is, a clear understanding of how quantities are measured—is also important for cost control when quantities, rather than percentage completed, are being used to measure work in place. There have been a number of efforts to standardize measurements, and in British countries there are methods which have the sanction of professional quantity surveyors. In the United States there are no standardized methods, although the highway departments have standardized their work to a considerable degree.

Each person believes that the name he applies is correct and definite; anyone else must be ignorant. If a manager gains experience in different areas and countries—even those using the same language—he realizes that every word not only has a regional meaning, but to some degree a personal meaning. Most political and sociological arguments are over definitions of words. A contractor's employees, in reporting costs, follow the contractor's lead in definitions. But as the business grows, new superintendents and estimators do not understand the nomenclature, and reported costs become confused. Does *footing excavation* include backfill, usually more expensive than the excavation itself? Gravel or concrete in soft sports? Compaction of the backfill under slabs? Overexcavation and sand backfill? Roots, which can be included in clearing? Are footing grade stakes for concrete put under excavation, layout, or placing concrete? The list is endless. More experienced men know even more reasons to classify an activity under one name or another, and become just as confused as the novice. Some contractors use "footings concrete volume" to include excavation, steel, forming foundation walls, pouring concrete, and rubbing foundation walls as a single estimating and cost item.

Proper identification begins when items are separated on the estimate, as this is the source of budget information. It is common in books, and somewhat less common in practice, to use the items in the estimate for the budget cost items without further identification. If this works for you, fine; if not, you can always return to the more detailed procedure. Many managers say they use the estimate identification as is, but in practice they usually modify them—sometimes quite extensively.

Materials and labor should be separate. The larger the quantity and the more uniform the cost of material, the less the manager is concerned with its frequent calculation for cost reports. The foreman calculates concrete quantities of each pour because he orders the concrete separately and must be sure the company is getting what it paid for. The foreman must also determine the amount of concrete to be ordered for the last delivery of the pour. There is no purpose in separating materials of slightly different prices or in separating different materials that are used together. For example, column concrete is often stronger and more expensive than floor slabs, but the money difference per cubic yard is small. In a wall, you may have brick of different kinds—mortar, block, and reinforcing—but all are one cost items. One can always recalculate material used to see if an estimating error was made; there is no need to encumber the weekly cost report with such details. You may notice that you immediately depart from most estimates when you combine material cost, as the estimator prices each item separately.

Now that labor is a separate item, designate *items that really have a purpose, are large enough that a small error, as one man-day, will not materially affect the result, and an item that the foreman or timekeeper on the job can readily separate from other items.*

A contractor sometimes needs unit costs of an item, even though it is very small, for use in estimating. In such case, the manager must give special attention to labor cost reporting on the item, bearing in mind that workers under observation will nearly always react in a way that will raise or lower unit costs—by dragging if they think the cost will be used to establish piece work, or more often, speeding up, to demonstrate what they can do. A superintendent acquainted with cost reporting can make the costs come out the way he wants them to.

The writer recommends the following guide for separating work into cost account items, or *work orders.*

1. The work order should be completed within four weeks of starting, rather than last a longer time. Fewer work orders are reported on at any time, and the total is obtained while the memory is still fresh. Work on longer items should be divided into two or more successive work orders. For example, excavation for a wall and backfill may be separated so that determining the cost need not wait for construction of the wall.

2. Each work order should be for work supervised by a single foreman, so that the efficiency of individual foremen can be measured. It is also important to recognize the foreman and allow him to demonstrate his ability by a separation of his work from that of others.

3. Within these limitations, work orders should include as many of the same kind of units of work as possible. For example, a work order to set door frames should include a large number so that the average cost is representative. An item should be for at least

five worker-days, if possible. Smaller units are unreliable because the time that workers change jobs is usually not precise.

4. When precise results are need for estimating, as the cost of beam sides separated from bottoms, or a comparison of two similar methods of doing the same work, spot records are useful. These are short-term records, with direct observation rather than merely recording the work done over a period of several days. These spot reports may be adjusted for the individual, the weather, condition of materials, and similar factors which are not observed in ordinary cost reports. Short-time items give imprecise results when they are not directly observed and adjusted for special circumstances.

In summary, the desired results must be compromised with the practical, requiring reports which may be readily obtained by current personnel and methods. The work included in an estimate item and that in a budget item may be made to coincide; but before you require that they do, you must consider the extent to which one or the other is made cumbersome compared to the cost of converting the items from estimate to budget items, on the successful bids.

Few contractors separate cost items in such detail as here described. Cost items often last for the duration of the job, and the work of each foreman is not separated. If you make a breakdown as described above, you can produce a CPM chart for the on-site work by the general contractor by showing on the work order for each item what item will precede and what item will follow it. If the dates for each work order are established to suit the work force, you have the advantage of including manpower planning in your CPM chart initially. If you know what you are doing, if the work is not too complicated, and if there is no problem with material availability or subs, you will have the same completion date that the CPM computer run will give you after clicking and flashing lights over the same data transcribed to tapes or cards.

2.11   THE WORK ORDER

The work order system described above and following is unusual in construction only in that most contractors list these descriptions on one sheet rather than use a page for each order (cost account), and pass other information to the foremen by other means—usually by requiring that the superintendent personally observe the work and instruct the foremen. The writer prefers to use a separate sheet of paper for each cost item, with a description of the item. In order that the simplest possible clerical work be required of job supervisors, these cost items are numbered consecutively. The term *work orders* is used in shipyard work very similar to the method here shown, and *job orders* are a corresponding term for shop work. The work order gives the foreman a description of the work to be done and charged to that number, and some of all of the following as well.

1. Authorization to proceed, which is important where changes are expected.
2. Verification of the latest drawing number and revision.
3. The construction method to be followed. Some firms use an order for this purpose alone, and have a detailed procedure for writing such orders.

4. Charges for extra work. All contractors use an order for work for which they must justify costs, usually termed *extra work order.*

5. Special safety measures, such as scaffolding, ladders, masks, or other protective gear, and any dangerous characteristics of the material to be used, such as asbestos.

6. Location of material, or purchase order reference to materials and their delivery date. Both the materials and the orders may also be keyed to specification paragraphs.

7. Designation of the trade to be used. This is particularly important when a pre-construction conference has established jurisdictional lines in advance.

8. Worker-day or cost estimate, and time scheduled for work to begin and end. This assists the foreman in his planning and gives him a measure of performance.

9. Priority in respect to other work, the ultimate purpose of labor scheduling.

10. Cross references to account numbers, CPM schedules, file numbers, and other information sources.

11. Corrections of plans, or details of plans where the work is shown.

By returning the work order form to the superintendent when the work is completed, the superintendent is notified that the work is completed for inspection by the superintendent or other inspector.

Work orders may be written at one time, or as time permits the examination of plans and placing of material orders. If orders are small notebook size, the foremen may keep time sheets and notes from the superintendent in his pocket. Work may be started with drawings incomplete or lacking, if the work order shows the name and telephone number of the person who can furnish the missing information. Nearly every contractor has had a disagreeable experience by transmitting information regarding plans or specifications verbally through several persons. If such information is on the order that the foreman uses to make out his daily time report, the superintendent knows that the foreman has received the special instructions or information before he starts work.

Work orders are sent down the line of supervision to the trade foreman. The superintendent may put instructions of his own on the order and may use it as a check list for establishing work sequence and assignment of workers to the various foremen. When the order reaches the foreman, it is an order to him from his immediate superior.

If changes are anticipated in part of the work, work orders on that part may be delayed. By checking his file or orders retained, the superintendent can promptly inform the architect or manager on request, if work has begun on any item.

Copies of work orders are kept by the clerk preparing regular cost reports. A change order or addition to the contract is handled as a work order, and billing for the change is prepared from the work order clerk's work sheet, or from the ledger page of costs for each work order. Since the foreman who charges time to work orders does not need to know if they are change orders chargeable to the owner, the owner or architect is assured that there is no temptation for the foreman to report time improperly on change orders.

An example of a work order is shown as Figure 2-1. Items shown are explained below.

1. *Slovak and Carp.* indicate that Mr. Slovak is the foreman in charge of the work and that the work is of the carpenter trade. If there are many foremen, the superintendent may

Acct: __0811__   Work Order No. __47__

Date __3/15__   Job __S.J. Hospital__

Foreman __Slovak__   Trade __Carp.__

Install sheetmetal frames (20) and leadlined frames (3) on second fl. This charge includes partition layout, bracing frames, handling, removing, storing and cleaning bracing, shotin studs to floor, nail to overhead.

Begin __3/19__   End __4/20__

Est. WD __20__

Material: SM frames   PO57-Pioneer
          Exp. Smith
          LL frames   PO46-Suraf
          Exp. Smith
          (555-2323)

Plans: Check Sh. 36 for Rev. 3
Superintendent's notes:

Man using stud driver must have card signed by me. Check finish schedules to figure wall thickness each frame —

B.J.

Figure 2-1
WORK ORDER

choose the foreman for the work. The trade is identification, but may also be a determination of jurisdiction if there is doubt about which trade is to do the work.

2. *Quantity and units* given in the instructions help the foreman check his material and make percentage completion reports by account when the quantity is small.

3. *Beginning and completion times* should comply with the CPM schedule when it is kept current, and the foreman gives priority to the earlier dates.

4. WD indicates worker-days. This estimate enables the foreman to plan well in advance the number of workers he needs and to adjust the size of his work force. This approximation does not need to conform exactly with cost item estimates.

5. PO indicates the purchase order for materials needed. With this number, the foreman can call the expeditor directly, or the job clerk can trace materials without asking information from the superintendent or consulting the files for information.

6. The *superintendent's notes* illustrate that the superintendent has checked the drawings carefully, foreseeing possible errors by the foreman. He need not find the foreman to give him oral instructions.

2.12 ACCUMULATION OF COSTS

The basis for labor cost accumulation is the time report, listing the workers by name, the hours worked, and the work they did. Some firms use a weekly report, usually with a carbon copy the foreman may keep when he submits the original. The foreman or a timekeeper may complete the report, or it may be completed by the foreman and the total time checked by the timekeeper from his daily records. The weekly report is preferred by many because less clerical time is involved; if a daily report is used, the weekly report may be required for calculating payroll.

The writer prefers the daily rather than the weekly report, for the following reasons.

1. Labor cost figures are available daily. It is rarely possible to figure the gain and loss on labor daily; but the overall figure, by work orders, shows the jobs workers are on.

2. An error on the weekly report will not be checked until the end of the week, when memories are cold.

3. Although the total clerical work is greater, most of the work is completed daily, leaving less for the end of the week. With the same labor force, it is therefore possible to pay sooner after the end of the workweek.

4. If foremen are forced to record the cost report part of the payroll each day, they are less likely to wait until the end of the week—an impossible task. In addition, if the foreman is off from work the last day of the pay week, a weekly report may be delayed until he returns.

5. If charges are reported on days with no charges on intervening days, this indicates that foremen are leaving jobs unfinished and must return to them again.

6. Smudging and folding of the time sheet makes it less legible if the foreman carries it all week, rather than turning one in each day.

7. The daily time sheet may be printed up only two or three days in advance, so the list of men is more up to date, and the foreman has fewer names to add to and deduct from the list. On the weekly sheet, he has a week's changes, or more. The office needs the old sheet to make the new one, but the foreman needs it the next day to record his time. Hence the foreman either has no sheet one day, or two weeks' sheets must be printed at once. This increases the number of changes that he must make.

8. A prompt report makes dishonesty on the part of the foreman, padding the payroll, more difficult.

A sample daily time sheet, as it would come from the foreman, is shown as Figure 2-2. This sheet is made up in the office from the previous records, but with only the names and badge numbers, if used, and pay rates. The hours, and workers added or taken off the foreman's crew since the list was made, are added by the foreman. After the time is completed by the foreman, one method of making calculations is as shown in Figure 2-3. If hours are to be accumulated for a weekly payroll, extensions can be rounded off to the nearest dollar, although pay *rates* will have to be kept to cents. The payroll data are calculated as shown; if an 8 ½ - by 11-inch sheet is used, the extensions may be made on the report itself. *Extensions* are arithmetical products. In payroll, they are the product of time and rate of pay; in estimating, the product of dimensions to produce quantities or, more often, the products of quantities and unit costs. Hence the estimating sheet where total costs per item are obtained is the *extension* sheet. It is also called the pricing sheet, since unit prices are placed on the items at this point.

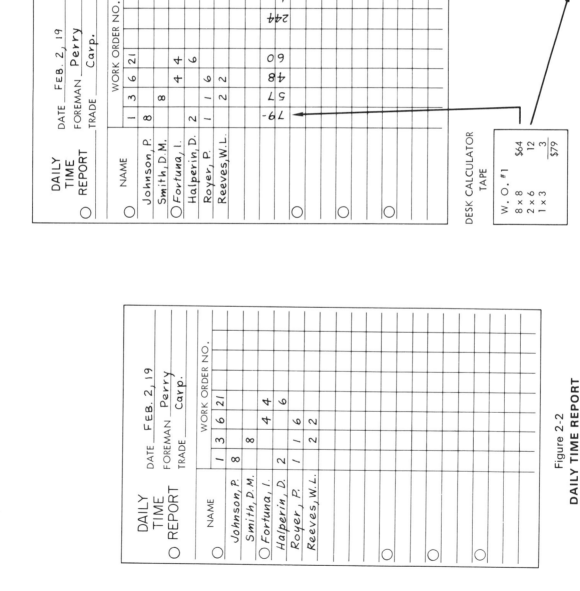

Figure 2-2
DAILY TIME REPORT

This report is made out daily by each foreman. It has the advantage of fitting in a pocket-size notebook. (*Courtesy of Lefax Publishing Co.*)

Figure 2-3
ACCUMULATION OF LABOR COSTS
(*Courtesy of Lefax Publishing Co.*)

If a small page is used, the page may be machine-copied onto a larger sheet which files easily, permits further computations, and allows the original report to be sent directly to the home office for permanent files or auditing. The daily charge to each work order is calculated, and matched to the daily payroll, $244 in the illustration. The amount charged to each work order is then made on a card, which is easier to handle than a ledger page; either may be used. The payroll data shown on the sheet may also be multiplied and added, with only the daily total recorded. The weekly payroll clerk must then go through the time reports for his hours, and an error at this stage is difficult to locate.

Figure 2-4 illustrates a simpler method of calculation, for weekly calculations. This method allows only two hourly rates, which is less precise, but it eliminates many operations.

The first step is to enter the payroll costs in the lower right-hand corner, as account No. 1, $1,126.00, is the total of payroll checks. Account No. 10, $219.06, is the amount withheld from payroll for income taxes. These are shown separately because they are available from the accounting records, but their sum is not. Account No. 11, $201.72, is the estimated cost of social security, insurance, and other costs; a result of the payroll amount. Some contractors handle this as an overhead account, in which case it is not distributed as job cost but added later as an overhead percentage. If total costs are desired, all these fringe costs are added each week, as shown here. In the example, the total payroll cost is $1,546.82.

Laborer hourly rate in the illustration is used here as a base, and carpentry hours are counted twice as much—that is, two "equivalent hours." This proportion may be any number. If several different pay rates are used, the use of a single proportion and wage rate gives the average for each classification of hourly worker. The calculation is simplified, and the result is sufficiently accurate for cost accounting purposes. The method divides the total payroll proportionally according to hours, which does not charge more than two pay rates, or any overtime, exactly.

The steps of calculation are as follows:

1. From daily time sheets, enter work item symbol (as D, H, J, G) and numbers of hours for each item on the corresponding line (as 8, 8, 2, 3, etc.) for skilled and unskilled categories.
2. Add up hours skilled and unskilled and enter in *total* columns (as 29, 9, 24).
3. Check total hours on job charges against total hours on daily time sheets for the week.
4. Multiply skilled for each job (as 29 hours for D) by 2 (or other ratio used) and add to unskilled (as 21), and show total in left column (as 79).
5. Add equivalent hours in left column (179) and divide into total payroll cost of average rate ($8.64). This number is left on the calculator, so no significant figures are lost—not important in itself, but it assures that the total charges will equal total payroll.
6. Enter total hours for each job (as 79 for D) in lower table.
7. Multiply hours for each job times rate (79 X $8.64 is $682.67) for the charge to each job.
8. Add costs for each job ($682.67, 224, 68, etc.) and check this total ($1,546.81) against total labor cost for the week. They should be the same.

By this method, one clerk can distribute the payroll cost for 20 men to 10 jobs in about 20 minutes, and the distribution total will always match the total cost in accounting records. If this method is used for billing cost-plus work, the customer must accept in advance billing on an

WEEKLY LABOR COST DISTRIBUTION

Week ending 12 - 10

Equiv. Hrs.	Job	Skilled Manhours	Tot. Sk. Hrs.	Unskilled Manhours	Tot. Unsk. Hrs.
79	D	8+8+2+3+8	29	8+7+2+4	21
26	H	4+5	9	4+4	8
72	J	8+8+8	24	8+8+8	24
2	G			2	2
			Total 62	Total	55
179					

D - 79	$ 682.67
H - 26	224.68
J - 72	622.18
G - 2	17.28
Total	$1,546.81

Sk. Equi. Hr.	124
Total Equiv.	179

1-	$1,126.00
10-	219.06
11-	201.76
Total	1,546.82

$$\frac{\$\ 1546.82}{179} \qquad \frac{\text{Dist.\$}}{\text{Equiv. Hrs.}} = \$ 8.64$$

Figure 2-4

WEEKLY LABOR COST DISTRIBUTION

62

average worker-hour cost rather than actual cost. In most cases, it is not suitable, as each customer wants an actual list of names of persons employed on his work, and their pay rates.

Many variations of the foregoing are possible. Many contractors do not charge payroll-based costs such as workmen's compensation hourly, as exact calculation is tedious. This writer charges these costs as a percentage, even though they are not, to an internal clearing account. This is an account used to approximate variable charges; the difference between actual and calculated charges is found, and it is charged to expense or income to balance the accounts—and keep the IRS happy. In this way, lump-sum contract costs are kept current in the same manner as are cost-plus charges, even though the actual fringe payments may not be completed or accurately known for months. By necessity, these fringes are approximated on cost-plus contracts.

Only the active cost accounts are kept in the card box, through which the bookkeeper looks for the work order required; consequently, on a job with 300 work orders, only 50 may be under way at any one time. Cost accounting may be mechanized by entering daily time reports on a computer; the time to make the entries will be substantially the same, since the same items must be recorded, but errors are easier to locate. There is no double-entry or self-checking feature for work order mischarges, regardless of whether a machine method of entry is used; that is, if a charge is made to #10 rather than #100, the accounting clerk will accept it—whether it is a mischarge by the accounting clerk or by the foreman. The writer recommends that the cost accounts for each foreman, as well as the summarized cost accounts, be given to the foreman. He can usually recognize his own charges and correct errors, and feels much more a part of the cost and entire management process.

Equipment costs are charged in a number of different ways. Figure 2-5 is a daily equipment report which shows the work on which the machine has been engaged, the payroll of operators, and complaints of the operator. An operator is unlikely to forget a report that is necessary for his own pay, but he may be careless about submitting a report that is separate from his time report.

The writer prefers equipment cost on an *operated* basis; that is, the work order is charged with an hourly rate just as though the equipment were being rented from another firm. The charges are then carried through a clearing account as for fringe payroll costs. The hourly *bogie* or *standard cost* used may be the actual hourly cost for the previous year. Several pieces of equipment may be charged to one clearing account, if no separate costs are needed. The work order to which operating expenses of the machine are charged may be included as a job cost account or may be an overhead cost account of the firm. Notice that for equipment there are two accounts—the work order to which it is charged, and the expense account to which costs of the equipment itself are charged. This is similar to considering a bulldozer an employee—its time is a cost to a work order, but the employee–bulldozer must be paid for its work. The bulldozer cost is to a firm if rented; if owner, to an account to which gas, repairs, and depreciation are charged.

The *daily equipment cost report* may be signed by the foreman ordering the work as well as by the operator. Most contractors keep equipment costs and operator's wages separate on cost accounts. For various reasons, including insurance reports, they must be separate on the overall accounts, but the job cost accounts need not be separate. The writer prefers that equipment charges be included on the same work order as labor, so the choice between labor and equipment, which must be made constantly on building construction, is made by the foreman on a cost basis. The foreman's costs then include his overall skill, not just his skill in direction of labor. He will also use equipment much less. This can be done, also, with separate work orders, depending on the type of work and size of work orders.

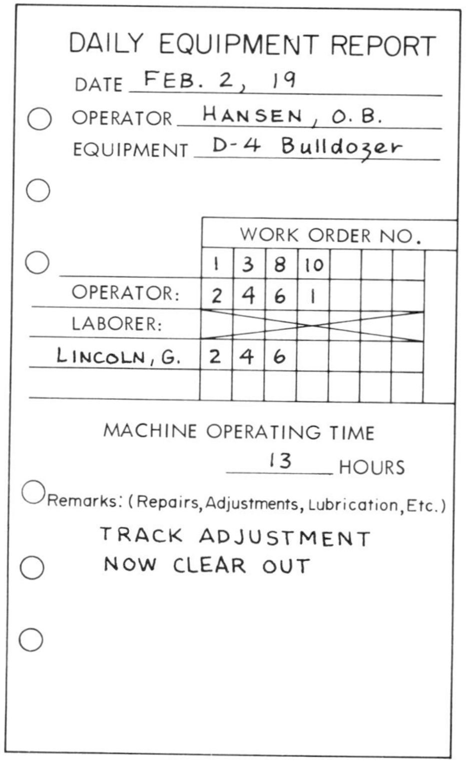

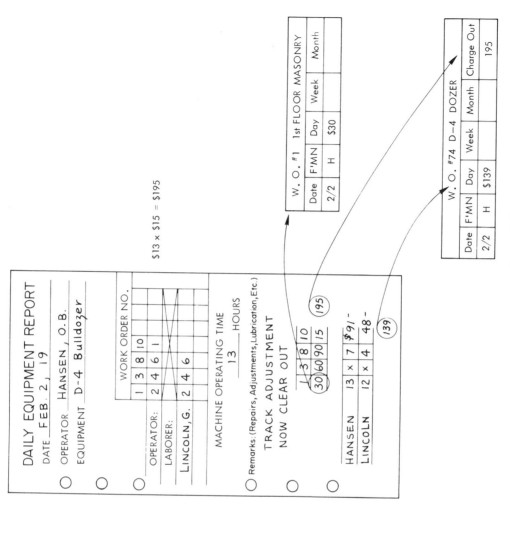

Figure 2-6
ACCUMULATION OF EQUIPMENT COSTS
(Courtesy of Lefax Publishing Co.)

Figure 2-5
DAILY EQUIPMENT REPORT
This report serves both for equipment accounting and as a payroll form for the operator and his assistant. *(Courtesy of Lefax Publishing Co.)*

A method of accumulation of equipment costs is shown as Figure 2-6. Equipment charges are calculated in the same way as labor, using an appropriate hourly rate, or standard hourly cost, here $15 per hour. Payroll costs are charged to an account of each machine or group of machines. The equipment cost may also include fuel, repairs, and depreciation, or these may be kept separate. The equipment cost account, an accumulated record of *all* hours, is a closing account in the regular profit accounting system. The equipment cost account is not an easily controlled cost, so it is relatively unimportant how often it is closed out; this may be done as infrequently as once a year. The importance of cost reports, remember, is to take action; if no action is probable even if costs are high, the job superintendent and cost accountant are not interested in it. Some contractors charge the job with *dry equipment cost,* so operator and running expenses are separate job costs; only depreciation and overhaul are included in the standard cost. This means that there must be an additional job cost account for fuel and operator; but the superintendent can control only the *usage* of the machine, not the hourly cost. The additional account therefore furnishes little information, except on quite long jobs.

2.13 SUMMARY AND COMPARISON OF COSTS

A weekly labor cost report is now well established in the construction industry, although the detail, the time taken to obtain it, and the form of the report vary considerably. Even firms who have no effective cost reporting system profess to a system. The ineffectiveness usually arises from the size and type of cost items, the accuracy of reporting, the report interval, and the time required to complete the report. The time may be months in some cases, which is not effective unless the jobs last many years.

Many firms attempt to obtain a labor cost summary by computer if other accounting work is done by machine, but there are many instances where the operations branch find a mechanized system supervised by the accounting department to be useless, and ignore it to establish their own manual reporting system. The fact that a firm has machine accounting does not, of course, describe what type of system it may have. One large company did its accounting manually and then printed its periodic reports by computer! The prestige of machine accounting was then available without the cost.

When using machine reports, there is a strong tendency to print out *all* cost accounts each week because it is easier than printing out only the information needed. Some firms have weekly printouts which are virtually useless, because they are made up of hundreds of pages. There is then no way to separate the significant data from the insignificant. Such reports are not necessary, and an adequate reporting system always includes a program of printing out data at various management levels, so that each manager receives data pertinent to his responsibilities. There is a tendency to print out material cost weekly in the same report as labor cost, which is another hindrance to interpretation of cost data. The most detailed report should be for the foreman, who can recognize a very detailed portion of an overall report.

Table 2-2 illustrates a portion of a typical weekly report. The work orders are listed on the report form; the blank form with all active work orders from the previous week may be copied, with new orders added and completed ones omitted, if the report is to be made manually. If but a portion of the work orders are to be listed, such as those with over $500 weekly change or the 50 largest (most active) orders, it is necessary that each report be completely anew each week.

Table 2-2. Weekly Labor Cost Report

Wilson Towers Job 234   Weekly Labor Cost Report   Week Ending October 26, 1980

	Work Order and Number	Budget Amount	Cost Last Week	Cost To Date	% Complete Last Week	% Complete To Date	Value of Work Last Week	Value of Work To Date	Cost Status Current Week	Cost Status To Date
6	2d fl. fms.	$20,000	$15,000	$21,500	70	100	$14,000	$20,000	−$ 500	−$1,500
7	" " concrete	2,000	–0–	1,600	0	100	–0–	2,000	+ 400	+ 400
8	3d fl. col. fms.	22,000	5,000	20,300	10	90	2,200	19,800	+ 2,300	− 500
9	" " concrete	2,000	500	1,200	20	90	400	1,800	+ 700	+ 600
10	1st fl. ext. mas.	27,600	10,000	22,470	30	80	8,280	22,080	+ 1,330	− 390
11	" " int. "	8,300	2,000	5,000	25	65	2,075	5,395	+ 320	+ 395
12	1st fl. doors	700	–0–	300	–0–	50	–0–	350	+ 50	+ 50
13	" " windows	650	–0–	350	–0–	50	–0–	325	– 25	– 25
	TOTAL	$83,250	$32,500	$72,720	.....		$26,955	$71,750	+$4,575	−$ 970

The person reviewing the weekly changes must bear in mind that the estimated percentage of completion for some items may be so inaccurate that the change shown by the report will be in the wrong direction; that is, reports can show that the percentage of work completed on an item is lower than for the preceding week, particularly when progress on the item is very low.

The *budget amount, cost last week, cost to date, percentage complete last week,* and *value of work last week* all come from the cost record for the work order or from the previous week's cost report. *Value of work* means the value according to the budget and is obtained by multiplying the budget value by the estimated percentage completion. This is sometimes called the *budget* or *estimate* amount. The difference between cost and value is the over- or underrun, also called the gain or loss. If no gain is shown on the total, the profit on the job presumably is as estimated. Other costs than labor, of course, will be over or under. The *percentage complete to date* is obtained by the foreman, superintendent, cost engineer, project manager, or whoever is delegated to make a physical inspection of the work at the end of each pay week or month. The cost status for the current week is the difference between the increase in *value* for the week and increase in *cost* for the week, and *cost status to date* is the difference between value and cost to date. The important figures are the totals under cost overall and for the week. But notice that if only active work orders are reported on, the weekly status change is obtained, but not the overall status to date. The labor cost report is also called the *red and black report,* since it shows if the job is "in the red" or "in the black."

Some variations of this report that you may encounter are as follows.

1. The size of the report can be reduced by omitting columns which need not be carried forward; such columns are still included in the calculations.
2. Columns can be added; for example, in Table 2-2, the columns for "cost for week" and "value added for week" are not shown directly, although they are used in the calculations.
3. Quantities and unit prices may be used in lieu of percentages. When this is done, the unit costs are also usually calculated.
4. Changes may be made to suit the machine method of calculation. The machine should eliminate useless information, not add it. Costs may be indexed according to the crew as well as the work order, doubling the number of cost figures; these results may be printed separately.
5. The reporting may cover only work for the week or work to date, rather than both.

2.14　DAILY COST REPORTS

A daily estimate of work done is necessary for daily cost charges. Fresh masonry can be distinguished from old, new concrete by quantity delivered, excavation by counting loads. The foreman may make the complete report, calculating not only the quantities but the cost amount ahead or behind for the day. One such form for foremen's use, in pocketbook size, is illustrated in Figure 2-7, *Masonry Cost Report.* The original estimate may have to be modified to provide data suitable for daily report forms, which will differ for each job. These daily reports may be kept on some items until summarized in the weekly report, along with items for which no daily cost reports are made.

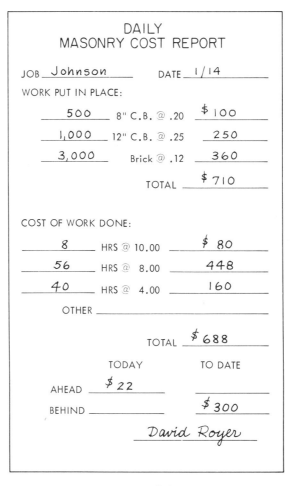

Figure 2-7
DAILY MASONRY COST REPORT

2.15   ACTION ON COSTS OVER BUDGET

If labor costs are going over the budget, what can you do about it?

Cost reports in themselves will usually reduce costs, even if no correction is made. Action will usually cut costs, even if the action is not justified—because the people involved will pay more attention to costs. The superintendent's action to reduce costs rests on the most qualified opinion, as he is closer to the work than anyone else. He may defend his operation and claim that the reports or the estimate is at fault, if he can see no unusual influence on the work. If he knows just what the reason is, he may not say so. He may know the reason and cover up; for example, one foreman's record had been good for years, but his costs began to run over. It was found that his child had leukemia, and the foreman was under high emotional stress. The person should have been transferred to a nonsupervisory position, both to restore the efficiency of the operation and to attempt to remove him from further emotional stress.

Unlike other industries, it is common to change a worker from superintendent to foreman to journeyman and back again, depending on the requirements of the work. If these changes are for short periods, his salary may be unchanged; for long periods, the salary will change to suit the work.

A foreman is seldom discharged for inefficiency that shows only on the reports and is not observable in the conduct of the job. The cost accounts show which supervisor to watch, and there is usually ample evidence of inefficiency. This is often unfair, as others of the same degree of ability get by because they are not observed. We have all too often discharged men for observable lack of ability, only to hire another who was even worse. Standards for supervisors in construction differ from those in industries where foremen and their supervisors are readily supervised and accurately rated on cost efficiency. People in construction are regularly demoted or discharged, even though they are hard working and careful in their work. When people cannot be closely supervised, and a high level of skill and ability is required, it is not possible to correct them by example and instruction, or to require they work "according to the book." There is no book, and each person is complete in himself. He must be accepted or rejected in total, not changed to eliminate his shortcomings. Under such conditions there is often injustice, and advance depends often on chance—as does the profit of his employer. This is a factor in construction employment which must be accepted by the young person; it is balanced by the equal chance of rapid advancement.

If a man is transferred and bad cost reports recur on his next job, it is reasonable to assume that a better supervisor can be found. You must be sure of your estimate figures and the circumstances in order to take action; few cost systems are well enough operated to ensure that a single work order is accurate. Weekly cost reports have a tremendous advantage over monthly ones in that considerable evidence of the direction a job is taking can be obtained in a much shorter time—although each report, in itself, is less reliable.

More often, the error is in the cost system itself; for this reason, each foreman should weekly review the cost reports of his part of the work. This review discloses errors; you may find that foremen will keep records of their own, if they do not think the firm's records are adequate or timely. The most severe test of a cost system's efficiency is whether it is accepted by the foremen and superintendents. *Never use a bad report in a discussion with a foreman unless he also sees the goods ones.* He will rightfully feel that his good work is unnoticed, and only his poor work noted. The manager *by exception*, who spends his time on work that goes poorly, sees only one side of the work and must be careful in this respect. If you do not give a foreman all the reports, you may observe and discuss his work from that angle. If you are not the foreman's direct superior, make sure that the superior is present when you talk to the foreman. Do not make any reference to reports which the foreman does not have, or you may cause the foreman to leave the job, with some of your best workers with him.

2.16   SUPERINTENDENT'S DAILY REPORT

The job superintendent often makes a general report to the project manager at the home office, or to the president of the firm. These reports are simpler than daily cost reports or completion reports for other purposes, and consequently they may be used if no other reporting system is used. Figure 2-8 is this type of report. Such reports have the dual purpose of notifying the home office of needs or problems encountered and also of providing a continuous record of weather and contractor activities, which is useful in case of claims for delay or disputes with subcontractors. The illustration shows generally what may be required, although details vary between firms and types of jobs. On a large job, the project manager may require daily reports from area superintendents. Such reports are not usually sent to the office; there are too many activities for one person to record, or for one person with other projects to read. A variation of this practice is a job diary, which keeps much the same information in a bound form on the job

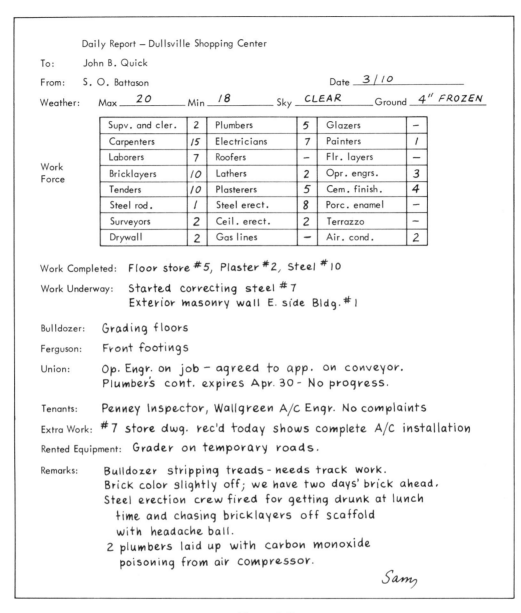

Figure 2-8
SUPERINTENDENT'S DAILY REPORT

where it may be referred to as necessary. In Latin America, licensed engineers may be required by law to keep such records.

2.17 ORGANIZATION MANUALS

Unfortunately, formal management policies and procedures, incorporated into an operating manual, are more of a hope than a practical reality. Such manuals, including descriptions of each person's duties and how they are to be done, are valuable for substitutes and replacements. Manuals should be specific, with examples, and should include the circumstances for disregarding the manual. Each field superintendent should be able to disregard the manual as far as it refers to his *subordinates,* but not in regard to himself. There should be a method defining which exceptions a person makes that should be reported to his superior.

An example of the description and responsibilities of one position, a materials clerk, is reproduced below.

Responsibilities of materials clerk. As job materials clerk, it is your responsibility to receive material deliveries. In general, this includes unloading or directing unloading in a proper location, accepting the shipping documents, forwarding them to the office to the Accounting Department, and notifying the person using the materials that they have arrived.

If materials have not been ordered by purchase order, it is your duty to see that the shipping documents show Accounting sufficient data to charge the client or job, as well as to pay for the shipment.

Delivery documents. You may receive one or more of the following documents. Note that no documents at all may be received. You do not refuse materials on this account, but you must then make out a *receiving report* with complete details.

Documents which may be received are:

1. *Trucker's or common carrier's delivery ticket.* This may show amount due, either for materials or for freight. The amounts may be payable or charged to the company account. You are not greatly concerned about the charges. Truckers are carriers who carry only one shipment, which you are receiving. *Common carriers* (who also may be truck lines) are those like Emery Air Freight, in the business of carrying and delivering freight to the public. All ship lines are common carriers, but we often hire private truckers to deliver material to the job. The difference is significant; conduct of common carriers is directed by statute, but we may make any arrangement with private truckers that we care to.

2. *Copy of invoice*, from shipper or vendor, may be combined with the above, or packed separately in the box. The invoice shows the amount due, and should also clearly indicate "pay this invoice." An information copy of invoice may be sent, which is not payable. It may not have prices, or it may be marked "do not pay this invoice." The original invoice for payment is usually sent by mail. Many accountants will not pay any but an original invoice. However, some vendors give more than one original, especially by photocopying; others give no original at all, keeping it for their own records.

3. *Packing slip.* This should be attached to the box or packed inside it. It lists the number of pieces of material, and description, usually by each box. You should check off each item, and sign as having been checked. This will usually go to Accounting, but you will need a copy on the job for future reference. If a copy must be made, make one by hand or photocopy.

4. *Ocean bill of lading, dock cartage slip,* and *forwarder's statement* or other papers may arrive with sea shipments. You are not required to recognize or check these, and they may be in any form. Be sure they go to the office; they may be quite important.

5. *Instruction books and parts lists* should always be packed with equipment. An original or copy must be sent to the office; often a copy is also needed on the job for operation or installation. If two copies are received, the foreman may keep one. Otherwise, make a copy or ask the office to do so, if the foreman or superintendent wants one.

   Note that a number of the papers described above are *inside*, so boxes should be opened within 2 days of being received. Some equipment must not be opened except in the presence of the shipper's representative. These will either be marked or you will receive instructions in advance.

6. *Purchase order.* Some vendors return a copy of our purchase order, which is handy for ready identification. As long as the papers are identified on the shipment elsewhere, the purchase order copy is not needed, but send it along to Accounting with the other papers.

Signature. Your signature is important. It is worth thousands of dollars, so should not be given lightly. You should give signing as much thought as signing your own checks. Never sign as a favor, or for merchandise you have not seen. Do not be a nice guy when your signature is involved. If you do this, you are worse than useless in your job; you are a potential disaster.

Form of signature. You name must include given family names. Do not use nicknames, as it must be identified by people not acquainted with you. If you follow Spanish custom, show your final initial or hyphenate your last two names as *Garcia-Alvarez*. If you cannot write legibly or must use a symbol as a signature, print your name. If you cannot print legibly, get a rubber stamp for your signature. Since this stamp is equal to your signature, you must keep it locked up or on your person when not in use. There is no reason to be embarrassed to use a stamp; this practice is much older than writing, and it is better to be clear than to be doubtful. Do not abbreviate your signature to anything less than what you expect it to be on your paycheck. Do not use initials, as they are likely to duplicate someone else's signature. When you get to be a bank president, you may sign as you please, but your job with us requires a clear and definite signature.

You may sign in one of the following ways.

1. Dick Johnson, 9/14/80
2. Received 9/14/80, Dick Johnson
3. Received but not opened, Dick Johnson 9/14/80
4. Received but not counted, Dick Johnson 9/14/80

Always add the date; it is part of your signature. Use a numerical system in keeping with the local practice, which in the United States is month/day/year. If persons on the job use differing methods, write the month.

Never sign *Approved, Accepted*, or similar words, either in your own writing or on a page where such wording appears. Your function is to *receive*; if the vendor requires more than evidence that he has made a delivery, refer him to the superintendent. If approved or accepted appears on the form given you to sign, cross out these words.

Refusal of shipment. On occasion, a carrier may refuse to deliver the materials unless a particular form of signature is used. In this case, refer him to the superintendent, and if the superintendent is not available, refuse the shipment. Do not be afraid of making mistakes by doing this; we would rather pay for a dozen re-deliveries than accept faulty or damaged goods without recourse.

If a trucker refuses to deliver a shipment when "Received but not counted" is the actual case, then he must wait until the shipment is actually counted. We must do this sometime anyway, so do not consider it a waste of time with common materials like brick or lumber. If a number of different sizes are shown, do not sign for the particular sizes (but you may sign the ticket) unless the sizes are verified. If you are reported for being uncooperative or ignorant, do not let it concern you; we will know you are doing your job. Capable carriers do not expect unverified signatures.

Inspection of shipment. You will be accepting shipments of materials with which you are not familiar. Consequently, you must sign "shipment received—count not verified," "shipment received, items not verified," or similar wording. If an exact verification is required, call on the superintendent or foreman to help you. An improper shipment is a source of more trouble for everyone than a delay in identification could possibly be.

Unidentified shipment. You may accept anything anyone wants to deliver to you; just sign *unidentified*. This indicates that the company does not necessarily accept responsibility. Notify the superintendent that you do not know what you have, and if he does not know, call the office. This should be the same day. If the shipment is fragile or subject to damage in the storage you have, refuse it.

Damaged shipment. If there is any apparent damage to crate or materials, note "damaged" and sign. Details are not necessary; from our point of view, they are better omitted. If the carrier wants the exact damage described, then do so. Otherwise, leave it as general as possible.

If in doubt, ask the superintendent. *If in doubt if you should sign anything, don't.*

2.18 QUESTIONS FOR DISCUSSION

1. "Many manufacturing and investment firms engage contractors on cost-plus contracts, rather than do work with their own forces, as much for reasons of internal organization as for lack of technical skill in the executive staff." Explain why this is so.
2. Give some reasons why a general contractor should, and why he should not, do his own subcontract work.
3. How does the custom of hiring men by the day affect job organization? If men are hired by the month, as on overseas construction, what are the differences?
4. Why do contractors not use a work order system as described in this chapter?
5. When cost reporting is not closely supervised, the early items to be completed are usually near the estimate, but savings and overruns show up later. Why?
6. What are the advantages and disadvantages of giving labor estimate amounts for each item to the job superintendent?
7. Most public agencies have a detailed budget made annually in advance, which limits payments to the exact amounts and items specified.
   a. Do you consider this a cost control system, as here described?
   b. What characteristics of a cost control system are lacking or weak, and how?
   c. If the budgeted amounts are all spent as budgeted, does this prove that the budgeting was correct? Why or why not?
   d. Such systems are considered efficient for running a university, but not for building a house. Why?

CHAPTER *3*

ESTIMATING

For which of you, desiring to build a tower does not first sit down and count the cost, whether he has enough to complete it? Otherwise, when he has laid a foundation, and is not able to finish, all who see it begin to mock him, saying This man began to build and was not able to finish. . .

Jesus Christ, from Luke 14:28–30
Revised Standard Version Bible

Estimating, which for the contractor is bid preparation, includes the methods by which a contractor arrives at a price for a construction proposal. An *estimate* includes takeoff, pricing, extension, summarizing, and assembly of the complete bid *(putting the bid together)*. A *takeoff* is the quantity survey portion of the work, which for a general contractor may be 60 percent of the cost of making the estimate. A specialty contractor will devote more of his time to takeoff, but pricing (which includes obtaining special prices from others) is also significant.

3.1 UNITS FOR ESTIMATING

Estimates are prepared by using some measures—*units*— which are adapted to the job and determining the numbers of such units in the job. The units are then priced at a single unit price for materials and labor, with individual unit prices for materials for labor, or using one kind of unit for materials and another for labor. Estimating books are based on actual material quantities. This so-called materials and labor unit pricing is widely used, but there are many variations.

Most contractors use a combination of labor units based on material and some sort of standard units, depending on the trade. Some of the variations used in lieu of material quantities are as follows:

74

1. Linear feet of trench in trench excavation.

2. For door and trim labor, a variety of units, such as percentage of the value of millwork material for the estimated labor, or price per door, ignoring hardware or parts of the frame and trim.

3. For roofing, square feet of finished roofing, neglecting sheetmetal work.

4. For floors of asphalt or ceramic tile, an approximation that ignores separate units for cement, mortar, trim, or toilet accessories.

5. For masonry walls, the square feet or the number of units, ignoring mortar and other accessories. Openings may be deducted partially or not at all, depending on their size, or may be deducted for material and not for labor, since the labor is greater than for solid walls, in most cases.

6. For scaffolding, and approximation by square feet of wall scaffold, among other possible variations, although the purchase, or lease, of scaffolding is figured by the component piece.

7. For plastering, by the square yard, counting openings solid, with no breakdown of materials or amount of returns, corner bead, and moldings.

8. For plumbing rough-in labor, by percentage of materials.

9. For electrical outlets and wire, lumping together without calculation of actual length of wire or breakdown of components.

10. For structural components such as outside walls, interior partitions, floors and roofs, and overhangs, tables of unit costs developed by contractors who do much work of the same kind of structure.

This list is only an indication of variations. The necessity for precision is balanced against the cost of estimates and the available time to make them. General contractors estimate plumbing by the number of fixtures or outlets, and electrical work by the number of outlets, but the bidding subs' estimates rarely are this approximate. A/E's estimates are less precise, usually made by the square feet of floor space.

3.2 TYPES OF BIDS

Since estimates are necessary for sales, the estimating department is the sales department for most contractors. Large firms engaging in cost-plus work may have full-time people on sales, but for most firms this is a duty of the manager, regional manager, or chief estimator.

An owner obtaining quotations for a construction job may use one of the following methods:

1. *Competitive lump-sum bid.*
2. *Negotiated lump-sum bid:* that is, a lump-sum bid from one contractor only.
3. *Irregular quotations* from several bidders, not necessarily comparable to each other.
4. *Cost-plus competitive bids.*
5. *Cost-plus negotiated bid.*

Competitive lump-sum bids are required for public work. Private owners, and sometimes public bodies, may accept bids from *invited bidders only.* For many architects and owners, this is done by formal prequalification; that is, a contractor may get on the invited list by submitting proper information to the A/E.

Lump-sum bidding requires that a complete set of plans and specifications be prepared in time for distribution to bidders, except on small or standard buildings, where specifications alone adequately describe the work.

Public bodies are subject to supervision of the courts if they show favoritism to a bidder; private firms are not. A contractor therefore never knows if a contract is really decided by price competition on private work.

Lump-sum bids are sometimes used as a maximum price, with a cost-plus contract for the basis for payment. Bids received under this arrangement often are not accepted solely on the basis of maximum price. Lump-sum bids are also received, for both design and construction, when the owner believes this is economical. A/E's have consistently refused to submit competitive bids for their own services, but recent antitrust action by the federal government has made any restruction or agreement which prevents such bids illegal.

A *negotiated lump-sum contract* is used when an owner wants a certain contractor and has confidence what the contract amount should be. In most cases, these contracts are for repeat orders of the same type of construction. This type of contract is the most sought after by contractors, as it combines high markup with low estimating cost.

Irregular quotations are any of the various owners' efforts to obtain low prices which the owner believes will be less than simultaneous bids, sometimes with a favored contractor.

One irregular quotation is a *closed* or *private bid opening.* That is, competitive bids are received for opening at a fixed time, but bidders are not allowed to attend. The owner may do this to award the contract to a favored contractor, but with the benefit of another low bid to set the price. Or the owner may take consecutive bids, exposing the previously low bid to each subsequent bidder. The owner may also do this without the appearance of simultaneous bids.

Cost-plus competitive bids may be received when the bid is for a stated lump-sum fee, percentage fee, hourly rate of equipment or merchanics time, equipment rental time, estimate maximum, or a combination of these. An advantage of cost-plus contracts is that the time of construction may be reduced while allowing a competitive bid to serve as a basis for contract selection.

A *cost-plus negotiated contract* allows work to proceed before plans are completed. Under competent management, it allows trade-offs of cost and time throughout the project, and allows orders for material without shop drawing approval, allows overtime, and enables construction to start with unfinished drawings.

3.3   UNIT PRICE CONTRACTS

Some types of work consist of relatively few kinds of work which can vary in quantity, but which can be measured. Bids are taken by unit prices, on quantities estimated by the A/E.

Highway and heavy construction are usually bid this way. The owner is allowed some changes at this unit price, and the quantities may vary according to final measurements. Where buildings and excavation are included in one contract, a lump sum for the building and unit prices for excavation may be used. Some bids have over a hundred unit prices, and most of them may require subcontract bids and summary sheets of their own. When unit prices cannot be calculated at the bidders' convenience, the bid is equivalent to scores of building bids at the same time, and the time for *putting together* (assembling all costs) *may be days for a unit price job, which would be hours if bid lump sum.* An estimator who has not made a bid of this type may fail to complete it in time for submitting the proposal.

The method of estimating varies with different kinds of bids. An estimate for a negotiated contract is different from an estimate for a competitive bid in the following ways:

1. The negotiated contract allows more time, usually, to prepare the estimate. It is therefore more nearly in the form in which it will be used in the budget and in material orders. The estimate takeoff requires more time for this added detail, and the preparation of budget and orders as a separate operation is indicated. The more careful estimate is less likely to have errors, and the errors will be smaller.

2. If a contract is being negotiated, subcontractors are less likely to get their bids in on time, as they believe there will be further negotiations (the less elegant word is "shopping") after the initial subbids (bids received from subcontractors) are received.

3.4 PRECISION AND ACCURACY OF ESTIMATES

Precision refers to the degree of exactness obtained by the estimate and is governed by detail of the plans (structural drawings are exact, but where excavation is involved or alternate materials are to be used, the drawings may be only approximate) and by the estimating methods used. The degree of precision of the estimating methods can be calculated by probability theory, but this is rarely done. If the estimator takes off larger items to a greater degree of precision than smaller ones, overall precision is improved. For example, an $100 item may be estimated with a probable error of 20 percent, with the same error of a $1,000 item estimated to 2 percent.

The degree of precision in estimates of material quantities is usually high; 1 percent is not uncommon. You may be wasting time by such precise estimating when the price of materials is subject to error of 2 percent and the price (unit cost) of labor is subject to error as much as 20 percent. Few contractors go broke because of a lack of precision—certainly not for 2 percent errors.

Accuracy refers to errors in the process which may be reduced by checking. There are few contractors who have not made embarrassing errors—and usually very costly ones—that appeared obvious in the estimate. For that reason, checking of estimates is directed almost entirely toward determining errors of extension and transfer of figures. If no extension in a bid is off over 50 percent in a large and complicated estimate, the contractor may consider that he has a good bid. Some estimators check only the large amounts on their bids—for example, those over $1,000. This method has the disadvantage of not detecting items which are given as $100 but should be $1,000 or $10,000.

An experienced estimator avoids useless precision. Estimates should be made more precise than the inherent error of the material; for example, if the average error of labor and material quantities is 5 percent, the precision should be closer than 5 percent. But if estimating expense is as limited as it usually is, 1 percent is more precise than necessary; probably 1 ½ to 2 percent is justified.

As for *checking figures,* from a practical standpoint there is no reason that sums and products be exactly correct. There is a reason why an error is made, such as:

1. The item is unusual or in unusual dimensions, so that the estimator's built-in warning system regarding the comparative size of cost items does not function. For example, the use of ceramic tile as full wall covering or as floor cover in only a few places does not produce a cost per unit which the estimator has seen before.

2. Items of unusual dimension, such as full-size lumber (marked in actual size or quarter-inches in an oversized convention), heating coils to an estimator who has figured only ordinary pipework, or slab forms with unusually long or short shores.

3. Large figures, not in proportion to the other products in the estimate. For example, if lumber quantities are in hundreds and thousands of board feet, an item of 100,000 or 30,000 board feet does not fit the space on a page, it is likely to emerge as 10,000 or 3,000 board feet. The same is true of materials which are two or three times as expensive as other substitute materials.

4. Figures which are not written well, or are cramped. If read in error once, they are likely to be misread again. Europeans use a definite flag on a "1" so that it appears like a "7," so there is no chance that it will be taken as merely a punctuation mark; the "7" is then marked with a cross piece on the vertical member to distinguish it from the "1."

3.5 QUALITIES OF AN ESTIMATOR

A number of qualities, some of them unfortunately contradictory, are required of estimators:

1. Patience for detail, technical knowledge of what to look for, and a memory which will keep in mind the details of one job without confusing it with a previous similar one.

2. A knowledge of construction and manufacturing processes in order to visualize how an item is made, and therefore the cost. A good estimator can *price* (estimate the unit cost of) anything if he knows the material of which it is made and the manufacturing process.

3. A knowledge and memory of the other subcontractors with whom he deals. He should gain the confidence of subcontractors' estimators, to the extent that he develops his own sources of information and at the same time can detect, not only by *what* he is told but also by the *way* he is told, what prices are low and who has them. The time sense is important too—when a bid is being put together, the time when quotations are received may give a clue to the direction of the bidding. That is, the meetings held and bids made during the last day before bidding, or sometimes the last hour, may follow a pattern as complicated as the mating ceremonies of the bird of paradise; the estimator should understand this pattern.

4. An ability to plan from the plans for a job, the *best method of making the estimate,* how long it will take, and how checks on errors can best be made. This is a greatly neglected skill, and because of its lack, estimating is more expensive than is necessary.

5. An idea of relative costs which will call large errors to his attention automatically wherever they may occur. The person who prepares an estimate may lose his

perspective of overall cost on that particular job, as the erroneous figures remain embedded in his memory. A second person should always read the figures for obvious errors of totals.

Since these qualitites are rarely to be found in one man, the best practice is to use more than one person on every estimate. The quality of patience in detail is not readily found in a person who also has general knowledge of processes, and some of the qualities listed above— especially the ability to plan the method of making an estimate—are acquired only by considerable experience.

3.6 QUALIFICATION OF BIDDERS

The use of an invited bid list for qualification has been mentioned. Many owners and A/E's are more formal in preparing this list, receiving experience and financial data continuously from prospective bidders, who *prequalify* (that is, qualify before plans are completed) either for a particular job or for all work of a certian type and size. The A/E also considers the reliability of the bidder in providing bids regularly, as a failure to obtain bids implies to the owner that the A/E is not informed on conditions in the industry, and delays the award.

On most *public work* bidders are qualified after bids are received. Since a low bidder can require that his lack of qualification be demonstrated, the most common method of refusing a low bid is to refuse all bids and readvertise for bids at another date. As public work is bonded, the owner depends primarily on the surety to investigate the bidders.

Generals have a more difficult problem qualifying subs, who may not be known until an hour or two before bid time, than A/E's have to qualify generals. One method is to call the sub's surety, A/E, or general contractor telephone references, although it may not be possible to reach them in the time available—especially with everyone tied up with other activities on bid days.

3.7 BID DOCUMENTS

Plans and specifications are provided by A/E's to Dodge rooms and builders exchanges without charge. General contractors may usually get two sets on returnable deposit; subs may pay for their sets, so they try to use drawings at Dodge rooms, exchanges, or general contractors' offices. These drawings are heavily used and often unavailable. When subs must buy drawings, they usually get only the sheets they believe contain their work, and often have incomplete bids because of incomplete plans.

The general contractor assists his subs by obtaining additional drawings; in many cases they are refundable, and if not, he can rotate them to several subs.

Some quotations may be obtained by reproductions of portions of prints and specs, with the general's quantities. Suppliers and subs who quote such incomplete data do not have sufficient data to quote other generals, and their bids are therefore especially valuable. It is worth the general's expense to obtain bids of this type whenever possible. This technique is particularly effective in obtaining quotations from equipment manufacturers who do not have local representatives, as on water and sewage plants and pumping stations. Exclusive sub-bids are the key to low bids.

3.8 COST OF "OR EQUAL" CLAUSES

Among the clauses used by some architects is that "approved equal" products, even on public work, must be submitted for approval during the bid period so that all bidders may use the alternative brand in their own bids. Such a clause practically disallows substitutions, as there is no incentive for a contractor to substitute a brand except to gain a bidding advantage. The specifications may even state that the *bidder* must obtain this approval, thereby ruling out the substitute manufacturers, who are the only persons who would gain by a substitution.

The distributor of a specified product may combine its installation with other work on the project, such as combining a specified roof decking with the structural steel bid. The breadkdown may be very high for the roof deck, eliminating competition not only for the steel erection but for the steel joists as well, as the steel erector-roof deck distributor-installer may be a distributor for a particular steel joist as well. In this manner, the original specified product ballons far out of control in added cost.

3.9 BID SUPPORT PAYMENTS

Contractors in some areas have agreed that the low bidder should make a lump-sum payment to unsuccessful bidders, to give compensation to unsuccessful bidders for their work done. If the bidders were in collusion, there would be no estimates made by unsuccessful bidders, and therefore no payments. However, the U.S. Justice Department filed suit in 1976 against the Lake County Building Contractors' Association in Illinois, declaring that such an arrangement conflicted with federal antitrust statutes. Under the Lake County plan, the payments were not made on a particular job unless all bidders agreed.[1] The charge grew out of a billing by a contractor to a public agency, in which payment to others for estimating the job was shown as a separate item. Evidently, the agency did not consider that cost of estimating work, by unsuccessful bidders, was a proper cost of the work.

On the other hand, there seems to be no objection to a contractor submitting a general bid on the same job on which he submits partial bids to all other bidders, as a subcontractor to them. The writer has a ruling on this only from the state of Florida, however.

3.10 COST OF ESTIMATING

It was mentioned before that the cost of estimating, for a general contractor, may be from 0.1 to 0.5 percent of the value of a job. This low cost is due to the fact that general construction work is rather simple and repetitive and includes large items that are sublet—and on which the general contractor gets estimates from the subcontractors. Estimates by specialty contractors are much more expensive and may be as high as 5 percent. On the work done by a specialty contractor, a 5 percent estimating cost is not unusual, in part because the specialty contractor's work, particularly mechanical and electrical work, is more varied and many items must be priced separately. In addition, subcontractors' estimators are seldom able to work in the luxury of their own offices but spend a great deal of time traveling to plan rooms and working in general contractors' offices.

[1] *Wall Street Journal,* May 20, 1976.

80

One measure of a contractor's future and the type of work he gets is a ratio between his estimating salaries and the rental on his office. A contractor with a $1,200-a-month estimator in a $1,000-a-month office is one who gets his work through an imposing front office—usually negotiated or cost-plus work. The contractor with a $3,000-a-month estimator and a $500-a-month office can be expected to get his work by competitive bidding.

The duplication of cost in estimating, which may amount for as much as 5 percent of the total cost of a building job and much more on a heavy construction or specialty job, has led to the establishment of independent quantity surveyors who sell their takeoffs to the various contractors. There are few contractors who depend entirely on these men for the basis of their bids, but many buy their takeoffs as a check on their own estimating departments. However, few estimators take off quantities the same way, and these differences make comparison of independent quantities with "in house" quantities difficult. For example, the definition of "footings" may vary. One estimator may take off the intersections of beams and columns as beams, another as columns. One shows all slabs on fill separately; another combines them.

In British practice, quantity surveying is a specialized profession, and surveys are provided by the owner along with the plans. Quantity surveying is taught as a professional course in the higher schools. If the quantity survey is in error, it is the responsibility of the owner, just as an error in the plans and specifications would be. The term *quantity surveyor* is British rather than American, but it is used by professional estimating firms in the United States. These firms also prepare complete estimates when requested, usually for architects but occasionally for contractors.

3.11 OVERALL ESTIMATE OF BID RANKING

An experienced estimator has a *feel* for an estimate and for his standing in the forthcoming bids. This judgment is based on the following factors:

1. The relative exactitude of the plans and their lack of dual meanings. When information is hard to find on the plans, it is usually probable that there will be omissions by the estimator, not detected because of the difficulty of obtaining clarifications from the architect.

2. The accuracy of the methods used and the reliability of the estimator who makes the quantity takeoff. Each person has characteristics, even when methods are specified (often they are not); some estimators are subject to overestimating, some to underestimating, and one estimator may normally be over on some items and under on others.

3. The reliability of the subcontract bids, and a knowledge of which competitors have the low bids. The distribution of subs' bids causes great variation between general contractors on building bids. If the estimator is unable to get a bid from a certain subcontractor but knows the sub has prepared a bid, the estimator needs to know the amount of the bid (even though he may not intend to use it). There are fewer secrets in business than most outsiders believe; the estimator may be able to get the bid of one sub from another, particularly if the bid he wants to know is high. These figures may not be available in numbers, but as relative ranking.

The estimator may find that a certain bidder has *none* of the local subcontractors' bids; he is usually a new bidder or from out-of-town. The estimator needs to know then, first, whether the out-of-town bidder is still in the race. This may be determined by a number of ways—the simplest way is to call the firm's estimator and ask. If he has dropped out, he will have no reason to keep it to himself. The architect may know, usually because of additional plans requested by the out-of-towner. The next step is to find out who the out-of-town man's subcontractors are (he must have them, since the local subs are not bidding him). At this stage, the estimator will already have a list of firms who took out plans from the architect. He will look for information next from specialty sub-contractors whose bids are indispensable and who have limited or no competition—such as heating controls vendors or some special item specified by brand on the plans. Then he goes through the wholesalers in the out-of-town bidder's home town and finally may have to resort to friends who are in a specialty business and who have bid against, or with, the out-of-town man in the past. Once he has found the missing subcontractors, he may still be unable to get a price from the out-of-town subcontractors, and the stranger's bid will remain a mystery until bids are opened. Such out-of-town bids on large jobs often break up a situation where the local unions, subs, and general contractors have managed to monopolize the construction in an area.

The estimator may also call the plan rooms. Subs may sign in the plan room or may check out drawings overnight; this serves as another source of information from which to find out which subs are bidding the job.

"Postmortems," or analysis of why the low bidder was low, are very useful. This requires a comparison both of possible direct material and labor variations, but usually more important, the source of competitors' subcontract bids. The competitors' subs are sometimes read at the bid opening; if not, they may be obtained from suppliers, other unsuccessful bidders, or eventually from the job personnel. If the sub is one new to you, he will ususally give you his bid. However, the low sub may have an exclusive agreement with the low general; that is, the sub gives his low price to no other general. Such agreements are usually well known in the trade, and make any separation of sub and general prices very difficult.

3.12   COURTESY BIDS

Sometimes the estimator discovers that there is no indication that *any* of his competitors have requested subcontract bids. This may prevent the estimator from putting together his bid, since he usually relies to some extent on such subs, who have access only to his competitors' plans. The estimator is then getting an indication that his firm will be the only bidder. Rather than inform the architect (who might postpone the bid date if he were aware of the situation), the various contractors who have decided not to bid may submit *courtesy bids* which they obtain from another contractor. So the estimator may find two or three other contractors calling him for a bid price (courtesy bid). The competitors want to stay on the architect's bid list, so they merely ask for a quotation from another contractor, knowing only that it will be higher than the low bid, but not how much higher. Courtesy bids are illegal in government work, because of the broad statement the bidder must sign stating that he has not consulted any other bidder for information to prepare his price. The architect usually is unaware that some of the bids he receives are courtesy bids.

In one instance, a bidder submitting a courtesy bid in a field of four or five bidders happened to call the contractor who turned out to be the low bidder, and was given a figure which came in a close second bid. The owner, for reasons of his own, chose not to award the contract to the low bidder. He called the contractor who had not figured the job, for data relating to negotiation of changes—which, of course, the second bidder did not have. The second bidder was glad for the opportunity to obtain a contract; like most courtesy bidders, he had refused to bid not because he had too much work but because he did not have time to estimate it properly.

The term *courtesy bid* is sometimes used to describe a sub's bid for which a general has no obligation. For example, a contractor bidding out of town may request a bid from one of his regular subs, although the sub does not want to work in the area. The sub donates this bid, which serves the general as a yardstick to judge the bids he receives. The sub gives it as a courtesy, in that he does not expect to be considered for the work. The sub may do this for a fee, but usually would rather have the general feel obligated to him.

3.13   COLLUSION

The writer has not encountered collusion between general construction bidders on any project, nor has the writer seen identical bids. There are various ways in which identical bids are rationalized, but they appear on their faces to be not only collusion but advertised collusion. In some cases, branches of the same firm have submitted identical bids to government agencies. Identical bids on general purchases by government agencies are common; in 1962, for example, 25.6 percent of procurement in Albany, New York, and 16.1 percent of procurement in Winston-Salem, North Carolina, was executed by purchases for which there were identical bids. The average for the country as a whole was 2.0 percent of purchases being made on advertisements which resulted in identical bids.[2] There were 37 identical bids received on construction contracts.[3] The U.S. Steel Corporation alone submitted 254 identical bids for materials supplied to government agencies—more than six times all the identical construction bids in the entire country![4] Only 10 of the 37 identical construction bids were low. Collusion is termed *bid rigging* in the trade.

On occasion, major manufacturers have been convicted of collusion on a national and continuing scale. A major effort is required to reach and repay purchasers, of whom there may be millions. In 1964 five manufacturers of plastic pipe fittings, including Celanese and Borg-Warner, agreed to repay $3,250,000 damages for collusion from 1966 to 1973.[5] This represented estimated payments on 80,000 homes and mobile homes, to people who could no longer be identified as having purchased the products.

In 1973 a larger settlement was made by U.S. Gypsum, Kaiser Gypsum, Flintcote, Fibreboard, Georgia-Pacific, and Celotex, in the amount of $68,000,000 for 1963–1967 in the form of an offer to unknown parties in 20 states, in a similar case.

The largest fines ever set in a felony price-fixing case were made against J. Ray McDermott and Brown & Root in 1978, when each of these firms was fined the $1 million maximum for collusion in allocating contracts for marine construction work. In addition, indictments were made against six executives for mail and wire fraud.[6]

---

[2]*Identical Bidding in Public Procurement:* Second Report of the Attorney General under Executive Order 10936 (Washington, D.C.: Government Printing Office, July 1964), p. 7.

[3]*Ibid.,* p. 10.

[4]*Ibid.,* pp. 99–100.

[5]U.S. District Court, Central District of California, *Stabler Construction Company, Inc. et al,* vs. *R. & G. Sloan Manufacturing Co. et al.,* Civil Action No. 71-1589-ALS and related cases.

[6]*Wall Street Journal,* December 15, 1978, p. 2.

In January 1979 the U.S. Justice Department filed indictments against 21 piping contractors in the Chicago area, charging that they regularly got together to discuss prospective piping construction projects and to pick which contractor would get the job; and that others stayed out of the bidding or submitted noncompetitive, collusive, rigged, and complementary bids so that the designated winner would have the low bid. Eighteen of the firms are charged with felonies, and the collusion is alleged to cover $100 million in work over two decades.[7]

3.14   TIME STUDY

The most difficult, and therefore most important part of the estimator's work is to establish labor costs. Time study, which is more commonly considered as an aid to production, is an aid in determining unit labor costs.

Frederick Taylor, a machinist of the 1800s, rose to shop superintendent and later became a consultant in the field of time study. He is credited with the organization of present time-study methods; there has been little basic change since he first designed them. In the terminology he originated, *scientific management* was basically an *attitude* of management. Now, however, other terms are used to designate a scientific approach to management. "Scientific," in this sense, means applying general principles gained from one industry or type of operation to another type. According to Taylor, managers assume responsibility for giving exact instructions to workmen as to what workmen are to do, how they are to do it, and how much they are to be paid for each item of work, rather than relying on the men to exercise their own judgment within the basic knowledge of the trade.

In academic management terms, Taylor was not a manager at all, since he emphasized obtaining data rather than just making decisions based on data. But in his own time, The American Society of Mechanical Engineers refused his paper on the topic, since they considered it to concern management, not engineering. Today technical management is still unrecognized as an academic study.

Taylor has been widely criticized insofar as application to the construction trade is concerned, because his methods do not comply with desires or rules of the trade unions. Union organization reduces the area of application of his principles and increases the time and effort necessary to carry out the savings possible by using his methods. Actually, the application of Taylor's methods have largely been absorbed by the labor unions and by the trades, so that it may be said with equal truth that Taylor's methods are not applicable to construction because the industry adopted them at an early date.

Today, development of improved methods is possible both in open-shop organizations and in union shops by using the union itself as an organizing vehicle. The unions today strongly resist piecework, but many smaller jobs are done by small subcontractors on piecework. Since there is free movement of workers from open-shop subcontractors to union labor, methods become freely adopted throughout the industry. If one studies workers in Central America, who have a history of development of the trades equally as long as workers in the United States, it is evident that the motions of the North American workers are much more efficient. Yet union members in Central America not only prefer piecework, but insist on it; union agreements specify piecework rather than hourly rates.

[7] *Engineering News-Record,* Vol. 202, No. 6, February 8, 1979, p. 17.

From a construction standpoint, Taylor's most important principle is that, under usual conditions, trade knowledge is a perogative of the tradesmen; that is, supervisors are unable to tell the workers what to do because the supervisors know less about the work than do the workers. The first step in improving management is to prepare extensive pamphlets or other information to include the knowledge of the trade. This trade information was not in book form (nor is it today), and Taylor complained that college professors did not consider such knowledge sufficiently exotic or unusual to be of importance. Taylor considered it of overwhelming importance; if the manager does not know the trade, he is unable to determine a day's work and therefore to institute a reasonable piecework pay scale.

Taylor's objective as a consultant was to develop improved work methods, apply them, and establish piecework rates. He repeatedly followed this procedure at various plants. The workers were expected to receive a 30 to 50 percent raise as a reward for increased production; this raise he determined by trial as the amount necessary to obtain their cooperation in working under detailed instructions. His usual procedure was to select some of the less intelligent workers first, to demonstrate that they could earn more money; the more intelligent (and influential) workers immediately saw the advantages.

The most striking feature of Taylor's work was his full realization of its difficulty; he expected that years of study would be necessary to accomplish savings and change over to an incentive system because one difficulty with such payment methods is that men improve much more than anticipated, doubling or tripling their previous wages. The owners then attempt to lower piecework rates, with immediate resentment by the men, even to the point of refusing to cooperate with further studies. Taylor believed five years a reasonable time to institute a system in a machine shop, and then only where the basic information was already available from other plants. Taylor and Thompson gathered data for fourteen years on concrete work before venturing into print with the data.[8] Taylor cited the work of Frank B. Gilbreth to show the time necessary to achieve results; the Gilbreths spent a year and a half studying bricklaying before they eliminated a single step of the bricklayer.

James F. Lincoln (Lincoln Electric Company, Cleveland, Ohio) established an incentive system by profit sharing in his shop, which over a period of many years resulted in an *annual increase in efficiency of 15 percent,* twice the industry average. By 1962 his sales per employee were double that of the industry, and his workers were paid nearly $12,000 a year—double the industry average. The efficiency of his workers was four times the industry average, and his turnover rate was 4 percent per year, which means that few workers left him except on retirement or death. According to Lincoln, he was discouraged from expanding from 1940 to 1960 because he would then dominate the industry and encounter antitrust laws.

Taylor wrote very little about his system but received considerable publicity from attempts to install his system in government plants, which resulted in an investigation by Congress. From the record of this investigation, it appears that a host of Taylor's imitators were purporting to sell the same service, but without proper study and without consideration of the workers. Taylor's primary objective was worker satisfaction, both during the changeover and afterward; he never had a strike during or after the changeover period in any plant.

Modern management consultants and theorists avoid revising the plant or workmen's work methods and concentrate instead on changes in accounting methods or corporate structure. This requires less knowledge of the actual work, and pays the executive a higher salary.

Taylor's methods, which require a considerable investment in the labor force, are well suited to a firm with a monopoly of a local labor force. For a contractor who can easily lose his

---

[8]See Figures 3-4 to 3-8 and the Taylor and Thompson reference in the Bibliography.

labor force and who cannot keep it at the same size for a long period, the advantages of intensive training would rapidly pass to his competitors. The previous applications of such organized management have therefore been made by large home builders in rural or near-rural areas. It is applicable to any similar situation, however, and may be a necessity to establish a local labor force in undeveloped countries. Our present apprenticeship programs are usually teaching technical knowledge, with neither time nor qualified teachers for work methods.

Modern time study. *Time study* requires that each task be timed, and that waste time be determined. *Motion study* is a determination of the minimum time required if the most efficient method (which must usually be determined also) is used. Since this text is devoted to management and not to methods, time study is treated only as it affects the estimator's cycle of obtaining information to be used for making further estimates. The time observations needed for this purpose are primitive indeed, consisting only of total time for completion of various portions of the building, and in such form that the foreman or timekeeper may keep such records in the time he has available, but more detailed time studies are made even for this purpose. Although the writer has not seen any modern studies which compare, in data provided, with Taylor and Thompson's *Concrete Costs,* further studies are being made by various agencies, particularly by the U.S. Department of Defense. During estimating of a job, an estimator may mimic the motions of a worker, in order to get some basis for estimating labor cost per item.

Time studies may be used for the following purposes:

1. To set piecework rates. It is particularly important that the individual be observed and that costs for different items which will be done by different workers be separated. Time studies for this purpose must be more detailed and precise than for cost accounting.
2. For future estimates.
3. To test various motion-study trials or gang size variations in order to determine the best working methods.

3.15 PERCENTAGE ALLOWANCES

Some cost items are not estimated in detail but are calculated as a percentage of detailed costs. Such approximation of overhead items for small jobs may be as much as 25 percent of the bid price. On larger jobs, percentages are figured for cost items such as those explained in the following paragraphs.

Payroll taxes and fringe benefits, although calculated as a percentage of payroll, are a detailed calculated cost. This is true also of *unemployment compensation insurance* and *liability insurance. Workmen's compensation insurance* is often calculated as a percentage of all payroll, rather than as it varies with the trade and work situation, and to that degree is an approximation.

Legal and auditing costs are sometimes calculated as a percentage of total project cost, which is, of course, an approximation, but it is better than including it in general office overhead, which results in larger jobs carrying more than their share of cost.

Payroll clerical costs can be calculated with some precision, if proper records or costs of services by others are kept, and the percentage is usually smaller on larger payrolls. The reverse

8800 JOB OVERHEAD SUMMARY

Building **WILLISTON CITY HALL** Listed by **BK** Sheet No. **1**

Location **WILLISTON, FLA.** Checked by **KR** Estimate No. **43**

Stories **3** Size **40×90** Cube **110,000 C.F.** Floor Area **10,000** Date **8/51**

Item	Class of Expense	Amount	Item	Class of Expense	Amount
8801	Equipment, Tools, Accessories			Brought Forward	
.1	Rental (See A. G. C. Rental Schedule)	500	8818	Telephone and Telegraph	
.2	Freight	40		(Regular, Long Distance: Connect. Fees)	75
.3	Hauling		8819	Rents	
.4	Loading, unloading, erecting, dismantling	100	.1	Offices	
8802	Job Organization		.2	Land, Unloading and Storage Facilities	
.1	Superintendence—(supt.; assistant)	3000	8820	Permits—Building, street, sidewalk, water,	
.2	Time and Material—(timekeeper; mat'l clerk)	1500		sewer, hauling over boulevards, use of equip-	
.3	Accounting—(bookkeeper; accounting)			ment, etc.	70
.4	Clerical—(stenographer, clerk, office boy)		8821	Insurance—Miscellaneous—	
.5	Shops—(blacksmith, machinist, tool man)			Fire, tornado, earthquake, riot, theft, boiler,	
.6	Safety—watchman; safety foreman)	300		plate-glass, automobile, payroll, etc.	1000
.7	Miscel.—(job chauffeurs; teamsters; waterboy)	500	8822	Petty Cash Items	400
8803	Light, Power, Water: Connections		8823	Interest—On deposits and job funds	1000
.1	Electricity—(light and power)	100	8824	Cutting and Patching for Trades	300
.2	Carbide, Gas—(lights, cutting, welding)		8825	Contingencies—Guarantees; (strikes, wages,	
.3	Gasoline, Heating and Illuminating Oils	75		labor, output, rains, freezing, floods, cyclones,	
.4	Coal and Coke—(power, heating, thawing)			earthquakes, material shortage and price, trans-	
.5	Water—(boilers, sprinkling, mixing, etc.)	75		portation, sub-soil conditions, finance, drastic	
8804	Supplies			and ambiguous contract provisions, supervising	
.1	Office—(stationery, time-books, forms, etc.)	50		personnel)	
.2	Job Shop—(steel, smiths coal, etc.)		8826	Cold Weather Expense	
.3	Equipment—(oils, waste, boiler comp., etc.)		.1	Thawing Materials, (Plant installation and	
8805	Traveling and Hotel Expenses			operation—See Item 8803.4)	
.1	Material—(expediting)	500	.2	Weather Protection—(window and door clo-	
.2	Labor—(procurement and transportation)	1000		sures, temp. walls, canvas, etc.)	
.3	Officials and members of permanent force	700	.3	Temporary Heat—(installing, maintaining).	100
8806	Express and Miscellaneous Freight			See Item 8803.4	
8807	Demurrage Allowance		8827	Repairs—Streets, sidewalks, property	100
8808	Hauling—hired for odd jobs	300	8828	Pumping—De-watering (If preferred, cover un-	
8809	Advertising—labor, material, equip.	300		der Excavation or Sheeting of Main Summary)	750
8810	Signs—company, warning, notice, etc.	50	8829	Final Clean Up—Windows, walls, floors, ceil-	
8811	Engineering, Surveys and Inspection			ings, fixtures, and premises	200
.1	Layout—(lines, levels, batter boards, etc.)	100	8830	Association Dues—Job share	300
.2	Public Inspectors—(boilers, wiring, etc.)		8831	Share of General Company Overhead	
.3	Inspection of sub-contract work			Expense of home office, shops and yard,	
.4	Lot survey			quantity survey, estimating, investigations,	
8812	Tests			dead time of permanent field force, taxes and	
.1	Soil—(test, pits, borings, bearing power)	75		all other company expense, not chargeable	
.2	Material—(cement, steel, aggregates, etc.)	200		to the specific job.	3000
.3	Structure—(floor loading)			SPECIAL ITEMS NOT LISTED ABOVE	
.4					
8813	Drawings—shop and setting				
.1	Drafting				
.2	Extra Prints				
8814	Photos	400			
8815	Patents and Royalties				
8816	Legal—Attorney and Notary Fees	200			
8817	Medical and Hospital Expense	200		Total Job Overhead	17560
	Carried Forward	10265		Equals 7 % of direct cost	

Figure 3-1

JOB OVERHEAD SUMMARY

(Courtesy of The Associated General Contractors of America, 1957 E Street, N.W., Washington, D.C. 20006.)

can be true, however, when a small payroll can be kept by personnel who spend the greater part of the week on other work.

Supervisory costs at the foreman level can be calculated fairly accurately by a percentage of labor; however, this cost varies widely with the kind of work done and should be calculated separately for each trade. The cost of one foreman for each eight men, for example, is a good approximation even if it is not followed in practice. If there are fewer foremen, there will be higher costs per foreman, that will be offset by higher efficiency; and more workers per foreman will be offset by lesser efficiency. The writer prefers to put the cost of foremen, as well as fringe benefits and payroll taxes, in the hourly rate of workers rather than as a separate item. This requires that these costs be separated again for the labor cost budget, as actual payroll rates are used for labor and cost reports. Labor contract bids and direct labor estimates may then be compared directly. When extra work or changes in the work, which require deletion of some costs and credit to the owner, are made, the clerical work is less (and it is also less subject to error) if fringe costs are all in the original extensions of labor costs.

Job overhead costs are those associated with direction of the job but not to any one trade. They include the superintendent, the project manager (even if he is responsible for other jobs as well, he is a cost to the job), clerical help, guards, and a host of other costs. The AGC job overhead classifications are shown in Figure 3-1. A number of items, such as temporary buildings, small tools, wrecking, some insurance, and taxes are separate items in the AGC breakdown and not considered as job overhead, although many contractors do not classify them in this manner. *Small tools* may be considered as job overhead or charged in the categories for which they are to be used. Small tools are generally any equipment which is allowed to be charged as an expense when bought, rather than depreciated over several years, by Internal Revenue Service (IRS) rules, and include hand tools or electrically powered tools. The wear under job conditions is high, and theft losses are great. Smaller contractors sometimes use a percentage of the total value for all job overhead costs; 5 percent is a popular figure.

Markup is the term for the amount added "at the bottom," after all calculated and allowed costs are summarized. This is sometimes termed *profit and overhead,* the overhead referring to home office overhead, or, for small contractors, including job overhead as well. As far as possible, all overhead should be calculated for the particular job, especially if the job is larger than or different from other jobs that the contractor has previously completed. Using a higher rate of overhead established for small jobs may keep a contractor from winning a large contract; or experience with complicated jobs may establish a high rate of overhead which will prevent a bidder from winning simpler jobs. Experience with lump-sum work provides an overhead rate too low for cost-reimbursable contracts.

3.16   OVERHEAD AS A DIRECT COST

Contractors universally look on overhead cost—that is, the salary of people not directly engaged in putting work in place—as a necessary evil. There is an implicit assumption that purchasing agents, secretaries, and project managers will do more work in less time if the job goes faster—that is, if they have less time to do it in. But this is not necessarily so; there is a certain amount of paperwork to do, and more often this delays the job rather than the paperwork just continues as long as the work lasts.

The contrary is more likely true—the job is done when the paperwork is finished. If there are too few people or too little effort in purchasing or in obtaining engineer or architect approval on changes or materials which can not be obtained, the work is delayed. It is a common observation that the delaying factor in construction work is more often comprised of the delays in buying, approval of shop drawings, and the like that establishes the completion date than it is due to the availability of labor. A shortage of supervisors delays the work and increases the cost. It is just as reasonable to assume that "overhead" is a fixed cost and that the work force will continue until the materials all arrive. Therefore, the work force is "overhead," being a fixed cost determined by the skill of the office force.

3.17   ESTIMATING MARGINS

Although a contractor bases his bids on his cost, his estimate of his competitors' bids is equally important. The cost estimate has two purposes: first, to avoid a loss contract, and second, to submit a bid under the competition. The amount left *on the table* (the difference between the low bidder and the next bidder) is a cash loss for the low bidder, even if the job is profitable as bid. An estimator's bid record can be shown by the total profit of the company, but the responsibility for this profit (or loss) is shared by the superintendent and buyers. The money left on the table is immediately obvious and is a fault of the estimate. It is not uncommon for a contractor to leave on a single job a sum as great as his year's profit. If the amount left on the table, for all jobs during the year, is several times the cost of operating the estimating department, as is quite common, it indicates that estimating is hurried and that the firm has too few people (or not the right people) in this department.Such a comparison should be made before a work-saving method such as computer calculations or separation of quantity surveyors from people who price units (who are higher paid) is introduced. Unit prices are not only greatly affected by the degree of repetition but also by the relative quantities of related work on the job. These points can be easily missed by an estimator who is pricing units and is not familiar with the details, particularly with how work of other trades will fit into the general contractor's work.

Estimators, who must by the nature of their work complete countless estimates only to see their work become useless on bid day, suffer from pressures to "get a job." Sometimes these are imposed by managers who are impatient with the estimator, sometimes by the requirements of the business, or sometimes because work has come to a standstill and the company must bid lower to survive. The manager should place such impatience "at the bottom" by lowering the markup, not by decreasing the unit prices, or he should *sharpen his pencil* (estimate more closely). Every estimator is affected by his emotional outlook as to whether he will remain "safe" in his estimates or make them as precise as possible. So long as his methods are consistent, the manager knows what to expect, and if the manager decides to take a loss job to keep working, he can tell what the loss will be. An attempt to pass this burden to the estimator will result only in uncertain cost estimates, and the markup will mean nothing at all. Contractors' offices are small, and causing an estimator to feel that he *must* get a job can result in bankruptcy.

Contractors do not agree on what a satisfactory estimate may be. Some estimators say their costs are safe; some will say they attempt to furnish a cost estimate which is high, in relation to costs, as often as it is low; that is, it has a 50-50 probability of overrun or underrun. Cost variation does not conform to a random pattern, however; the possibility of an overrun occurence may be 50 percent, but if it does occur, it is generally greater and may be *much* greater than underruns usually are.

3.18   ETHICS

Some consider a contractor unethical if he bids low on a job on which he knows that he must make a profit from changes which will be necessary to correct the plans. Such bidding is not mentioned in the AGC code of ethics, and this writer sees nothing unethical in it. The A/E prepares the contract documents, and any errors made by the contractor because he has not checked or verified critical portions is borne by the contractor. Bidding on an expectation of profits in changes results in lower bids to the owner; there is no reason to think a contractor who has a profit in the job before changes will accept any less for changes on this account. The contractor who checks the drawings or site more thoroughly can reasonably expect a payment for this service.

There is also a popular conception that a contractor will do a less careful job if he has very little or no profit in the job. This is sometimes used as a justification to avoid competitive bidding, and some owners are considerably influenced by this possibility, to the extent of negotiating fixed-fee contracts rather than invited competitive bids. There is no doubt that a contractor will take every conceivable path to avoid bankruptcy, and the contractor with little capital will make every effort to avoid loss. But the larger and well-capitalized contractors do not attempt to avoid their obligations. A large firm has difficulty in attempting to vary the management process from one job to another. There is no gain in doing so. A manager is expected to cut costs as well as he can, and the same limitations apply to him on a profitable job as on an unprofitable one. The tendency will be to be more generous with the owner who has repeat business in sight, regardless of whether the particular job under way is a profitable one. Cutting costs on a job which was bid too low implies that the manager could have cut costs on other jobs but did not bother. To the extent that this happens, it is poor management; a manager should put his time where the highest returns are made, not on a job where the saving will be more noticeable than on another job. The owner who expects to pay more to encourage better quality or more speed in the work should put his requirements in the contract documents, not in a contract which could have been negotiated at less cost.

3.19   RISK AND MARKUP

The object of a business venture is to "buy cheap and sell dear." The difference between direct costs and receipts is used to pay indirect costs; any money then remaining is profit. The difference between direct costs and total receipts is the *gross profit*; indirect costs are *overhead*. The gross profit is the markup, if the estimate is correct. However, the markup is not the gross profit, for a number of reasons. A contractor rarely is able to put his profit into an estimate accurately; the profit is a remainder after any number of unknown events occur.

Contractors use many methods to determine markup, sometimes by comparing one method with another, sometimes adding several factors of risk. Eight methods are shown on Figure 3-2. The numerical values in the figure are common, but the actual values used depend on the cost definition (what is included in cost), and the type of job.

1. The percentage of total job cost method is the simplest. When the jobs being bid are substantially the same, this percentage, which is the end result of other methods, is sufficient. Many contractors use this method alone, but always tempered by judgment based on consideration of the other factors.

2. The percentage of *labor and equipment costs* combines some of the other key factors. The labor and equipment cost has a direct connection to capital, supervision, and overhead cost. It is common to use one factor for percentage of labor and equipment, and add also percentages of material and subcontract cost.

```
┌─────────────────────────────────────────────────────────┐
│                    BASIS OF MARKUP                        │
│  On:                                                      │
│  1. Total Job Cost (Buildings) ...................... 7%  │
│                                                           │
│  2. Percentage of Labor and                               │
│       Equipment Rental ............................ 50%   │
│                                                           │
│  3. Subcontracts .................................. 3%    │
│                                                           │
│  4. Sub Risk – Cost of Losing Two                         │
│     Largest Subcontract Bids                              │
│                                                           │
│     Based on actual calculations, example:                │
│                                                           │
│                      Electrical        HVAC               │
│        Low......      $8,000          $12,000             │
│        2nd......       6,000           11,000             │
│                      ────────         ────────            │
│        Difference     $2,000          $ 1,000            │
│             Total risk ........................ $3,000    │
│                                                           │
│  5. Probability                                           │
│     Estimated cost, of which the probability              │
│     of exceeding is 30%, (see Figure 3-3),                │
│     is estimated to be 3% above the estimate              │
│     cost amount ................................... 3%    │
│                                                           │
│  6. Based on known competition.                           │
│                                                           │
│  7. Exposure on own materials and labor ........... 25%   │
│                                                           │
│  8. Cash requirements                                     │
│     You have          $30,000                             │
│     You must have to                                      │
│       live during the                                     │
│       period the above                                    │
│       money is in use   15,000                            │
│     Required return .............................. 50%    │
└─────────────────────────────────────────────────────────┘
```

Figure 3-2
**METHODS OF DETERMINING PROFIT AND
OVERHEAD MARKUP**

3. The *percentage charge on subcontracts* reflects the overhead cost of administration with capital and risk required. This percentage is much smaller than the percentage markup on the firm's own purchases and personnel.

4. *Subcontractor risk* is a more rational way of calculating the markup on subcontracts, based on the probability of loss by subcontractors who fail or refuse to accept the work at the bid as interpreted by the general contractor. In the illustration, failure of the lowest electrical bidder to accept the job would result in a $2,000 cost increase to the second lowest bidder. And the refusal of the low HVAC (heating, ventilating, and air conditioning) bidder would result in an additional cost of $1,000 to give the contract to the next bidder.

5. The *probability of increased cost* is explained in the pages following. The cost, for which a 30 percent probability of exceeding, may be calculated at 3 percent greater than the estimate cost used. A 30 percent probability of exceeding a cost is the same as a 70 percent probability that the work will be done for less than that cost figure. The probability may be applied to the job as a whole or to parts.

6. The *competitive estimate* is one based on the contractor's opinion of how his competitors will bid, based on past experience. Some contractors are consistently low on concrete and high on finish work, for example. Usually, a contractor will be high on work at which he has the least experience.

7. *Exposure* on materials and labor is another way of saying the markup is based on the part of the work which is most risky. Contractors often speak of exposure as a matter of judgment to compare with markup they have determined from other methods. Exposure may be based on any part of the work, but is most commonly used to refer to materials and labor.

8. Consideration of *cash requirements* is inherent in all methods above, but can be determined by the actual dollar amount, rather than an approximation by percentage. If the job being bid is large in respect to the firm's assets, cash becomes a very important factor. A contractor without adequate credit for his operations views the return on cash as an absolute requirement; if he cannot attain a minimum ratio, he cannot continue in business regardless of other considerations.

The *probability of increased cost* is always on a contractor's mind, and is the single item most strongly affecting markup. It is the *contingency* or risk factor, although contractors often say: "If I allowed for contingencies, I wouldn't get a job." *First*, a contractor's intention is to *avoid loss* of capital; second, to make a profit. By allowing for the former, the latter takes care of itself.

An estimator who prepares a large number of bids, all identical in type and amount and therefore equally risky, would find that the cost of work in comparison to his estimate could be plotted on a curve similar to that shown in Figure 3-3. Some of the jobs would cost more and some less, if he is successful in selecting the point at which the probability of greater or lesser cost is 50–50. In the illustration, heavy construction shows a greater variation than does building construction. This is the normal situation, since building construction, in addition to being more uniform and depending less on natural conditions is mostly bid by subcontractors, of which the lowest is selected before the general construction work (carpentry, concrete work, and other work for which the general contractor uses his own labor) is bid. That is, the subcontract work is even more indefinite in cost than overall bid, but the lowest of a number of sub bids is used by the various contractors when the overall bid is made up. The variation is therefore primarily in the work the general contractor does himself.

This variation is not symmetrical; that is, the chances of greater cost are not equal to the chances of lesser cost for every job. The estimator has one of the hundred jobs in this illustration, all equal so far as he can tell, but he does not know which of the hundred he has. The best he can do is get the right class—that is, an estimate of the probability that he is high or low. In the illustration, the probability of overrun over $108,000 (8 percent) is 100 to 80, or 20 percent. If $100,000 is used as a bid price, including profit, the chance of losing money on the job is 20 percent. This variation is not gradual, in some cases; the cost may depend on a single decision or happening, such as whether a certain subcontractor or vendor will be satisfactory or will be able to furnish work to specifications, or it may be such a chance as a dike being overtopped,

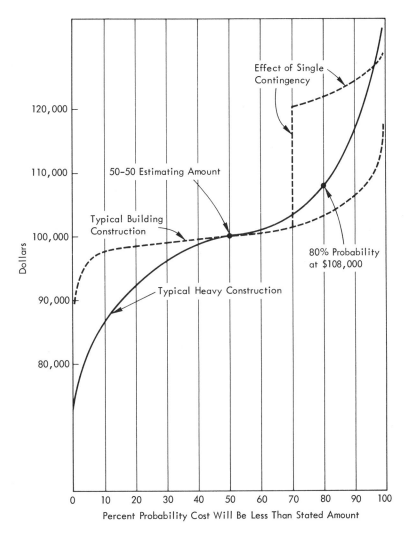

Figure 3-3
ESTIMATE PROBABILITY GRAPH
Percentage of Probability That Cost Will Be Less Than Stated Amount.

with damage to the work in progress. The single-contingency effect shown in Figure 3-3, has a 70 percent probability that it will *not* happen; but if it does, the added cost is estimated at $121,000 less $102,000, or $19,000. A prudent contractor may include this cost in his bid, but with a markup on the job which will equal this cost if it occurs, to the extent of eliminating or even bidding with a negative profit margin for this contingency. That is, he may decide to take a loss on the job if this contingency occurs, but not the whole $19,000. Since he knows that the other contractors, in most cases, are facing the same probability, he may bid $120,000—which would give him a small loss if the event occurred, but a good profit if it did not.

Contractors do not state the situation in this way; they like to believe that they will make a profit on every job or they will not bid it. They do think in this manner and recognize that a series of "unlikely" events (events with less than a 50 percent probability of happening) may happen, and, if they do, the job will lose money.

Each contractor gets the jobs he is low on; that is, his high estimates are above the low bid, and another contractor gets the job. Hence the contractor profits or loses on his low estimates but makes no profit—or loss—on his high ones. Therefore, the estimator's errors in the low bids are the only ones for which the contractor learns the real cost. He cannot tell if he could have made a profit on the jobs he did not get.

It is not uncommon for a whole series of bids for a project to be loss jobs, because of a shortage of work at the time. This occurred in the 1930s on the All-American Canal in California. A contractor with equipment payments to make and no work, who is being forced into bankruptcy, will take any job which gives him a chance greater than zero and gives him a paying job for himself for a few months.

Note that the estimate probability analysis diagram (Figure 3-3) is not one which is used for preparation of a bid, but an explanation of how the bid figure is judged. Since there are no identical jobs at bid time, and the final information is not available for all jobs in any case, an estimator's judgment on the various points of this curve cannot be evaluated.

3.20 BID ANALYSIS

Table 3-1 illustrates the comparison one contractor made between his own bids, the average bids, and the low bids. One proposed method of analysis is that a contractor may determine his own probability of having the low bid at any selected markup, and then, by statistical analysis methods, determine what his markup should be for any sales figure; from this, he can determine his bid for maximum profit. Following the same procedure, but for the purpose of analyzing bids made by others, he can determine the difference between the low bid and the bidder's cost, the estimated markup in the last column of Table 3-1. However, an inspection of these figures indicates that the low bidder must have had a different cost estimate than Lenox, the contractor who is making the comparison. The third job, for example, yields the bidder 0.7 percent markup; as a matter of judgment, this is corrected to 2 percent and so entered in the last column, but underlined to indicate that the figure is an estimate, not a calculation, and that there is probably a difference between the two estimates—Lenox and the low bidder—of 1.3 percent. The reason for the difference was not known to Lenox.

The fifth job, the low-rent housing, happens to have been won by an out-of-state bidder who had his own subcontractors. Consequently, there is every reason to believe that the low cost estimate was quite different from Lenox; 3 percent is used, but circumstances make this comparison useless. The sixth job, St. Paul's Church, was in an entirely different set of bidders. It was a church job and a member of the church was the low bidder; in addition, it had a large amount of millwork, with consequent uncertain labor costs. The 5 percent assigned to this job as the low bidder's markup is not comparable to other jobs during the period. Job 7, Seminole High School, was bid from out of state by Lenox against bidders usually not in the Lenox competition. Lenox had its own subcontractors but did not use a critical sub's bid; the job also was made up of several alternates and the apparent markup on different parts varied between bidders. The 2.5 percent assigned as the low bidder markup is therefore only reasonably correct.

During this period, Lenox bid $8,622,000 in work at an approximate cost of $20,000, or about 0.23 percent. With an average 13.6 bidders (the weighted average rather than the numerical average), Lenox should have won about 7 percent of the work, or $620,000 worth. Lenox could have done this by lowering its bids 0.9 percent; this would have reduced profit on Job 8 from 5 to 4.1 percent, a reduction of $6,000 on this job, which would have been the only

Table 3-1. Summary of Bids Submitted by Lenox Construction Company

Job	Lenox Bid Standing	Low Bid*	Lenox Bid*	Average Bid*	% Above Low Bid	% Below Average Bid	Low Bid % Over Lenox Cost	Low Bid Estimated Markup %
1. Phillips Hwy. Plaza Jacksonville, Fla.	2d of 5	$1,590	$1,605	$1,622	1.0	1.1	3.0	3.0
2. Jet Engine Rep. Shop Jacksonville, Fla.	3d of 14	1,048	1,077	1,103	2.8	2.4	1.2	1.2
3. Pharmacy Wing, Univ. Of Florida, Garnesville, Fla.	3d of 17	1,680	1,736	1,746	3.3	0.6	0.7	2.0
4. Post Office & C.H. Ocala, Florida	2d of 23	1,076	1,093	1,155	1.6	5.7	2.4	2.4
5. Low-Rent Housing Warner Robins, Ga.	8th of 20	866	907	935	4.7	3.1	0.3	3.0
6. St. Paul's Church Atlanta, Ga.	3d of 6	222	248	255	11.7	2.8	-1.7	5.0
7. Seminole High School Sanford, Fla.	4th of 12	1,459	1,513	1,565	4.0	2.8	0.0	2.5
8. Govt. Emply. Exch. Atlanta, Ga.	3d of 12	680	686	715	0.9	4.2	4.1	4.1
Total		$8,622	$8,865	$9,095	2.8	2.6		

Average no. of bidders 13.6
Lenox markup 4% over $1,000,000
              5% $500,000–$1,000,000
              10% under $300,000

*In thousands.

job won. With an average field of 13.6 bidders and an estimating cost of 0.23 percent the average bidder must have a 2.9 percent markup for estimating costs to break even; but if Lenox takes the one job, its job profit would be 4.1 percent × $680,000= $28,800. After estimating costs for all jobs of $20,000 is deducted, the net profit will be $8,800 or 1.3 percent of work done for the period. With this number of bidders, it is nearly impossible for *all* of them to make a profit at these prices.

If a 1 percent cut had been made in all bids by Lenox, the first job would have been won, with a profit of $47,700 on the Phillips Highway Plaza, more than offsetting the estimating costs; in addition, some of the later jobs would not have been bid at all, and, if so, the markup later would have been higher, since the firm would not have needed the work. However, Job 1 was for a private owner who asked for bids a second time on changes, and it is doubtful if the contract would have been obtained at this low figure. There were a number of later extras, however, which should have added to the profit.

In the foregoing discussion, it has been assumed that the cut in bids was made without consideration of the number of competitors. It is generally conceded that an increase in the number of bidders lowers the markup of the low bidder, but in this particular bid record that was not the case—the two jobs Lenox most nearly won attracted 5 bidders and 12 bidders, and Lenox's second-worst ranking was on Job 6, with 6 bidders.

It is asserted that a contractor, knowing the cost of the job, can derive the probabilities of winning a job at any markup, given the number of bidders. The number of bidders is usually known from the number of general contractors who have taken out plans, although there will be a few who drop out without notifying the architect. Added to this analysis, which assumes random distribution of bids, there must be a judgment for the known attributes of competitors. But often the known competition is such that the number of other bidders is unimportant. For example, suppose that a hospital, with which local bidders are inexperienced, is to be bid by three local union contractors and one open-shop outside contractor who is a specialist in hospitals. The locals can pretty well judge the other local bids, but the outside man is an entirely unknown quantity.

This analysis, however, is a way to judge the available information after all possible corrections for known variations have been made, but few contractors bid a sufficient number of similar jobs that consistent data can be accumulated. The competition not only varies with seasons and amount of work available for bid, but also changes from the outcome of the preceding bid. Only very large, or very aggressive, contractors will continue to bid at the same markup after they have had one or two successful bids.

Most bids are at least partially subject to analysis based on information from subs which was not available before the job was bid. For example, let us take three bids on the same job:

Royer	$73,000
Williams	73,811
Trieste	75,274

From Royer's point of view, the cost was $66,228. He knows he had two low sub prices which the others did not have;

Plumbing	$1,600	low
Roofing	900	low
Total	$2,500	low

As Royer received a late electrical bid $1,000 below the one he used, but which was available to the others, this is a lower cost that his competitors had. So Royer estimates the competitive cost estimates were

Royer cost estimate	$66,228
Plumbing and roofing bids	+ 1,600
Electrical bid	− 1,000
Royer's estimate of competitors' cost	$66,828

So Royer estimates that Williams markup on this basis was 9.0% and Trieste had a 11.1% markup. They, on the other hand, believed Royer's markup, without further investigation, to be $73,000 − 67,728, $5,272, or 7.8 percent (although it was actually 10.2 percent). The low bids are unknown to those firms who did not have them. The firm that gets the greatest number of sub bids not only will have the low cost, but will have the best information on the other bidders' costs.

So Royer's competitors, by using any analysis based only on competitors' bids (rather than costs), would add 7.8 percent markup to beat Royer. But in this case, the named low sub bidders were read with the low bid, and Royer's competitors could determine the other sub prices. The probability that subs, unknown to some of the bidders, have *prices out* at bid time, is not random; it depends on out-of-town bidders, the location of the work, the type of work, and whether out-of-town bidders already have relationships established with local subs.

These figures, which this writer believes to be typical circumstances for buildings show:

1. Naming subcontractors when bids are opened reveals in part the low bidder cost, since these subs will usually reveal their bids on request of an unsuccessful bidder.
2. Contractors can often determine why their bid was high, and therefore make more accurate projections of future bids than can be done by a method which assumes that bids are subject to laws of probability. A probability analysis, rather popular among authors, can be made when other conditions are unknown. There are not many bid results available in which such conditions existed.

3.21   ERRORS IN ESTIMATING

Failures of construction firms are most often blamed on "incompetent management" or "poor judgment." These errors usually show up in estimating—selling too low—or in construction—buying too high. In estimating, more than in any other operation in the construction business, a contractor can lose a large amount of money as a result of a few minutes' work, with very little chance of gaining a compensating amount, no matter how well he does anything else. Most errors are of the type which can be classified as "clerical," such as omission of items from the plans, neglect of reading pertinent specifications or of applying them to the estimate, failure to add or to include numbers in the addition and misplacement of decimal points in extensions. Estimating requires a high degree of experience and management ability; but, at the same time, people who have these abilities are often unable to attain the unthinking concentration re-

BUILDING ESTIMATE SUMMARY

Building _____ Stories _____ Estimate No. _____
Location _____ Size _____ Date _____
Owner _____ Floor Area _____ Listed by _____
Architect _____ Cube _____ Checked by _____

Item	Class of Work	Amount	Item	Class of Work	Amount
				Brought Forward	
100	Wrecking, Moving and Clearing Site		4800	Tile, Mosaic and Terrazzo	
200	Temporary Construction—Office, Sheds, Etc.		4900	Composition, Cork, and Patented Floors	
300	Excavation, Grading, Backfill and Special Fill		5000	Kalamein and Other Fire Doors, Windows	
400	Foundation Support—Piles, Caissons, Cribs		5100	Hollow Metal Doors, Windows and Trim	
500	Shoring, Sheeting—Temporary, Permanent		5200	Steel Sash and Partitions	
600	Underpinning—Temporary and Permanent		5300	Caulking	
700	Water Lines, Gas, Drains, Sewers, Conduits		5400	Weather Strips	
800	Paving, Curbs, Sidewalks and Drives		5500	Glass and Glazing	
900	Concrete—Except Items 800 and 2700		5600	Priming and Painting	
1000	Concrete Surfacing and Cement Work		5700	Decorating	
1100	Concrete Forms—Wood, Metal; Metal Arches		5800	Plumbing and Gas Fitting	
1200	Reinforcing Rods and Mesh; Metal Inserts		5900	Electrical Wiring	
1300	Concrete Block		6000	Electrical Fixtures	
1400	Common Brick		6100	Heating and Ventilating; Boiler Setting	
1500	Face Brick		6200	Elevators, Dumb-waiters, Hoists; Controls	
1600	Fire Brick; Flue Lining and Special Brick		6300	Elevator Doors and Enclosures	
1700	Hollow Tile—Load Bearing and Partition		6400	Mail Chutes	
1800	Terra Cotta Fire Proofing—Arch, Furring, Etc.		6500	Pneumatic System	
1900	Gypsum Block—Lintels, Furring, Book Tile		6600	Vacuum System	
2000	Water and Dampproofing		6700	Refrigeration System; Refrigerators	
2100	Architectural Terra Cotta		6800	Sprinkler System	
2200	Artificial Stone		6900	Tanks	
2300	Cut Stone—Lime, Sand, Blue, Etc.		7000	Vaults and Vault Doors	
2400	Exterior Marble and Granite		7100	Revolving Doors	
2500	Structural Steel and Iron		7200	Awning, Shades and Blinds	
2600	Metal Lumber		7300	Fly Screens	
2700	Stacks—Metal, Brick, Concrete		7400	Floor Coverings	
2800	Ornamental and Miscellaneous Iron		7500	Special Equipment not otherwise listed	
2900	Iron Doors, Gates and Shutters		7600	Special Interior Fixtures	
3000	Vault Lights		7700		
3100	Metal Store Fronts		7800		
3200	Bronze and Other Art Metal		7900		
3300	Lumber, Wall Board and Rough Carpentry		8000		
3400	Insulation and Sound Proofing		8100		
3500	Millwork and Finished Carpentry		8200		
3600	Special Cabinet, Panel, Stair and Door Work		8300		
3700	Artwood Floors—Parquetry		8400		
3800	Rough Hardware		8500		
3900	Finished Hardware		8600	Landscape—Leveling, Sodding, Planting	
4000	Roofing—Non-metal, Slate, Tile, Composition		8700	Job Overhead—Equip., Fees, Permits, Etc.	
4100	Sheet Metal Work, Plain; and Metal Roofing		8800	Compensation, Liability, Contingent Insurance	
4200	Pressed Sheet Metal—Ceilings, Panels, Etc.		8900	Taxes, including any sales and use tax, tax on	
4300	Metal Furring, Lathing, Corner Beads, Grounds			payrolls, etc.	
4400	Plastering—Plain, Ornamental; Plaster Board			Total Cost—Labor, Material, Sub-bids	
4500	Stucco			Margin and Profit	
4600	Imitation Marble and Stone, and Scagliola		9000	Bond	
4700	Interior Marble, Slate, Stone				
				Amount of bid	
	Carried Forward				

Figure 3-4

BUILDING ESTIMATE SUMMARY

(Courtesy of The Associated General Contractors of America, 1957 E Street, N.W., Washington, D.C. 20006.)

quired for the arithmetic. The best quantity surveyor is not the experienced superintendent or manager; the latter cannot prevent his mind from wandering to how something would be done, when the immediate problem at hand is to determine dimensions and quantities.

An efficient checking system is essential. It must be assumed that errors *of all kinds* will be made; and the contractor should plan a system which will locate them, starting with the largest errors and proceeding to the smallest as far as time and money allow. The first, of course, is to check the major categories (trades) of work for omissions. Every possible kind of error has been made, including forgetting to put overhead and profit on a job. Most estimators use check lists. Since the item forgotten is more often a well-known item (unusual trades tend to stand out in the memory), these lists must be of a type that are adapted to the job. The *Building Estimate Summary* (Figure 3-4) of the AGC may be used for the check sheet.

3.22   NATURE OF ERRORS

An error is made because the mind puts an extraneous particle of information into a calculation; at a critical time, one writes 7 for 70, 9 cubic feet to a cubic yard of concrete, or 27 square feet to a square yard of plaster. This often occurs when an operation is interrupted before it is completed. The mind will not go back to the incomplete item but remembers it as complete. When checked, the entire operation or the important part of it may be recalled; but the mind slides down "memory lane" and follows the same erroneous operation it did the first time. This, in turn, reinforces the memory, so the next time it is even easier to repeat the error. One can check the same operation 10 times without discovering an error. A factor may be used in error once and the mind will then keep drawing it from the recent memory, as this requires less effort than drawing it from the older (and correct) memory. In this way, a number of identical calculations are made, all wrong. Once the estimator sees the error in any of the calculations, he will find it immediately in the others, since he has substituted a more recent factor in his memory.

Because of this characteristic, checking more than once by the same person is unproductive; the arithmetical errors on a computation sheet can be checked and discovered much more easily by another person. To locate errors not shown on the sheet, those occurring between the plans and the sheet, requires another complete estimate. Time can better be spent on overall checks—determining the average amount or cost of concrete per floor, number of board feet of lumber per 100 square feet of floor space, checking one floor total against another, and using a positive checking system (that requires a mark on each number transferred to another page) for transferred calculations. Calculations should *never* be made on sheets which are then thrown away, but products may be accumulated on a calculator and the tape fastened to the sheet. If another person is not available to check calculations, "sleep on it" and check your own work the next day when your erroneous memory has faded.

3.23   RIGID SPECIFICATIONS

One cannot ignore an unreasonable specification just because *everyone knows* it will not be required. For example, a U.S. Corps of Engineers specification required *"Forms shall be true to line and grade."* *Everyone knows* that forms are not perfect; when tolerances in finished work are stated, as in this case, forms to that tolerance are the contractor's responsibility. Yet the contracting officer insisted on perfect forms, and the Board of Contract Appeals refused extra payment. The contractor did collect in the Court of Claims, but this took 8 years, and put the contractor out of business.[9]

How can one avoid bidding a specification which is evidently in error? Some of the methods used are:

1. *State exception or interpretation in the bid.* There is always the probability that the low bid will not be allowed if it contains any exception on public work. On private work, this is acceptable.

2. *Letter of inquiry to the A/E.* This is the method usually specified. However, there is the probability that the A/E will not answer. It is the practice of many A/E's to consider any letter frivolous if it appears obvious to the A/E, and they may avoid an answer if their own error is involved. They do not like to mention errors in writing. The most dangerous possibility is that the A/E will not respond, but later use the letter to demonstrate that the bidder is aware of the rigid specification, and not having received an answer, should have assumed the more rigid requirement. Since the other bidders are not exposed, the letter writer may write himself out of the job.

3. *Letter to A/E, requesting explanation but stating that the less rigid requirement will be assumed.* On a public job, this obliges the A/E to answer or be open to the charge that he knew that one bidder was using an assumption the others were not. The A/E will naturally resent this type of letter; it is useful only on appeal, and consequently many contractors will take their chances by doing nothing at all. This is a decision which should be made by the responsible manager of the bidder's firm.

3.24 ESTIMATE SHEETS

The three processes in estimating, usually done separately, are quantity takeoff (quantity survey), pricing, and summary (recapitulation). A number of publishers and associations provide printed sheets, or ordinary accounting sheets may be used. It is important that transfers from one sheet to another be properly marked. The writer prefers that each kind of sheet be numbered in a different series and that the pricing sheet show the page in the calculations from which the quantities were taken. On the takeoff, the totals transferred should be circled in red or otherwise marked; transfer of totals from the takeoff to the pricing sheet can then be verified.

This use of three kinds of estimate sheets is important. One may combine all three kinds of work on one sheet, but do not price on some takeoff sheets unless you price on all of them, for the same trade item on the summary sheet, or some items priced on the takeoff sheet, may not be transferred to the pricing sheet. To be sure that all quantities or costs are transferred from any sheet, place them underlined against the right margin and circle them when they are transferred, and wherever they are transferred to *must* result in another transfer to the right boundary. Any uncircled number on the right boundary then represents an omission. Use of color is preferred. If the transferred items are quantities of material, the sheet is a takeoff sheet; if the quantities are dollars, the sheet is a pricing sheet.

Experience in estimating small jobs can be detrimental when an estimator goes on to large ones. Since all items on the takeoff of small buildings can be remembered, the estimator does

---

[9]*Kenneth Reed Construction Corp.* v. *The United States,* U.S. Court of Claims, No. 144-70, March 16, 1973.

not learn to record the work he has calculated. If the job is complicated—for example, a job with $10 million in concrete work alone—and irregular, such as a water treatment plant, some marking system must be used to show on the estimate what work has been calculated, and a marking system must be used on the plans themselves, with the system fitting that particular job. Calculations to analyze methods of determining unit prices are separate, belonging to neither takeoff, pricing, nor summary sheets.

3.25 COMPUTER ESTIMATES

Since the extensions should be verified by some other person than the original estimator anyway, many estimators list takeoff dimensions for multiplication and extension by others. The dimensions may be listed so that multiplication is performed by computer, and if unit prices are carried in the computer, the entire operation is done by machine. By one method, unit prices are fed into the machine by code numbers, and the printout then contains the extensions and the name of the item.[10] Unit prices may be adjusted by using several code numbers for the particular job being bid, which are keyed to unit prices stored in the machine. Also, an overall correction factor for labor or material may be fed into the machine. The same numbers and figures then become the budget from which cost control printouts are made. This method is justified only for special circumstances, as when a contractor does a large number of estimates of similar jobs. Use of a computer only for multiplication and addition is like shooting a fly with an elephant gun. Coding each item separately according to the price for that particular job is of no advantage, as the price can be given directly with no more work. Nevertheless, some large firms are estimating in this way, since they have the computer for other purposes. If a computer is to be used, the maximum benefit is obtained by a large estimating department, which may utilize its memory for storing information. The computer is valuable chiefly for its storage capacity, as estimating calculations are simple processes.

 The trust placed on machine-stored data varies from estimators who use it as a matter of routine to those who consider even one unchecked unit price (that is, one supplied from the machine's memory) as too risky.

3.26 CORPORATE LICENSE IN ANOTHER STATE

A corporation must have a license in each state where it operates. If it does not have one when a contract is made, it usually may not sue on the contract later, and for public work a bid may not be acceptable if the bidder is not properly licensed. The licenses required to bid work vary with the state. There may be three types of licenses required of one corporation, as follows:

1. A corporate license to do business—that is, a recognition of the corporation by the state, usually by the secretary of the state. Many firms are incorporated in Delaware, as the laws there are lenient, but their Delaware office is the office of an attorney who acts only as a legal agent to receive notices and may also represent a hundred other corpora-

---

[10]Allen Bros. and O'Hara, Inc., used this system successfully on Holiday Inns throughout the country. In their case, the structures were very similar and the prices were not made in competition, which permits low estimates to balance high ones.

tions in the same way. They comply with Delaware laws for corporate organization, but with the laws of the states where they operate for business operations.

2. A license as a contractor—that is, a professional acceptance of competency, issued by a contractors' license board of the state, county, or city in which the work is to be done.

3. An occupational license from the county, city, or both. This is obtainable through a tax usually paid at the tax collector's office.

3.27 LABOR UNIT COSTS

Labor unit costs are calculated by dividing an item of work previously completed, for which the total cost is available, by the number of units of work. The units are usually square feet or yards, cubic yards, or pieces. The accuracy of such unit costs depends on the similarity of the previous item of work to the new item being estimated. Since the work is never exactly the same a second time, some adjustments are used to compensate for the expected difference in new work. The limiting factor in determining actual unit costs is the feasibility of separating the labor cost of small parts of the work; as a result, many firms use quite gross measures, as square feet of building area. Square feet of concrete form area is nearly as approximate, since labor cost varies with the cuts, (labor in making the edges) and therefore with the size of individual areas, more than with the size of the entire area.

An industrial engineering approach to this problem, as it would be handled in factories, is illustrated for concrete formwork. Each operation is analyzed in detail, either from separate time studies or by interpolating time for the operations not actually observed. Figure 3-5 shows the results for 252 operations, for cutting pieces of lumber which may be necessary for concrete forms. Figure 3-6 shows 180 operations of making up forms on the bench, with variations of speed of workmen. Figure 3-7 shows how the results of these operations are used to make an estimate for a column, and Figure 3-8 combines similar operations into setting and removing a column form. These last two calculations are then combined to make a chart of the labor total for columns of various sizes—the actual sheet which may be used for estimating (Figure 3-9).

These charts were part of those derived by Sanford Thompson and Frederick Taylor over a period of 17 years and are not practical for an individual contractor. However, the methods can be followed for estimating where repetition of one type of construction warrants it, and time can often be obtained in a similar way for more simple work. For example, in placing concrete by crane, the fixed times, such as for dumping and loading, may be timed separately from time in transit, so the results may be adjusted for work when the time in transit is greater.

Considerable work in time study has been done at Stanford University, particularly with time-lapse photography. Time-lapse pictures are motion pictures which are taken very slowly—as slowly as once each 3 seconds. When shown at ordinary speeds, the action is greatly speeded up; observations can be obtained in a fraction of the time (since the camera does not require an operator while taking pictures) that would be necessary for direct observations, and the films shown to foremen and superintendents also demonstrate the lost motion in their crews. Time-lapse photography is principally advantageous for gang operation, determining the best arrangement of the work so that all men are kept busy. Individual workmen usually move around too much to keep a camera on them, if the camera is close enough to observe motion in detail. Some contractors are following up on this work, using their own cameras. Such films may help materially to cut labor costs.

TABLE 161]

SAWING LUMBER

Times include placing and removing lumber from saw.

Times assume saw working continuously with no allowance for rest and delays.

Lumber before cutting is taken as 12 feet long, but values are approximately correct for other lengths.

Use "2 cuts per Length" column for squaring ends or sawing into two lengths.

Use "6 cuts per Length" column for sawing into 6 lengths.

Use "12 cuts per Length" column for sawing into 12 or more lengths.

In any case multiply the unit time by the number of cuts.

DIMENSIONS OF LUMBER	ITEM NUMBER	ON MILL-SAW TIME PER CUT						ITEM NUMBER	BY HAND TIME PER CUT					
		AVERAGE MEN			QUICK MEN				AVERAGE MEN			QUICK MEN		
		2 Cuts per Length min.	6 Cuts per Length min.	12 Cuts per Length min.	2 Cuts per Length min.	6 Cuts per Length min.	12 Cuts per Length min.		2 Cuts per Length min.	6 Cuts per Length min.	12 Cuts per Length min.	2 Cuts per Length min.	6 Cuts per Length min.	12 Cuts per Length min.
1"×2"	(1)	0.17	0.11	0.08	0.11	0.07	0.05	(22)	0.38	0.33	0.31	0.25	0.22	0.21
1"×4"	(2)	0.18	0.12	0.09	0.12	0.08	0.06	(23)	0.49	0.42	0.41	0.33	0.27	0.27
1"×6"	(3)	0.19	0.13	0.10	0.13	0.09	0.07	(24)	0.60	0.53	0.50	0.40	0.35	0.33
1"×8"	(4)	0.24	0.16	0.12	0.16	0.11	0.08	(25)	0.70	0.62	0.60	0.47	0.41	0.40
1"×10"	(5)	0.28	0.19	0.14	0.19	0.13	0.09	(26)	0.81	0.71	0.67	0.54	0.47	0.45
1"×12"	(6)	0.32	0.22	0.16	0.21	0.15	0.11	(27)	0.96	0.82	0.76	0.64	0.55	0.51
1½"×2"	(7)	0.24	0.17	0.11	0.16	0.11	0.07	(28)	0.43	0.38	0.36	0.29	0.25	0.24
1½"×4"	(8)	0.26	0.18	0.12	0.17	0.12	0.08	(29)	0.60	0.53	0.50	0.40	0.35	0.33
1½"×6"	(9)	0.31	0.22	0.15	0.21	0.15	0.10	(30)	0.76	0.68	0.66	0.51	0.45	0.44
1½"×8"	(10)	0.36	0.26	0.18	0.24	0.17	0.12	(31)	0.92	0.80	0.77	0.61	0.53	0.51
1½"×10"	(11)	0.41	0.29	0.20	0.27	0.19	0.13	(32)	1.10	0.95	0.92	0.73	0.63	0.61
1½"×12"	(12)	0.46	0.33	0.23	0.31	0.22	0.15	(33)	1.31	1.12	1.09	0.87	0.75	0.73
2"×2"	(13)	0.31	0.22	0.15	0.21	0.15	0.10	(34)	0.49	0.42	0.41	0.33	0.28	0.27
2"×4"	(14)	0.33	0.24	0.17	0.22	0.16	0.11	(35)	0.70	0.62	0.60	0.47	0.41	0.40
2"×6"	(15)	0.39	0.29	0.20	0.26	0.19	0.13	(36)	0.92	0.80	0.77	0.62	0.53	0.51
2"×8"	(16)	0.46	0.34	0.24	0.31	0.23	0.16	(37)	1.17	1.02	0.99	0.78	0.68	0.66
2"×10"	(17)	0.53	0.39	0.28	0.35	0.26	0.19	(38)	1.41	1.28	1.25	0.94	0.85	0.83
2"×12"	(18)	0.60	0.44	0.31	0.40	0.29	0.21	(39)	1.66	1.46	1.41	1.11	0.97	0.94
3"×4"	(19)	0.32	0.22	0.16	0.21	0.15	0.11	(40)	0.92	0.80	0.77	0.61	0.53	0.51
4"×4"	(20)	0.41	0.29	0.23	0.27	0.19	0.15	(41)	1.17	1.02	0.99	0.78	0.68	0.66
4"×6"	(21)	0.60	0.44	0.31	0.40	0.29	0.21	(42)	1.66	1.46	1.41	1.11	0.97	0.94

Figure 3-5

SAWING LUMBER TIME STUDY

(*Source: Frederick W. Taylor and Sanford B. Thompson, Concrete Costs [New York: John Wiley and Sons, 1912], p. 662.*)

TABLE 162] MAKING COLUMN, BEAM, GIRDER, SLAB, AND WALL FORMS [UNIT TIMES ON BENCH READY TO PUT TOGETHER

For explanation of this table see p. 627.

No allowance made for rests and delays.
For ordinary construction add 50% to times for unavoidable delays.
For large well organized work add ⅓ to times for unavoidable delays.
For small jobs with inexperienced builders add 100% for unavoidable delays.
Add 50% to total of making when old lumber is used.
Times based on form side 24 inches wide by 12 feet long.
For lengths over 12 feet increase time proportionately; for lengths under 12 feet add 20% to proportionate times.

Item Number	Description	Columns	Beams	Girders	Slabs	Walls	Unit	Average Men 1-inch Form Lumber min.	Average Men 1½-inch Form Lumber min.	Average Men 2-inch Form Lumber min.	Quick Men 1-inch Form Lumber min.	Quick Men 1½-inch Form Lumber min.	Quick Men 2-inch Form Lumber min.
(1)	Rip △ strips for corner beading on mill-saw	Cl-2-3	B	G	S	W	Corner	1.47	1.47	1.47	0.98	0.98	0.98
(2)	Rip △ strips for corner beading by hand	Cl-2-3	B	G	S	W	Corner	6.70	6.70	6.70	4.45	4.45	4.45
(3)	Carry² 1" × 2" cleats 50 feet to bench³	C	B 2	G 2	S	W 1–2	Cleat	0.03	0.03	0.03	0.02	0.02	0.02
(4)	Carry² 2" × 4" cleats 50 feet to bench³	C	B	G	S		Cleat	0.12	0.12	0.12	0.08	0.08	0.08
(5)	Carry² 4" × 4" cleats 50 feet to bench³	C	B	G	S	W 1–2	Cleat	0.30	0.30	0.30	0.20	0.20	0.20
(6)	Place 1" × 2" cleats on bench	C	B 2	G 2	S		Cleat	0.09	0.09	0.09	0.06	0.06	0.60
(7)	Place 2" × 4" cleats on bench	C	B	G	S		Cleat	0.13	0.13	0.13	0.09	0.09	0.09
(8)	Place 4" × 4" cleats on bench	C	B 2-3	G 2-3	S	W1-2	Cleat	0.17	0.17	0.17	0.11	0.11	0.11
(9)	Rip form boards for pockets on mill-saw	C					Side	1.24	1.40	1.56	0.83	0.93	1.04
(10)	Rip form boards for pockets by hand⁴	C					Side	1.70	2.15	2.70	1.13	1.43	1.80
(11)	Rip 12 ft. form boards on mill-saw⁵	C	B	G	S		Side	2.94	3.32	3.72	1.96	2.21	2.48
(12)	Rip 12 ft. form boards by hand⁵	C	B	G	S		Side	8.70	12.68	17.34	5.80	8.45	11.56

Figure 3-6
FORM MAKE-UP TIME STUDY

(*Source: Frederick W. Taylor and Sanford B. Thompson, Concrete Costs [New York: John Wiley and Sons, 1912], p. 634.*)

TABLE 162] MAKING FORMS—Continued   [UNIT TIMES]

#	Item	C	B	G	S	W 1-2	Type	See Table 156, page 655.					
(13)	Carry² form boards 50 feet	C	B	G	S	W 1-2	Board	0.40	0.47	0.55	0.27	0.31	0.37
(14)	Place form boards on bench	C	B	G	S	W 1-2	Board	0.20	0.20	0.20	0.13	0.13	0.13
(15)	Wedge form boards together using iron clamps⁶ (see p.491)	C	B	G	S	W 1-2	Cleat	0.30	0.30	0.30	0.20	0.20	0.20
(16)	Wedge form boards together with wood wedges⁶	C	B	G	S	W 1-2	Cleet	0.30	0.30	0.30	0.20	0.20	0.20
(17)	Measure and mark for nailing⁴	C	B	G	S	W 1-2	Cleat	0.05	0.05	0.05	0.03	0.03	0.03
(18)	Nail 2-12 ft. form boards to cleats⁷	C	B	G	S	W 1-2	Cleat	0.42	0.48	0.54	0.28	0.32	0.36
(19)	Nail 4-12 ft. form boards to cleats⁷	C	B	G	S	W 1-2	Cleat	0.83	0.96	1.08	0.55	0.64	0.72
(20)	Nail 6-12 ft. form boards to cleats⁷	C	B	G	S	W 1-2	Cleat	1.25	1.44	1.62	0.83	0.96	1.08
(21)	Nail 2 form boards to 1" batten⁸	C			S		Batten	0.37	—	—	0.25	—	—
(22)	Nail 4 form boards to 1" batten⁸	C			S		Batten	0.74	—	—	0.49	—	—
(23)	Nail 6 form boards to 1" batten⁸	C			S		Batten	1.11	—	—	0.74	—	—
(24)	Turn form over for clinching nails	C			S		Side	0.90	1.10	1.30	0.60	0.73	0.87
(25)	Clinch nails 2 form boards per batten⁸	C			S		Batten	0.15	—	—	0.10	—	—
(26)	Clinch nails 4 form boards per batten⁸	C			S		Batten	0.30	—	—	0.20	—	—
(27)	Clinch nails 6 form boards per batten⁸	C			S		Batten	0.45	—	—	0.30	—	—
(28)	Carry² 1" × 4" pocket pieces 50 feet (see p. 655)	C	B	G	S		Side	0.09	0.09	0.09	0.06	0.06	0.06
(29)	Place 1" × 4" pocket pieces on form	C	B	G	S		Side	0.30	0.30	0.30	0.20	0.20	0.20
(30)	Nail 1" × 4" pocket pieces to form	C	B	G	S		Side	0.74	0.74	0.74	0.49	0.49	0.49
(31)	Mark and square across end of form	C	B	G	S	W 1-2	Board	0.10	0.10	0.10	0.07	0.07	0.07
(32)	Saw off end of form 24" wide by hand	C	B	G	S	W 1-2	Board	0.23	0.41	0.60	0.15	0.27	0.40
(33)	Throw form aside on pile	C	B	G	S	W 1-2	Side	0.53	0.68	0.82	0.35	0.45	0.55
(34)	Extra for making clean-out hole	C					Col.	2.20	2.65	3.10	1.47	1.77	2.07
(35)	Level and mark across top of column	C					Side	See Table 163, Item 23.					
(36)	Cutting beam pocket	C		G			Pocket	12.45	14.60	16.75	8.30	9.73	11.17

¹ For example, for any column select proper items from those marked C choosing from items marked C1 for type 1, see Fig. 48, p. 491, from C1-2 for both types 1 and 2 see Fig. 49, p. 493. Those marked C only apply to all types of form construction.

² Can be done by laborers—figuring costs at laborers' rates.

³ For cleats over 4 ft. long see Table 156, p. 655.

⁴ This operation finished in erecting, see Table 156, p. 655.

⁵ This considers ripping 2 side edges of each form, if one side only is ripped off take ½ of item.

⁶ Based on 8 cleats per form.

⁷ Based on 2 nails per 6-inch board per cleat using 8d nails for 1 inch boards, 10d nails, for 1½-inch plank and 12d nails for 2-inch plank.

⁸ Based on 2 nails per 6-inch board per batten using 6d nails.

Figure 3-6 (Cont.)

(Source: Frederick W. Taylor and Sanford B. Thompson, Concrete Costs [New York: John Wiley and Sons, 1912], pp. 664-665.)

Description	Table Number	Item Number	Number of Units	Unit Time	Time per Column
				min.	*min.*
Rip △ strips for corner beading........	162	(1)	4	1.47	5.88
Saw 2″ × 4″ cleats.....................	161	(14)	4×7	0.24	6.72
Carry 2″ × 4″ cleats 50 feet to bench.	162	(4)	28	0.12	3.36
Place 2–2″ × 4″ cleats on bench........	162	(7)	28	0.13	3.64
Saw off form boards, both ends......	161	(3)	32	0.19	6.08
Rip form boards for pockets on mill-saw	162	(9)	8	1.24	9.92
Rip 12-foot form boards on mill-saw..	162	(11)	4	2.94	11.76
Carry form boards 50 feet—Table 156, p. 655....................	162	(13)	16	0.30	4.80
Place form boards on bench..........	162	(14)	16	0.40	6.40
Wedge form boards together with wood wedges....................	162	(16)	16	0.30	4.80
Measure and mark for nailing	162	(17)	16	0.05	0.80
Nail 4–12 foot form boards to cleats.	162	(19)	28	0.83	23.24
Throw form aside on pile............	162	(33)	4	0.53	2.12
Extra for making cleanout hole.......	162	(34)	1	2.20	2.20
Saw 1″ × 4″ pocket pieces...........	161	(2)	12	0.12	1.44
Carry pocket pieces 50 feet to bench..	162	(28)	4	0.09	0.36
Place pocket pieces on form.........	162	(29)	4	0.30	1.20
Nail pocket pieces to form..........	162	(30)	4	0.74	2.96

Total time to make 1 column form complete.............	97.68 min.
Add for unavoidable delays, ineffective work, etc...........30%	29.30 "
	126.98 "
Add for foremen, making benches and contingences due to weather, etc. (see p. 125)............................27%	34.28 "
	161.26 "
Total overall "Make Time" =	2.69 hrs.

This total coincides with the time given direct in Table 134, p. 630.

If work is scientifically managed the carrying items will be practically eliminated and other items reduced.

HAND SAWING vs. MILL SAWING

Example 2: How much longer will it take to make the above form if all the sawing except △ strips is done by hand instead of on a mill-saw?

Solution: Mill-saw items used above: Items (14) + (3) + (2) from Table 161 + Items (9) + (11) from Table 162 = 35.92 minutes.

Replace these by hand saw items which amount to 90.00 minutes.

Difference is 90.00 − 35.92 = 54.08 minutes = 0.90 hours net.

Adding percentage as above, gross time is 89.28 minutes = 1.49 hours.

Total time if sawing is done by hand is 2.69 hours + 1.49 hours = 4.18 hours, or an increase of 55%.

Figure 3-7
FORM LABOR CALCULATIONS
(*Source: Frederick W. Taylor and Sanford B. Thompson, Concrete Costs [New York: John Wiley and Sons, 1912], p. 649.*)

TIME PLACING AND REMOVING COLUMN FORMS

Example 7: Figure time placing and removing column forms, first time.

See Example 1 for details of form.

Solution: Take items from Unit Time Table 163, pp. 668 to 670.

DESCRIPTION	TABLE NUMBER	ITEM NUMBER	NUMBER OF UNITS	UNIT TIME	TIME PER COLUMN
				min.	*min.*
Locate, line, and brace columns..................	163	(1)	1	61.60	61.60
Carry sides of forms 50 ft.	163	(3)	4	2.40	9.60
Place 4 sides of form on horses..................	163	(4)	1	1.75	1.75
Measure sides for 4 △ strips, mark and ready to place them..........	163	(5)	1	8.00	8.00
Place new △ strips on 4 sides of column........	163	(6)	1	20.20	20.20
Get wedging boards 50 feet away..................	163	(10)	4	0.30	1.20
Mark cleats and nail on wedging boards.........	163	(9)+(11)	7	1.60	11.20
Place and drive wedges, get, place, and tighten iron clamps.............	163	(14)+(19)+(20)	7	2.00	14.00
Nail 4 sides of column together................	163	(12)	1	1.04	1.04
Carry 3 sides to place, 50 feet....................	163	(15)	1	7.20	7.20
Carry 4th side to place, 50 feet....................	163	(16)	1	2.40	2.40
Lift and place 3 sides (attached), place 4th side..	163	(17)+(18)	1	3.00	3.00
Square form, place temporary braces. Place piece at cleanout hole, change to next column..	163	(21)+(22)+(24)	1	16.80	16.80
Make and carry wood wedges 50 feet..........	163	(29)+(30)	1	10.00	10 00
Oil forms...................	163	(33)	1	5.00	5.00
Saw off 1 × 2 inch strips ..	161	(22)	8	0.38	3.04
Clean out rubbish from column.................	163	(35)	1	2.75	2.75
Remove forms............	163	(36)	1	69.60	69.60
					248.38
Add for unavoidable delays and ineffective work (see p. 625)			30%		74.51
Carried forward to next page........................					322.89

Figure 3-8

PLACING FORMS TIME STUDY

(*Source: Frederick W. Taylor and Sanford B. Thompson, Concrete Costs [New York: John Wiley and Sons, 1912], p. 651.*)

LABOR ON COLUMN FORMS [TIMES

TYPE 3.—BOLTED CLAMPS—Fig. 49A, p. 493

For costs see opposite page   *See pp. 613, 623 and 648*

BEFORE USING THIS TABLE, OPEN FOLDING PAGE 653

SIZE OF COLUMN	TIME IN HOURS PER COLUMN											
	MAKE FORMS for Different Story Heights*			PLACE AND REMOVE FORMS 1ST TIME†			PLACE AND REMOVE FORMS AFTER 1ST TIME (Same size col.)			REMAKE, PLACE AND REMOVE FORMS		
	6 ft. hr.	12 ft. hr.	18 ft. hr.	6 ft. hr.	12 ft. hr.	18 ft. hr.	6 ft. hr.	12 ft. hr.	18 ft. hr.	6 ft. hr.	12 ft. hr.	18 ft. hr.
1-inch Lumber (Nominal)												
8″ × 8″	1.5	2.5	3.8	7.1	9.5	12.2	6.0	7.5	10.1	8.1	11.1	14.8
10″ × 10″	1.6	2.6	4.0	7.2	9.6	12.6	6.1	7.8	10.4	8.3	11.3	15.1
12″ × 12″	1.7	2.8	4.2	7.5	10.1	13.1	6.3	8.0	10.8	8.5	11.6	15.5
14″ × 14″	1.8	3.0	4.5	7.8	10.5	13.5	6.6	8.4	11.3	8.7	12.0	15.9
16″ × 16″	1.9	3.2	4.6	8.1	11.0	14.1	6.9	8.8	11.8	9.0	12.3	16.4
18″ × 18″	2.0	3.3	5.1	8.4	11.3	14.7	7.2	9.2	12.3	9.3	12.7	16.9
20″ × 20″	2.1	3.5	5.4	8.7	11.7	15.2	7.5	9.5	12.8	9.6	13.1	17.4
22″ × 22″	2.2	3.7	5.7	9.0	12.0	15.6	7.8	9.9	13.3	9.8	13.4	17.9
24″ × 24″	2.3	4.0	5.9	9.3	12.5	16.2	8.1	10.3	13.8	10.1	13.8	18.3
26″ × 26″	2.5	4.2	6.5	9.8	13.1	17.0	8.4	10.7	14.4	10.5	14.4	19.1
28″ × 28″	2.8	4.7	7.2	10.2	13.8	18.0	8.9	11.3	15.1	10.9	14.9	19.9
30″ × 30″	3.0	4.9	7.6	10.5	14.3	18.5	9.1	11.6	15.6	11.2	15.3	20.5
32″ × 32″	3.1	5.2	7.9	11.0	14.6	19.1	9.4	12.0	16.1	11.4	15.6	20.7
34″ × 34″	3.2	5.4	8.0	11.3	15.0	19.5	9.7	12.4	16.6	11.7	16.0	21.2
36″ × 36″	3.4	5.6	8.3	11.6	15.5	20.1	10.1	12.7	17.2	12.0	16.4	21.8
2-inch Lumber (Nominal)												
8″ × 8″	2.0	3.3	5.0	8.1	11.0	14.1	7.2	9.2	12.4	10.5	14.4	19.2
10″ × 10″	2.1	3.4	5.3	8.4	11.3	14.6	7.6	9.5	12.7	10.7	14.7	19.6
12″ × 12″	2.2	3.7	5.5	8.7	11.7	15.2	7.8	9.8	13.1	11.2	15.1	20.0
14″ × 14″	2.3	3.9	6.0	9.0	12.2	15.8	8.1	10.3	13.8	11.3	15.6	20.7
16″ × 16″	2.5	4.1	6.3	9.3	12.6	16.4	8.4	10.7	14.4	11.7	16.1	21.3
18″ × 18″	2.7	4.4	6.8	9.8	13.1	17.0	8.8	11.2	15.0	12.1	16.5	22.0
20″ × 20″	2.8	4.6	7.1	10.1	13.5	17.6	9.1	11.7	15.6	12.4	17.0	22.6
22″ × 22″	2.9	4.9	7.4	10.4	14.0	18.2	9.5	12.1	16.2	12.7	17.4	23.2
24″ × 24″	3.1	5.1	7.8	10.8	14.4	18.9	10.1	12.6	16.8	13.1	17.9	23.8
26″ × 26″	3.4	5.6	8.5	11.3	15.6	19.7	10.3	13.1	17.6	13.6	18.6	24.8
28″ × 28″	3.7	6.2	9.5	12.0	15.9	20.7	10.8	13.7	18.4	14.2	19.4	25.8
30″ × 30″	3.9	6.5	10.0	12.3	16.5	21.5	11.2	14.1	19.0	14.7	19.9	26.4
32″ × 32″	4.1	6.8	10.4	12.6	17.0	21.6	11.5	14.6	19.8	14.9	20.3	27.1
34″ × 34″	4.2	7.0	10.6	12.9	17.4	22.2	12.0	15.1	20.5	15.2	20.5	27.6
36″ × 36″	4.4	7.3	11.1	13.4	17.9	23.0	12.3	15.6	21.2	15.5	21.3	28.3

* If old form lumber is used add 90 % to "Make Forms."
† Values increased 50% as labor is generally inefficient on first set up.

Figure 3-9

COLUMN FORM LABOR CALCULATION

(*Source: Frederick W. Taylor and Sanford B. Thompson, Concrete Costs [New York: John Wiley and Sons, 1912], p. 634.*)

3.28 SUBCONTRACTORS' BIDS

A general contractor is not bound legally to give a job to a subcontractor who submits a low bid, unless the general has expressly agreed to do so.[11] This promise can be oral, although an oral agreement, especially by telephone, is not only difficult to prove but for long jobs (over a year) may not be valid. A subcontractor, however, is held to his bid, even though the general has made no specific promise.[12]

A subcontractor's liability for his bid was clearly defined in a 1978 Massachusetts case, *Loranger Construction Corporation* v. *E.F. Hauserman Co.* Hauserman quoted $15,900 for steel partitions for a school, and refused to honor the quotation. Loranger purchased the partitions elsewhere for $23,000 and sued Hauserman for the difference. The appelate court stated that the subcontractor, under the theory of *promising estoppel*, was barred from reneging on his quote once Loranger submitted a bid in reliance on it. The Supreme Judicial Court used a different reasoning: that a jury could have found that the proposal was accepted (a) when it was made, (b) when the general used it in submitting his bid, or (c) when the general requested the subcontractor to perform. All three courts—trial court, appelate court, and State Supreme Court—decided against the subcontractor.[13]

The legal situation where a sub makes an error is not directly comparable to that when a general makes an error. The contractor has acted on the sub's bid, to the contractor's detriment, and has often submitted a bid bond to commit himself, but the owner has suffered no damage if the general's bid is withdrawn. However, the *general* does suffer if the sub's bid is withdrawn; he can avoid the contract only by claiming the bid of the sub as an error. Logically, the sub's release on the grounds of error should depend on the general's release on the same grounds. Even then, the general has suffered the loss of the cost of the bid—often a substantial sum.

However, the enforcement of the sub's obligation to honor his bid is nearly impossible. The contractor must prove that:

1. he had received a clear and definite offer from the sub;
2. the subcontractor could expect that the contractor would substantially rely on the offer;
3. the contractor actually and reasonably relies on the offer; and
4. the contractor relies on the offer to his detriment and this reliance must be reasonable and justifiable.

In a 1975 Illinois case the subcontractor who was substantially low claimed release from his bid on the grounds of a mistake. With the usual wide variation of subcontractor bids and the few bids normally received, the low bid can quite often be shown "unreasonably low," and therefore not be relied on by the contractor.

[11] *Northeastern Construction Co.* v. *Town of North Hempstead,* 121 App. Div., 187, 105 N.Y. Supp.581.

[12] *R.P. Farmsworth & Co., Inc.* v. *Albert,* 50 F. Supp. 27.

[13] *Loranger Construction Corp.* v. *E.F. Hausermann Co.,* 384 N.E. 2d 176 (S. Ct. Mass. 1978) as reported in *Engineering News-Record,* March 29, 1979.

This case illustrates two other defenses available to the subcontractor. Since the contractor had accepted the bid *subject to architect's approval*, the court held that the contractor had not accepted the bid until the architect approved the subcontractor. The contractor could not have avoided this qualification except by omitting the qualification and claiming later that this was a specification requirement implied in the sub's bid. It illustrates how important an unqualified acceptance may be.

The last defense allowable to the subcontractor, pointed out by the court, is even more general. Since the subcontract referred to so many matters which were absolutely essential to the performance of the proeject, it was extremely unlikely that the contractor and subcontractor intended to bind themselves contractually until the written subcontract had been completed and signed, as there were many matters that went beyond the bid price.[14]

This presents two dilemmas to the general contractor. Trade opinion and professional ethics maintain that he is obligated to the sub whose bid he *used*, yet the sub has given no enforceable bid. Is the general therefore justified in considering all prebid obligations as tentative?

The second dilemma is that the general must, in many cases, name his subcontractor in the general's proposal. He can, of course, later claim that the sub would not verify the sub's bid. Some contractors refuse to name their subs, stating all subcontractors "to be determined." This author has found no legal cases on this matter.

3.29 BID DEPOSITORIES

To prevent "bid shopping" by general contractors, specialty contractors agree to give copies of their bids (or in some cases, the original bids) to a neutral agency such as a builders' or contractors' association, to be opened at a later time. Bids in depositories are usually opened at the same time general contractors' bids are opened. Since each subcontractor then knows the competitive bids, he cannot continue negotiations without the other subcontractors being aware that he is cutting the low bid.

The chief disadvantages of bid depositories are as follows:

1. Since his prices are exposed, a subcontractor is placed in an embarrassing position if he has given different prices to various contractors—a common practice. If a contractor is to realize a gain for the efforts he makes in financing and organizing a job, it must be through lower prices from his subs whose costs are reduced.

2. There are many ways for the subcontractor to avoid the depository limitations; a simple way is to give one figure to the contractor and another to the depository. Unless the depository bid is an original proposal, it may be lower; otherwise, the depository bid may be higher. Of course, this exposes the cost cutter, but never without some doubt, since large contracts may be made in different ways by including or excluding some items. If more than one subcontractor does this, the bids in the depository become confused.

---

[14]*S.N. Neilson Company* v. *National Heat and Power Company, Inc.,* Appelate Court of Illinois, October 8, 1975, as reported by Marvin Schechter in "Legal Matters," *Constructor,* Vol. 58, No. 8, August 1976.

3. The contractor may be obliged only by direction of the architect to let the job to the bidder who was low at bid time. Any of the subcontractors can cut their price afterward, the action that the bid depository was set up to prevent.

4. The effectiveness of the depository depends either on the cooperation of the subcontractors or on their control of the market. If they would cooperate to the extent of making firm bids, there would be no problem without a depository. If they control the market, they may be subject to antitrust action. The market control may be through the municipality, which could refuse to permit outside contractors by means of a rigged licensing examination, or through the unions, who might refuse to work for other contractors and refuse to allow them to work nonunion. Sometimes the material men refuse material to an outside contractor, or quote a higher price. All such actions, of course, raise the price of construction.

5. An outside contractor, not a party to the local agreement, may underbid the local people after the bids are exposed—making a bid very simple for him.

6. With a subcontractor specified at bid time, the low bidder will not pass on savings he may later realize by cuts in the price of materials which are part of the subcontract bid. This saving is not gained by the owner anyway, if the contractor has a lump-sum contract, but it would be gained by the owner under a fixed-fee contract. These price cuts are prices the sub did not have at bid time.

7. Cooperation is also required of the material suppliers to get prices in before the time that bids are scheduled to go to the depository. The subcontractor who has his bid already in, and who then receives a material price cut, will be anxious to transmit this cut to his bid. This is especially true if the sub concerned suspects that his competitor has used either a lower material price or has submitted a lower price, anticipating some bid shopping of his own materials. The larger firms—such as Westinghouse and General Electric—are as slow to get final prices in to the subcontractor as are smaller suppliers.

3.30 FOUR-HOUR PLAN

In some areas, general contractors have agreed to accept no bids after a certain time—such as 2 or 4 hours before the time for submission of general contract bids. This allows the general contractor time to determine who really is the low subcontract bidder. Otherwise, he really does not know in many cases, because of the complexities of the number of items which may or may not be included in the bid. The general also has time to check his bid properly for errors. This procedure works well when general contractors cooperate. This cooperation fails under the following conditions:

1. Contractors from areas without such a rule bid against the local people. These outside contractors may bring their own subcontractors and suppliers, who are not accustomed to bidding under the local practice.

2. The general contractor requests clarification of bids from subs who did get their bids in on time but not in detail. A sub can put in a bid on time but later raise or lower it by adding or deducting what he includes in the bid, in which case cooperation fails.

3.31 DETAIL OF SUB BIDS

The subcontractors' bids may vary widely as to what is included in the same trade. Generally, the most definite bid specifies "everything mentioned in Sections 2, 3, and 4 of the specifications." Or it may be "all specified," or "all work of our trade specified." These are different bids but are all capable of computation. Many bids do not mention sections but say, "all work on electrical drawings," which is less definite.

Everything mentioned may include not only the work of other trades incidental to a trade, such as bases and fences around transformers for the electrical trade, but also totally unrelated items mentioned in a section, such as "heating and cooling controls for the ventilating systems" in the electrical section. The ventilating system would not be considered under the electrical contractor's work simply because it is mentioned. The electrical section may mention "wiring the bathroom ventilating fans," which is less definite. Medicine cabinets with fluorescent lights are a kind of electrical fixture but might not be furnished by the electrician. Often work is mentioned in more than one section of the specifications. Any reference to the specifications *only* runs the risk of omitting work shown on the drawings but not in the specifications.

All specified in the particular section of the specifications is a little more definite. However, work may be specified in a section but not required—specifications are descriptive rather than definitive anyway. This may be as ambiguous as "everything mentioned." For a long time, the United States government included description of starters and safety switches for pumps in the electrical section of the specifications, with a note in the mechanical section, "see electrical specifications," or vice versa. Consequently, an item may be *described* in only one section of the specifications but *specified* in two. Painting is often not only specified in the painting specifications but also in the section for the item being painted. This may lead to confusion since some millwork items, particularly kitchen cabinets, are prefinished. The painter may not include painting piping because he does not know the extent of it; piping is often shown only in a schematic manner. Electrical or electronic controls are sometimes confusing, since the electrician often does not do this work; it may be included in the electrical work, the mechanical work, or in a separate contract. *All specified* may omit items on the drawings, especially on the electrical or mechanical drawings, which are not included in the specifications for these trades. Again, this may be concrete work for transformers, boiler foundations, concrete fill around electrical conduit, or wood blocking for some of these trades. Often these details depend on an item to be included in the proposal which is an option of the electrical or mechanical contractor. Only the mechanical or electrical contractor knows just what is needed.

All work of our trade is probably the most definite of all, if tied down to union working rules. The unions have separated the work by trades for many years, and there are generally decisions—or there will be—for all items. Under these circumstances, the concrete items and blocking are given to the general contractor. Not all doubt is removed, but this provides a reasonable way to determine the differences. There may continue to be disputes over the plywood backing for electrical switchgear (which is the electrician's work), cuts in wood finish for electrical fixtures (which may be either carpenters' or electricians' work), suspended ceilings with panels set in (which is the electricians' if there are fixtures above it, but not otherwise), supports for bathroom fixtures (which are the carpenters', unless they are metal, but this may depend on local practice).

All work on plumbing drawings leaves out items which may be specified, or shown on architectural drawings, but which must be done by the plumber. How about a plumbing fixture shown on the architectural drawing but not on the plumbing drawing?

3.32   SUB BID SHEET

When sub bids are transmitted, most subcontractors have their estimator, rather than a clerk, call the general contractors. These bids should be received by the estimator for the general contractor, for others are not familiar with the job and do not know what questions to ask about the items in the bid. The general's estimator should make a list, checking the specifications for all doubtful items or for items mentioned more than once in the specifications. He then receives the sub's bid by telephone with this list in front of him. Another person can take the bid under these conditions, but not nearly so quickly; the name of an item may not be enough to identify it. The subcontractor often does not check all the specifications, and he may not even have all the drawings and specifications for the job; the person discussing an item needs to know what drawing shows it. The sub—mechanical or electrical, or even for exterior work, as utilities or paving—will usually order only the sheets showing his work. These are often identified by a special prefix of drawing number, so the sub may not know how many sheets there are, although the sub's estimator should look over the rest of the drawings just to find such things (as the general contractor's estimator looks over the specialty drawings). There are sometimes large pieces of work required by a vague clause in the specifications, such as, "Construction of the elevator shaft shall be as shown and in accordance with the Elevator Code." Such a clause, appearing in the elevator specifications, may require concrete tie beams or other structural work in the building, not in the elevator contractor's work. Such items are left vague in the specifications because the requirements vary for different equipment that might be used, and the architect, by showing the details, might be drawing an item which would not apply to the equipment used, or he might be restricting the item to one equipment manufacturer.

3.33   SAMPLE BID

A telephone conversation receiving a bid may go like this:
"Smith of Hotshot Electrical with a figure for Cutem Hospital."
"Johnson here. Shoot."
"All sections 3,4,5, and 6 but not painting, 156,400."
"Okay, $156,400. Have you light on water tank?" (As water tanks are built by specialists, they may have entirely separate drawings made for the A/E by a contractor.)
"Didn't know there was one. What is involved?"
"Navigation light and float valve on two circuits."
"OK, we'll include the light circuit . . .let me see . . .yes, we have the float valve in the control man's bid."
"Excavation and concrete for conduits?"
"We don't dig and fill, as always" (of course, how would you know if you have not done business with him before?) "but we'll take the concrete."
"Safety switches and starters?"
"We'll install, mechanical man furnished."
"Bond?"
"Included." (He probably would have said no after he got the bid, if he was low. Now you've got a 1 percent better price than you otherwise had.)

TELEPHONE BID

PRACTICAL FORM 512

ESTIMATE NUMBER **45**

DATE **8/10/**

JOB **Cutem**

LOCATION **Boot Hill, Ga.**

FIRM **Hotshot Electrical**   BY **Smith**

ADDRESS   PHONE

CLASS OF WORK **Electrical**

WORK INCLUDED	AMOUNT OF BID
Sections 3-4-5-6	$156,400
Incl: Circ for water tank	
Float valve "	
Conc for conduits	
Bond	
Safety Sw's & Starters	
Temp wiring to 100 ft of rooms	
Disconnects only old wiring	
(Noncommittal — probably high).	
	TOTAL BID

EXCLUSIONS AND QUALIFICATIONS

No painting.
No excavation

ACKNOWLEDGEMENT OF ADDENDA:

DELIVERY:

TAX	
EXCLUDED	
INCLUDED	

MFD. IN U.S.A.   FRANK R. WALKER CO., PUBLISHERS, CHICAGO

RECEIVED BY: **K.R.**

Figure 3-10
TELEPHONE BID
(*Courtesy of Frank R. Walker Company.*)

"Temporary wiring?" (This may not be in the specs at all; if it is, it is probably in the General Conditions.)

"Included to within 100 feet of rooms." (This means he will give an outlet every 200 feet throughout. The other subs can pull their own extensions from there.)

"Removal of old wiring?" (Wrecking is also not in trade specifications and need not be done by trades.)

"We'll disconnect only." (The general does not usually know what is required to take hot wires from a building; it may be something an electrician can do in a few minutes.)

"How good is this bid?" (If you know the other estimator well, you can tell from his tone, rather than his words, how much—and how little markup—he has on the job.)

"We're always good." (Which can mean anything, but probably would mean just the opposite of what he says. That is, it is just another job, and they are not after it very hard, or they are giving preference to another general.) The results of the conversation would be recorded on a telephone bid sheet as shown in Figure 3-10.

Depending on your relationship with the other estimator, you may try to get some useful information on other bidders—which ones he is bidding, if he is giving them the same figure, if some bidders have dropped out, if there are other electrical bidders whose figure you do not have (he would know through his suppliers who other electrical bidders were, and you could find out in the same way). If he does not give you his competitors' names, he knows he may be denying you the job with no benefit to himself. A subcontractor's estimator usually knows his competition better than you do. He may know of subs who are bidding only one general, and who that general is. This in turn offers you an opportunity to try to convince the other sub that you should have his figure. You have an advantage in getting *all* sub bids which are *out*, that is, bids which have been submitted to any general contractor. If each competitor *does not* have one low bid that you *do* have, this may give you the job.

3.34   CUTTING SUBCONTRACTOR BIDS

From getting all the bids, it is only a step to believing there are low bids you do not know about, and lowering your bid to use the bids you do not have. You are now a *bid shopper* because you have to look for a sub you did not have, or a price you did not have, to make a profit on the job. You can rationalize this by thinking that your subs would not have gotten a job anyway if you had used their bids—the next step is that your sub does not know if there were lower bids, so why shouldn't he cut his price? Then all the other generals do the same thing you are doing, and now you are in a full throat-cutting competition. This situation exists in some areas, not in others. This is the unethical practice most resented by the subs and is not in accordance with the AGC code of ethics. Nevertheless, it is hard to stop a general from accepting a lower bid if it is later offered, or if he knows it will be offered, from relying on it in advance.

3.35   STIPULATED SUBCONTRACTORS

In many cases, A/E's require that the name of principal subcontractors, whose bids were used, be named in the bid. Presumably, the low bidder will be protected in this way against bid shopping, as the subcontractor may be changed only with permission of the A/E. The A/E is primarily interested in getting subs he knows are qualified and is also interested in a basis for negotiation if he must allow an extra for a change of subs. When the architect directs a higher price sub's proposal be used in a contract, he allows an extra payment to the contractor for his added cost.

There are several practices which may avoid the A/E's requirement that the sub be named.

1. Writing "undertermined" in the space provided.
2. Naming a subcontractor known to be unacceptable to the owner, giving the contractor time after the bidding to obtain a better price from one of the reputable subs.
3. Naming a friendly subcontractor who will refuse the job after the bids are taken (and who may not have estimated the job at all). The subs are not bound by the general's bid, and the manner in which most subs' bids are given is so vague that enforcement is very difficult.
4. Offering the owner part of the saving made by changing subs after the contract award.

3.36 IMPROVING INDUSTRY RELATIONSHIPS

The basic difficulties in enforcing any type of anti-bid-shopping plan are as follows:

1. Subs' bids are received in a manner that does not constitute a contract if accepted, and are received so late that an acceptance is very difficult.
2. Sub and suppliers will settle for a lower price when a firm contract, rather than just a chance at a contract, is offered. When a sub is low with a general contractor, he still has no assurance that the general will get the job. A bid depository for *all* subs of a trade, if followed, would solve this problem, but would still leave the question, "Is the owner going to build at all and, if so, will he accept the low contractor?" Without a firm commitment by the owner, which, in turn, requires a reasonable accurate architect's estimate when drawings are started, contractors will continue to incur losses by estimating jobs not built. This is less of a problem with public than with private work.
3. The owner, when offered an unethical saving, is strongly tempted to take the unethical purchase when it is a one-time or rare transaction. Many owners will build a major job only a few times a generation. They have to live with the higher price for the life of the project, but they will probably never again do business with the contractor.
4. A contractor must maintain the goodwill of his subs in order to get prices in the future.

3.37 CONTRACTORS' ALTERNATES

A contractor may know of ways in which the drawings or specifications may be changed to provide substantially the same construction at lower cost. In such a case, it is not uncommon that he will bid an alternate at the same time as bids are received under the engineer's conditions: he may submit two bids, one in accordance with the invitation, and one in accordance with the alternate construction.

In such a situation, the contractor would not submit his alternate early because (1) subs' prices would not be available, and (2) the A/E would probably ask for other contractors' bids on the same alternate. To submit it later than the bid time would be unethical of the contractor. Very often, the contractor will bid only the alternate, since if he were high on the regular bid and low on the alternate, the owner (1) will assume that the difference between the two represents the cost saving of the alternate, and (2) will offer the low bidder a price on the alternate, using this difference as the basis of negotiation. With a lump-sum bid on the alternate only, the owner has no such basis for comparison.

Such alternate bids cannot be considered in public work when complete plans and specifications are provided; the contractor may put in a low bid on the invitation, hoping to change to the alternate later (which is allowable, and, in recent years, is called *value engineering*).

Value engineering is the review of plans and specifications with the goal of making cost-cutting substitutions. Most architects and engineers feel that this is their duty, and that such review considers only small parts of the design, rather than fully considering the design as a whole. It is the engineer's duty to save money. In recent years, the US government has offered a contractor part of the saving resulting from the contractor's design changes; 20 years ago, resident engineers were forbidden to grant such credits (but had no way to determine the savings, and shared with the contractor anyway). The Navy, which already had a similar, much older department of cost reduction, in a typically bureaucratic procedure established a value engineering department in addition to the one already existing. The federal government's program of value engineering requires consideration of the design at three stages:

1. Lectures to design personnel in the early stages of design.
2. Review of plans on completion of design.
3. Review by contractor after contract is let.

Under the contractor's review provision, savings by contractor's alternates are to be shared by the contractor and owner. A/E's have always agreed to alternates, but as they have inadequate cost data to check the contractor's proposed credits, they often discourage substitutions after the contract is signed and the owner no longer can obtain another price on the option.

On private work, the owner may accept such alternates; but he may also elect to ask for bids a second time, with several contractors bidding the alternate. The value of such alternates is therefore less valuable to the contractor if he clearly describes them. The contractor may give a price, however, contingent on unspecified modifications being accepted. In some circumstances, this may be successful. A/E's resent such proposals, which imply that their plans and specifications are unnecessarily expensive. An A/E is particularly embarrassed if the alternate voids substantial portions of the plans which the contractor proposes to replace at no charge. For example, a plan showed driven step-taper piling for foundations with a comparatively short depth to rock bearing. The contractor proposed to use Franki piling—a much larger concrete pile with no casing—and to redraw the foundation plans, thus offering a substantial saving. The owner immediately recognized that the engineer had charged him for extensive drawings now of no value, as well as selecting the wrong foundation. The more experienced engineer will probably shrug this off and say, "You can't win them all," but the less experienced designer may feel he has done an inadequate job and resent the contractor who made the improvement. If the contractor depends on the A/E for negotiated work, where the designer's recommendation carries considerable force, he may avoid submitting bid alternates of this type. If a contractor is low, he can negotiate these changes privately with the A/E, and then the two can present the alternate to the owner as a joint effort. The AGC code of ethics does not forbid unsolicited alternate bids, but some A/E's consider them a shady practice if not previously discussed with the A/E.

3.38 QUESTIONS FOR DISCUSSION

1. Your employer says, "I don't use anybody's bid. I take the bids, and they are under no obligation to me. If a sub is low with me, I give him first refusal of the job, but not necessarily at his bid price. If it appears he is too low, I tell him so before he signs a contract, but I don't tell him how much—I just tell him he appears low and should check his estimates. The bids I go in with are mine, and I don't hold anyone responsible if he's wrong." You note that he puts no profit on his bids and has an average bidding record.

   Do you think he is ethical? If not, would you continue to work with him? Do you think this policy is wise from a business standpoint? Why or why not? What might possible results be?

2. You were the estimator who took the telephone bid shown in Figure 3-10, one hour before the bid was due at the architect's office. You were unable to figure out whether the bid was low in the time you had, so you did not use it. You town has no 4-hour bid rule. You are the low bidder, and 2 hours after bids are received, both Hotshot and another electrical contractor, Sparkwell, whose bid you used, call you back. Both say other contractors told them they were low on the job, and asked if you used their bids. What would you tell each of them?

3. Two days before bid time, you find that a sheet in your drawings evidently is not for the job you are bidding, and the sheet is not listed in the index of plans. There is the name of another project on the sheet but it is not in a conspicuous place and, since it fits in well as an additional small building, you believe the other bidders will have included it in the estimate, if they also have the sheet. What would you do?

4. The circumstances are the same as in Question 3, but the sheet is definitely part of the bid drawings; however, the layout (site) drawing indicates "Not in this bid—future construction" on the building. What would you do?

5. You have a close relationship with an architect and have been given negotiated work on his recommendation. You are one of four bidders on a $100,000 building, and an hour before bid time you discover that two bidders have asked you for courtesy bids. The lack of interest of the fourth bidder, as shown by your contacts, makes it appear certain that he has dropped out; in fact, it has been repeated to you that their estimator said, "I think we'll have to pass this one up." Would you go ahead with your bid as planned?

6. Continuing the situation from Question 5, you have bid the job and find that the fourth bidder did drop out, and the three bids received were actually those you figured.

   What would you tell the architect? When would you tell him? When would you wait to see what the owner is going to do?

7. You have prepared a bid for a moderate-size job in another state and have it nearly ready a week before bid date, when you discover you cannot bid without registration. Examinations for registrations are conducted once a month, and another month is needed to obtain a license. What course is open to you if you want to bid the job?

CHAPTER *4*

CONSTRUCTION CONTRACTS

When a friend deals with a friend—let their agreement be well penned—that they may continue friends.

Benjamin Franklin

A contract may be compared to a painting; the attorney supplies the frame, the canvas and the brushes, and the manager does the painting. The responsibility for drafting the essential parts of a contract and executing it rest with the project manager, not the lawyer. The function of the attorney is to advise what dangers may be incurred in the use of a certain clause, or, in some cases, what may be the legal result of proceeding without a formal contract. Unfortunately, the latter advice is seldom obtained, since delay may have been the result of the lack of advice obtained from the attorney whose advice on the consequences of delay you are now seeking!

4.1 ESSENTIALS OF A CONSTRUCTION CONTRACT

Legally, a contract must have two competent parties, a matter to contract, and a meeting of the minds as to what the contract intends; the intent is a matter on which you can consult your attorney if there is any doubt. In this chapter, you may expect to see the *general provisions* you may find and how they may be interpreted. Part of the contract documents is known as the *general conditions,* or, in the Uniform Construction Index, the *general provisions.* Often contractors use the term *general conditions* as a cost classification to designate job overhead on the summary sheet. But *results* are obtained from a contract or subcontract by having the following:

1. A clear understanding by the parties of the meaning of each part of the contract. Legal language confuses and causes delay, since the other party will seek his own legal opin-

ion when he might not, if he understood the intent of each part before legal language was used. "Trick" clauses, which have a definite legal meaning but which are not clearly stated in layman's language, may gain legal rights for a party but do not result in getting work done. A party will not seek ways of avoiding a well-understood document, unless he is pressed into heavy losses or bankruptcy. In such case, the contract will be a sad experience for both parties, regardless of its provisions. Subcontractors resist a printed standard contract less than a typed one, as they feel the printed form has been accepted by others.

2. A contract written with a board of arbitration in mind. Many contractors, and especially small subcontractors, have only a rudimentary knowledge of law and even of specifications. If you make such a contract clear, you must include a legal education in the contract itself; this is not only lengthy but may, in itself, scare off prospective subs. Consequently, the contract should be written so it may readily be understood by professionals in the construction industry; it will then be understood by arbitrators.

3. The legal teeth of formal phrases only to the extent necessary. The attorney will use less legal phraseology if you give him a clearly written first draft. You may consider the *standard contracts* well enough established so that the legal terms are generally known. The contract includes all the plans and specifications, as well as usages of the trades; an attorney must depend on the manager to prepare the technical portions, which necessarily cause the greatest difficulty in enforcement.

   Standard contracts are those approved by the AGC, AIA, ASCE (American Society of Civil Engineers), or other national societies and associations.

4. Proper qualification of the contractor or other party requiring a legal business license. In states where a professional license is required, a contractor usually cannot collect payment without it. Any attempt to extend one's license over a nonlicensed contractor is both illegal and poor business. A 1970 California decision established that a mason with no contractor's license, representing a licensed contractor, could not collect for his work, nor could the licensed contractor. It appears, although not stated in the summary, that the mason was attempting to use another's license, or appeared to do so. The mason did not reveal either that he had no license, nor that he was working for a licensed contractor. The mason could not collect because he was unlicensed, and the contractor could not collect, as he did not disclose himself as the principal who contracted the work.[1]

   However, when the contractor acts with good faith and the work is satisfactory, but the owner invokes a technicality to avoid payment, the courts will stretch the point. Contractors who were unlicensed for a short period, or who did work for which their licenses were not normally adequate, have collected for their work.

These requirements do not represent differences, but the end result sought by the manager is achieved if these guides are observed. The first two categories are not the work of an attorney; for example, unless he is a construction specialist, he would not know what knowledge is expected of the subcontractor or of a possible arbitrator. If the manager will arrange the obligations of each party to a contract in the simplest possible language, an attorney can provide the form and grammar which will make the intent legally binding. If the manager attempts to write the entire document, he must be complete and clear; otherwise, the attorney attempting to han-

[1] *Muth* v. *Leineke,* 88 Cal. Rep. 1 (1970) quoted in *Building Design & Construction,* October 1975.

dle language that is unclear to him (for he is usually not a construction specialist), will distort the meaning intended. A contract is a joint effort of manager and lawyer.

Three language definitions are involved in a construction contract: construction usage, common usage, and legal usage. The manager speaks the first two; the attorney, the last two. Avoid terms that may have one meaning for construction and another in common usage, because the meaning in common usage is often taken at law over the construction one. This responsibility is the province of the manager. The attorney, on the other hand, must avoid words which have different common and legal meanings. Words with more than one meaning should be defined in the contract itself.

As an example, "sophisticated" has a common and a technical meaning, neither of which has yet appeared in some dictionaries, where there are several prior meanings nearly the opposite of present-day usage. "Cleaning" applies to several construction processes, which bear only a casual resemblance to the common meaning.

4.2 CLAUSES TO AVOID

In April 1972, a subcommittee of the estimating and cost control group in the Construction Division, American Society of Civil Engineers (ASCE), reported on cost-increasing contract provisions. The group objected to specification clauses which are intended to protect the engineer, usually at the cost of the contractor, and which do not comply with the standard contracts. A number of these are the following:

> *The use of multiple contracts which make it difficult for the general contractor to properly plan and coordinate the overall project.*
>
> *The lack of set arbitration or appeal procedures, or (use of) finality clauses which make the owner or the engineer the sole arbitor of disputes.*
>
> *The use of hold-harmless agreements . . . which require the general contractor to assume the legal liabilities of the owner . . . . Even though many such agreements are unenforceable, resulting litigation is a cost-increasing factor.*
>
> *The lack of a clear statement of the owner's position on acceptance of risk of damage arising from force majeure, war, civil riots, and so on. (Force majeure refers to an event beyond anyone's control; an act of God, as wheather.)*
>
> *The use of clauses requiring construction to be performed to meet local codes even though design has not been performed to meet this requirement.*
>
> *The requirement of material and equipment guarantees from the contractor which exceed the guarantees that the manufacturer provides.*
>
> *The use of clauses requiring that the general contractor warrant the sufficiency of the design although the general contractor has had no part in its preparation.*
>
> *The overt or covert use of closed specifications, providing unscrupulous manufacturers or suppliers with the opportunity to bid items covered at much higher prices than would be the case under competitive conditions.*
>
> *The specification of untried and unproven materials, or the use of vague and ambiguous technical specifications which also require at the contractor guarantee the results.*
>
> *The general lack of performance-type specifications combined with incomplete and insufficient design detail.*
>
> *The provision for the use of standard specifications such as the American Society for Testing Materials (ASTM), American Concrete Institute (ACI) Standards, or the like, when, in fact, such standards have been modified by the designer.*

> *The use of localized standard technical specifications in areas other than those for which they were intended.*
>
> *The use of vague and/or incomplete drawings with notations such as "The contractor shall visit the site and make appropriate provision for any work which is required to adapt these drawings to site conditions."*
>
> *The use of the term "true to line and grade" without specifying allowable and reasonable tolerances.*
>
> *The use of the term "as directed by the engineer" in place of clearly stating what is required.*
>
> *The use of designs which do not allow the maximum reuse of concrete forms . . . .*

A/E's are at times rather naive about the purchasing process and assume that the price of the same item is the same under all conditions; hence the use of "overt or covert" specifications which restrict purchase to one supplier. "Covert" refers to the practice of using specifications which, although no manufacturer is specified, make it impractical for more than one brand to be used. Such specifications are usually provided by the manufacturer's agent at the A/E's request. The restricting clause may be some inconspicuous description of a particular, and usually patented, detail of manufacture. The A/E often does not know that he has inserted a restricting specification. This practice cannot be considered unscrupulous since virtually all manufacturers do it when they have an opportunity. Who wants to write a free specification which allows your competitor to sell the job? When there is a legitimate reason for requiring a particular brand, it is more economical for the owner to take open bids and negotiate the brand wanted with a change order before the contract award.

Some architects pride themselves for specifying brand names, as they consider that this assures quality for the owner, by not allowing the contractor to substitute a cheaper product. Such a practice makes it unnecessary for the architect to be acquainted with competing brands, as he would have to be in order to select an "equal" brand. Federal specifications or commercial standards are often considered too lax by A/E's; actually, there are few manufacturers whose products consistently are superior to the minimum standards of the industry, as in roof deck and aluminum windows; but for an overwhelming number of products, there is no substantial difference. In these cases, where the better product is needed, most manufacturers can be eliminated by specifications which give metal thickness or similar restrictions.

4.3 COMPETITIVE BIDDING

The desirability of competitive bidding as a general concept is seldom discussed among contractors. Each favors the operation under which he is successful, and nearly all are interested, in varying degrees, in reducing their competition. Sometimes the method of reducing competition is to require an initial capital and stability which, if required in earlier years, would not have allowed the presently well financed contractors to have started in the business. The competitive nature of the business, reputed to be one of the reasons for the high failure rate, has also prevented monopolization of the industry by large firms, as has occurred in other large industries. The giant contracting firms are large predominantly because the individual jobs are large, not because they have displaced small contractors.

It is claimed that competitive bidding (and unless it is unrestricted, as in public work, it is not really competitive) is not really economical for the owner or profitable for the contractor. There are many failures; the cost of duplicated effort in bid preparation occupies the efforts of many of the people who have great talent for management; work on the early parts of projects is delayed for months while plans are completed; plans and specifications must be more detailed than otherwise would be necessary, and the A/E has a greater obligation to examine submittals carefully, since the contractor will use the cheapest possible product. All this is true to some extent. Nevertheless, subcontractors are usually bid competitively even if a fixed fee or negotiated contract is used, so the conditions do no change materially in respect to the larger part of building work.

Some contractors thrive on the competitive bid market, usually on public work where contractor's qualifications are enforced by performance bond requirements, and standards of ethical conduct (of the owner, at least) are enforced by auditing agencies or the courts. Public bodies are obliged to give impartial treatment to bidders. Some contractors prefer to do no *cost-plus* work, as there is a tendency for the people in the organization to lose interest in cost reduction. Costs either go up or down; if not strictly watched, they gradually rise, and the contractor is no longer able to do competitive work at a profit.

The competitive market also serves as a qualifying race for contractors who become owner-builders. Costs are later kept under control by hiring managers who have recently survived competitive conditions. Competitive construction contracts are *hard money* contracts; advertisements for managers often require hard money experience.

4.4 COST-PLUS CONTRACTS

Cost-plus, or *cost-reimbursable,* contracts, poorly written or not fully understood between the parties, are a source of dissatisfaction and litigation. Some contractors refuse them; some contractors refuse any other type of contract. Most contractors get all the cost-plus work they can.

The advantages of cost-reimbursable contracts to the contractor are:

1. Hopefully, he eliminates risk in regard to his responsibility for the cost of materials and labor.
2. If so planned, he reduces or even eliminates capital requirements.
3. Although never admitted by the contractor, he makes profitable use of equipment and personnel not otherwise needed.
4. Requirements for cost accounting and managerial decisions are reduced; consequently, he can use less skilled managers.

The disadvantages of cost-reimbursable work to the contractor are:

1. Increased clerical work to properly account for and to collect charges. If the contractor's methods are not adapted to cost-plus work, this objection alone may make cost-plus work unsuited for him.
2. Delays in purchasing and subcontracts, in order to obtain client (owner) approval and comply with client requirements.
3. Exposure of otherwise private cost records to the owner.

4. Decreased efficiency throughout his organization, which is later applied to lump-sum contracts.

A lump-sum contract is much simpler to administer than is a cost-reimbursable one. The following items should be clearly defined in the contract:

1. The amount of the contractor's fee and how paid.
2. Definition of reimbursable costs, particularly insurance, interest, fringe benefits, management salaries, and travel.
3. The extent to which the contractor's funds will be used; that is, the amount and time of payment.
4. Who is to authorize purchases and purchase terms; that is, must they be on competitive bids?
5. How contractor's equipment and materials is to be priced, if sold or rented for job use.
6. To what extent, if any, the contractor is responsible for faulty workmanship, errors in the work, or other mistakes.
7. If the contractor is to furnish a performance bond, or be responsible for total cost at completion time, his duty to manage the work must not be hampered by preceding restrictions.
8. The method of termination of the contract shall be defined.
9. Responsibility for later claims arising out of the work should be defined. If men later recover for underpayments in wages, for example, does the owner remain liable?

   Unfortunately, some contractors assume that a cost-plus contract gives them the right to do as they please at the client's expense, and they find they have an uncollectible claim for their work.

Contractors often charge cost-plus contracts with the Associated Equipment Distributors *Green Book* costs, which reflect average retail equipment rental rates. Such rates are usually unjustified for this application, as they do not reflect actual contractor costs. *Green Book* rates are set by including costs of operating a retail establishment, advertising, and the other incidental costs of obtaining customers, which is not a cost to the contractor on his own equipment.

Also, the contractor is not a neutral party in that he will use his own equipment much more lavishly than he would use rented equipment. He has not only a different cost structure but a different cost for idle equipment; that is, he will not be renting it out to others if idle. The Associated General Contractors' equipment cost book is a much more valid index of contractor cost; it is based on actual monthly costs, based on new cost.

Rarely can a contractor assume the actual cost of late completion of a construction project. Consequently, when time of completion is important, the owner waives the security of lump-sum contracts and engages a contract under such terms that the owner pays costs without a guarantee of the maximum cost and the construction may proceed immediately after the plans for parts of the project, without waiting for completion of all the plans. Owner-built projects also include houses and apartments, where the extent of the work depends on sales, so the progress of plans and construction varies as the work proceeds.

4.5   LETTER OF INTENT

In many organizations, the manager who negotiates a contract is not legally authorized to execute it, or a contract may have to be approved by several persons in an organization. In such cases, a contractor may be authorized to proceed with the work under a letter of authorization before he receives a signed contract. The contractor himself may or may not have signed it. Such authorization may take the form of a *letter of intent,* which signifies the intention of the owner to enter into a contract for the work. Such forms are a method of protection against a clumsy owner's organization which would otherwise delay the work. The procedure is a poor one, but a manager is often part of such an organization, which he cannot circumvent in any other way.

The letter of intent is usually used by public bodies. It says that a contract will be entered into and that the contractor may proceed. However, because of different legal requirements, the letter of intent from a public body may not be considered to have legal significance, and the situation is risky for the contractor. *A private owner, on the other hand, should never use a letter of intent,* as he can be held to have completed the contract by the letter, in lieu of the formal contract. The owner cannot then cancel the letter, and if the contractor objects to a stipulation of the contract, the owner is in a poor position to require it. The contractor may require the payment of his profit plus damages if a contract is refused, under breach of contract, and such a claim would probably not be subject to arbitration, since only matters under the contract are negotiable; if the owner insists that no contract was made, he cannot logically apply a term of the contract.

A letter to proceed, on the other hand, may be written as a separate contract that will terminate when the final contract is executed. For example, such a letter could read as follows:

> *Gentlemen:*
>
> *You are requested to proceed with work on the Boment Dam in accordance with plans and specifications furnished you. In the event we do not agree on the final contract, you will be paid for work done as we may agree, or as an arbitrator may determine in accordance with the Construction Industry Rules of the American Arbitration Association.*

A letter of intent should contain:

1. An order to proceed with certain work.
2. A statement making clear that the parties have not yet agreed to any other contract.
3. A statement of how the agreement may be terminated, usually by execution of the final contract.
4. Some methods to determine the amount due if the contract by letter of intent is terminated, whether or not a formal contract is executed.

It is expected that such a letter will be in force only a few days and is to be used as an authority to order material. It should not be written and accepted unless the parties have agreed to all contract terms.

4.6 LEGAL CASES AS AUTHORITY

Construction law includes technical and custom considerations with which few attorneys are informed. Like other occupations, attorneys do much similar work most of the time and do not study cases outside their specialty. The writer as an arbitrator has seen a number of attorneys in action on construction cases, and in most instances, it was immediately evident that one of the two attorneys had no case, either because the facts were not to his liking or because the attorney was not informed. From the cases on the record, it appears that judges also cannot understand construction law. Even the apparently obvious cases are appeals from the A/E, a lower court, and sometimes an appeals board, all of whom have been proven wrong by legal standards on final decision. *A contractor who knows no law can neither represent himself, nor obtain adequate representation,* or properly provide records to document his case.

Also, it must be remembered that law is a political expression, made by legislatures and interpreted by political appointees. The shorter the period of service for a judgeship, the more rapidly decisions change. The National Labor Relations Board, with five-year terms, changes more rapidly than others; its decisions are identified with who was president, and appointed its members, at the time. Other courts change more slowly, but their decisions are still identifiable in many cases with the interest of their patron.

Legal cases are not indexed according to their application to construction. As far as this author is aware, textbooks relative to construction law are intended for general use of contractors, not for attorneys, and are therefore not detailed as a legal textbook would be. Also, they are written, but not read, by attorneys. If you choose an attorney available to you, the chances are very great that he will never have read a book on construction law; he starts with the assumption that he is dealing with a branch of the law he already knows, and therefore chooses general cases. For these reasons every effort should be made to use arbitration, and to stress the use of construction professionals, not attorneys, on appeal boards for construction contracts.

Judges determine law and juries determine facts; but the judge declares what is law and what is fact. Consequently, evidence must be carefully presented in order to get to the jury at all.

An attorney should be able to find ample citations to support either side of the disputes which occur in construction. Records of judgments are published only for decisions of appeals courts, supreme courts, and the U.S. Court of Claims (to which U.S. government cases are appealed). Countless judgments in small claims and circuit courts are not published; hence lower court cases not appealed are not in the available records.

The law does not necessarily have to follow precedent, except in the same state. Some courts consist of a number of judges, and the decision cited may be reversed a following year by a change of one judge in the make-up of the court. The cases cited as precedents in many instances were decided differently in the lower courts; but *usually* only the final appeal is quoted. You should realize also that the cases cited do not necessarily prove the law to which you may be subjected.

The writer does not agree that no one should be his own lawyer; businessmen, and particularly project managers, are making legal decisions every day. By the time an attorney is consulted, it is often too late to remedy a situation; and if an attorney is consulted for each decision, the decisions are made too late to be useful. The manager must often compromise the wording he would like to have with a wording which he believes is ample but which an attorney believes is doubtful. Psychology, not law, is the governing factor in the majority of contracts, since they are not disputed in court.

4.7 ORAL MODIFICATIONS TO WRITTEN CONTRACTS

Often contractors engage in extended oral negotiations before a contract is signed. During this time, understandings are reached, and offers and counteroffers are made. A written contract, which does not include all these details, is then prepared. Of what value are the earlier verbal agreements?

The general rule is that no evidence of conversations prior to the signing of the contract is admissible in court unless the following conditions apply:

> *(1) the verbal agreement is a collateral one;*
> *(2) it must not contradict express or implied provisions of the written contract; and*
> *(3) it must be one that the parties would not ordinarily be expected to embody in the writing.*[2]

Agreements made orally *after* the contract is made may constitute a new contract. An oral contract is usually as good as a written one but is more difficult to prove; but if oral modifications are inadmissible as evidence, they may not be presented in court at all. However, they may be accepted in arbitration. Of course, merely being admissible as evidence does not make them valid. These situations most often arise with subcontracts, as the agreements are less formal than are general building contracts.

For example, a written contract for laying a floor did not include the contractor's oral promise to polish the floor and to guarantee the permanency of the colors. This promise could not be introduced as evidence, since the court ruled that it should have been included in the written contract.[3] Likewise, when an owner verbally agreed to give a builder some old materials in addition to cash payment, the court held that the contractor could not offer the conversation as evidence, because the materials were not mentioned in the later written contract.[4] A contractor who agreed orally to remove stain saturations in stonework and who later in a written contract agreed to "clean down thoroughly all the front and side of building" was not obliged by the court to remove stains.[5] In this case, it appears that the court interpreted "clean" in the common dictionary definition, "to rid of dirt." In arbitration, it would be expected that the word would be given its meaning in construction usage for the trade, which in this application includes acid cleaning to remove all stains which were *caused by* the contractor.

4.8 AMBIGUOUS CONTRACT PROVISIONS

A clause may be judicially declared ambiguous—capable of two meanings. A jury may not decide. Not only may oral evidence be used to clarify the meanings; but also the manner in which the parties have worked under the contract and under previous similar contracts may be considered. The intention of the parties is drawn according to the following:

[2] *Mitchill* v. *Lath,* 247 N.Y. 377.
[3] *McKeige* v. *Carroll,* 120 App. Div. 521, 105 N.Y. Supp.342.
[4] *Abramson-Engresser Co.* v. *McCafferty,* 86 N.Y. Supp.185.
[5] *Krauth* v. Harris, 194 N.Y. Supp.525.

1. *Where contracts are partially written and partially printed, and there is a conflict between the two parts, the written portion prevails.*[6] This situation often occurs in printed subcontracts. For example, a printed contract may specify "furnish all labor and materials" and the (type-) written portion, "furnish labor only." If the written portion says "furnish labor," the subcontractor may still have to furnish the materials; it may be interpreted that repetition of the word "labor" does not refute "materials" elsewhere in the contract.

2. *In case of conflict or ambiguity in the agreement, it is to be construed most strongly against the party who drew it, and any promise is to be construed in the way that the promisee had good reason to understand it.*[7]

3. *Words intended to exempt a party from liability because of its own fault are to be construed strictly against it.*[8] Hold harmless clauses are difficult to enforce because of this principle.

4. *A contract will not be so construed as to put one party at the mercy of the other.*[9] Many contracts contain such phrases as "work shall be done in accordance with superintendent's schedule," which give the general contractor almost absolute power to terminate the subcontract for the subcontractor's failure to keep up to a progress schedule which the subcontractor has no part in making. This delay is usually because of neglect by the manager, as it is normally possible for him to set some dates—at least, in terms of the number of days after notification—which the subcontractor may be held to, and more important, to plan his own work.

5. "Specific provisions of a contract or statute prevail over general provisions."[10]

4.9   ANTICIPATED PROFITS

Anticipated profits may be claimed by a contractor if the owner fails to comply with his portion of the contract. When claiming anticipated profits, the contractor may not submit in evidence subcontracts made after the original contract. This is because the most accurate evidence of the profit expected would be the contractor's latest estimate; but it is assumed that such subcontracts represent favorable purchases after the original contract was signed and therefore were not contemplated by the contractor when he signed the contract. Logical or not, it opens to the contractor an opportunity to make a claim based on his anticipation at the time the contract was signed—which may be much more than the profit he anticipated by the time the contract was abandoned. The legal measure of damages is the difference between the contract price and what it would have cost to complete the structure, or, as the contractor's option, the reasonable cost of the work done. The admissibility of evidence, in the case of the contracts signed after the work was done, does not necessarily mean that the contractor's claim on those grounds will hold. Anticipated profits are speculative at best, and the determination of the amount is a jury matter; but the admissibility of evidence is a judicial matter. If the decision does not conform to the evidence admitted, the judge can set aside the jury verdict.

For example, while the Grinnel Company had a contract in progress, the owner went bankrupt. Grinnel sued the owner for anticipated profits and then contracted the completion

---

[6] *Wilson & English Construction Co. v. New York Central R.R.,* 240 App. Div. 479, 269 N.Y. Supp.874.
[7] *Ibid.*
[8] *Ibid.*
[9] *Ibid.*
[10] *Ibid.*

of the work with a later owner who purchased the property. The trial court awarded Grinnel six cents damages on the ground that Grinnel had made its original profit because it finally completed the work. This verdict was reversed on appeal; the appeal court, judging that the second contract had nothing to do with the first, admitted no evidence about Grinnel's actual profit on the second job. The appeal court's reasoning was that Grinnel suffered a loss of profits by the cancellation of the first contract and was therefore entitled to collect for the loss. Of course, it was collecting from a bankrupt estate, which may have been a very small payoff. According to the court, the second contract had no more connection with the first than did any other contract Grinnel made. The company had to go to the same effort, in a competitive bid or by convincing the owner, to get the second contract as it did the first. Grinnel could have used this effort to get another contract, rather than the one for this building.[11]

4.10 SPECIFIED SUBCONTRACTORS

The owner may take separate bids from subcontractors, then contract with a general contractor to assume these subcontracts. In one instance where a general contractor had failed, the owner agreed with a second contractor, Norair Engineering, to complete the work. Norair agreed to use the subcontractors already on the job, as a condition necessary for Norair to obtain the work. Although the owner specified the subs to be used, the contractor was responsible to the owner when the work of an owner-specified /contractor was defective; it was held that *since the general had agreed to allow the owner to specify subs, the general was responsible for the work of these subs.*[12]

4.11 METHODS VERSUS RESULTS

If a specification sets forth a detailed description of how work is to be done and also requires a guarantee by the contractor of the results obtained, it cannot be enforced. For example, a waterproofer was required by the specifications to guarantee that a cellar of a building be absolutely watertight. He did the waterproofing, using materials and methods required by the engineer. Later, dampness developed in the cellar. It was held that the guarantee was unreasonable and that the court must give a reasonable interpretation; namely, that the guarantee applied only to workmanship and that the contractor had agreed he would make the cellar watertight, so far as the plans and specifications permitted. As the work was done under inspection, it was assumed that the workmanship was good, and the owner had not claimed otherwise.[13]

If the contractor has chosen the materials and methods, the situation is entirely different even though the specifications embody the contractor's specifications. A painting contractor contracted to paint a house with a certain brand of paint, and he guaranteed his materials and workmanship for 15 years. He was held to his guarantee, since he had proposed the brand used. Nevertheless, many contracts specify both methods and results. For example, on water plants all work is carefully specified, but the contractor must guarantee the structure against leakage.

[11]*Grinnell Co.* v. *Voorhees et al.,* 1 F. 2d 693.

[12]*Norair Engineering* v. *St. Joseph's Hospital,* Court of Appeals of Georgia, October 2, 1978. From *Civil Engineering,* Vol. 49, No. 5, May 1979, p. 14.

[13]*MacKnight Flintic Stone Co.* v. *Mayor, etc. of the City of New York,* 160 N.Y. 72.

4.12 WITHDRAWAL OF ERRONEOUS BIDS

It is generally held that if a bid is so low that the person receiving bids "knows or had reason to know because of the amount of the bid, or otherwise, that the bidder has made a mistake, the contract is voidable by the bidder."[14] On one occasion, when a government official insisted on accepting a bid which was low even though the discrepancy was obvious, the court commented, "It is interesting to consider what would have been the attitude of the Government if the shoe had been on the other foot, and the Government had made a typographical error in the contract or in payment under it."[15]

It is generally conceded that in a case where a single mistake in the bid has resulted in an obviously low bid, the bid may be withdrawn if detected at the time bids are opened. The contractor can usually recover his bid bond, but municipalities in particular often attempt to collect on the bond. The fact that a contract has been entered into does not, in itself, control; when the bids are received, an offer and an acceptance, which constitute a valid contract, may exist.

If the bid is not obviously low, a contractor's right to withdraw is doubtful. Also, error in such a case may not be immediately detected, and later the owner may suffer real damages if the contract is voided.

A contractor may be low because of errors of judgment, such as accepting a subcontract bid from a firm unable to do the work, or having taken an overly optimistic view of labor costs. As such errors do not establish a "mistake" by law, efforts are sometimes made to change estimates after the bid has been submitted. A contractor may make an entirely new set of pricing and summary sheets, with the purpose of accumulating the error in one place, and then claim the new set to be originals. One contractor would have a telegram delivered to him after he entered the bid room, just before each bid opening; it said, "Have discovered error. Raise bid $100,000," and was signed by his estimator. If he was too low, he would then take the message from his pocket, open it, and show it to the owner, pointing out his own mistake in failing to read the telegram. Of course, he would have to do this before he actually knew of the low bid was in error. Many contractors are willing to withdraw their bid immediately if it appears too low, without locating an error in their estimate.

In one instance, when a contractor refused to contract a job after it was bid, the U.S. Comptroller General withheld from the contractor a sufficient sum of money, due him on another contract, to cover the difference between the bid withdrawn and the next highest bid. The U.S. Court of Claims decided that the contractor had a right to withdraw the bid without penalty.[16] The mistake was a misunderstanding of a sub's bid—one of the more common errors.

On a U.S. Bureau of Reclamation dam, the low bidders claimed a $1 million mistake on a $5,700,000 bid for the keyway and foundation contract. The government estimate was $10,000,000. The Bureau released the low *three* of nine bidders, and readvertised the job.

Most owners, especially U.S. governmental agencies, realize that a contract with an unwilling contractor is not to be desired, and make no attempt to enforce an erroneous proposal. One government agency, which has many small contracts to let, inserted an additional step in the contract procedure for this reason. After proposals with bid bond were received and award decided, the low bidder was asked to confirm his proposal before a contract was mailed to him. The default rate was quite high; sometimes the low two or three bids would be in error.

[14]*Saligman et al. v. United States,* 56 F. Supp.505 (1944).
[15]*Ibid.*
[16]*Alta Electric & Mechanical Co., Inc. v. United States,* 90 Ct. Cl.466.

The examples of erroneous bids cited all have in common that the error was large enough that the low bid appeared too low, and the error was discovered in a day or two after bids were received. When either of these conditions is not met, the contractor may be held to his bid. On numerous occasions, the low bidder has attempted to raise his price after bids were opened; this is universally denied on public work. On private work, it is of course an individual matter. Attempting to raise a bid after becoming low bidder is referred to as *taking a second bite of the apple*.

Although a court may always make a different opinion, the requirement that a bid bond be forfeited, or that a contractor be responsible for an erroneous bid, does not appear to be enforceable—provided that the error is discovered in time that there was no harm done to the owner.

In a 1949 case, the conditions for withdrawal were described as follows:

1. The mistake must be so serious that to enforce the contract would be unconscionable (without conscience, unreasonably excessive).
2. The mistake must relate to a vital feature of the contract.
3. Reasonable care must have been exercised by the party making the mistake.
4. Rescission (withdrawal) must not prejudice the other party (that is, the owner), except by loss of the bargain.

Some cases are as follows:

1. A sheet of computations have been overlooked, resulting in a bid of $143,171, $50,000 lower than intended. The error was discovered before acceptance, and the contractor relieved of his obligation.[17]
2. In 1970, a contractor made a $10,000 error on a $34,000 job in transferring amounts from work sheets to the total sheet. The mistake was discovered the day after bids were opened, but the owner had already awarded the contract. The judge decided that the owner had suffered no serious prejudice, and released the contractor.[18]
3. A federal court in 1969 made a similar decision for a contractor who recorded a sub's bid as $22,000 rather than $330,000, which made the contractor's bid over 10 percent low. The owner attempted to accept the bid, knowing that it was a mistake.[19]
4. A 1955 case was decided the same way on different grounds. The contractor had omitted $34,000 in subcontract bids, and found the error before award. The court chose to invalidate the clause, barring withdrawal of the proposal.[20]

Most specifications allow a cashier's check (the modern equivalent of a certified check) to be submitted with the bid, rather than a bid bond. The contractor is occasionally tempted to submit a check rather than the bond, when waiting on a bid bond means that the proposal could not be made. This is a very dangerous practice, both because of the lack of recourse if there is an

[17]*Conduit & Foundation Corp.* v. *Atlantic City,* 64A32d 382, 1949.
[18]*Cataldo Construction Co.* v. *County of Essex,* 265 A2d 842, 1970.
[19]*M.J. McGough Co.* v. *Jane Lamb Memorial Hospital,* 302 F. Supp.482, 1969.
[20]*Peerless Casualty Co.* v. *Housing Authority of the City of Hagulhurst, Georgia,* 228 F2d 376, 1955. This and three preceding cases as quoted by R. E. Vasant, *AE Concepts in Wood Design,* March–April 1975.

error, and because the surety may refuse or delay the performance bond. The contractor still has the right to a refund in case of an erroneous bid, but is in the position of having to sue for it. This may take years and expenses greater than the money at stake.

A certified check is a contractor's check for which the bank has set aside funds from the contractor's account. A cashier's check is a promise by the bank itself to pay an amount, and is paid out of the contractor's account, just as if he had purchased a money order. The charge is nominal in either case. A cashier's check is *certified* by the bank, and therefore acceptable when a certified check is required. A *letter of credit* is often used in international bids; the bank promises to pay as with a cashier's check, but it may depend on some stated condition. A letter of credit cannot be deposited as can a cashier's check.

4.13 INSURANCE

The AIA standard contract provides the owner

> *shall purchase and maintain property insurance upon the work at the site to the full insurable value thereof. This insurance shall include the interests of the Owner, the Contractor, Sub-contractors and Sub-subcontractors in the Work and shall insure against the perils of Fire and extended coverage and shall include "all risk" insurance for physical loss or damage including, without duplication of coverage, theft, vandalism and malicious mischief.*

Damage to work caused by accidents, but not identified as due to a particular person, come under the heading of vandalism; the damage may be accidental rather than malicious. Furthermore, the owner is required to furnish a certificate of insurance to the contractor, and if the owner does not do so, the contractor may obtain insurance and collect the cost from the owner. From the wording of the AIA contract, it would appear that if neither the contractor nor the owner obtained such insurance, the contractor would have a difficult time collecting from the owner for damages caused to the building. The ASCE/ACC contract has a similar insurance requirement.

It is not economical for large organizations with many separate risk exposures to obtain insurance on buildings—or on anything else. The normal losses on uninsured risks are less than the premiums (or there would be no insurance companies), and large losses by such a firm are extremely improbable. Such firms are therefore often self-insured, and the insurer may be a separate, affiliated corporation or merely an expense account. Most states have definite requirements as to the manner in which a corporation may be self-insured to protect the workers for accident compensation benefits. Few construction companies are large enough to be self-insurers, as their risks are high in proportion to the size of construction operations.

Unless a contract states otherwise, a contractor is liable for delivering a completed building, which means that the contractor must replace damage due to accidental destruction or loss due to fire or storm. The contractor should allow for an insurance premium to cover these risks if the insurance responsibility is not mentioned in the contract, although he may be able to share this cost with subcontractors.

An ideal insurance policy, from an insuror's point of view, is one that avoids all probable liabilities, and leaves a way for the insuror to avoid improbable liabilities in case they should occur. The contractor is not on equal basis in attempting to buy adequate insurance, in view of the insuror's legal talent and knowledge of statistics. The courts and legislatures attempt to counter this situation by interpretation of insuror-written policies against the insuror, and statutory definitions of policies.

In one instance one of the largest insurors in the country wrote a professional liability policy for an A/E on a "claims made" basis; that is, the insuror was responsible for claims made only during the period of the policy, regardless of when the event occurred. In April 1974, an explosion occurred for which the *insuror was notified in November 1974* of potential claims. The following month the insuror canceled the policy because it terminated operations in the state, and refused to defend the claim on the ground that *no claim had been made in November 1974*. The insuror was upheld on this point. However, the policy itself was declared unconscionable—that is, grossly unfair.[21]

The National Council of Compensation Insurance has proposed *wrapping up* workers' compensation and general liability insurance on projects with premiums of $500,000 or more. *Wrap-up* insurance is the type obtained by the owner, prior to the start of construction, covering all contractors on the job. The AGC (Associated General Contractors) opposes such insurance as an infringement on the general contractor's right to select his own insuror.

4.14 EXTRA AND ADDITIONAL WORK

The differentiation between *extra* and *additional* work is important in public work, as the engineer acting for the public agency usually has no authority to order work done beyond the scope of the contract. This places the contractor in a difficult position if he believes work to be outside the contract; if the contractor proves to be right after the work is already done, then the engineer had no authority and the contractor cannot collect damages! While still trying to protect the owner, in this case the public agency, against the action of its own agents, the courts have taken a lenient view and have allowed the cost of such work as damages for breach of contract on the part of the municipality.

The distinction between extra work (outside the contract) and additional work (a modification of the work included in the contract) is illustrated by a case where the engineer required (1) that portland cement be used in an area where specifications were ambiguous, and (2) that an elevator be installed to permit public officials to drive through a sewer, with the sewer lighted for their inspection. The court held that the contractor could collect for the cement, since there was reason to believe the material could have been required by the contract, but he could not collect for the car elevator and lighting because the contractor should have known that such work was not required by the contract.[22]

The authority of the A/E is usually much less restricted in private contracts, as compared with public contracts, so the question of his authority and work outside the contract is less likely to arise. It is important that the contractor carefully examine the A/E's authority as expressed in the specifications.

In ordinary business language, *additional* work is seldom referred to; all work for which added payment is sought is called *extra* work, regardless of the legal basis.

Claims for extra or additional work are the usual reason for application of the disputes clause, paragraph 2.2.12 of the AIA General Conditions (Appendix A). This provides that any appeal from an architect's decision must be made within 30 days, if the decision so states. Other contracts usually have a similar clause, without the necessity of an A/E specifying the time in the appeal. The contractor must use great care that his appeal is both timely and *complete*. Sometimes the architect's decision is not clearly stated, forcing the contractor to appeal. If the architect then verifies or clarifies the decision, another appeal may be required.

[21] *Heen & Flint Associates* v. *Travelers Indemnity Co.* Supreme Court of New York, December 8, 1977. From *Civil Engineering*, July 1978, Vol. 48, No. 7, p. 17.

[22] *Borough Construction Co.* v. *City of New York*, 200 N.Y. 149.

For example, an architect ordered a concrete floor on grade removed, and steel cables temporarily installed to replace the floor, to carry arch action. The contractor appealed the decision that required "removal and replacement of the floor," but did not mention the steel cables. The appeal judge ruled that the appeal applied only to the floor removal, not to the cable installation. Presumably, the contractor could have recovered for the slab, but not for the cables.[23] If the cables had not been mentioned, the contractor would have designed them anyway, and could have collected with the slab.

4.15 SUBCONTRACTS

Generally, the A/E may not inquire into a general contractor's liability to his subcontractor; the general is entitled to be paid for extra work, regardless of whether an extra payment is required from the general to the sub. When *damages* are involved, however, the general may not collect for the sub if neither the general nor the sub suffered damages; but in the court case in which this verdict was stated, the sub was doing the entire contract and the prime contractor was merely collecting an 8 percent commission.[24]

The general contractor should understand and supervise the work he subcontracts, and his profit from it should depend on his efficiency. For example, a subcontractor will often bid to the general with qualifications, if unable to obtain a needed clarification from the architect or a necessary price from the supplier. The general contractor then takes a risk on what the final cost will be and whether the general will assume a risk or qualify his bid also. A contractor often has special knowledge of the job—such as extras or time delays he knows will occur—and may either modify the subcontractor's quotation or delay subcontracts, or both, until a matter is settled. The general in this way acquires a claim the sub does not have, and has taken a corresponding risk; the general can reasonably expect to make a profit from this. A/E's often assume that the subcontract terms are the same as those in the general contract, when this is not the case. Unfortunately, many generals act as brokers for their subs, rather than supervising them.

4.16 OWNER'S LIABILITY

The owner of property has no liability, in general, for damages to persons or property caused by construction operations on his premises; the contractor and the architect, as independent agents, assume this responsibility. Likewise, the general contractor is not responsible for the negligence of his subcontractors. Each firm or person is responsible for its own negligence. None acts as agents of the others. The owner is responsible for negligence only to the extent that he or his agents personally directs the work, which is not usually the case.[25]

There are certain exceptions when a person employing a contractor is liable:

where the employer personally interferes with the work, and the acts performed by him occasion the injury; where the thing contracted to be done is unlawful; where the acts performed

[23]Author's appeal to State of Florida.
[24]*Degnon Contracting Co.* v. *City of New York,* 235 N.Y. 481.
[25]*Burke* v. *Ireland,* 166 N.Y. 305.

create a public nuisance; and where an employer is bound by a statute to do a thing efficiently and an injury results from its inefficiency. . . . Supervision, however vigilant, if confined to enforcing the terms of the contract or subcontract, does not constitute participation such as would render the owner or contractor liable for the negligence of the contractor or subcontractor.[26]

It has also been held that in order for the owner to be liable, it must appear (1) that he intended the work to be constructed in the manner that proved to be improper, (2) that he knew or ought to have known that such manner of construction was improper, and (3) that such improper construction caused the injury complained of.[27]

In another case, it was stated that:

a builder or contractor is justified in relying upon the plans and specifications which he has contracted to follow unless they are so apparently defective that an ordinary builder of ordinary prudence would be put upon notice that the work was dangerous and likely to cause injury.[28]

Previous rulings do not prevent suits from being brought and settlements from being made to avoid court costs, nor do they prevent interpretations by the judge or jury which are obviously not in agreement with the opinions of a contractor, architect, or other person acquainted with trade practice.

Large corporations often have large technical staffs, even though they hire architects and engineers for new construction projects. These large corporations may furnish the A/E's detailed engineering criteria and even standard designs, so the owner may reasonably become more liable than an owner normally would be. Also, a damaged party will *sue the deep pocket;* that is, he sues the party with the most money.

In a number of cases the owner has been found at fault under a general doctrine that the owner is responsible when the operation is "inherently dangerous," such as demolition adjoining city streets or open excavation required in traveled streets or sidewalks. He may also be responsible for hiring an incompetent contractor (which may be merely a contractor who has not done that type of work before), or responsible under a duty to keep his premises in such state that "invitees" are not unduly exposed to danger.[29] Decisions following such trends in recent years have led to suits against nearly everyone connected with the site of an accident. The person usually at fault—the general contractor—is least likely to pay for accidents to his workers, as he has the protection of the workmen's compensation laws, which restrict his liability to that payable under these laws.

4.17 HOLD-HARMLESS CLAUSES

A *hold-harmless clause* is one under which the contractor agrees to assume responsibility (that is, to hold others harmless from litigation) for all claims for damages brought by others, as a result of the work. The contractor assumes a legal obligation, when so stated, that he will insure the owner and A/E against their own negligence. A/E's justify such clauses on the grounds that

[26]*People* v. *Gaydica,* 122 Misc. 31, 203 N.Y. Supp. 243.
[27]*Herman* v. *City of Buffalo,* 214 N.Y. 316.
[28]*Ryan* v. *Feeney & Scheehan Building Co.,* 239 N.Y. 43 (1924).
[29]*Janice* v. *State,* 107 N.Y.S. 2d 676 (Court of Claims of New York, October 18, 1951).

everyone is suing them, and they shouldn't be sued; so let the contractor be sued. The extent of the contractor's liability varies with different clauses. Hold-harmless clauses are unenforceable in some states on the grounds that it is against public policy that a contractor, rather than an insurance company, insure another against liability due to the other's own negligence.

The use of hold-harmless clauses have been disputed between contractors and architects for many years. A contractor normally carries liability insurance for his own operations; that is, if he becomes liable, usually because of negligence, for damages to other persons, the insurer will defend the suit and pay such damages, within the limits of the policy. This insurance, however, does not cover the liability the contractor acquires *by contract* (as opposed to liability inherent in his operations). If a contractor agrees to guarantee a third party protection from claims due to the negligence of the same third party, the insurer will not pay the contractor under a general liability policy. The insurer is insuring the liability of construction operations, not underwriting a contractor who enters the insurance business by assuming to underwrite the liabilities of others. *Since these clauses are often ambiguous, it may be necessary for the contractor, before bidding the job, to submit the clause to his insurer, to determine if he has or can obtain coverage.* Some such liabilities covered by hold-harmless clauses are as follows:

1. The contractor may be required to provide insurance that will protect the owner from any damages. This is close to the second situation listed here but may be open to the defense by the contractor that no such insurance is available.

2. The contractor may be required to protect (hold harmless) the owner from all claims. Since this is a contract to protect the owner from the consequences of his own negligence, even though not so stated, the courts may refuse to honor it. "Contracts will not be construed to indemnify a person against his own negligence unless such intention is expressed in unequivocal terms." [30]

3. The contractor may be required to protect the owner against claims arising out of the contractor's or subcontractors' negligence.

4. The contractor may be required to defend suits (or pay the defense costs) of claims arising out of the work, but he may not be responsible for damages paid by the owner. This liability is substantial, as the costs of defending a suit are often greater than the claim itself.

5. The contractor may be required to pay to third parties the cost of property damage and personal injury arising from the work.

The hold-harmless clause contains no provision that it applies only to claims for which the contractor is otherwise liable; the contractor is insuring passers-by, abutting landowners, trespassers, and low-flying aircraft against injury, regardless of whether he would otherwise be liable or not. One such clause reads, "injury to persons or damage to . . . the property of others caused by the act of the contractor." [31] This type of clause permits an action under the contract by third persons—persons injured even though they have no connection with the work and could not otherwise collect for damages. It may also make the contractor liable for damages resulting from the design of the structure.

The AIA standard contract requires the contractor to hold harmless the owner and architect only for damages resulting from actions of the contractor and for actions for which the owner and

---

[30] *Coley* v. *Cohen*, 289 N.Y. 365, 45 N.E. 2d 913.
[31] *Ibid.*

architect may have partial responsibility with the contractor. The ASCE/AGC engineering contract requires the contractor to maintain insurance "as will protect . . . unless otherwise specified, the Owner, for claims . . . arising from operations under this Contract whether such operations be by himself or by any Subcontractor. . .," which appears ambiguous and is the type of clause which would not be held to protect the owner. In the case of the AIA contract, it is more likely that the architect would be protected from his own negligence than the owner would be, since the action of the architect is carried out by the contractor, who must clear himself of contributory negligence before the architect can be held responsible to the exclusion of the contractor. The clauses, of course, do not protect the architect or owner from claims but are designed to require the contractor to reimburse them. The AIA clause relieves the contractor of the architect's liability arising out of his plans and his orders on the work, with some qualifications. A claim on the architect could, therefore, be the subject of a court action by a third party, and also of an arbitration to determine if the claim against the architect is payable by the contractor under the hold-harmless clause. The AIA clause also provides that the contractor is liable *to the fullest extent permitted by law,* since the clause as stated is not allowed or enforceable in many states.

A/E's sometimes require contractor liability for their own protection, when they are partially at fault. It is the practice, when an accident or failure occurs, for the person damaged to sue everyone connected with the work—architect, owner, contractor, subcontractor, and manufacturer of materials. The A/E's attempt to put hold-harmless clauses in the contract to cover this type of liability. They also try to avoid responsibility for inspectors who fail to require proper work, or to use adequate safety practices, and are willing to state the contractor has no design responsibility. It is important to the contractor that he avoid design responsibility unless he intends to fully check the design. In 1978, 30 percent of *all* A/E firms were sued for faulty design and construction, and insurance now costs 2 to 10 percent of a design professional's gross income.[32]

Generally, a contract for one person to assume the liability of another is unenforceable, as against public policy. For example, in an Illinois case, an injured worker sued an architect for negligence. It was held that if an architect were relieved from the liability for his own actions, the worker would be deprived of his constitutional rights. In this state, as in several other states, including California, Michigan, and New York, hold-harmless clauses in construction contracts are void by statute. In the case described above, the statute was upheld.[33]

This writer has paid an insurance rate of $0.87 per thousand (0.1 percent) of total contract cost to cover the 1970 AIA General Conditions liability, which is virtually the same as the 1976 clause. At an architect's fee schedule of 6 percent, this means an effective increase in cost of the architect of 1.5 percent of his fee, which eventually must be paid by the owner.

Under a hold-harmless clause, it is possible for a contractor to pay for the same accident twice, instead of not at all. For example, an employee is injured due to an A/E action or design. He is entitled to workmen's compensation from the contractor's insuror, and also a common-law judgment from the A/E or owner, which the contractor is also required to pay. Without the hold-harmless clause, the A/E would be responsible to the contractor's insuror as well as for his direct damages.

The hold-harmless clause of the standard AIA (American Institute of Architects) General Conditions has been of continued concern to the AGC (Associated General Contractor), who prevailed upon AIA to withdraw their 1966 edition. The 1978 edition (see Appendix A) re-

[32] *Wall Street Journal,* December 22, 1978.
[33] "Misplaced Crane Strikes a Blow for Individual Rights," Robert A. Brown, *Heating/Piping/Air Conditioning,* July 1976.

quires that the contractor assume responsibilities of the owner to the extent that the claim is due at least in part to the contractor's operations and is because of injury, or damage to property. Such claims are usually made against both contractor and architect. The AGC tries to persuade A/E's who use more rigid requirements to modify them.

4.18 BREACH OF CONTRACT

Generally speaking, a contractor may claim payment, not as stated in the contract, for any of three reasons:

1. For *breach of contract,* in that the owner has not complied with the contract;
2. For *additional* work, caused by the *alteration* of work required to be done in the contract; and
3. For *extra* work, which is work related to that shown by the drawings but which is an addition to, rather than a modification of, the work contracted.

A *breach of contract* occurs when a party refuses to act as required by the contract. An owner may breach the contract by refusing to make a payment, refusal to comply with his own architect's decisions, or refusal to furnish plans; assuming, of course, that he has a clear contract obligation to perform these acts. A contractor may breach his contract by refusal to continue with the work, refusal to abide by the architect's decisions, or in numerous other ways. Generally, contract requirements state that the contractor must continue to do disputed work even if the architect has not agreed to pay for it; a refusal to proceed would be a breach of contract by the contractor. If the owner breaches the contract, the contractor has the option to forgo any further work on the contract and claim his payment for work done. He cannot claim a profit on the entire job, however, if he elects to claim payment for work done.[34]

An oral agreement in which a general contractor agreed to use the subcontractor's bid and award the subcontract to him if the general contractor was successful is a valid contract, and a subcontractor has collected anticipated profits on such a contract. The extent of damages under a breach of contract is most difficult to prove, since appeal judges differ in what evidence is admissible to the jury and how the jury is to be instructed.

When an owner has breached a contract, the contractor may proceed with the contract without giving up his right later to claim damages arising from the breach of contract.[35] That is, when the owner breaches the contract, the contractor has an option to do likewise, or to continue work under the contract; no matter which of these options the contractor takes he retains the option to demand damages suffered by the breach.

A contract, in general, may be canceled (*breached,* when no cancellation clause is inserted in the contract) by either party at will. This is often the best way out of a dilemma, especially for material orders. The offending party may lose much less by paying for profits or damages than by continuing the contract, and sometimes one party is unable to continue because of lack of money. It is important that disputes be settled quickly, either by direct negotiation or by ar-

---

[34]United States for use of *Susi Contracting Co., Inc., et al.* v. *Zara Contracting Co., Inc., et al.,* 246 F.2d 606.
[35]*Blair* v. *United States,* 147 F.2d 840.

bitration, rather than be taken up in the courts, which are ill-equipped to handle such technical matters. At times, court do require performance of a contract, but these are almost always personal service contracts not applicable to construction.

A contractor who accomplishes *substantial performance* of his contract does not breach the contract by subsequent abandonment of the job but is responsible for the cost to complete minor details. There is no clear definition of "substantial performance," but it has been held that substantial performance had not been given when faults were scattered throughout the structure, even though correction of these faults would cost less than 1 percent of the value of the building, in a case in which the contractor had willfully departed from the specifications. One definition of defects which a structure could have, even after *substantial completion,* was this:

> *They are such as that the work needed to make them what they should be is no other in degree of difficulty, disturbance or inconvenience, than such as in the ordinary repair to buildings from year to year made needful by wear and tear and decay.*[36]

4.19 CHANGED OR CONCEALED CONDITIONS

The contract clause allowing extra payment is referred to in engineer contracts and those of the U.S. government as *changed conditions.* The AIA General Conditions uses the term *concealed conditions.*

Concealed conditions refer to the existence of physical conditions at the site, normally below ground, which in AIA contract paragraph 12.2.1 are as follows:

> . . . *concealed conditions encountered . . . at variance with the conditions indicated by the Contract Documents . . . unknown physical conditions . . . of an unusual nature, differing materially from those ordinarily encountered and generally recognized as inherent in work of the character provided for in the Contract . . .*

Such concealed conditions give rise to extra claims, changes in plans, or sometimes abandonment of the work. All standard contracts contain a provision that the contractor must promptly notify the A/E (the AIA contract allows 20 days for such claim) of changed conditions, and the engineering contracts provide that the contractor must do so before proceeding with any construction which would make it difficult or impossible to determine accurately the conditions found.

Because of the broad description of such clauses and the difficulty of proving the conditions, the contractor must move promptly not only to notify the A/E, but also to document his claim properly with photographs and consultants such as geologists, hydrologists, and soil mechanics engineers, in accordance with the type of problem encountered. Claims are complicated by the fact that many engineers furnish few borings, often include inadequate information, and state in the contract documents that such borings are for information only and are not guaranteed. It is not unusual for the engineers to fail to include data which would be discouraging to the contractor, or to refuse to allow the contractor to speak to the engineer who made the borings. The latter prohibition is based on the duty of the engineer to give all contractors equal

---

[36] *Woodward* v. *Fuller,* 80 N.Y. 312.

information; information is usually given at a tour of the site to all interested bidders at the same time. The boring engineer may participate in these meetings. Unless a contract states otherwise, a contractor assumes the risk for underground conditions;[37] many A/E's specify that the contractor assume such a risk. The decisions are not uniform, but it has been held that the contractor is obliged to assume "the care and prudence with which the businessman of average caution would exercise in serious undertakings."[38] Some would say that a business man of average caution would not be engaged in the construction business.

In the New York case, it was held that a contractor could not rely on information shown on drawings, where it was stated that the information was not guaranteed and that the records of rock depth were not part of the contract.[39] In another New York case, it was held that a bidder may be obliged to rely on available information, even though the contract states otherwise, when it would be impractical for him to determine the true conditions, and where the state had made an honest mistake in locating rock levels. The court held that a mutual mistake had been made and that it might have been grounds to void the contract for lack of mutual understanding. However, the contractor did not so plead at the time; and, once the work was completed, the contractor had accepted the levels as he had found them and could not recover the added cost.[40] A contractor who encountered more rock than borings anticipated, on an unclassified unit price contract (payment by the cubic yard, regardless of whether rock or earth was encountered), was disallowed a recovery, although the state legislature passed a special act for the benefit of the contractor.[41]

The contractor is not required to give up gains he has made because underground conditions were less difficult than anticipated. At the Baltimore Ashburton Water Plant, for example, borings for the main structure showed the rock was encountered at an irregular depth, but generally not far below the surface of the ground. The engineers did not attempt to determine if the rock was a hard structure or isolated boulders, which could have been determined only by machine excavation. As it turned out, the borings had been stopped by boulders which were readily excavated by machine shovel at a cost several hundred thousand dollars less than had been anticipated.

Some examples of conditions in which the contractor was granted extra payment under federal government contracts are these:

1. The ground surface was shown 3 feet too high on plans, and the building site was actually under water.

2. A hard layer of material, which could be ripped and which the government knew to exist, was not shown on drawings.

3. A soft rock bottom rebounded materially after excavation; that is, it flowed upward and had to be reexcavated.

4. Borings indicated solid ledge rock, which turned out to be boulders. The contract was canceled and the project redesigned.

5. A buried taxiway not shown to exist on available drawings was encountered on an airfield job.

---

[37]*Niewenhous Co.* v. *State*, 248 App. Div. 658, 288 N.Y. Supp.22.

[38]*Ibid.*

[39]*T.J.W. Corp* v. *Board of Higher Education of the City of New York et al.*, 251 App. Div. 405, 296, N.Y. Supp. 693, aff'd 276 N.Y. 644.

[40]*Foundation Co.* v. *State of New York*, 233 N.Y. 177.

[41]*Weston* v. *State*, 262 N.Y. 46, 186 N.E. 197.

6. Pile driving in a Navy yard struck an old submarine which had been junked and forgotten, and which had to be excavated and removed in pieces.

The contractor was unable to recover added costs in the following situations:

1. Boulders were encountered in glacial drift; recovery was denied because several borings showed rock, and the type of material was one in which boulders were to be expected.
2. Rock was shown as "shattered," but ledge rock was encountered. The government held that foundation conditions did not cause an increase in cost.

Misrepresentation of data by the owner is a breach of contract. In federal work, this breach of contract technically places the matter outside the authority of the normal agency procedures, but it has been recognized as changed conditions.

The federal government may not pay a contractor for delay due to changed conditions; to get relief, the contractor must put such costs directly in his proposal for added costs. Other owners, not restricted by statute, may be required to pay for delay. When a contractor knows that conditions on a job will be more difficult than are shown on the plans and specifications, he may be actually at a disadvantage in the bidding. The changed conditions clause essentially pays the contractor's additional cost due to circumstances *unknown to the contractor*. The contractor may notify the engineer of such conditions, in order that the information will be transmitted to other bidders; but if the engineer chooses to disregard it, the contractor is at a disadvantage. *If the contractor bids his knowledge, he will be higher than other bidders; but if he ignores his knowledge and bids low, he may not be able to collect for the extra costs he, unlike the other bidders, knew about.* It has been suggested that, in this case, the contractor seal a statement, postmarked, to the effect that, in the bid, the information he has not used because he assumed that the engineer had better information. The inclination of contractors to avoid a market after learning of its difficulties is well known. A road-grading contractor, it is said, does not bid a *second* golf course (after having done one).

4.20 LIQUIDATED DAMAGES AND DELAYS

The contract clauses under which a contractor bids provide for completion times in one of three ways:

1. If a contract specifies no completion date, the contractor must make a reasonable effort to complete the project, but *any* claim against him for damage due to late completion is very difficult to collect.
2. A contract may have a specified delivery date, with a *Time is of the essence* clause. In this case, the contractor is obliged to complete on the specified date, subject to any extensions allowed for in the contract, and failure to do so is a breach of contract. If he does not finish by the contract date, his contract may be canceled, or the owner may collect

damages incurred because of delay. The contractor must have been aware of the extent of such possible damages when he bid the work.

3. A contract may have a specified delivery date and a provision that the contractor will pay specified damages on failure to complete. This provision will hold, subject to extensions of time as provided for in the contract and provided that the specified (liquidated) damages are reasonable.

Liquidated damages provide for the contractor to pay a stated amount for a project completed later than the calendar date specified; this is in lieu of paying other damages to the owner caused by late completion of the project. The amount is usually less than the actual damages, since an overstated damages clause may be interpreted by a court as penalty rather than damages and be uncollectible. The ASCE/AGC and ASCE/APWA (APWA is the American Public Works Association) contracts include a liquidated damages clause as well as clause which rewards the contractor for early completion. There is no necessity for a reward clause; the purpose of the liquidated damages clause is to reimburse the owner for damages suffered. The owner is seldom in a position to profit by early completion. The AIA contract has no liquidated damages clause but includes a *Time is of the essence* clause with specified dates.

The difference between liquidated damages and daily reward clauses for early completion is largely psychological; it forces the contractor to set his own most probable completion date rather than adding a few days to the A/E's estimate, unless ample time is provided. The contractor, in assuming the obligations of liquidated damages, usually allows a contingency for some delay in his estimate.

Unless the contract provides otherwise, delay may not be proportioned between the contractor and the owner. That is, if the owner delays the work, the contractor is relieved from the liquidated damages clause altogether.[42] The liquidated damages clause may also be considered waived if additional change orders do not contain appropriate extensions.[43]

In an arbitrated case, a contractor building a hospital accepted a large number of change orders, receiving appropriate additions in contract price and extensions of time. At the completion of the job, the contractor asked for additional reimbursement on the grounds that, although each change order separately could have been completed in the time specified for the change order, the effect of the whole—which delayed the project for over a year—caused additional overhead costs which could not have been foreseen at the time the change orders were accepted. The contractor won an award. In this case there were three arbitrators: a contractor, a salaried employee of a contractor, and a government engineer. The engineer was the most generous to the contractor, and the salaried employee in the middle. The contractor, when choosing an arbitrator, may do better to find an engineer or an architect rather than another contractor.

The liquidated damages clause usually carries a waiver of claims by the contractor for owner-caused delays. In such cases, only where delays are so serious as to virtually annul the contract—that is, when delays are considered beyond the scope of the contract—does the contractor have the option of collecting extra for them or ceasing work claiming a breach of contract. A delay of 175 days, for example, was held not to be covered by the delay clause when the owner was making attempts to redesign large parts of the work.

There are many faults with the method of specifying completion time in construction contracts, with liquidated damages for late completion. The collection of liquidated damages

---

[42]*Village of Argyle* v. *Plunkett*, 175 App. Div. 751 N.Y.
[43]*United States* v. *United Engineering & Contracting Co.*, 234 U.S. 236, 58L. ed. 1294.

constantly introduces a specification difficult to enforce in court, since there are nearly always causes of delay beyond the control of the contractor; in addition, it has been found on one occasion that the owner was not entitled to collect liquidated damages if he did not actually suffer damages. The clause *Time is of the essence of this contract* has been held to make the contractor liable, provided the contractor was aware of the damages which the owner would suffer by late completion, and provided the owner did actually suffer such damages. These damages are nearly always much greater than those in liquidated damages clauses. Few contractors would accept a contract with the damages to the owner spelled out in numbers, but they commonly accept this liability when it is stated as a "Time is of the essence" clause. Many contractors are well acquainted with the damages and delays clause of the United States government and are inclined to assume that such a clause is the law in other cases. This is not true, and lacking other exemptions, the contractor can become an insurer of the completion time and of the owner's damages resulting from delay; that is, there need be no relationship between the contractor's liability and whether delay was beyond the control of the contractor. An insurer has an absolute liability; a life insurance policy would be worthless, for example, if the insurer was exempt from paying because the insurer did not cause the death of the deceased. *The contractor, as an insurer, assumes the liability not only for his own acts, but for strikes, acts of God, wars, changes in the type of manufactured items, and any other cause; he is exempted only by acts of the owner.*

The liquidated damages clause is a relief to the contractor from possible higher damages. The owner becomes his own insurer under a liquidated damages clause—for an often petty amount, he assumes the liability for damages caused by delay beyond the amount of the liquidated damages. The contractor's losses for delay are fixed as a maximum by liquidated damages; he is granted a variety of causes for delay (he often has not such releases under a contract which specifies "Time is of the essence"); and if the work is not actually used by the owner, the contractor may claim that no damages were suffered by the owner and may be relieved of liquidated damages.

A contractor can collect damages suffered by delay caused by the owner under certain circumstances. These may be because the owner failed to make the site available, failed to complete drawings or approvals as required by the contract, or did not furnish materials or materials and labor for which the owner has taken the responsibility for supplying. In the same manner, a contractor can collect damages from a subcontractor for delay, if the subcontract so states in sufficient detail, and the extent of the damages (which need not be liquidated damages) is known to the subcontractor as well as to the contractor. Damages for delay are difficult for a contractor to collect from the owner, as the owner usually makes sufficient exceptions in the contract, as to reasons for delay, that the delay seldom is caused by a reason not covered in the contract. If the contract states that the owner will extend the completion for delays caused by the owner, it has been held that the contractor is barred from claims for his expense due to delay. A contractor has collected damages for delay on the part of the engineer, who was lax in approving submittals and making drawings, even though it was a public contract in which the engineer had no authority as agent of the city. It was held that since the city made the work of the engineer preliminary to the construction work, the city was responsible for the delays.[44]

It is important to remember that under a "Time is of the essence" or liquidated damages clause, or under both, when the completion date is specified in the contract, the contractor is liable for delays. He is relieved only for the reasons for delay specifically stated in the contract. *When one enters into a positive agreement, he is not relieved of that duty or absolved from liability for failure to fulfill the covenant by a subsequent impossibility beyond his control, even though it is an act of God.*

[44]*Litchfield Construction Co. v. City of New York,* 244 N.Y. 251.

Many contractors are not aware of this requirement and assume that they are relieved from performance for circumstances beyond their control. *This is true only if stated in the contract.* The writer has prepared scores of contracts with absolute completion dates and has rarely had an objection from a contractor. Such clause may be ineffective not because they lack legal teeth, but because the contractor does not understand them before signing. When he does understand, he doesn't sign. Many contractors believe they have no liability for damages unless there is a liquidated damages clause in the contract.

The practice used by public agencies of letting separate contracts, particularly for mechanical and general construction work, often increases costs because of the inability of one of the contractors to keep up with the work. This delay is always present for subcontract work also, but it is more likely to occur on public work, where directors of the work are not qualified to manage it, that is, not qualified for direction of work in the sense of the contractor's project management. This is not a public official's normal function, and public officials' qualifications do not extend to such detail. This difficulty is reinforced by the tendency of public agencies to pay low salaries, which does not attract able personnel or retain them after they are hired; but the security of position in such agencies, with the result that more incompetence is tolerated; and by their lack of ability to choose business managers.

When a contractor is late on a job under a contract with a liquidated damages clause, he may prefer not to resolve whether delays are authorized. If extra work is authorized and no additional time is given (and the contractor, presumably, must *not* agree that no extension is necessary), the liquidated damages clause of the contract may not be enforceable by the owner.[45] This depends somewhat on the conduct of the owner, however, and the cases may vary. That is, the contractor must not agree to inadequate delay, but he need not complain if the matter of delay is not considered.

When a contract requires that requests for delay be made promptly, the owner can refuse to honor late requests, even though the extension is otherwise justifiable, and even though the owner has been accustomed to honor such requests previously.[46] Many times material is late, weather is bad, or other delays occur in part of the work. The contractor may not believe these will delay completion, but he must report a delay or sacrifice liquidated damages in the future.

If the contract requires reporting the cause of a delay *when it occurs,* you may find that an owner refuses to allow delay because the original happening, not the delay, was unreported. For example, a contractor orders a steel building for June 1. Foundations have been delayed, and are not ready for the steel building until July 1. But the building is not delivered until August 1. Just after July 1, the contractor reports that there will be a delay for the steel building. The owner may answer that "the delay began" on June 1, although the building could not have been installed until it arrived. Therefore, the contractor is a month late in reporting the delay when the delay began![47]

4.21 PERFORMANCE AND PAYMENT BONDS

A *bond*, in construction usage, means a *surety bond*, or agreement of one party (who is presumably rich) to guarantee the financial ability of another party (the contractor, presumably poor) to complete a construction job and to pay for it. There are also general surety

[45]*Haffner & Taylor* v. *Perloff,* 174 La. 687, 141 So. 377.

[46]*City of Houston* v. *Gribble,* Court of Civil Appeals of Texas, September 30, 1978, rehearing denied October 28, 1978, as reported in *Civil Engineering,* Vol. 47, No. 5, May 1977, p. 14.

[47]Author's experience with State of Florida.

bonds made to cover certain operations of the contractor for a period of time; for example, a contractor who does work on public property, even for a public owner, may post a bond to cover possible claims of the city or third parties. A contractor who installs privately owned public utilities, or who builds driveways for private persons on public property, may furnish the city with such a bond.

Some types of bonds required of a contractor are the following:

1. A *performance bond*, under which the surety (bonding company) guarantees to an owner that the contractor has the financial ability and the skill to complete the job. The contractor need not default on his payments to creditors to cause the surety to take over completion of the work, although the failure of the contractor is the most common reason.
2. A *payment bond*, to protect all creditors on the job. This protects the subcontractor who is unable to put a lien on public property, or who is under various other restrictions. Generally, a payment bond is preferred by creditors to lien rights.
3. A *payment bond*, to protect the owner. This pays liens to which the property may be subjected as a result of failure of the prime contractor. It does not, however, protect subcontractors who do not have a direct claim against the owner. For the subcontractor, it insures the financial ability of the owner to pay for his own liabilities. It does not insure the general contractor's obligations, and it is important that the subcontractor distinguish between the two. A provision in the bond that the general contractor shall pay all bills for labor and materials is sufficient to allow action by third parties.

Public agencies usually require general payment bonds, which are combined performance bonds and payment bonds to all creditors. If in doubt about whether the bond will cover liabilities to him, a subcontractor may qualifty his bid (since he may not know the language which will be actually used in the general contractor's bond), or he may consult an attorney after the sub has obtained a copy of the general's bond.

Since the funds of many corporations may be withdrawn by the officer-stockholders at any time, sureties require that the manager-owner sign both as an officer and in a personal capacity. He may also be obliged to post collateral, such as a mortgage on his own house.

A monthly status report, termed by the surety in Figure 4-1, *Schedule of Position on Uncompleted Work,* is the form for reporting work in progress by the contractor to the surety. It is prepared by the contractor and sent to the surety monthly, preferably as soon as payment requests have been sent to the owner and the previous month's invoices have been received.

Notes on Figure 4-1:

1. Job No. 80-2 has been completed, but since the last report. This report is the last time it will appear.
2. This is not a cash or liquidity statement. Cash on hand and retained percentages are not shown. The excess of billings over costs is not accounts receivable, because of retention.
3. The important amount is $470,000, estimated cost to complete. If the surety's bonding limit is $700,000, then they will at this time write a bid bond of up to only the difference of $230,000.

Name of Contractor __KING ROYER, INC.__

Date __JUNE 10__

'SCHEDULE OF POSITION ON UNCOMPLETED WORK
(Please Include Any Contracts Completed Since Last Schedule)

Job. No.	Contract Price	Amount Billed to Date	Costs to Date	Excess of Billings Over Costs	Estimated Costs to Complete	Total Estimated Cost	Estimated Gross Profit
80-1	$173,026	146,029	130,732	15,297	30,000	160,732	12,294
80-2	30,080	30,080	26,417	3,663	0	26,417	3,663
80-3	456,700	423,732	431,036	(7,304)	10,000	441,036	15,664
80-6	206,000	103,246	71,000	32,246	110,000	181,000	25,000
80-10	75,000	17,400	18,032	(632)	50,000	68,032	6,968
80-11	327,000	43,426	22,000	21,426	270,000	292,000	35,000
	1,267,806	763,913	699,217	64,696	470,000	1,169,217	98,589

Comments

King Royer, Pres.

" COSTS TO DATE " INCLUDES ALL ACCOUNTS PAYABLE.

Figure 4-1
SCHEDULE OF POSITION ON UNCOMPLETED WORK
"Cost to date" includes all accounts payable.

CONTRACTOR'S SUPPLEMENTAL STATEMENT

NAME ___ King Royer, Inc. ___

SUBMITTED TO ___ Associated Surety Co. ___

CONTRACTS IN PROCESS OR TO BE
CLOSED OUT IN CURRENT FISCAL YEAR

March 1, 1981
DATE

CONTRACT Indicate with Whom, Whether Lump Sum, Fee, Etc. Percentage of Retention, and Brief Description of Job	Total Contract Amount, Including Extras	Amount Being Subcontracted	Per Cent Completed	Estimated Completion Date	Cost of Work Performed	Total Amount Billed	RECEIVABLES				Anticipated Total Profit on Completion
							Amount Billed and Now Owing, Exclusive of Retainer	Amount of Retainer	Estimated Job Profit to Date		
80-1. LS 10%, Alachua County School Board, School	$ 173,026	$122,000	84	3/1/81	$130,732	$146,029	$131,426	$14,603	$10,327		$12,294
80-2. LS 5%, Florida National Guard, Barracks	30,080	10,000	100	1/15/81	26,417	30,080	28,576	1,504	3,663		3,663
80-5. LS 10/5%, U. S. Dept. of Agriculture, Buildings	456,700	360,000	93	2/15/81	431,036	423,732	400,897	22,835	14,568		15,664
80-6. Fee 5%, Sqartout Coil Co., Plant renovations	206,000	110,376	50	6/15/81	71,000	103,246	98,084	5,162	12,500		25,000
80-10. LS 10% Franciscan Stationers, Store	75,000	40,200	23	4/1/81	18,032	17,400	15,660	1,740	1,603		6,968
80-11. LS 10%, U. S. Dept. of Fisheries, Houses	327,000	302,426	13	7/15/81	22,000	43,426	39,083	4,343	4,550		35,000
Totals	1,267,806	945,002	--	--	699,217	763,913	713,726	50,187	47,211		98,589

This interim form is submitted for the purpose of providing supplemental information in connection with establishing or maintaining credit with the above named bank. The information contained herein is true and correct to the best of my knowledge and belief.

King Royer
_____ Signature
King Royer

_____ Title
President

Figure 4-2
CONTRACTOR'S SUPPLEMENTAL STATEMENT

147

4. The cost to complete may be reduced by bonded subcontract amount included in it.

5. If your past reports on estimated profit have been correct, the surety may deduct this amount from cost to complete. Both of these deductions (5 and 6) increase the contractor's bidding and bonding capacity.

6. This schedule includes all contracts under way, not just bonded contracts.

7. The *excess of billings over costs* is not profit. Until the job is nearly completed, there may be no relationship between the excess and the profit. As the job approaches the end, draws ahead are eliminated.

8. The actual cash billing is the amount billed to date less retentions included, plus final retention only, on the final billing on the job.

9. A cash excess of billings over the cost does not mean that the contractor will get paid in advance of expenditures, since he must continue to make payments, and does not know how just when the payment requested will be received.

Figure 4-2, *Contractor's Supplemental Statement,* is a more elaborate version of the *Schedule of Position.* It is supplemental in that the surety requires annual operating statements (profit and loss statements) and statements of condition (balance sheets), and the monthly statements are supplementary to these. The form in Figure 4-2 shows the amount being subcontracted, which is a reduction in exposure; the percentage completion, which gives an idea of the reliability of the anticipated profit; and the receivables, both current (exclusive of retainer) and deferred until completion of the work (retainer, or retention). Neither of these reports shows a key figure—the contractor's receivables, exclusive of amount of these receivables owed to others. The contractor should always calculate this amount, as it shows his solvency or lack of it—the surplus cash available.

The manager of a firm that contracts public work must understand and supervise this report. Since it is not a book of account and does not "close out" or affect the books of account, a bookkeeper may not understand just what figures are wanted, and may inadvertently show that the contractor has more outstanding work than is actually the case.

The Schedule of Position (Figure 4-1) shows billing in excess of cost, which is "front-loading." This practice, rather than being condemned by sureties and lenders, indicates to them that the contractor is carefully managing his cash and therefore requires less working capital than otherwise. The same is true for an item on the balance sheet "deferred income taxes" or "reserve for taxes"—which may be a liability account (an amount owing) for payment of taxes next year on this year's profit, and is necessary to reconcile two accounting systems used, one for the surety, the other for tax reporting. Both these items—excess draw and deferred taxes—are liabilities which represent interest-free loans to the contractor. Of course, at the end of the calendar year, a contractor's statement should show the taxes due the following April 15 as a liability actually due—not a reserve.

On the other hand, costs in excess of draws, or underpayment requests, indicate to the surety not only poor management but possible job loss.

4.22 COMPLETION BY SURETY

If a contractor is unable to pay his debts on a bonded job, or does not continue with the work as specified, the owner will serve notice to the surety to complete the contract. If the contractor cannot pay his debts, he cannot continue; but he may not stop work and still be able to pay his debts. The surety may then take over the contract and, in so doing, may acquire not only the construction company but the home and other private property of the contractor. The surety may arrange for the work to continue under another contractor or person hired for the work, or it may finance the contractor to continue work with his own forces under the existing contract. The latter route is least expensive if the surety has been called in early, before the contractor has gotten so far into debt on other contracts that an attempt to rescue him would be more expensive than completing the job. Suppose a contractor on a $100,000 job has completed $50,000 worth of work and has been paid this amount. Suppose further that the work can be completed for $50,000, that the contractor has no debts other than those secured by accounts receivable, and that he has no cash. In this case, the contractor may be in good financial condition but merely out of money. A $10,000 loan from the surety may enable him to complete the job and repay the surety. The surety would not profit from taking over the job completely and then trying to get back an overrun from the contractor.

But suppose now that the same $50,000 worth of work has been completed, that $70,000 worth has been paid for by the owner, and that the contractor has $100,000 in other debts against which he has no receivables or other assets. The surety can look at a $50,000 outlay to complete $30,000 in work—a $20,000 loss. But the surety wants no part of the contractor's other $100,000 in debts, so it would not try to save the contractor. These are two extreme cases; intermediate cases would be a matter of the surety's judgment. If the surety took over the job but not the contractor's corporation, it might hire the contractor as a manager to complete the work. Judgment is required not so much to decide the proper course of action with a known balance sheet as to decide how much confidence to put in the contractor's balance sheets. It is difficult to determine the financial condition of a construction firm under the best of conditions—that is, when the people are honest. It is much more difficult when books are not properly kept, either because of ignorance or fraud.

It is not uncommon for a general contractor to discontinue all operations for years, his assets being held by a surety depending on the outcome of a few claims. In heavy construction, a single extra claim may be greater than the total capital of a contractor. When such a claim is denied by the owner, the contractor cannot remain in business while he waits for the outcome of court action or appeal.

Although the surety guarantees the overall owner against the overall loss, in the event of failure the surety's first responsibility is to complete the work. This is because

1. the owner, if left to complete the job, will take a longer time and incur greater costs, and
2. until the work is completed, the liability of the surety is not known.

The primary function of sureties is service, and secondarily it is financial liability. They are service companies, but guarantee their services by financial responsibility; in this respect they function like a contractor in a joint venture.

A surety is not an insuror, although many of them are subsidiaries of insurance companies. A surety differs from an insuror in that:

1. An insuror accepts all customers, knowing there will be losses. A surety accepts only principals which it has evaluated as being of minimum risk.
2. An insuror does not directly prevent the risk against which it insures; a surety directly intervenes in the operations of a contractor before the liability is claimed.

3. An insuror considers only the risk it insures, and is not involved in the contractor's other risks. A surety by necessity must often direct other construction on which it receives no fee.
4. An insuror protects its policyholder against a risk; a surety protects a third party, the owner, and has no responsibility unless the principal (policyholder) is insolvent and cannot benefit from the bond.

When considering a new principal for bonding, the surety will consider previous causes of default in other firms, as

1. lack of experience or management experience
2. lack of capital
3. lack of depth of management, particularly lack of persons to take over management in the event of death or disability of the manager
4. spending habits of the principals of the firm—that is, are they spending profits or reinvesting them?
5. extent to which the firm has other interests. If a contractor does public contracts and is also a homebuilder, for example, he may tie up his capital in unsold houses. To the extent that the firm's available capital is in unbonded work, the surety assumes an unpaid risk.
6. extent of contractor's unbonded work, for the same reasons as (5). The surety is bonding only part of the work, but the firm's capital may be tied up in unbonded jobs.

In case of a contractor's inability to complete a bonded job, the surety will act as soon as it knows of the situation. The surety does not wait on the owner. The surety may:

1. Install its own accountant to audit, observe, or authorize the contractor's expenses, while the surety's money is supporting the business.
2. Endorse (guarantee) the contractor's notes so that he may obtain additional bank loans.
3. Loan money direct to the contractor.
4. Install its own or its recommended engineers, project managers, or superintendents in the contractor's organization.
5. Employ consultants to assist the contractor.
6. Allow the contractor to go bankrupt or to relinquish the job, then completing the work with the contractor's forces, new personnel, or letting the work to another contractor on a lump-sum or cost-plus contract.
7. Allow the owner to take over the work and complete it. In this case, the surety pays the overrun to the owner.

4.23 AIA FORM, PARAGRAPH 13.2, CORRECTION OF WORK

The 1976 revision of the AIA Standard Conditions made indefinite warranty by the contractor a requirement. Contractors have generally assumed that they were not responsible for faulty materials and workmanship for more than a year after substantial completion, but even under the old (prior to 1976) AIA conditions the courts have generally held the expiration of the guaranty period does not terminate the contractor's responsibility for defective work.[48]

The correction period is the time during which it is assumed that responsibility for failures rests with the contractor; the contractor is legally presumed to be at fault, unless he can refute it. For example, a contractor was called to clean a sanitary sewer which repeatedly blocked up, in a school. The contractor found that the blockage was caused by blueprints stuffed into water closets. The sewer was not designed for such disposal, so the contractor was able to refute the presumption of fault. However, he was obliged to make the "trouble call" under circumstances he could not collect.

In another instance, aluminum screen doors on a barracks were not in accordance with the specifications, and had to be replaced within a year. The contractor's files were able to show he had repeatedly asked for clarification of the specifications, which were based on an unknown proprietary specification. Also, he had contacted over a score of manufacturers, and secured an alternate sample; the architect's representative had approved the sample, and the superintendent so recorded on his daily report. The architect had no records at all, and did not further claim that the doors be replaced.

After the year's *correction period*, formerly termed the *warranty* period, defective work is not presumed to be the fault of the contractor, but of the owner. The owner must then prove the fault lies with the contractor, which is usually difficult.

4.24 ARBITRATION

According to Article 15 of AIA Construction Contract Form A107, *"All claims or disputes . . . shall be decided by arbitration in accordance with the Construction Industry Arbitration Rules of the American Arbitration Association."* As a general rule, such an arbitration rules out a suit at law. When the matter is not subject to federal law and the state has no arbitration statute or precedent, a party intending to refuse arbitration should consult an attorney when an arbitration notice is received.

The nature of arbitration is often misunderstood. Arbitration is a substitute for litigation, and the acceptance of the AAA Construction Industry Arbitration Rules effectively forbids the contractor from placing a friend or acquaintance as an arbitrator who should be neutral. Under the Arbitration Rules, the arbitrators are chosen from a panel of prequalified, impartial experts. Arbitration is not mediation or reconciliation; it is a private court which need not technically comply with the law. Unlike a court of law, which must decide on the basis of evidence presented, the arbitrator may include his special knowledge of the trade or business. The

---

[48] *Omaha Home for Boys v. Stitt Construction Co., Inc.*, 238 N.W. 2d 478, 1976; *Baker Crow Construction Co. v. Hames Electric*, 566 P2d 153, 1977; *City of Kennewick v. Hanford Piping*, 558 P2d 176, 1977; *Board of Regents v. Wilson*, 326 N.E. 2d 216, 1975. From *Concepts*, January-February 1979, p. 4 ("Vansant's Law," by Robert E. Vansant.

panel of arbitrators is judge, jury, and expert witness; in addition, it is the final appeal as to law and fact except on such narrow grounds as bias, corruption, fraud, or exceeding contractual authority.

A lawyer is accustomed to pleading in a very special way—one side presents its case, then the other; argument is restricted to opening and closing statements; and evidence is introduced by one attorney or the other. Attorneys are also accustomed to delays when agreed to by the parties. An arbitrator, on the other hand, may require all evidence on a point to be presented at once, rather than hearing all points of one side and then all points of the other; he may accept any evidence, including hearsay; he may question and cross-examine not only the witness but the parties and the attorneys, at any time; he may recognize law and precedent as he wishes—he does not even have to accept as truth that on which the attorneys have agreed. He can make up his procedure as he goes along and is not obligated to accept a delay, since the arbitrator in most cases is unpaid and is often from another city or state. From the attorney's standpoint, the verdict is completely unpredictable, but a jury decision is also not predictable. An attorney who does not understand a technical point or trade practice can understand how it will be presented to another lay person, that is, to a jury, but he does not know how it will sound to another technical person with experience and ideas of his own.

Procedure in arbitration. The verdict of an arbitrator can be set aside only by "external" irregularities—if an arbitrator has been partial to one side, for example, through proven connections by business or blood, which were known only to one party, or if he was bribed. An arbitrator need not make the reasons for his decision known, and the proceedings may or may not be recorded.

Some kinds of evidence admitted in arbitration which would not be admissible in court are:

1. Letters or other documents without authentication and cross-examination of the writer.
2. Evidence relating to the conduct of the parties not directly related to the matter at hand.
3. Parol (oral) evidence regarding a contract which later was executed other than as discussed between the parties.
4. Opinions from persons not established as expert witnesses.

Since arbitration proceedings are intended in part to avoid attorneys, it is to be expected that attorneys will be somewhat suspicious of arbitration. A competent contractor or engineer may therefore do better without an attorney, although this procedure may be risky if an attorney is chairman of the board of arbitration.

The writer has observed that contractors on arbitration boards are likely to be more partial to the owner than are owners and architects. This tendency of professionals to be harder on those in their own profession than on those not in it also occurs in other professions. Undoubtedly, arbitrators act more in accordance with their own beliefs, as opposed to defending their own occupations, when the verdict is private and does not reflect on the profession as a whole as it does in public hearings.

In arbitration, ignorance of the law *is* an excuse, particularly if *both* parties were ignorant of the law, or if one party appeared to defraud the other. An arbitrator makes a moral and ethical decision, even to the exclusion of a legal one. If an attorney presents a clear-cut picture of the law, he is performing a valuable service for his client; but if there appear to be two legal opinions, the arbitrator is entitled to disregard them both.

Various public bodies have appeal bodies, who give decisions in a process that is not arbitration. A state appeal board, appointed by the governor, can bury an appeal by deliberately extending the hearing with trivia, which the contractor must pay for typing before a court appeal will be considered. An appeal board hearing can be much more costly, and just as long as a court suit. A judge, having no stake in the outcome of a trial, will not allow indefinite material evidence; an appeal board may do so. It is this author's opinion that unless one is familiar with the appeal board, it should be considered as a proceeding to complete before the matter is decided in court.

An attorney who addresses an arbitrator as he would a judge—implying that the attorney and the arbitrator have some special knowledge denied to the litigants—may do a disservice to his client. An arbitrator has the right of any citizen; he can ask any question, but he cannot require an answer. The arbitrator may give little or no weight to a refusal to answer. If the attorney addresses the arbitrator as he would a jury, he is also frustrated, for the technical part of the case, which the attorney would explain to the jury, is more familiar to the arbitrator than it is to the attorney. The arbitrator depends a great deal on impressions the parties make (as does a jury), and he may be adversely affected by a party's not taking the stand in behalf of his own case. In some states, arbitrators have the right of subpoena and may require a party to testify.

The arbitrator also has a duty which a court does not consider—as far as possible, he wants to leave the parties in a position in which they will continue to do business with each other. This is an objective of arbitration and, for that matter, is the aim of justice. Arbitrators establish no precedents, since their decisions are not available, and they can freely tailor their decision to the special case as they see it—not as it would be understood by an outsider or by participants in a later dispute.

Under AAA rules, the arbitrator must decide the claim as presented in the demand for arbitration; if he should accept any additional claims, the decision could be questioned as partial to one party who had an opportunity to prepare for it.

In popular usage, *arbitrator* is connected with *labor arbitration;* the popular press is inclined to use the work "arbitration" for proceedings which are often mediation. The results of mediation are practically always compromises, so arbitration matters are erroneously also considered compromises. It is true that a compromise is usually dictated by the claims of the parties in arbitration; that is, one party asks for more than he expects (or may previously have offered to settle for) and the other party will file a counterclaim, hoping to settle for a zero claim. However, the arbitrator is by no means limited to a settlement compromise.

4.25 ADVANTAGES OF ARBITRATION

The advantages of arbitration are the following:

1. Delays are reduced and the date when the proceedings will end is stated before sessions begin.
2. Legal costs are reduced, both because the sessions are shorter and because the parties concerned can state their own case without the difficulties in admission of evidence they

would encounter in court. The lawyer, if used, spends less time in preparation of the case. In a technical case, an attorney must spend a great deal of time preparing the matter so it is apparent to a lay jury; this preparation is unnecessary in arbitration.

3. Key issues can be disposed of so that minor issues may be settled by the parties. Some parties do not want to concede small matters until they know how the big ones will come out; and, in many cases, they can foretell the probable outcome of small issues by hints dropped during the proceedings involving the larger issues.

4. The verdict is final, as compared to the verdict in a court of law. Although a court may be required to enforce the verdict, either by directing a payment or by refusing a suit of collection by the losing party, the grounds on which an arbitration proceeding can be overturned are so few and difficult to prove that arbitration is rarely appealed. Court enforcement is a factor partially balancing the no-appeal factor of arbitration. For this reason, arbitration is not entirely free of court procedures.

5. Arbitration proceedings are private. This is particularly important to professional contractors, who do not want to publicize disputes, particularly if the other party is a religious institution or a business with a good reputation in the community.

Powers of an arbitrator. An arbitrator has a great deal of power, but he may not improperly determine jurisdiction; that is, if the matter is not subject to arbitration in the first place, the arbitrator's decision is not valid. On a three-man board, the chairman is usually an attorney, and one may expect that the procedure would be much more formal under such conditions of procedure resulting from an attorney presiding.

The privacy of arbitration proceedings has the disadvantage that few people, even the arbitrators and the administrators of the AAA (who do not act as arbitrators), must rely mainly on knowledge from personal experience rather than on precedent because the record of proceedings, when kept, is distributed only to the parties. The arbitrators are usually unpaid and consequently serve in few cases—sometimes one each two years.

A board of arbitration seldom explains the reason for its verdict, as some parties, arbitrators, or writers fear that an explanation may cause a verdict to be vacated (voided). "The arbitrators are only required to render an award in conformity to the submission and an award need contain no more than the actual decision. . . . The means by which they reach the award is needless and superfluous."[49] An arbitrator can be overruled for excluding evidence from the case, as being partial to one party, but he cannot be overruled on the ground that he admitted evidence which should not be heard. Consequently, it is to the advantage of the arbitrator to admit all evidence which may be useful in the case. An arbitrator seeks to complete the case as soon as possible; if he does not, the advantage of arbitration is largely lost. He therefore will not be interested in proper authentication of written matter which requires out-of-town or otherwise difficult-to-obtain witnesses. Because of the necessity for prompt action with limited cost, an arbitrator will not engage in the kind of legal examination of a brief which is expected of a judge, as he has no legal clerk to do so, and no access to a law library. The law as presented to an arbitrator should be complete. A reference is not suitable; the text should be provided. It is customary for an attorney to include in his brief a reference to cases rather distant from the one in question; the judge must then determine if it parallels the one under consideration. An arbitrator has not time for such study, even if he has the talent.

---

[49]*Gary Excavation Co.* v. *Town of North Haven,* 279 A. 2d, 543 (1971).

4.26   AN IMPROPER DECISION

An arbitration which was voided because the proceedings were not in compliance with law was as follows: A subcontract provided for unit price charges for additions to and deductions from the subcontract. The subcontractor attempted to overthrow this method of calculation for changes in the work and won an arbitration award. Since evidence of variation in the circumstances of the work could have been the only reason for the action even though it was not mentioned in the court's decision, it may be assumed that there were valid reasons to question the units; any competent estimator can demonstrate such variations.

The court held, however, that the arbitrators exceeded their authority and changed the contract; since their authority was to decide disputes under the subcontract, they could not change the subcontract itself.[50]

This distinction is a fine one to construction managers, who often fail to distinguish between claims under the contract and claims not within the arbitrator's jurisdiction. Such a distinction must be carefully preserved, however, if an arbitration decision is to be valid. Assuming that the arbitrators were correct—that the circumstances had changed and did justify an extra payment—there were a number of errors on the part of the subcontractor and the arbitrators:

1. Unit prices for changes are dangerous to guarantee, and the subcontractor should not accept such contracts. If unit prices may not be avoided, such a clause as "The unit prices shall control for work not differing materially from other work in this contract" should be added. This clause virtually nullifies the unit-price clause, as "materially" could conceivably be a dollar, but will give the owner some sort of base for changes. It amounts to saying that the owner, not the contractor, is bound but that other prices must be carefully prepared as extra claims.

2. The subcontractor calculated extras separately from the unit prices, in effect denying any validity to the unit prices in the contract. He could have avoided any clash with the working of the contract by a claim of change in the type of work contemplated in the units.

3. The subcontractor's counsel had said, "Certainly it will possibly be 'scrapping the contract,' " and this statement was used to prove the counsel's awareness that the subcontractor was avoiding the contract; and the subcontractor claimed the contract was abandoned by both parties.

4.27   THE A/E AS ARBITRATOR

Some contracts state that the architect or engineer is the final authority for interpretation of disputes. An example of such a statement is this:

> The Engineer shall in all cases determine the amount, quality, acceptability and fitness of the several kinds of work and decide all questions . . . His determination and estimate shall be final and conclusive on the contractor.

[50]*Strange et al.* v. *Thompson-Starrett Co.*, 261 N.Y. 37, 184 N.E. 485.

Such clauses have generally been unenforceable, with the courts holding that the engineer's decision is not final in regard to interpretation of the contract, since that is a legal matter. However, since virtually every decision of the engineer is an interpretation of the contract (he has no other base of authority over the contractor), it follows that his other decisions could also be appealed to the courts. In one case, a jury was instructed to decide, as a matter of *fact*, whether certain underpinning of buildings was necessary to prevent collapse.[51]

The engineer is given no such final authority under the ASCE general conditions, although the AIA contract maintains the final power of the architect on matters of artistic effect. It can be assumed that the contractor will not object to the architect's artistic effect but may dispute the cost; the question then becomes, "Who decides what questions can be arbitrated?" Obviously, the Board of Arbitrators. The matter reduces to "I'll play your game, but you let me be the referee." As a practical matter, the items which result in dispute over artistic effect are still subject to arbitration. The contractor rarely purchases art work; one exception is murals, which may have to be provided with the building. If a contractor used an artist other than the one specified, the finished work might be refused on artistic grounds.

When the A/E is specified by a contract to be the arbitrator in a state lacking an arbitration law, the courts will disregard this as they may disregard any arbitration clause in a contract. If the state has such a law, however, the contractor should avoid a contract which gives the A/E power as an arbitrator. Such a clause would probably not stand, as being inequitable, but it may be used. Few A/E's follow the standard contracts exactly (in the standard contracts, this is no problem), and there is a strong tendency for many of them to attempt to give themselves as much power as possible. Some of these clauses simply amount to, "The contractor shall do all work to the satisfaction of the architect, with no recourse," and these will be overthrown more readily if arbitration by the A/E is not mentioned. Some contractors, and especially subcontractors, will sign virtually any paper placed in front of them, if the description of the work and the price are correct. Inequitable clauses may be eliminated only by the refusal of contractors to bid work under these clauses; very few contractors go broke on jobs they do not do.

When a party has the choice between arbitration and suit at law, as is true when one party proposes arbitration even though no arbitration clause is contained in the contract, the admissibility of evidence may be an important consideration. An attorney whose case rests on evidence which he believes may be ruled inadmissible may do much better at arbitration.

A *quasi-arbitrator* is a person with some—but not all—of the powers of an arbitrator. The resident engineer may be held to be an arbitrator, so far as questions of fact are concerned, but not for questions of law.

The federal agencies administering construction contracts have fixed boards of quasi-arbitration, or more properly, boards of appeal, since the contractor has no voice in selecting them; the boards consist of government employees assigned to that duty. Usually, one of them is an attorney, and the board decisions are exposed to legal review (by the Court Claims), as arbitrators' decisions are not. These appeal boards make long explanations of the verdicts they decide, and these verdicts are excellent sources of information regarding how future cases may be handled. The verdicts are available at various places, depending on the agency. Usually, no extra copies are kept long after the proceeding.

4.28   CITATIONS OF ARBITRATION LAW

For those with access to a legal library, The American Arbitration Association publishes a free list of pertinent cases regarding the legal background of arbitration, with cases on the topics listed below.

---

[51]*Dock Contractor Co.* v. *City of New York,* 296 Fed. 377.

1. The parties may limit the time within which a difference may be referred to arbitration. This is 30 days in the AIA General Conditions, and it may not be changed by an arbitrator. An arbitrator *interprets* the contract, but may not disregard or *change* it.

2. Generally, cases may not be combined. For example, a subcontractor may make a claim on the general, and the general makes the identical claim on the owner, but it can be made one case only by agreement of all three parties.

3. A public body may be obligated to arbitrate, even though it lacks authority to pay an adverse ruling.

4. A surety or other party which takes over a contractor's claim, assumes obligations and rights to arbitrate. Others may also be restricted by reference to the general contract, as subs who incorporate by reference allusions to a general contract with an arbitration clause.

5. Participation in litigation may waive arbitration rights. However, filing a lien or suing on the lien may not affect arbitration rights.

6. Arbitration proceedings may apply against foreign countries.

7. The arbitrator, not the court, initially determines what are arbitrable issues.

8. Ambiguous matters are construed to be arbitrable.

9. The invalidity of a construction contract is for decision of the arbitrator.

10. An action at law for defective design is not barred by an arbitration agreement providing that "questions of fact arising under the contract are subject to binding arbitration."

11. The issue of the existence of the agreement to arbitrate is for the courts.

12. Fraud in the inducement of the arbitration clause itself is a question for the court.

13. Locale may properly be a question for the AAA.

14. Adjournment is discretionary with the arbitrator.

15. Independent investigation by the neutral arbitrator of the premises constitutes a valid ground to vacate the award.

16. Where the agreement specifies allocation of the arbitrator's fee, the arbitrator is required to adhere to the terms of the agreement.

17. An arbitrator does not have the power to award punitive damages.

18. An arbitrator may not be required to testify for the purpose of impeaching his award.

19. A "completely irrational" award may be vacated (changed) on the statutory grounds that the arbitrator "exceeded his powers."

Determining if an arbitration clause is applicable in a particular case is not a simple matter. In Montana in 1979, Westech, Inc., refused to pay Palmer Steel Structures $50,000 for a building, and demanded arbitration. Palmer refused, filing suit to enjoin arbitration procedures because the arbitration clause was not valid in Montana. Although the trial court upheld the obligatory arbitration clause, the State Supreme Court said that it was void.[52]

---

[52] *Palmer Steel Structure v. Westech, Inc.*, 585 P.2d 152 (Montana Supreme Court, 1978), *Engineering News-Record*, December 14, 1978.

4.29 STATES WITH ARBITRATION STATUTES

As of 1972, the states below have statutes or court decisions upholding arbitration procedures similar to those provided for by the AIA clause.

Alaska	Hawaii	Minnesota	Rhode Island
Arizona	Illinois	Nevada	South Dakota
Arkansas	Indiana	New Jersey	Virginia
California	Louisiana	New Mexico	Texas
Colorado	Maine	New York	Washington
Connecticut	Maryland	Ohio	Wisconsin
Delaware	Masschusetts	Oregon	Wyoming
Florida	Michigan	Pennsylvania	District of Columbia

You should be familiar with the laws relating to arbitration in your own state. Not all the states listed above necessarily recognize all cases, and other states not listed recognize some cases. All the states listed have a full arbitration statute, but the District of Columbia does not. Arbitration proceedings have a limited application in the twenty states that lack an arbitration statue (and lack also supporting judicial decisions). However, courts will recognize the Federal Arbitration Statue as controlling on United States government construction (under bonding and payment requirements of the Miller Act) and on work where interstate commerce is involved. Since it is virtually impossible to construct even a small project without interstate construction being involved in some way, the application of the Federal Arbitration Statute is always probable. It has been held, for example, that an architect was engaged in interstate commerce when his drawings were sent for approval to an owner temporarily out of the state.

The factors discussed above apply when arbitration is agreed to in a contract. Voluntary arbitration on *specific disputes after they arise* is enforceable in all states.

4.30 AMERICAN ARBITRATION ASSOCIATION

Under the previous AIA arbitration procedure, the parties each appointed an arbitrator and these two appointed a third; if the two were unable to agree, the third would be appointed by a local court. This procedure led to disputes over the appointment of the third arbitrator and almost automatically assured that the arbitrators would be persons acquainted with each other and with the parties. Local people with no technical guidance on arbitration procedures or experience in arbitration cases were selected and were usually paid a fee. In 1966 the AIA adopted the clause which refers all cases for arbitration to the American Arbitration Association, and dropped their previous recommendation, which provided for a choice of the AAA procedure or the old procedure.

The American Arbitration Association is a nonprofit organization with public membership, formed for the purpose of providing assistance to parties who desire to use arbitration. The AAA provides typical clauses for contracts to make arbitration enforceable and clauses for agreements to arbitrate, does necessary administrative work to provide arbitrators, and if the parties cannot agree, makes decisions relating to selection of arbitrators and to location of hearing. The AAA fee for these services varies and is calculated by a decreasing percentage of the award granted by the verdict. There are no overall statistics on arbitration awards, but a sample of 30 typical cases in 1969 showed an average award of $5,775 and a median award of $4,770. That is, there were as many awards in amounts above $4,770 as there were awards in amounts smaller than $4,770. The smallest award was $315 and the largest $20,590. The average AAA fee for these cases was $254. Of the 30 cases, eight had three-man panels and six of the chairmen of these panels were attorneys. The average time for completion of the cases from first filing was 5.5 months, but a number of the cases were unusual in that arbitrators resigned or parties failed to approve lists of arbitrators.

Under the Construction Industry Arbitration Rules of the AAA, the demanding party mails the regional office of the AAA a copy of his demand, with the filing fee and two copies of the agreement to arbitrate (usually the construction contract). The AAA mails copies of the demand to the other party, who may within seven days file an answer to the AAA and to the other party but is not obliged to file an answer. If the parties have not appointed their own arbitrator, the AAA mails them lists of arbitrators available in the area and qualified in the matter at hand. Each party crosses off unsatisfactory arbitrators and lists the remainder in order of preference. If the parties cannot agree on eligible arbitrators, the AAA appoints them without further reference to the parties. The location of hearings, if any, is agreed to by the parties; or, if they do not agree, the AAA decides the location. The number of arbitrators is agreed to by the parties; if they do not agree, the number is determined by the AAA. Usually, only one arbitrator is appointed in cases under $50,000, unless the parties ask for three arbitrators. Once the arbitrators have been selected, they are in charge of the proceedings, including setting the time and place of hearings, determining who may attend (although parties may be excluded, witnesses may be, and the public always is excluded), setting a time for examination of the site, if necessary (since the parties must be advised and have an opportunity to attend). The chief restriction on the arbitrators is that they must not take any evidence, whether oral, by questioning persons on the job, or by physical inspections, unless the parties are present or have been invited to be present. The ruling is made by a majority of the arbitration board.

The writer considers that the presence of an attorney as chairman will make the proceedings rather formal, and therefore more difficult for a party who is presenting his case without an attorney; but the opposite might occur with the attorney-chairman feeling obligated to represent the party who has no lawyer. Arbitration proceedings take more time than they should, largely because of the time consumed in writing rather than telephoning. Some correspondence must be written, but there is no reason why the parties could not telephone their objections to the panel and settle a date by telephone; and communications of the AAA with the arbitrators could be by telephone. The thirty cases mentioned earlier were prolonged because arbitrators resigned before the hearings, perhaps because the arbitrators could not make appointments as far ahead as was necessary under mailed procedures.

The use of attorneys by the parties in arbitration appears to be overdone; this slows the procedure and increases the expense, and arbitration is designed to avoid such expense and loss of time. It is surprising that contractors who will settle a contract bid by flipping a coin for $10,000 will not proceed with a $10,000 arbitration case without an attorney once the contract is settled. But then maybe these coin flippers also use this method to settle their disputes!

To preserve the "poor country boy" defense, the writer is inclined to go without legal representation to arbitration with an attorney for chairman if he is trying to break the other party's contract. On the other hand, if a party is defending his own language drawn by an attorney, he should employ an attorney.

A party can exercise a certain amount of control over the profession of the arbitrators by marking off members of the profession he does not want, without consideration of other qualities. The reaction of arbitrators is very difficult for one of the parties to predict, and usually has no connection with any prejudice of their own profession. Since any acquaintances of either party are not usually accepted as arbitrators, neither party has personal knowledge he can use in selecting names on the AAA list. *Any* connection with a party or with either attorney, not merely business connections, must be reported by the arbitrators to the parties. The writer was once disqualified as an arbitrator because one of the attorneys several years before, then in another city, had prepared some routine papers for him, not connected with the construction business. The attorney involved did not even recognize the name, but the opposing attorney refused to accept this connection as impartial.

The AAA recommended contract clause to establish arbitration is as follows:

> *Any controversy or claim arising out of or relating to this contract, or the breach thereof, shall be settled by arbitration in accordance with the Construction Industry Arbitration Rules of the American Arbitration Association, and judgment upon the award rendered by the Arbitrator(s) may be entered in any court having jurisdiction thereof.*

The version used in the AIA contract form is substantially the same.

4.31 FINAL PAYMENT

It is well established that a final payment clause like the following is enforceable. *If the contractor accepts the final payment on a contract, the owner is relieved from further claims under the contract.* It is immaterial whether the contractor has made previous written claims for additional work. However, the contractor may accept final payment if the owner agrees to release him from his obligation to drop all claims. The person making the final payment to the contractor may not have the power to allow him to exempt claims not paid at that time, since the release of the contractor from his obligation to drop all claims must be by the body or person authorizing the contract (not necessarily signing it), in the case of public contracts. The AIA and ASCE/AGC contracts, as well as the federal construction contract, allow the contractor to reserve claims at the time he accepts final payment, or allow the contractor to continue with claims he has already presented in writing. It has also been held that the release does not relieve the owner from a claim for damages for breach of contract (as distinguished from claims arising under the contract) even though the contract may not make any exception for pending claims.[53] That is, the final acceptance of payment may exclude claims arising out of the contract but will allow claims for breach of contract.

4.32 UNIT PRICES

Unit prices in construction contracts appear in two situations; in engineering contracts for roads and similar work and for changes in work provided for by unit prices.

Engineering contracts for roads, paving, and similar uncomplicated work are awarded on the basis of an engineer's estimated amount, with a unit price for each kind of work. So far as possible, these units intend to cover different kinds of work in detail. For example, a sewer contract may have a unit price for each size of pipe and several unit prices for excavation, based on the first unit price for excavation to 6 feet, another for the next 2 feet, and so on, so that a

[53] *Faver* v. *City of New York*, 222 N.Y. 255, 261.

single trench may have four or five different kinds of work upon which unit prices are calculated. Manholes may be designated a similar way, according to depth, and materials to be excavated may be differentiated by hardness. In unit price contracts, the engineer usually measures the work for payment after it is done. On building contracts, the site work—such as grading, paving, and utilities—may be contracted by units and the building itself by a lump sum. Where a building is the principal item and the contract drawings are prepared by an architect, site work is usually included with the building.

Unit prices of this type seldom cause difficulty but when changes are made under a unit-price contract, the question arises, "Is the new work comparable to the old?" The ASCE/AGC contract states: *"The number of units . . . is approximate only . . . [In the case of] Changes in the work . . . that cannot be classified as coming under any of the Contract units . . . either the Owner or the Contractor may request a revision of the unit price for the item so affected. . . ."* The contract also provides for maximum variation in the number of units which may be inserted in the contract without adjustment of the unit price.

Changes in the work may be provided for by unit prices for electrical outlets, feet of pipe, square feet of windows, or other kinds of units in building contracts. It is not best practice to use unit prices in this way; the AIA contract does not provide for them, and many contractors refuse to bid such units. The actual cost of such units varies widely, and their use is a constant source of controversy.

Units are often used to make changes in contracts where exact plans are not practical. Golf-course sprinkler systems, for example, are bid this way; unit price adjustments are provided for changes in the number of sprinklers and linear feet of pipe. The contractor knows that if the number of holes is not changed, the changes in the system must be minor.

Contracts which provide for variation in the units or admit changes in the plans using the same units should be entered into by the contractor with full knowledge of the risk he is assuming. In one instance, unit prices were contracted under a clause which stated, *"The state reserves the right . . . to make such additions to or deductions from such work or changes in the outcome of case plans and specifications covering the work as may be necessary . . . no claim shall be made by the contractor . . . because of any such change. . . ."* The court upheld the changes as within the scope of the contract.[54] In this instance, the length of a canal was reduced from 3 ¾ miles to 3 ¼ miles, and a lock was changed to a location not in the original contract, due to poor foundation conditions.

If no provision is made in the contract for change in unit prices for change in the work, the engineer may not be empowered to give relief on public work. This arises because the unit prices are part of the contract, and the engineer is able to interpret, not change, the contract. He cannot arbitrarily change unit prices (or a contract price, of course) without specific authority in the contract itself, and this authority may be omitted. In a lump-sum contract, there is only one price, and the A/E is empowered to change it by clauses regarding payments for extras.

The extent to which changes can be made in a unit price contract without negotiation of new prices is not clearly defined. The kind of change, rather than the value, seems to be the criterion; courts have allowed changes in quantities as great as 41 percent when the change did not affect the general extent of the work, and it has been held that a change as little as 2 ½ percent could not be made. Incidental items can generally be omitted, but the description of the work as contained in the contract may not be changed.

If a case in arbitration, as previously cited, involves a claim that the unit prices are not applicable, the arbitrator must be careful that his decision does not state that any change was made in the contract.

---

[54] *Kinser Construction Co.* v. *State of New York,* 204 N.Y. 381.

4.33 CHANGE ORDERS

Often a contractor or subcontractor requests additional payment from the owner because of conditions which do not appear changed to the owner. Each such request must be based on a valid change, or consideration, in the work. A contractor who requires the owner to pay for overtime or higher wages to complete the work cannot collect such increase unless it is shown that the contractor had a valid change in the work. A contract, written or oral, which obligates the owner to pay an additional amount is not valid, and the contractor will not be able to collect the additional amount if it is demonstrated that no valid consideration existed. A contractor is occasionally faced with this situation when a subcontractor refuses to proceed with his contract unless granted an increase in price. The contractor's agreement to pay an increase is not collectible if made only to obtain completion of the contract.

Standard contracts require that change orders be authorized in writing. The A/E often does not have authority to give a change order at additional cost, unless he has specific authority from the owner, written or verbal; if the owner gives a verbal order, it may have to be proved that he was aware of the added cost and accepted the progress of the work as ordered. Although added costs for such verbal orders are usually collected, contractors attempt to have such orders verified in writing as early as possible. The verbal order of the architect is less dependable, since it must also be proven that he has authority to give the order. It often occurs that the contractor is misled by the apparent harmony between the architect and the owner to believe that the architect has full agency powers and the confidence of the owner; but, on completion of the job, the owner later discovers that the work is more expensive than he was led to believe, so he refuses the contract's claims and disclaims the architect's authority to authorize them. The owner has the advantage over the contractor, since the owner can refuse to pay the contractor; but the architect, by this time, has collected almost his entire fee. The architect can then be of little help to the contractor; written and verbal change orders alike, issued by the architect, may be disclaimed by the owner.

Many change orders are billed by the contractor on the basis of cost as the work is done, under a cost-plus percentage fee arrangement. On federal contracts, however, contracting officers are prohibited by statute from contracting cost-plus work; the criterion is the *value of the work* as determined by the contracting officer. The contracting officer may accept the contractor's cost figure as the value of the work, and he will usually accept the contractor's detailed estimate as the value of the work. Prompt submittal of detailed extra claims avoids delays in payment on federal work, reduces bookkeeping required of the contractor, may enable him to collect more for the work, and reduces disputes over the charges.

When an extra claim is prepared, the writer recommends the use of any standard cost figures, particularly labor costs, which may be available from publications, rather than using the contractor's actual estimate figures on his work. Such standard costs are usually adequate and are more readily accepted. Since the markup is limited, the time of estimators or supervisors may be included as direct cost in such an estimate; if the claims were based on actual cost later, the contractor would have to have a cost accounting system reporting such items in detail, to justify the claim. Few contractors keep such records.

Claims for extras usually must be made in writing within a certain number of days after the condition causing it is discovered. A court has held that this time period starts when the work is begun, not when the order is given.[55] A verbal claim may be sufficient, if it can be shown that

---

[55] *Fleischer Engineering & Construction Co.* v. *United States,* 98 Ct. Cl. 139.

the person to whom the claim was made was aware of it.[56] In federal contracts, the claim must be made to the head of the agency within thirty days after an adverse decision of the contracting officer. Since the contracting officer has no authority to waive such a requirement, failure to make the appeal will void the claim; fairness or justice has nothing to do with it, since the public (via the agency head) is being deprived of its right to decide if it wants to spend the money. It has been held by the Corps of Engineers appeal board that a contractor was entitled to payment for the portion of an extra claim which was completed *after* he complied with formalities, but not before.

The contractor may assume that the A/E representing a private client has an agency power that he would not have when representing a public agency, since the authority of public officials is assumed to be known to all.

A 1975 decision by the Armed Services Board of Contract Appeals, quoting a Court of Claims decision, denied a contractor's claim for extra work on different grounds, *the interpretation of a contract . . . before the controversy . . . is . . . of great, if not controlling weight. . . .*[57] In other words, agreement when work is *performed* that the work is included in the contract virtually precludes a later claim to the contrary. It is therefore of paramount importance that the job superintendent not agree to work without written protest and complete it, before it can be reviewed by the person in charge of the business aspects of the contract.

A contractor should authorize his superintendents to proceed with such work using a routine form by the superintendent, countersigned by the A/E, that proceeding with the work does not waive any rights the contractor may otherwise have.

4.34 LIENS

A contractor, supplier, or laborer on a construction project has a right, in general, to claim what is due him for the work directly from the owner or from a later owner of the property. That is, if the owner is insolvent, refuses to pay, even though he has sold the property, the claimant, through the mechanism of his lien, may force the property to be foreclosed and sold and the proceeds used to pay the claim. If a contractor has been paid and has not paid his subs, the subs may collect the amount from the owner, depending on local law and other circumstances.

There are numerous variations in the mechanism of these claims, and restrictions on the claims by statute. Since state laws vary widely, every contractor should have a copy of the lien law in the state in which he is operating. These are usually available from the secretary of state of the state concerned.

Lien laws are intended to give the subcontractor or laborer an improved claim for collection of debts. The general contractor may also claim a lien, but the complications do not usually arise in his case, since he deals directly with the owner and the owner is aware of the amount due. On the other hand, the owner may pay the contractor, who fails to pay a subcontractor or laborer, and in this way the owner may be obliged to pay more than his contract agreement for the work. The owner may have no knowledge of the names of contractors and material suppliers who have claims against the general contractor. Typically this situation occurs when the prime contractor or a subcontractor is bankrupt.

[56]*McGovern* v. *City of New York,* 202 App. Div. 317, *Empire Foundation Corp.* v. *Town of Greece,* 31 N.Y.S. (2d) 424.

[57]Michael S. Simon, in November 13, 1975, issue of *Engineering News-Record,* p. 69, quoting appeal of Forsberg & Gregory, Inc., ASBCA No. 18069, 75-1, BCA 11,177 (March 19, 1975).

Some typical requirements of lien laws in the various states are as follows:

1. There must be a minimum amount of money due. The minimum is as small as $10 in some states.

2. The owner may be required to record the construction contract with the local land record office, or he may be required merely to state in the same manner that construction has begun. This notice is independent of the building permit, which is issued by another agency. If an owner fails to take this action, he may lose protection from claimants unknown to him.

3. The lien is usually honored in the order it is filed; that is, the prior mortgage is ahead of the lien in being paid off by the foreclosure. However, there are many variations, depending on the state and the manner in which the various parties have given required notices. The lien may date from the time construction was begun, not when the lien was filed; it may apply only against the value of the improvements, not against the value of the land. A direct labor lien often has priority, but there may be more stringent time requirements and lesser amounts for which the labor lien is applicable.

4. Various notices may be required between the claimant, owner, and prime contractor, to protect lien rights. It may be necessary for each subcontractor to give notice to the owner when he signs a contract, to preserve his rights; and a material supplier may be treated differently from a subcontractor. The owner may be required to post his name and address on the job.

5. After a lien is filed, proceedings for enforcement must be commenced in a stated time, which may vary from 60 days to 3 years.

6. A subcontractor or contractor may waive his lien rights by provisions in his contract. A provision in the general contract included in a subcontract or purchase order merely by reference will not obligate subs, suppliers, or laborers. For example: a statment that "general conditions of the general contract apply to this subcontract," is included in a subcontract. The general conditions state that the contractor may not file a lien, but this reference does not prevent the subcontractor from filing a lien.

7. The amount of a lien may be limited to the amount of the general contract; that is, an owner may not be required to pay out more than his contract amount, and he may even reduce this amount by the amount necessary to complete the work of a defaulting contractor. This restriction may be general, or it may apply only if the owner or the subcontractor has taken or failed to take specified action by giving notice. This limitation may make a claim worthless when a general contractor has gone bankrupt, which is the case against which the subcontractor usually seeks protection.

8. The A/E and the land surveyor may have lien rights against the property as do contractors.

4.35   OTHER COLLECTION RIGHTS OF SUBCONTRACTORS AND SUPPLIERS

A subcontractor has a claim for recovery of amounts due from the general contractor or the owner if the money has been set aside for that purpose by a clause in the mortgage commitment or otherwise. In some states, contractors who receive payment for their subcontractors' work hold this money in *trust* for the subcontractors; that is, money paid to the contractor for one job must be held for that job and may not be used to pay subcontractors on a previous job. Such diversion of money is a common practice of insolvent firms. It is a crime in the states requiring the money to be held in trust; and individuals, not companies, are called to account for crimes. In New York State, such a diversion constitutes larceny, which is important to remember if you sign checks, even though you are an employee. On United States government work, payment bonds are required by the Miller Act. When collections under payment bonds are made in federal court, the United States Arbitration Act applies, making the AAA arbitration clause applicable in all states.

In case of bankruptcy, money in trust must be paid to those for whom it is held in trust, rather than shared with creditors. Generally, withholding taxes and social security payments which have been deducted from workmen's wages are held by the contractor in trust for the United States government.

Payments of any kind should state on the check the invoice or job for which payment is being made, so there is no later confusion regarding payment due from bonded jobs. If a general contractor is behind with payments, it is to the subcontractors' and suppliers' interest to insist on payments on an older, unbonded job rather than on a new, bonded job. The creditor can then claim payment from the surety if the contractor goes bankrupt or otherwise fails in his obligations on the bonded job.

4.36 APPEALS IN DISPUTES

Federal contracts require that appeals of the contracting officer's decision must be made within 30 days, and it has been repeatedly held that the merits of a contractor's claim have no bearing if he fails to comply with this requirement. Although the detailed estimate is usually submitted before the contracting officer makes a decision and is therefore available at the time of appeal, there is no requirement that the estimate be made when appeal is made. It has been held by the Chief, Corps of Engineers, U.S. Army, that "intention to appeal" was sufficient, and it has even been held that the contractor established his claim by a request to the contracting officer for additional time to prepare an appeal. This concession should not be relied on, of course, by the contractor, but serves as a ray of hope when the contractor has already neglected to make an appeal. The manager often is given a job in progress or completed, in which the claims for extras have been neglected during progress of the job, and he must do the best he can with it.

4.37 MATERIAL AND WORKMANSHIP GUARANTEES

The AIA contract provides for the contractor to correct faults of material and workmanship for one year after completion of construction, unless there are other specific requirements for portions of the work. There are similar requirements of builders on U.S. government insured housing, and United States government contracts may require such a guarantee. According to the standard clause of U.S. government contracts, the contractor is responsible for latent defects and further clarification is needed in the particular project specifications for a time guarantee. Responsible contractors expect to maintain a guarantee for a year; some of them have service crews for this purpose, accepting as their responsibility any fault which cannot be demonstrated to be due to the occupant. Usually they provide also for nonguarantee service, billing the occupant for work not included in the guarantee, such as broken glass, stopped-up toilet fixtures, or repair of a garbage disposer after a fork has been dropped into it. A contractor is entitled to include a contingency account for this work when completion is reported for income tax purposes; it shows as a cost of the job but if not spent within a year, becomes a profit for the following year.

Maintenance obligations are also included as part of the construction contract, usually for mechanical equipment and roofing, for which the services are available from the subcontractor. The roofing guarantee, which is paid for by the contractor, is in the form of a surety bond written upon the roof, not upon the roofing contractor. This bond does not ordinarily reimburse the owner for damages caused by water leakage to the interior of a building. The roofing guarantee, which is practically a maintenance guarantee, covers materials and workmanship, but it is assumed that in the absence of mechanical damage, the surety will make good on the damage. Roofing guarantees are for 10 to 20 years; whereas air-conditioning guarantees will usually be for 5 years. The roofing contractor is responsible for a lesser period for the cost of repairs due to faulty workmanship—2 years for Johns-Manville and GAF—and is reimbursed for all repairs after that time for the life of the guarantee, or surety bond.

In one case, a contractor furnished a 10-year bond for a roofing guarantee, which specifically stated that the contents of the building were not covered by the bond. Subsequently, leaks developed, and the owner sued the contractor not upon the bond but upon the original contract. The contractor pleaded that the 10-year bond superseded the contract, so that he was responsible only under the bond. The court held, however, that the contractor was responsible for faulty construction and therefore must pay for the interior damage as well as for the roof, under an implied guarantee that the roof would do what would reasonably be expected of a roof—that it would not leak. In the case cited, the time for completion of the roof to the damage was not stated, but the use of the ten-year guarantee in the court's decision implies that the contractor, rather than being relieved by a bond he took out to limit his liability, may have been interpreted as describing the roof as a ten-year roof, thereby extending his own guarantee.[58] Since many bonds do not cover flashings—and never cover flashings installed after the original bond inspection—the contractor may be exposing himself to a considerable liability. This liability is extended even further by the failure of sureties to include flashing which were not listed in the bond. Ordinarily, construction contractors' guarantees extend only one year from completion.

The implication above that a contractor assumes as additional, not a lesser, liability when he furnishes a maintenance bond (manufacturer's bond) is important because it has been held, in some areas, that a contractor is not responsible for defects in materials—that he is only responsible for ordinary care in purchasing and examining them. In Missouri, warping of joists after construction caused extensive damage to a house, and it was decided that the contractor was not responsible.[59] It was even held that when a garage collapsed because of faulty welding, the contractor who repaired the damage could not collect from the firm which manufactured the welded rods—although the contractor might have been responsible had he welded the rods himself! The court's reasoning was that the owner could not collect from the contractor because the contractor had exercised due care in examination of the rods. Although the contractor was not liable, he contributed the cost of the repairs when he proceeded with them; but then he was unable to collect from the firm who did the faulty work.[60] It appears that this firm had better lawyers than welders.

---

[58] *Rowson* v. *Fuller,* 230 S.W. (2d) 355.

[59] *Whaley et al.* v. *Milton Construction Co.,* 241 S.W. (2d) 23 (St. Louis Court of Appeals, Missouri, June 19, 1951).

[60] *Flannery* v. *St. Louis Architectural Iron Co.,* 194 Mo. App. 555, 185 S.W. 760.

In recent years, there is a tendency to *sue the deep pocket* when an accident or failure occurs in a building. That is, the party sued is the one best able to pay. In an accident involving the failure of a temporary elevator, the contractor, the architect, and the manufacturer of the wire rope were sued. Liability of this kind does not terminate a year after a job is completed; there seems to be some doubt that it ever ends. The contractor, therefore, has a variety of liability insurance policies to choose from, according to the time they remain in force; for current operations, for one year after the work is ended, or completed contract insurance, which extends indefinitely.

4.38 QUESTIONS FOR DISCUSSION

1. In the Grinnel case (Grinnel is a fire-sprinkler installation firm), do you agree that it should collect twice for one job, in accordance with the court decision? Should the court first find out how much of the first award Grinnel would get? (Since the firm was bankrupt, the full award was not necessarily paid.)

2. A union contractor overruns his time limit in a contract with a *Time is of the essence* clause. The delays clause of the AIA had been modified in this particular case, deleting as grounds for delay "or by labor disputes" and "any causes beyond the contractor's control." There was an extended labor dispute, but the owner refused to grant an extension. The owner invoked AIA paragraphs 3.4.1, and 8.2.1, canceled the contract, and engaged an open-shop contractor to complete the work. Was the owner's action justified, or was it a breach of contract?

3. In a situation similar to that just described, the AIA paragraph 8.3.1 (delays) was used without change. There was a shortage of labor but not labor dispute. Could the owner suspend the contract?

4. Suppose that you are on a board of arbitration in the third case given here. What questions would you ask, and of whom, to determine how you would confirm the statements made above, pertinent to your decision?

5. You are a contractor who has incurred delay, under the standard AIA contract, because of unusually wet and cold weather conditions. Can you get a time extension on such grounds? How do you justify it? Is your extension limited to the number of days of bad weather?

6. You are employed by the Hilo Piping Company, which has $5,000 due from the Sinkfaster Construction Company for work on the Jetville Baseball Club (all completed, but $2,000 represents Hilo's retained percentage). The same Company owes you a $10,000 progress payment for work on the Golden Rule School. The Jetville payment is 60 days overdue. You receive a $10,000 payment on the school, but Sinkfaster says he cannot pay the older bill until he gets the retention from Jetville. The Jetville contract is not bonded, but the school is bonded. Based on these facts, do you advise your employer to accept the check? What courses of action are open to him? If he refuses the check, can he also refuse to sign the month's release on Golden Rule?

7. The facts are as given in Question 6. Your firm has a line of credit at a local bank. How might you find out, and what would you want to know, to make a better decision above? List information needed and possible sources.

8. The situation is the same, except that the Jetville bill is 4 months old. How does this affect your decision?

9. How does the current contract on the Golden Rule School affect your decision?

CHAPTER 5

PURCHASING AND SUBCONTRACTING

When purchasing materials, the buyer considers the following factors:

1. Price
2. Delivery when needed, and the arrangements for transportation to the job
3. Quality, in compliance with specified minimum standards
4. Ethics, the consideration of how present purchases affect future prices
5. Payment terms

Price is entirely variable; once the decision to use the particular material has been made, it must be purchased regardless of price, and it must be purchased at the minimum at which other requirements can be met. *Meeting the budget* (buying at or lower than the estimate price) is considered at times to have some magical quality, but there is none. A buyer should pursue the *bottom dollar* (lowest price) just as strongly when the price is below the estimate as when the price is above the estimate.

To arrange satisfactory delivery, the buyer needs not only the vendor's statement but also an independent determination, so far as possible, of the vendor's reputation and commitments. If shop drawings are to be submitted, the time allowed for approval by the contractor or the architect should be specified; many vendors use lack of approved shop drawings as an excuse for nondelivery, even though the delay occurs in their own organization. The buyer may specify that fabrication may not be delayed by lack of approval of shop drawings when there is not a specific requirement for the drawings. Because of added transportation costs, the point of delivery must be considered for its effect on delivered price. Usually, materials can be accepted at any point and a corresponding correction made to competitive prices for added freight. Transportation costs, for example, may put an out-of-town vendor's lower price above a local vendor's higher price.

There is a considerable difference between buying materials *FOB jobsite* or *freight allowed jobsite*, both of which mean that the material will be delivered to the job and that the buyer unloads. The cost of the delivered material is the same, but under a freight-allowed quotation, the buyer must pay for the merchandise and claim any damages from the carrier. Also, under a freight-allowed purchase, the seller is not obligated to replace material damaged en route; and when such material is reordered, he is not obligated to replace it at any particular time. Although the carrier is obliged to pay for the replacement of material damaged en route, including the freight cost of the material damaged, he is not obliged to pay the transportation costs for the replacement shipment. The difference may be considerable. The freight costs on the damaged material are based on the large shipment (perhaps a carload or truckload); the replacement quantity is likely to be small and the freight rate which the buyer must pay for this small shipment will be much higher than the freight rate the carrier pays on the damaged material. Vendors who have a monopoly or near-monopoly are those who are the least helpful in replacing shipments, and who have the most strict quotations regarding delivery. At times, steel companies give a quotation which says, in effect, that they will deliver materials when they please at the price they decide to charge at the time of delivery. Their quotation is not so worded, of course, but that is the legal effect of the proposal. It is not unusual for some distributors or manufacturer's agents to make a quotation which is not firm, being subject to approval by their home office, but then they may not notify the buyer on the response of the home office, and he must find out whether the quotation has been accepted.

Quality of material to be purchased is not usually difficult to determine; in many cases this is covered by the architect's specifications. The buyer should be familiar with standard quality specifications, such as ASTM standards, state highway department specifications for aggregates, Commercial Standards, and standards of manufacturing organizations. Federal specifications are useful even for nongovernment work, as they are not uneconomically rigid and are comprehensive. Each vendor will claim superiority for his product, but a higher price is no guarantee of quality. When the actual ingredients are unknown, as is often the case for patented products such as coatings, it is logical first to try the least expensive.

Ethics is less often a problem in material purchasing than it is in subcontracting. If a vendor's prices are passed on from the buyer to another vendor, it is to be expected that the first vendor will not continue to quote or will give higher prices. There is often an obligation to a vendor who has given the lowest quotation before a job is bid, but this is usually not considered except when the product is a custom item requiring special effort on the part of the vendor to provide the quotation. The buyer is more likely to try to impose eithical standards on the seller in the case of material purchases, than vice versa, as many sellers are in a monopoly or trade agreement position. Rarely does a buyer forego an opportunity to buy directly rather than through a distributor. On occasion, however, it is preferable to give a large order to your supplier of small quantities, rather than to another dealer with a lower price; you are paying the small dealer for services on small orders. The large contractor does not have to give repeated orders to get service, as he buys from several vendors and will buy in large quantities. The contractor who at times buys by the truckload and at other times needs small orders may get better service by staying with one supplier. On the other hand, he also cuts himself off from the low-cost high-volume supplier in this manner, and makes bidding and completing large jobs more difficult. A growing contractor may be obliged to change suppliers when his requirements reach larger quantities than his previous supplier is accustomed to selling.

Payment terms for materials purchased are important. If suppliers are identical or nearly so in other respects, the contractor short for cash may extend his contracting ability by negotiating for longer terms. It is much better to arrange long payment terms than to default; the contractor preserves his reputation for prompt payment, and if his credit is good, many suppliers may actually prefer to make the interest on delayed payment. It is much better to make arrangements for credit you don't need than to fail to make payments when due. The writer has arranged for payments delayed as long as six months, with a previously agreed interest rate, without having the material price raised as a result. On occasion, vendors may agree to warehouse items on the job on a *consignment* basis; that is, the items belong to the vendor and the buyer incurs no liability until he draws them from the warehouse. This is particularly desirable on foreign work or when it is necessary to receive materials well in advance of use in order to assure availability. Spare parts, for example, are necessary on isolated jobs, but the date they will be needed is indeterminate. The consignment warehouse may be on the job itself; on large jobs where a variety of small hardware is needed, the cost of interest on unneeded items is less than the labor cost of obtaining needed items from even a short distance. Items shipped on consignment are returned if they are not used; they need not be purchased.

Some vendors will not compete with each other for price, but will offer *free loan of equipment* or other attractions. Oil companies, for example, will furnish free tanks and service pumps, and sometimes even lubrication apparatus for greasing heavy equipment.

Payment made by one business to an employee of another firm to obtain the second firm's business is *commercial bribery,* according to many state laws. They are quite common in countries other than the United States. A more charitable term is *kickback.* Under federal tax laws, such payments are not allowable business expenses (and therefore cost the firm twice as much) if against state or federal law. However, if the state law is not being enforced, the practice may be permissible, as in the case of liquor for entertaining, which was considered permissible in spite of a state statute to the contrary.

5.1 SOURCES OF INFORMATION

To purchase an item at the lowest price, it is axiomatic that you need to contact as many sources as possible. You may find that the salesmen who beat a path to your door represent firms with the greatest number of salesmen, and often the highest sales expense. No matter how many offers are delivered to your desk personally, look as far as possible for inconspicuous vendors, particularly if the purchase is of substantial size. Established firms usually deal with the most economical suppliers, but even here a new project manager may restore relations with a vendor who has received no business from the firm for years and has therefore stopped providing prices.

Since manufactured goods are almost universally sold through distributors or manufacturer's agents, you should contact all distributors, although many firms will sell direct in areas where they are not represented. You will therefore also want to contact manufacturers direct. Larger jobs are quoted directly by manufacturers, although the local distributor presents the quotation and often makes the takeoff from the plans.

Some vendors may be reached through the following sources:

1. Contact all manufacturers shown in *Sweet's File of Manufacturers* and *Thomas' Register of American Manufacturers* (see the Bibliography), preferably furnishing them with a takeoff (quantity survey) of a pending purchase for quotation.

2. Contact manufacturers of rated products of any kind through the association making the ratings. For Underwriters' labeled products, lists of approved manufacturers are

available. For firms with products rated by the manufacturers' associations, as lumber, windows and doors, obtain membership lists of the associations.

3. Contact all local distributors and subs from telephone yellow pages, membership lists of builders' exchanges, local municipalities' lists of licensed contractors, and for specific jobs, from local material suppliers. Most suppliers will be glad to give you the names of contractors who install their products, either as subcontractors or labor subcontractors.

Do not be concerned that many of your inquiries return to the same firms. It will do them no harm to know you are looking for all possible suppliers.

Many suppliers are not interested in making detailed takeoffs from plans in your area, and you may find the best sources are those to which you must furnish details of quantity, sizes, and specifications rather than avoid this expense by buying only from firms who give a complete quotation.

Types of vendors include the following:

1. *Dealers* buy wholesale and sell retail. The designations "wholesale" and "retail" are vague; a contractor is considered a wholesale buyer by some manufacturers and a retail buyer by others. There are three pricing levels for many products; manufacturer, wholesale jobber (who sells to retailers), and retailer. The contractor may buy at any of these three levels, depending on the product and quantity. The wholesale jobber may be a subsidiary of the manufacturer and sell only the manufacturer's products, or may sell competitive products as well.

   A dealer owns his stock and sets his own sale price; usually he sells to the general public as well as to contractors.

2. *Distributors* buy directly from the manufacturer and sell either to contractors or jobbers. A distributor may have an *exclusive area*, in which other distributors of the same manufacturer do not compete, or he may compete with others. Often a distributor will sell to contractors in his area regardless of where the job is located, or to jobs in his area regardless of where the contractor is located. Many distributors—for example, for roofing and acoustic tile—install the products they sell; other subcontractors must buy from them. For this reason, a change in material after the subcontractor is selected may result in higher cost for an equal product.

   Distributors, like dealers, own the products they sell, but they do not usually stock large quantities. The usual dealer order, therefore, requires a factory order. Distributors may also be dealers.

   A distributor cannot prevent you from buying from any distributor you want to, nor can a manufacturer prevent a distributor from selling in any territory the distributor chooses. A distributor in Florida, for example, may sell to a contractor in Louisiana for shipment to a job anywhere. A distributor may not advertise in another's territory, however. A distributor will often try to make you believe you must buy from him. A manufacturer who "sells" territories by means of a franchise fee or other charge also may not fully inform his customer–distributor.

3. *Manufacturers' agents* are similar to distributors, but they do not own or stock the product. They usually represent a number of little-used products, selling them on a

commission basis, and bid their products on a job, by making a takeoff or by quoting only unit prices.

There are also manufacturers' representatives who have no authority to sell to the contractor. Cement and drywall manufacturers have such representatives, whose primary duty is to promote the product and furnish specifications to architects and engineers. These are salaried employees of one manufacturer, who on occasion will indicate distributors who will act as "paper intermediaries," charging a minimum markup for materials they never see and in which they have no investment. In cement, for example, where the price to the distributor is fixed by agreement, 1 or 2 percent is paid by the contractor to the distributor.

4. *Manufacturers* of some products will sell directly at the distributor's price. This situation occurs either when a manufacturer has no local distributor or agent, or when the customer is a very large—and usually continuing—buyer. If one customer has established a relationship, he may buy small quantities at low prices as well. Since prices to distributors also vary for large shipments, the price to the customer for steel joists, roof decking, heating equipment, or hollow metal doors may be less than the usual distributor price.

5.2 ACCEPTANCE OF OFFERS

The manager is often requested to write an acceptance of an offer made by a vendor. Some vendors, particularly large corporations, make a very detailed proposal (offer), and an unqualified acceptance by the construction manager commits him, as it is a contract. Since these terms may void any promise of price or delivery, the manager should send his own purchase order; the statement on a purchase order, *Unless otherwise stated, this order is not acceptance of any proposal,* makes the order a counterproposal, not an acceptance. Acceptance of a price and specified product is not necessarily an acceptance of an offer. Most firms accept the purchase order, realizing that it differs from their proposal. They would like to have their standard terms but may not insist, particularly for large orders.

On the other hand, the manager may receive an offer which is low but lacks detail, and he may be anxious that the vendor does not have the opportunity to withdraw his offer. If the manager sends a standard purchase order, or adds any details lacking in the original proposal, he does not accept the order but may void it. In this case the vendor has made an incomplete proposal; the manager should call the vendor for the missing data and write an acceptance of the offer (his purchase order), adding the details discussed and specifying that such details are as discussed by telephone. His action may void the vendor's offer instead of accepting it as intended, and the vendor is then released from his price by the manager's purchase order, which amounts to a counteroffer. It is bad practice to accept *any* bid which is low by mistake. However, bidders often obtain additional prices shortly after the job is bid, and the low bidder may then want to raise his price. Usually, a bidder will hold his price unless he is released by a counteroffer.

When a general contractor has used an erroneous low bid from a reputable subcontractor, the general is entitled to demand that the sub stand by his price. "Reputable" in this case means a sub who *will* stand by his price and has the financial resources to do so. A general should never be caught between a sub who is low and unable to complete his contract, and an owner who demands performance at the losing price. Some contractors will take the loss themselves, as a matter of prestige; such conduct is unrealistically commendable.

Other common but doubtful practices may give considerable trouble. For example, a vendor gives a proposed price to a contractor; the contractor answers by writing a purchase order with additonal terms. It is immaterial what the additional terms are; the order becomes a counteroffer merely by the addition of any terms. These terms may even be the vendor's ordinary way of doing business, as "FOB jobsite," or "2% 10th; net 30 days" (for payment). The vendor receives the order and proceeds to order or manufacture the materials. The vendor may reasonably claim that his proceeding with the work was an accpetance, even though the contractor was not so informed. The contractor, not having an acceptance of the purchase order, may proceed to purchase the material elsewhere. Each has a valid reason to believe that a contract exists (in the case of the vendor) or does not exist (in the case of the contractor). Careful firms do not allow this to happen; the vendor acknowledges the order, and the contractor requires such acknowledgment. If the material is stock and may be delivered in a few days, or if telephone follow-up is used for all order, this formal exchange is not so important. Nevertheless, the failure of vendor and buyer to agree on time of delivery is an important cause of delays.

Some large companies who quote through local distributors do not allow the local distributor (or manufacturer's agent) to make a binding proposal. You may therefore receive a proposal which states that the *proposal is binding only when verified in the vendor's home office.* The firm wants a voidable contract, to which they may hold the contractor without being bound themselves. Since the contractor has an opportunity to reopen the price for *all* vendors on a bid if the low bidder has this qualification, vendors rarely refuse to honor the agent's proposal, attempting to give the quasi-proposal the status of an actual proposal. The difference between such a "maybe" low bid and the next higher bid is a contingency item in preparation of the project estimate.

A great number, if not all, proposals received by the estimator who is making up a competitive bid will be irregular in some particular. Some firms, particularly structural and miscellaneous iron vendors, will give a very detailed proposal well in advance of bid date but without the price. The price is given later, either by telephone or personally by a salesman. If the salesman is not authorized to sign the bid, can his statement of the price be accepted as valid? Of what value is a signed incomplete proposal? Such proposals would probably be enforced, since it can be shown that this is the customary way for these firms to do business, but the price still must be proven when a telephone quotation was received. The subcontractor is subject to the same errors in preparing his bid as is the general contractor; cases involving subcontractors' proposals rarely show up in the courts, probably because the contractor has so little evidence to prove the existence of a valid contract.

5.3 THE PURCHASE ORDER

Many firms use no purchase orders, placing orders on suppliers' proposal forms or by telephone. There are firms with millions of dollars' worth of construction a year, who work by memory and by routing bills to the person placing the order. One firm with billings of nearly $1 million a month, with a score of jobs under way at one time, uses no formal method of verifying deliveries of materials and sends no purchase order copy to the accountant that material the company has been billed for was actually ordered (and received). No particular method of purchasing is necessary for profitable operation—it is only important that some uniform method, which protects the firm against overpayment, is being followed. Written memoranda of purchase and subcontracts are economical in that they save time for managers, protect the firm if a person dies or resigns, and reduce opportunity for fraud.

Like most managers who entered the construction business through the front door (the office), the writer recommends that a written record should be made of every decision which authorizes the spending of money. This is standard in medium-sized and large construction firms.

The simplest form of purchase order is as shown in Figure 5-1; these orders may be purchased from an office supply firm in tablet form, with a carbon paper insert. The buyer writes his order in longhand, sending the original to the vendor and keeping a copy for the file. With copying equipment, copies can be made as necessary. Stock order forms (forms bought from supplier's stock) are available in a variety of forms and sizes, some with a number of clauses relating to delivery and payment.

Custom order forms. Contractors usually use a purchase order form with a printed letterhead and *logo* (short for "logotype," a symbol used with the company name, usually a design which includes the initials of the firm); consequently, the printed name is expected by suppliers, as it affects handling of credit approval and whether the supplier will accept the order before credit is investigated. Also, contractors prefer copies marked to suit their own office system and incorporating their own terms, both on the face and the back of the form. One such form is shown here from C.H. Leavell and Company of Dallas, Texas, predominantly a building contractor.

Leavell does not title its form "Purchase Order" because the same form is used for contract changes. Usually, purchase orders are numbered consecutively by the printer; accountants feel a little more secure if they can account for all numbers, knowing there are no orders outstanding about which they have no information. The purchase number need not be consecutive for each job, and they may be combined with the cost account number. The purchase order number is needed to identify shipments in the field and to identify invoices to be matched with purchase orders and receiving reports.

As many as three addresses may appear for the firm sending the order; the address at which the order is written, the address to which the shipment is to be made, and the address to which the invoice is to be sent. Leavell leaves the form blank for all three addresses, since the order

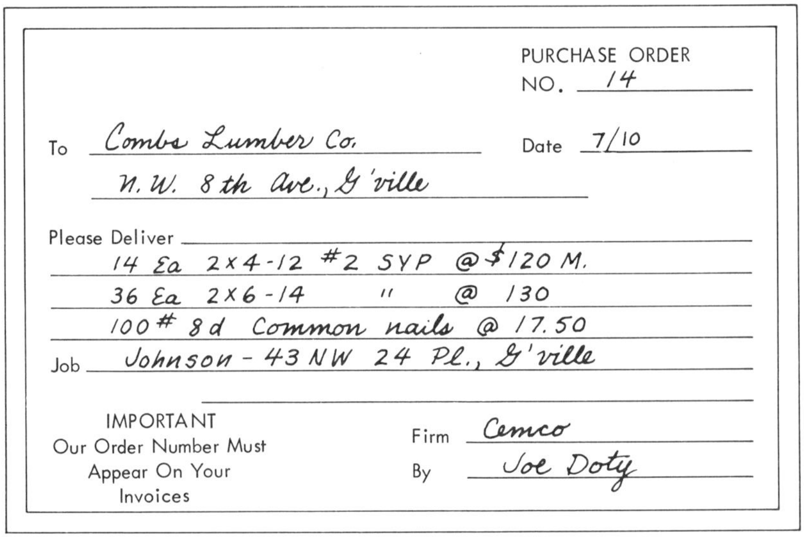

PURCHASE ORDER
NO. 14

To Combs Lumber Co.
N. W. 8th Ave., G'ville
Date 7/10

Please Deliver
14 Ea 2×4-12 #2 SYP @ $120 M.
36 Ea 2×6-14 " @ 130
100# 8d Common nails @ 17.50

Job Johnson - 43 NW 24 Pl., G'ville

IMPORTANT
Our Order Number Must
Appear On Your
Invoices

Firm Cemco
By Joe Doty

Figure 5-1
STOCK PURCHASE ORDER

1. Discount period shall begin from date bill is received or goods are received, whichever is later.

2. Material shall arrive in such time that three men may unload before quitting time, or vendor will be responsible for overtime or demurrage to lay over until the following day.

3. Purchaser assumes responsibility for merchandise delivered when no representative of purchaser is on the job, unless delivery has been requested otherwise, but purchaser must be informed within 24 hours that such delivery has been made.

4. When a delivery date is specified, we may in addition to any other legal remedies refuse to accept delivery at a later date.

5. We are not obliged to pay for shipment of less than all items and quantities shown, unless partial shipment has been ordered.

6. When no delivery date is specified, delivery date is assumed to be no later than three weeks after order is received by vendor.

7. Time is of the essence of this contract.

8. When request of acceptance is made, such acceptance must be returned within 10 days of its receipt by the vendor, or we may consider our offer rejected.

9. This purchase order constitutes a proposal to purchase, and unless reference to a quotation is made, the conditions of the quotation apply only at the option of the buyer.

10. If this order is for equipment rental, operator shall have duplicate time tickets signed daily by our timekeeper. No time not actually operating is to be paid for except as stated herein, and premium time for operator is at vendor's expense unless otherwise agreed.

11. Any disputes arising under this contract or the breach thereof shall be submitted to arbitration in accordance with the Construction Industry Rules of the American Arbitration Association.

Figure 5-2
PURCHASE ORDER (back)
(*Courtesy of King Royer, Inc.*)

GENERAL TERMS & CONDITIONS

1. No charges allowed for boxing, crating or packing, unless so stated on reverse.

2. Goods subject to our inspection on arrival, notwithstanding prior payment to obtain cash discount.

3. Goods rejected on account of inferior quality or workmanship will be returned to you with charge for transportation both ways plus labor reloading, trucking, etc., and are not to be replaced except upon receipt of written instructions from us.

4. The Seller agrees to pay and to accept exclusive liability for the payment of any payroll taxes or contributions for unemployment insurance, or old age pensions or annuities, which are measured by the wages, salaries or other remunerations of his employees, as well as any and all sales or use taxes, or taxes measured by receipts in connection with this Purchase Order.

5. Upon request by the General Contractor (Purchaser), the Seller agrees to furnish waivers or releases from his materialmen or other suppliers for the purchases covered by this order.

6. The Seller shall hold and save the General Contractor (Purchaser), its agents, servants and employees, and successors and assigns, as well as the Owner, harmless from liability of any nature or kind, including costs and expenses for and on account of any patented or unpatented invention, process, article, or appliance manufactured or used in the performance of Seller's obligation under this agreement, including their use by the General Contractor (Purchaser), its successors and assigns, as well as their use by Owner.

7. Insofar as they are not inconsistent with the terms and conditions of this order, the General Conditions and Provisions of the General Contract, for which the material covered hereby is to be supplied, are incorporated herein by reference and made a part hereof as fully as if written herein. The said General Conditions and Provisions of the General Contract are on file at the office of the General Contractor (Purchaser), and available at any and all times.

8. Claims for extras positively will not be allowed unless ordered in writing.

9. Time is the essence of this Contract and if the Seller shall fail to deliver any of the materials at the time or times specified herein, the General Contractor (Purchaser), upon written notice to the Seller mailed to the address noted on the reverse hereof, will have the right to procure said materials elsewhere and the Seller hereby agrees to pay any additional charge, cost or penalty that the General Contractor (Purchaser), may incur thereby. The Seller further agrees to indemnify the General Contractor (Purchaser) for any loss the General Contractor (Purchaser) may incur through delay in the completion of the General Contract where the breach of this contract causes said loss.

10. Delivery of materials will be in accordance with the construction schedule established by the General Contractor (Purchaser). In addition, materials or equipment to be supplied shall be within accepted standards for workmanship and quality, and shall be satisfactory to and meet the approval of the Contracting Officer and/or Architect having jurisdiction and the General Contractor (Purchaser).

11. The Supplier agrees not to assign this purchase order, or any money due or to become due to the Supplier hereunder, without the written consent of the General Contractor (Purchaser) first had and obtained.

12. All questions arising on this Purchase Order shall be decided according to the laws of the State of Texas with reference to which this contract is made.

13. No references to conduct, course of dealing or usage of trade are to be made outside of this purchase order to supplement or vary its terms.

Figure 5-3
PURCHASE ORDER (back)
(*Courtesy of C.H. Leavell & Company.*)

may be made at project office. Leavell calls for *multiple invoices,* for various offices, an indication of centralized payment procedures, and requires *return of copies* signed to indicate acceptance.

The clauses on the back reflect the firm's experience and type of work. The Leavell terms require the seller to pay sales taxes. In many states, materials may not be subject to sales tax if used on public contracts, but the contractor must furnish proper documentation of the use. Consequently, a contractor accustomed to these conditions would provide for the sales tax on the face of each purchase order, while a contractor accustomed to paying the tax would include it as part of the order. Vendors often do not bother to return copies of purchase orders, and the failure is readily overlooked by the buyer. Shipment of an order, in general, signifies acceptance, and many orders are shipped as soon as the acceptance would be returned—sometimes sooner.

5.4 CLAUSES RECOMMENDED
FOR PURCHASE ORDERS

For simple orders, where no extensive fabrication is involved or where the restrictions of a government owner are not in effect, the writer recommends the following clauses in the purchase order:

1. *The discount period shall begin from the date the bill is received or goods are received, whichever is later.* Bills and goods are often sent by separate departments, and the billing people are sometimes more prompt.

2. *The material shall arrive in such time that three men may unload before quitting time, or the vendor will be responsible for overtime or demurrage to lay over until the following day.* Materials delivered "FOB Jobsite" are unloaded by the buyer. Many truckers have their own equipment and cannot drop the trailer, since there is no tractor locally to return it to the point of origin. Lacking such a clause, the trucker can reasonably claim he has fulfilled his duty when he has delivered *FOB* (free of freight charges, on board the carrier); it is not his fault if you are unable to unload it.

3. *When a delivery date is specified, we may, in addition to any other legal remedies, refuse to accept delivery at a later date.* This choice is inherent in the combination of the time clause with a specified date; the purpose is chiefly to impress the vendor with your choice rather than to add a legal alternative. Legal action is usually waived when delivery is accpeted.

4. *We are not obliged to pay for shipment of less than all items and quantities shown, unless partial shipment is ordered.* A common saying in construction is, "The first item shipped is the last one needed." Vendors often ship as is convenient for them; for example, column bars without ties, or partial shipment of brick where the last shipment may or may not match in color. Requiring full delivery before payment gives the shipper an incentive to ship the entire order.

5. *When no delivery date is specified and shipment cannot be made within three weeks, please notify us of the delivery date, or we reserve the right to cancel the order.* This is a convenience clause, to avoid the necessity of writing a delivery date on each order. An order available from local or regional stock may be delivered in three weeks; beyond

that point, the vendor is often unable to give a definite delivery date. This clause does not substitute for a delivery date on items important enough to merit it, but it is intended to prevent vendors from holding purchase orders for long periods of time, shipping materials when they are no longer needed.

6. *Time is of the essence of this contract.* This clause is necessary to make effective any delivery dates or other statements regarding time.

7. *When request of acceptance is made, such acceptance should be returned within ten days of receipt, and no contract exists until we receive an acceptance.* This is a substitution for the return copies of the other purchase orders illustrated. The face of the purchase order is modified for the circumstances when acceptance is necessary; this clause puts a time limit on the return of acceptances.

8. *This purchase order constitutes a proposal to purchase, and unless reference to a quotation or proposal is made, is not an acceptance. When reference to such quotation is made, this is a conditional acceptance and supersedes the proposal where they conflict.* As a practical matter, it is often timesaving to use the vendor's proposal, especially when it is lengthy. This clause strives to avoid the consequences of error when such proposals are accepted without full understanding of the contents. If the vendor's proposal is not acceptable in some way, the purchase order should be specific on the exceptions. If the vendor's proposal *is* being accepted as noted in Chapter 5, this clause may be deleted.

9. *If this order is for equipment rental, operator shall have duplicate time tickets signed daily by our timekeeper. No time not actually operating is to be paid except as stated herein, and no premium payments will be made except as agreed in writing.* Equipment is rented from many firms who have only a few machines, or from contractors who use the machines for their own work most of the time. Their arrangements for billing are quite irregular. Very often they charge machine time for all hours for which the operator is paid, although the operator is given extra time daily for maintenance work. Travel time is customarily paid for by the person renting the machine, but the time should be prearranged, as it is often billed at a lower rate. In any case, the equipment owner should not charge *each* customer twice—for moving in and moving out— if the machine goes directly from one customer to another. Field repair time is often charged, as is coffee time and premium time. On large machines the owner saves enough on the fixed charges to pay the extra half-time for the operator; consequently, the customer may get this overtime at the straight time rate. Since much of the machine cost continues when the machine is idle, the owner's cost on overtime is no more than for the regular eight-hour day, even though the operator gets time and a half. The Leavell purchase order tends to avoid disputes by its statement *No references to conduct, course of dealing or usage of trade are to be made outside of this purchase order to supplement or vary its terms.* The recommended clause does not apply, of course, to purchase orders other than those for equipment rental.

10. *No back orders will be permitted. If any item cannot be delivered, delete it and notify us at once to determine if the balance of the order is acceptable.* This is really a restatement of clauses 4 and 5 in regard to partial orders. Distributors often cannot determine if entire orders can be shipped from the factory; this clause puts them on notice that they must request verification from the plant before partial shipment is made. Partial shipment of some goods is worthless; the stretchers in a tile wall, for example, are useless without the matching base, since color varies not only between manufacturers, but for different shipments from the same manufacturer. *Back order* refers to the

order placed by a distributor to the manufacturer when he finds he has most of an order but lacks certain items. At the plant, some items may be manufactured at different times, and back orders wait until the next production run of the missing items.

11. *Any disputes arising under this contract or the breach thereof shall be submitted to arbitration in accordance with the Construction Industry Rules of the American Arbitration Association.* Material orders are as likely to result in court cases as are subcontracts and are even more difficult to settle, since they are normally not subject to the detailed clauses of the general contract. If the general contract is included by reference, as is done by some generals, the problem remains both of applying the contract and of interpreting it. Many materials bought for construction are not incorporated in the final structure, and even when they are, suppliers often do not know the purpose to which their materials are to be applied.

5.5   PURCHASE ORDER COPIES

The number of copies of purchase orders required grows rapidly as a company installs more complicated control systems; the use of seven to nine copies is not unusual. Copies may be used for some or all of the following:

1. The *accounting department* uses copies to keep a record of outstanding orders, to keep a record of orders against each cost account budgeted item, and to authorize payment of the invoice when it is received.
2. The *cost accountant* uses a copy of the purchase order because it has the cost account number to which he will charge the invoice when it is paid.
3. If a stock of materials is kept as a separate account, rather than charging materials directly to the cost account when they are received, the *warehouse* will want a copy to know what is expected, and when to reorder.
4. The *superintendent* needs to know when materials are coming, from where, and how they are being shipped.
5. A separate *expediting department* will use a copy for follow-up on outstanding orders.
6. On a cost-plus job, the *owner* will receive a copy to show how his money is being spent. Even on lump-sum jobs, some owners—particularly the U.S. government—require copies for expediting and material priority control, and for assurance that the materials have been ordered in accordance with the contract—not only for quality, but for source and labor conditions.
7. The *purchasing agent* usually wants to retain a copy for price and source reference in the future.
8. The *supervisor of the purchasing agent* and the *superintendent* or *project manager* may each want a copy for his file, to see what purchases have been made.
9. A *chronological reading file* is often maintained as a master file where any item can be located by date.

5.6 RECOMMENDATIONS FOR COPIES

Each job and company has its own requirements. This writer has found that the following copies are usually sufficient, but this depends on the locations of the various offices, if the work is cost-plus, and the complexity of the expediting necessary.

1. The *originator* copy serves as the permanent reference both for the purchasing agent and the supervisor.
2. The *expediting* or *pending file* copy serves as the copy to receive the goods and is sent to accounting with the delivery record (and with the invoice if invoices are received on the job). The cost accountant receives this copy with the invoice.

Other copies are made for special requirements. It is assumed that no record of outstanding commitments will be kept, except at the jobsite. Outstanding commitments are subcontracts and purchase orders for which the work is not billed or materials billed. Outstanding commitments records on any but very large jobs are insignificant for purchase orders, since orders are delivered within a month after the purchase order is written. If the job is complicated (as a process plant) and the orders take months to fill, this record is important. If a manager other than the buyer wants to review copy, the same copy may go to the reading file afterward.

5.7 SIMPLIFICATION OF PURCHASING PROCEDURES

Control is used loosely by executives to indicate their approval before execution of purchase orders, contracts, and checks. This is really spot auditing since no overall total or systematic checking is involved. Some managers consider *control* to be approval by as many people as possible before a purchase is made, whether they are informed on the matter or not. This kind of purchase control may supersede control of price, sales, and progress, with disastrous results. Simplification of purchasing procedures carries with it a lack of "control" by some of the people otherwise involved. But to have information readily available for two people, and to have them act on this information (otherwise it is of no value), requires an added cost in clerical and executive time. Two persons cannot do twice as well what one can do alone; and as specialization brings added information to the attention of each executive, with considerable time lost at the executive level to communicate between specialists, the result is not only higher overhead costs but more important, loss of supervisory control because of the contacts necessary between persons. Any system must balance loss of control and resulting errors against expense and speed of reaction of that same control. The federal government (or any government, for that matter) is used as a classical example of bureau ossification, where the number of persons for any task keeps increasing as the efficiency keeps dropping, until the actual work must be accomplished outside of the organization if it is to get done at all; the bureaucracy can then continue to exist without hindering essential operations.

5.8 PURCHASING SUBCONTRACTS

Let us suppose that the estimator–manager has received a proposal from a subcontractor prior to bid time, that the price is low, that the estimator has used it and later wants to sign a subcontract. What is his procedure? The subcontractor may never have seen the general's form of subcontract. The estimator may know his employer will insist on a particular form of subcontract. This subcontract may contain some clauses which the estimator is sure the subcontractor has not expected—an unusually high insurance coverage, for example.

The estimator has several options:

1. He may send the subcontractor a contract, signed by the general, for the sub's signature.
2. He may send the sub an unsigned contract for the sub to sign and return.
3. He may send an acceptance of the proposal in letter form, with a contract enclosed.

The first method is rarely used, as it constitutes a counteroffer by the general, which may not be accepted by the sub. If the general has not signed the contract, the estimator can send it in such a way (option 3) that the contract is an offer made *after* acceptance of the proposal—an important point.

If the proposal is a good one and the general does not want to give the sub an opportunity to withdraw it, acceptance of the sub's proposal must be unqualified. A letter may do this. The acceptance may not be made contingent on the signing of a contract, however, but may request signature of the contract "formalizing our agreement." There is no agreement at this point, no matter what the parties *say,* since there is no "meeting of the minds" on terms unless the proposal was definite on contract terms or unless the parties had previous contracts. If the proposal is not one which is of particular value, the general may be less careful—he may, in negotiating the contract, deny that a contract exists, which gives the sub a chance to withdraw the offer. If the general sends an acceptance, he cannot be said to have refused the proposal; after the acceptance, counteroffers may be made to *modify* the contract.

Because of the importance of preserving the original proposal until a substitute is found, a buyer often first approaches the second or third bidders. If the low bidder knows—and can prove—that others are continuing to discuss the work, he may be able to withdraw his proposal on that account, as it indicates that the general is not accepting his bid. If you are a subcontractor and are asked by the general to reduce your bid, you can reasonably well assume that you are not low, particularly if the general says he has a better price. The general would be very unwise to tell the low bidder he was not low, thereby releasing the low bidder, unless he had another quotation close to the amount of the low bid.

The procedure described above is not considered ethical by many managers. Although the general may avoid trying to undercut the low bid when he is submitting a competitive bid with a public opening (which the AGC code of ethics requires), he is often obliged to bid in accordance with the owner's whims, and to go back to his subs because the owner has not accepted the low bid. It may be noted that the AGC code of ethics requires the same as legal care—that the low bidder should not be informed that he is not low. According to the AGC Code of Ethical Conduct, *"The Contract should preferably be awarded to the lowest bidder if he is qualified to perform the contract, but if the award is made to another bidder, it should be at the amount of the latter's bid."* However, it would be unusual to find a contractor who would pay another sub more than the low bid, if the contractor had a choice.

5.9 CLERICAL PROCEDURE

Purchasing is the first step in a procedure which is intended to furnish a structure of authorization for expenditure of funds in the company. The critical step is the purchase; following steps are concerned with determining that the delivery and payment of the item are properly carried out. The accounting department is most closely concerned with this chain of events, but other persons make the critical decisions; the accountant is responsible for determining that the framework is designed to protect the company against fraud and unnecessary expenses. The functions here outlined may be administered by one or several persons, one or several departments; essentially, it is the operations that are necessary. Cost control, previously outlined for labor cost control, is a part of this procedure. Other cost accounts are equally important but more often will be handled entirely by the accounting department because time is not so critical, authorization of payments may be simultaneous with or closely related to the payments, and detail is not so useful or necessary.

Accountability, one of the purposes of proper clerical procedures, has a number of meanings, including the following:

1. Responsibility for *physical custody of materials.* A superintendent or foreman, or a workman, is given tools for his use. He is *accountable* for returning these tools in good order; sometimes he may be required to pay for lost items, but such rules are usually only an incentive to provide reasonable care for them.

2. Responsibility for *custody of money.* A clerk may be responsible for a petty cash fund, or the accountant may be responsible for all funds, including checking accounts. An accountant's responsibility is more direct than is a supervisor's responsibility for physical goods, chiefly because it is easier to keep track of money; money requires signatures of the receiving and delivering persons every time it changes hands.

3. Responsibility for *authorization of expenditures;* that is, obtaining proper value for money spent. This is the primary job of the manager of the company, and he delegates authority to each person in the company. Misappropriation by dishonesty can be detected, but the more common fault is lack of good judgment. Judgment is relative, and it is difficult to tell if a person has spent money wisely. An important aid is to classify expenditures so that each person can determine if the person subordinate to him is spending money in the way that was estimated, which is the goal of a cost control system.

To assure that expenditures are made by the person authorized, and that he gets what he should from the agreements he makes, each step of the purchase and spend procedure must be verified. The steps are as follows:

1. *Purchase authority* in the form of a signed document.
2. *Receipt of the materials or labor,* verified for the quantity and quality ordered.
3. *Payment,* properly made—that is, paid only after receipt of the materials or work as agreed, in an amount not exceeding the agreed sum.

Figure 5-4, a flow chart for cost control and accounting illustrates a typical procedure. On the basis of the estimate, a job budget is prepared for guidance of the manager. The work is then purchased as (a) materials, by purchase order, (b) installed work, as subcontracts, or (c) direct

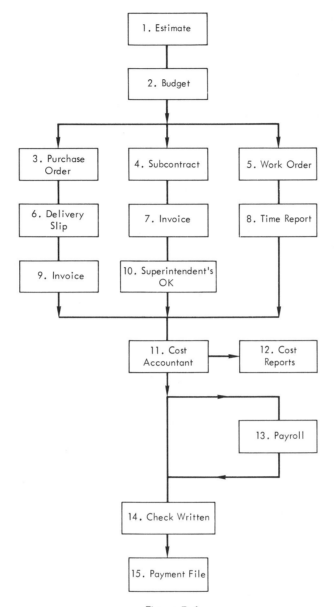

Figure 5-4
FLOW CHART FOR COST CONTROL AND ACCOUNTING

labor, by issuing a work order or cost account number for the guidance of the superintendent. The three vertical paths on the chart represent the three kinds of purchases. After payment is authorized, it is recorded according to the cost accounting system and the cost reports are used to determine current costs of work done. The invoice (or payroll) then passes to the accounting department for payment, profit accounting, and permanent record.

Any step of the preceding may be done informally. Every contractor follows this process, although it may be greatly simplified by avoiding detailed written breakdowns, cost reports, written authorizations and records, leaving this information in the memory of the manager. The advantage of written records and specialized help is that one manager may extend his ability further, and not be required to carry records in his head. The entire system may require only purchase orders, on which appropriate notes are made when material is ordered, delivered, and paid; the purchase order itself can be the cost account number. For small jobs this simplification is economical.

As detailed below, the procedure is suitable for jobs from $500,000 to $10 million, with appropriate differences in the number of cost items. Since the accounting system is basically by cost account number rather than by job, a small job may have only one cost account number.

Cost account numbers may be designated by any convenient system, such as the Uniform Index already described. An integrated whole-number system is favored (for example, showing 1809 rather than 18.09), as it may be used for files, cost accounts, and purchase orders.

The preparation of the *estimate and budget* is discussed in Chapter 6. The budget, a list of updated cost accounts from the estimate, serves as a guide for the numbering of cost accounts.

5.10 COST CONTROL AND ACCOUNTING SYSTEM

The flow of paperwork has been shown in Figure 5-4. After the *estimate* (1) and the *budget* (2) are prepared, the job is bought by *purchase orders* for materials (3), *subcontracts* for work bought in place (4), and by *work orders* or *assignment of cost accounts* to the labor items (5). Any of these authorizations may be verbal, and the authorization to the superintendent may be no more than providing him with a list (although sometimes this is also oral) of the work he is *not* to do—that is, the subcontracted items.

Receipt of material on the job is acknowledged by the signature of the superintendent or the materials clerk on a paper delivered with the materials. This may be a *delivery slip* (6), a list of materials contained in each shipment from the vendor. This list is called a packing slip for crated items and is detailed by crates. Some firms send an *invoice copy* as a delivery slip; it may be a copy without price, an extra copy not intended for billing purposes, or the invoice copy for payment (9). If no delivery slip or other paper comes with the shipment, the clerk prepares a *receiving report,* listing the materials received and the vendor on a report form he has been given for that purpose or on a blank sheet of paper. Some firms use receiving reports whether a delivery slip comes with the order or not.

The *time report* (8) is the equivalent of the receiving report for labor; it is the superintendent's admission that the firm has incurred an obligation to pay. Few firms have a corresponding report for subcontracts; the *subcontractor's invoice* (7) is substantiated by the superintendent's approval, the inspection of the work by the project manager, or a telephone confirmation to the superintendent by the project manager.

Invoices for materials (9) may be sent directly to the job, to the project manager, or to the accountant. Many accountants feel that all incoming invoices should pass through their office when first received by the company, regardless of where they may go for approval. Usually this is because they distrust other offices and believe the invoices may be delayed or lost elsewhere. A creditor calling about payment of an invoice will invariably first call the accounting department, and the accountant feels embarrassed if he does not know anything about the matter. Managers and superintendents have long been accustomed to telling vendors, "Well, I sent your bill to Accounting two weeks ago. I don't know what's holding it up."

Where the invoices go first is not of importance, but prompt handling is. Anything that can happen to an invoice elsewhere can also happen in accounting, with one additional hazard—it can be improperly paid. Once the invoice is received, there are two alternatives. The invoice may be sent to the manager's or superintendent's office, where delivery slips are kept and reconciled with the invoices as they come in, or the reconciliation (matching of shipments and billings) may be done in the accounting office. In the latter case, delivery slips are sent to accounting as they arrive, and accounting can pay invoices as they arrive without further approval by anyone on the job. The person who matches invoices and delivery slips should be familiar with the materials involved, but it is not important what department he is in. If in the home office, he is usually more closely supervised technically; if he is on the job, he can much more readily check for errors or irregularities in deliveries.

To the *cost accountant* (11), all three kinds of expenditures (materials, labor, and subcontracts) arrive for summarizing (posting) in cost accounts and for preparation of *cost reports* (12). Here again there are several options:

1. *The labor cost accounts may be kept by a clerk* before the cost accountant receives the time reports, so the cost accountant need only record the total payroll cost. Material and subcontractor charges may be entered either from the invoices or from the check stubs.

2. *The cost accountant may also charge labor cost accounts.* These must necessarily be charged in much greater detail than by checks, but if the number of cost accounts is not too great, labor charges may be listed on the check stub for the payroll. The use of the check stub as a payment voucher, or the same machine entry for checks as for cost accounting entries, saves time and labor for the accounting department but results in delay in cost reporting.

3. *Material and subcontract cost accounting entries may be made from approved invoices* or from the stubs of checks used to pay the invoices. The difference is substantial to the project manager but is inconsequential from the accounting standpoint. Many invoices are approved for payment and accumulated until the end of the month or week, to avoid writing many small checks and to reduce cash requirements. If the cost entry is made from checks, there will be a delay in the cost report. If cost entries are to be made from each invoice, machine accounting may be impractical for cost accounting purposes; consequently, there is a conflict of interest between the accounting department and those persons who need cost account data on which to base their actions. This conflict is aggravated because by the time accounting calls upon the project manager with questions on costs for the previous month, he no longer has records or recall of data to answer adequately.

A project manager who is expected to explain cost overruns should supervise the cost accounting. He should also be required to approve invoices for his job; otherwise, it is much easier for the accounting department to receive a matched purchase order, delivery slip, and invoice for payment—all fraudulent, and none of which have been seen by the project manager. Central control of purchase orders and signature by a purchasing agent does not assure protection; the purchase orders are not difficult to procure, either by sending a forged requisition to the purchasing department, stealing a purchase order blank from the purchasing department and forging an order (not difficult if the third or fourth carbon copy is being received by accounting anyway), or ordering oneself a supply of purchase orders from another printer. Printers do not require confirmation of a manager to print purchase orders, and they prefer that they be picked up rather than delivered to the customer. Any of these forgeries can be accomplished without access either to incoming mail in the office or to the office where invoices are approved.

Requisitions are requests for materials made by one person to another in the same organization. When a separate purchasing department is maintained, requisitions are made by the project manager to the purchasing agent. If the firm maintains a warehouse, materials are issued from the warehouse by requisition. A requisition to a purchasing department contains the same information as the purchase order, but it will usually not have the name of vendor and price. If the vendor and the price are included, it is for all practical purposes a rough copy of the purchase order.

5.11   BILLING METHODS

Many accountants will pay only on an *original invoice,* to avoid the possibility of payment both on the original and on a copy. This delays payment when the vendor is accustomed to sending a copy of his invoice with the materials and to keeping the original indefinitely for his own records. If invoices are properly checked for delivery, the copy should not pass through after the original has been paid; but there is always a possibility that it will.

A *statement* is a summary sent to each buyer at the end of each month, listing purchases and payments for the month. As machine accounting has become more common, less information is given on the statement; usually the date, invoice number, and amount are shown on machine-prepared statements. The buyer must then ask for copies of any invoices that are missing.

There are two common methods of billing:

1. Delivery of an invoice, usually the original, with materials, and a statement with copies of all invoice–delivery slips at the end of the month. This is done by smaller dealers, major oil companies, and many credit-card companies.
2. A delivery slip with the material, an invoice mailed later, and a statement listing only invoice numbers and amounts at the end of the month; as an alternative, the invoice may go with the materials instead of a delivery slip, with a statement of the same type at the end of the month.

The statement should go directly to the accountant, should be marked "statement" if there is any doubt, and should not be approved for payment as an invoice. When a firm sends invoices with goods, there may be no way to distinguish a monthly billing from a monthly statement. These invoices are often marked *This is your invoice,* and if no statement is to be sent, *You will receive no further invoice.* The modern trend is to send no monthly statement. The invoices are approved for payment, either as they come in or at the end of the month. In many cases, statements are in error; they may not include credit for payments, returned materials, or materials not actually delivered. Some firms use a central machine-billing system in another city; invoices and statements must be checked carefully, as corrections are often delayed for weeks. Usually, the invoice is sent to the central office for billing before the material has been delivered so errors caught at the time the material is received do not get to the central billing office. The use of central machine accounting by concerns with offices in several locations has slowed down corrections and has generally made identification of invoices more difficult.

5.12 DELIVERY SLIPS

Delivery slips or receiving reports are kept on file at the job or at the office, depending on where the invoices are received. These may be kept alphabetically in an accordion file, and serve as an *accounts payable* file; the clerk refers to this to determine in advance what the amount of bills at the end of the month will be. Delivery slips are not payable and have no amount. They do represent the materials which will be due for payment no later than the tenth of the following month. This is important in order to make arrangements to have money to pay them, or to make arrangements for delay. The accountant has the bills actually received and unpaid by him, but he has no record of materials received on the job which have not yet been invoiced (unless *he* is maintaining this accounts payable file).

It is important that the accountant receive all statements. If a superintendent or a clerk is holding invoices because of discrepancies in receipts, or just because he doesn't know what to do with them, the costs they represent remain unknown. Failure to transmit invoices causes a complete collapse of cost control, and the result can be that the company goes bankrupt. Some vendors are very slow to follow up on unpaid invoices, and invoices may accumulate for months.

When the clerk receives material, he pulls the job copy of the purchase order and checks against the delivery slip or the invoice received with each delivery to determine the following:

1. Whether the purchase is authorized, according to the purchase order.
2. Whether the price is correct, as indicated on the purchase order, or as authorized by the superintendent if the purchase is within his authority.
3. Whether the materials delivered are of proper quantity and quality, determined by physical inspection and the delivery slip.
4. Whether the materials are properly stored and the person needing them is notified of their arrival. The latter point is particularly important on a large job with a central warehouse.

Obtaining ready-mix concrete delivery on time is one of the general contractor's chief delivery problems, as he must have men waiting before the trucks are loaded. Also, an hour's delay not only causes an hour's overtime to complete the pour, but lowering temperatures, darkness, and weather changes cause further losses. Concrete firms routinely charge for truck time if unloading is delayed, but refuse to pay for contractor's added cost if the truck is delayed. Truck delays are usually accepted as the contractor's losses, but good and detailed records of time of order and of deliveries reduce late deliveries, and if the vendor makes contract commitments, can be used to justify claims on the vendor.

S. J. Groves, in a 1978 case in Federal court, was able to collect $55,000 damages for damages due to late concrete delivery. The principle in contention was the extent to which Groves was required to mitigate (reduce) his losses. Groves had continued to use the same vendor, rather than try other rather doubtful sources; the appeals court held that Groves' action could not be questioned in hindsight, in that it was not the best method to reduce damages.[1]

[1]U.S. 3rd Circuit Court of Appeals 77-1802, argued February 23, 1978.

5.13 SUPERINTENDENT'S PURCHASES

The superintendent needs authority to make small purchases on his own, for which he usually has petty cash, a reimbursable amount of money on the job. The petty cash account represents both an authorization and a method of payment. He may also have to use cash to pay for items for which a home office purchase order has been issued; and he should have a written authorization as to what type and size of purchases he may make for payment at the office. The superintendent may be issued a purchase order describing the general type and maximum amount of his authority, rather than a letter; he then forwards the bills or expense account statements marked as a charge against that purchase order. For the superintendent's use in ordering, some firms give him a *field purchase order* form, usually smaller than the conventional order, to be handwritten by the superintendent. A copy is sent to the office and invoices are paid as on regular purchase orders. These field purchase orders may carry a maximum amount on the face.

5.14 INVOICE APPROVAL

Approval of invoices, which the receiving and purchasing procedures are intended to safeguard, is the most important action in a firm. Invoices should be approved by the project manager or by a person with a full knowledge of the situation on the job and of the kinds of materials purchased. The accounting department also examines invoices before payment, but fake or duplicate invoices are much more difficult for them to detect. If a clerk in accounting checks the invoices against receiving reports and purchase orders, he often also writes the check and sends it to the officer of the company for signing; the company officer signs these checks by the dozen and has no time to check them. In most cases (certainly in any case where an intelligent clerk has made even the least credible fake papers), the company officer cannot recognize a fraudulent payment.

For invoice approval, a stamp similar to that shown in Figure 5-5 is recommended. Sometimes signatures are used without any explanation; this is not recommended because of the difficulty of determining just who signed the invoice and why. The project manager or the superintendent, for example, may have signed an invoice as a delivery slip, meaning that the material was received, and the signature may be accepted by someone else as meaning approval for payment. This situation arises when invoices are not stamped until received by the accountant for payment. Previous signatures, such as those acknowledging receipt, approving price, or giving the manager's approval can then become confused. The stamp provides for approvals as explained on page 189.

```
P. O. ____432_____
Cost ____L.R._____
Chckd ___L.R._____
OK to Pay $1,500.   KR
Check No. ___456_____
```

Figure 5-5
APPROVAL STAMP FOR INVOICES

P.O. The purchase order number, for verifying proper authorization. Often purchases are made without a purchase order; this is noted by the clerk so the approving authority will understand that the authorization and the price have not been checked previously. He will therefore examine it more closely. Items without a purchase order are usually utility bills, taxes, permits, and items which are paid for in cash at the time of purchase. Some firms write purchase orders for such items after the invoice is received. In such cases the original (vendor) copy of the purchase order should be destroyed, and the copies marked "For internal use only." As an alternative, all copies of the order are marked "Confirming telephone order." Otherwise, a purchase order may find its way to the vendor, and a second shipment or billing may be made. The purchase order copies are sometimes useful, however, for the persons who receive information copies of the order. Late purchase orders are also made to conceal purchases not made as prescribed by the manager and indicate that the subordinates are trying to hold up a weak manager.

Cost. A clerk indicates to what account the payment is to be charged. Often the cost account may be indicated on the purchase order; it is then copied from the purchase order. The account to which the material will be charged may be unkown when the purchase order is made, usually because the material will be charged to different orders when it is delivered, or there may be changes from the material arrives. At the time the invoice is approved, it is good to confirm the cost account number from the delivery slip.

In any case, the cost number should be placed on the invoice *before* the manager's approval, so he may correct it if necessary. This usually requires that the cost number be assigned by the field clerk. The often-used practice of allowing the accounting clerk to number cost account charges leads to mischarges and confusion, as the accounting clerk does not know the materials or the status of the job. Materials charged to job overhead or to the small-tools account merit special attention, as the manager's check of stolen goods is assisted by proper knowledge of expenditures in these areas.

Chckd. This indicates that an overall check has been made of receiving reports against the purchase order and the invoice to verify that quantities and charges are correct. If no purchase order was issued, the receiving report is checked against the invoice; in the case of temporary utility installations, it may consist of calling the superintendent to verify that certain work has been performed, since the superintendent ordinarily does not submit a receiving report. The checking may be done by a clerk under the manager, or it may be done in the accounting department. In any case, it should be clearly understood who is to check the extensions on invoices; sometimes both the job clerk and the accounting department check extensions, unknown to each other; sometimes neither does it, assuming it is the other's duty.

OK to Pay. This is the approval of the manager, who has authority to write purchase orders and receive materials. Many firms delegate the paying function to a nonmanager, in order to reduce the requirements on the manager's time. A project manager can readily approve bills personally for projects under $1 million per month, although the number of bills depends on the manner in which the job is bought, rather than on the money value. It is preferred to allow subordinates greater authority in the purchasing procedure and to exercise direct control by approval of invoices. This does not, of course, prevent some unwise purchases from being made, but it does assist in developing managers and supervisors with greater personal attachment to their work. Some firms delegate purchasing to a subordinate, who buys in accordance with requests (*requisitions*) submitted to him by the supervisors. This procedure has the disadvantages of the supervisors' buying, since the purchasing agent has no alternative but to buy as directed, and it also has the disadvantage that supervisors will not be careful in their purchases since they do not feel it to be their responsibility to get the least expensive item that will do the job. This is one of the reasons that large organizations, who employ people specialized in the mechanical aspects of their profession, become less efficient with time.

If the amount to pay is shown as well as the approval, the chance of error in payment is reduced, and the manager can approve invoices for a lower, corrected amount if an error is found in the billing. *Only the last page should be approved when an invoice of several pages is summarized on the last page.* A check may erroneously be written for each page of a long invoice, where each page total is the sum of previous pages. Sometimes a billing may have separate amounts for materials and for installation, with no total. Some amounts on invoices do not represent the actual amount due; the manager, by marking the amount clearly, is assisting the accounting clerk, who may not understand a bill (especially when written in longhand) as easily as does the manager.

It is fundamental that the manager, not the accounting department, establish the cost account to which an item is to be charged. If there is an obvious error, the accounting clerk should notify the manager when the charge is made. For effective cost control, costs authorized by each person should be charged to an account by which that person's ability can be measured; he should therefore be the authority on where the item is budgeted. This is not at all a simple matter, and very often cost accounts are rendered useless by mischarges. Of course, no manager should charge expenditures to the account of any other manager, which can readily happen unless the procedure prevents it.

A check offered in full payment of a debt, and if the check is noted as payment in full, usually discharges a debt in excess of the amount of the check when the check is cashed. If you receive a check of this type, and do not want to accept it, you should return it. If you keep it, you may be held to have accepted it; holding a check for an unreasonable length of time becomes an implied acceptance.[2] In some states, notifying the debtor that the check is unacceptable denies this implication; in others it does not.

5.15 SUBCONTRACT PAYMENTS

If the supplier also installs his materials, a *subcontract* is prepared. Usually, an officer of the corporation is required by the firm's regulations or by-laws to sign contracts, but practically, the authority of the purchasing agent or any project manager—even the superintendent—is seldom questioned by the subcontractor. A copy of the subcontract is required for each party, but there is often a third copy made for the superintendent. The amount may be deleted on the superintendent's copy.

Figure 5-6 is a stamp of approval of subcontractors' invoices. Subcontractors use a wide variety of forms of invoices. Ideally, they should show the total contract, a statement of items completed and their value, previous payments, and the amount due. Some contractors furnish an invoice form, with these data to be filled out by the sub, for the information shown in Figure 5-6. The *balance due* on the stamp is not the amount to be paid, but the amount not yet paid

---

[2] 156 S.E. 2d 467; *The Businessman and the Law,* 1977 undated, *Man and Manager,* 799 Broadway, New York, N.Y. 10003.

```
COST NO. ___1804___
FILE NO. ___1806___
TOTAL CONTRACT  273,000 -
PREV. PAYMENTS   62,500 -
BALANCE DUE     210,500 -
OK TO PAY        50,000
                    KR
```

Figure 5-6
STAMP FOR SUBCONTRACTOR'S PAYMENT INVOICE

on the contract. Since the contractor wants to assure that a large enough amount remains to complete the job, the amount still unpaid is a key figure in deciding whether to accept the sub's billing. The amount after *OK TO PAY* clarifies, for the accounting department, just how much is to paid on the current invoice; invoices are sometimes confusing in this respect. The file number is a convenient reference for the manager and the superintendent, as a continuing file must be kept for subcontractors to determine the amount due; this is not necessary for material suppliers. The *previous payments* statement is necessary on each approval as bills are sometimes confused or in error, and when an invoice is submitted, the sub may not have included in the amount due a previous billing which has not yet been paid.

Many firms demand invoices in duplicate, so the manager may keep a copy, although the duplication itself can lead to duplicate payments. For subcontractors, however, a duplicate of invoices in the manager's file is needed so that the manager may check and correct future invoices. Subcontractors' accounts often get confused because one subcontractor may have more than one subcontract, may have extras to the subcontracts, and may be selling some materials to the contractor as a vendor. The project manager needs an invoice copy with all notes and markings, preferably a photocopy of the invoice as it *leaves* his office.

For small subcontracts, a purchase order is often sufficient and should have the requirement that the subcontractor carry proper insurance. Generally, a contract form is preferable if the subcontractor cannot be paid for his work all at one time. Many subs will work on a job less than a week, and some kinds of work may take less than a day; blown insulation, for example, is practically a delivery process except that the unloading is done with special equipment. Verbal subcontracts are satisfactory only for the smallest work, and then only with the firms who are accustomed to working with each other. Some firms have such a relationship that either will automatically comply with anything the other considers fair, but even here it is best that such goodwill not be diluted by ambiguous contracts—as are verbal agreements.

5.16 SUBCONTRACTS

Subcontracting is a much more complicated form of purchasing than is buying materials, and the purchasing agent, if his duties are separated from the project manager's, must rely on the estimator or the project manager for detailed requirements, obligations assumed by the estimator, and the subcontractor's performance and ability. Obtaining subcontract prices has already been discussed. Usually, after the subcontract is prepared by the general contractor and mailed to the subcontractor, the sub signs and returns two copies or brings them to the contractor's office for discussion, changes, and mutual signing.

Firms with centralized management controls often require a contract to be approved by the corporate officers. If both the contractor and the subcontractor have such requirements and the various officers feel it their duty to comment on the contract, weeks may pass while inconsequential details are being cleared up. It is preferable that the contract be negotiated and signed by the same person, to reduce delay and the use of the manager's time.

The importance of specific description of the subcontract work, rather than incorporating a sub's bid into the contract, was illustrated in a 1979 case in Rochester, New York, between Flower City Painting and Gumina Construction Company, where the contract was held not to exist and the sub could not be held to knowing the custom of the trade because it was the first project Flower had undertaken.

In this instance, Flower quoted painting various apartment units for $98,500. Flower's bid did not specify the actual work to be performed, and was incorporated into a subcontract, along with the prime contract and the plans and specifications.

Flower then wrote that it had contracted only the interior of the units, not the exterior, and claimed extra payment. Gumina canceled the subcontract and Flowers sued. The U.S. District Court held that Flower breached its contract by submitting an invoice for extra work, and could not recover. On appeal, the decision was confirmed, but for different reasons; the appeals court found that the two parties never actually had a meeting of the minds with respect to the functions covered by the subcontract, since the contract was ambiguous on this point.

The finding that the custom of the trade does not apply where a sub is ignorant of the custom of his own trade comes very close to a determination that lack of understanding is a defense against a written contract, and enforces the necessity that a contractor must be certain the subcontract is clear to the sub.

5.17 SUBCONTRACT FORMS

The subcontract is prepared on a form used by the general contractor, modified as necessary for the particular project. One contactor's form and two standard forms are included in Appendix D:

1. AIA Document A401, the Contractor-Subcontractor Agreement published by the American Institute of Architects.
2. The subcontract form of the AGC, which was prepared jointly with a number of subcontractor organizations.
3. A contract form used by Lenox Contracting Co., illustrating special adaptations for the particular state and project. This form is technically a contract form for specialty contracts, where the contractor is also the owner and the architect does not supervise the work.

At least two sets of drawings and specifications should be included in the subcontract documents (the AIA contract requires three for the general contract). Each of the parties initials one set and delivers it to the other. This ceremony is considered unnecessary by many contractors, who consider that a reference to the plans and specifications of a certain date is sufficient. In most cases this is true, but some errors may occur, as explained below.

Missing pages or sheets. Although most specifications list the number of pages, many do not, and a sheet can be missing from the set the subcontractor figured. Borings are often not part of the bid documents but may be desirable to include in a subcontract.

Changes in drawings or specifications. Good practice requires that a revision note be made on each tracing on which a change is made, but such notes are sometimes neglected. The original cannot then be identified when compared with a revised drawing, and a substitution may be to the advantage of one party. Since the contractor provided the drawings, the burden of proof is on him, but how can he prove which is an earlier drawing? As the subcontractor usually returns plans to the general after he makes an estimate, many estimators initial the prints they use and insist that these become contract drawings or be returned for comparison with contract drawings. With the increasing practice of delivering tracings or sepias to a field office, a number of people have an opportunity to change drawings. *Sepia* originally referred to the color in which transparent prints are made, and then to the prints themselves. It now is loosely used to indicate any type of transparent print. *Never* use marked prints for contract drawings, as the marks cannot be dated. Unsuccessful bidders return their prints to the A/E after bids are read, so the general should request these prints; some of his subs will have figured the job from them.

Few contractors use either of the standard subcontract forms with modifications. Most contractors have their own form, based on their own experience. Such forms change from time to time, as do the standard forms. As a firm grows larger, a change in the form becomes a matter to be approved by several officers, who are not familiar with daily problems, and the contract form becomes rigid. In one instance, a hold-harmless clause requiring insurance which was difficult or impossible to obtain remained in a contract form for years although project managers made no attempt to enforce it.

5.18 AIA AND AGC SUBCONTRACT CLAUSES

The time of completion clause includes *Time is of the essence* and leaves the description of time limits to the writer. These time limits are difficult to schedule, and more difficult to enforce; specifying the number of workment to be furnished is one (seldom-used) alternative.

Articles 11.1 and X(1) set forth that the Subcontractor assumes the Contractor's relationship to the Owner (AIA) or the Contractor's responsibility to the Owner so far as it falls with the subcontract (AGC). The AGC adds a nondiscrimination clause, probably because the contractors had to comply earlier with nondiscrimination requirements, since they were included in contracts for government financed construction. Actually, some contractors have used such a clause for 20 years. The AGC clause on the Subcontractor's responsibility is superior to the AIA clause, as it admits of less possibility of conflict.

Articles 11.8 and X(2) give the Subcontractor responsibility for monthly billings. It is customary for the subcontractors and the suppliers to bill the general before the end of the month for the month's work, and based on their bills, the general bills the owner. The owner should pay before the tenth of the following month, whereupon the general pays the subs, and the subs their suppliers. It is not unusual for the contractor and his subs to have their checks already written before the owner's check is received. The checks may then be mailed on the same day on which the deposit is made in the banks to cover them. Well-financed contractors can pay their bills without receiving the owner's check, but even in their case, they will normally make such early payments on only a small proportion of their jobs. Payment by the tenth often yields a 2 percent discount on materials and is the critical point by which a firm is rated for credit as "slow pay." Failure of the general to pay the sub, however, is not considered important if the owner's check is late. Payment to the general is computed on a percentage completed on major items on which a breakdown has been furnished to the A/E.

Articles 11.2 and X(3) require a *receipt from the sub's suppliers* (which is a release of future lien to the owner) that the suppliers have been paid. Before requesting another payment, the subcontractor must pay the *previous* month's bills from payment received from the contractor for that purpose; and the sub must furnish certificates from creditors to that effect. If the contractor is not solvent and is unable to make these payments (having used the money to make payments on earlier obligations), the creditors may be induced to sign a release, as their opportunities of collecting their balances are improved by doing so.

"Insolvency" here indicates "without cash to pay current bills." Many firms manage to become solvent after having been insolvent, without becoming bankrupt. Many others merely terminate operations without a legal proceeding, the assets being informally distributed among creditors, or the assets being too small to justify a proceeding.

Occasionally, the contractor and the subcontractor are required to pay outstanding bills *before* payment is made by the owner. In this case, it is not uncommon for the supplier to give a "courtesy" release before receiving payment, as shown in more detail in the cash requirements discussion in Chapter 6.

Articles 12.6 and X(10) are identical in that the contractor may cancel the subcontract if the sub's progress is unsatisfactory. Even under the best of conditions, proof is difficult. In a 1974 Michigan case, a paving subcontractor failed to make proper progress and was discharged. A lower court found that the subcontractor was properly discharged, but ordered the contractor to pay the sub's expenses nevertheless. On appeal, the general succeeded in reducing the sub's recovery to the contract amount less the cost of completion, stating that the *object of the measure of damages is to put the injured party in as good a position as he would have been in if the performance had been rendered as promised.* But the contractor had to appeal, as well as go through a lower court hearing, to establish this presumably obvious principle.[3]

Articles 11.4 and X(15) require the subs to *participate* in the preparation of *coordinated* drawings, chiefly where mechanical work is crowded together. The clauses are almost identical. Physical conflict of piping, ductwork and electrical work is common on complex work, as the A/E avoids the expense of detailed drawings and usually lacks the experience for determining the most economical methods; mechanical contractors differ on the best methods. It is customary for these subcontractors to make their own details, and shop drawings they prepare separately may have drains or sprinkler lines passing through each other or through ductwork. In some cases interferences cannot be avoided in the available space, and the contract drawings must be changed. An extra is usually authorized only if there is no possible solution within the scope of the contract drawings; that is, longer lines, more fittings, or larger spaces must be provided. The owner is not responsible for extra work caused if one contractor must install work which would not have been necessary if details had been worked out in advance.

Article 11, the hold-harmless clause of the AIA subcontract, makes the subcontractor responsible for damages caused jointly by himself and the contractor and is the same as a

[3]*Maraldo Asphalt Paving, Inc.* v. *Harry D. Osgood Co., Inc.,* 220 N.W. 2d 50 (Div.2-1974) quoted in *Engineering News-Record,* October 16, 1975, p. 53.

similar clause in the general AIA contract. Presumably, if an architect's error is so obvious that either the contractor or the subcontractor should have recognized it, the liability finally rests with the subcontractor! The AGC and the other contractor and subcontractor organizations have consistently opposed such a clause, which makes one party liable for another's legal obligations toward third parties, and it does not appear at all in the AGC subcontract.

The 1978 edition of the AIA standard subcontract form A401 was approved by the AIA and the American Subcontractors' Association, but it was rejected by the AGC and by the Council of Mechanical Specialty Subcontracting Industries (CMSCI). The CMSCI objected to the hold-harmless clause, which held subs responsible for all claims resulting from their work, as the clause is illegal in many states and creates legal complications. The AGC objected to the provision that the general must pay the sub even if the architect hasn't approved payments to the general, and required interest payments on overdue progress payments to the subcontractor from the general contractor. "They take the architect off the hook and leave us holding the bag," said Campbell Reid, director or the AGC's building contractors division.[4]

Articles 12.1 and X(19) require the contractor to represent the subcontractor toward the Owner. The AGC subcontract is nearly the same, with emphasis on the obligation between contractor and subcontractor rather than on the contract documents which are written for the general contractor-owner relationship. This clause gives the subcontractor a right which the general contractor will normally give him in the general's self-interest. The contractor is obligated to make almost any claim on the owner that the subcontractor makes, as the contractor may otherwise become liable himself. However, the A/E is influenced by the way the contractor handles these claims, and if the contractor makes no investigation and does no more than forward the claim (as is usual), the A/E will consider the claim less important.

Articles 12.4 and X(21) permit the subcontractor to obtain information directly from the A/E regarding payments for the subcontractor's work. The sub can usually get this information anyway, but without this clause the contractor could claim that confidential information was being requested. The contractor often fails to pay the sub his amount due when the amount has been received from the owner, so the sub must find out from the A/E if the payment has actually been made to the contractor. It is, of course, to the owner's advantage that such money is properly paid, in order to avoid lien claims by the sub. As noted elsewhere, in some states retention of such payment on the part of the contractor, and diversion to other payments, is a crime on the part of the person making the payments. A later clause in the contracts permits the sub to stop work if such payment is not received within seven days.

Articles 12.3 and X(24) are identical and specify that the *Contractor shall give no instruction to other than the Subcontractor's superintendent.* This is to avoid work being ordered of which the subcontractor is not aware. In case of emergency, this procedure need not be followed, and in some cases—particularly in unsafe work—the contractor may assume a legal liability by failure to warn the men directly. This rule, therefore, like any other provision of the subcontract, cannot void the contractor's legal obligation either by statute or other contract. If a sub's workman, for example, is causing harm to other work by his actions, the contractor's superintendent will take prompt action.

Articles 12.5 and X(25) prevent the Contractor from holding the Subcontractor liable for liquidated damages not specified in the contract. It applies only if liquidated damages *are* named in the subcontract, but it could be interpreted to exempt the subcontractor from damages due to delay. Since the delay and date clause is not standard or printed in the form, the person preparing the subcontract must carefully review this clause with the special clause being inserted to assure that the two are consistent.

---

[4]As quoted in "Specialty Contractors Oppose AIA Subcontract," *Engineering News-Record,* Vol. 188, No. 23, June 8, 1972, p. 53.

Articles 12.5 and X(26) are identical *backcharge* clauses, which require that such charges be presented to the sub within the first ten days of the following month; that is, with the next billing.

A *backcharge* is a charge for work done by the contractor for the subcontractor and therefore is deducted from subcontractor's contract. Such charges may be for the loan of workmen when the sub has little or no need for regular men on his own payroll. The sub may ask for tool boxes, shanties, temporary wiring, materials which the general has in abundance, or may ask the general to do work in the sub's contract but not in the sub's trade, such as concrete bases for equipment. Many of these charges are disputed; the general may charge for cleaning, shared watchman service, or other shared services. Many subs insist on clauses in their subcontracts which prohibit their being charged any prorated charges without prior approval by the sub.

The sub's superintendent on the job is usually authorized to order items to be backcharged, and in well-organized firms the superintendent makes such request in writing with a copy to the sub's office. In many cases, a superintendent fails to write such requests and the sub receives the bills, toward the end of the job, which he could have avoided had he known about them. This subcontract clause assures the sub that such bills will be presented promptly, while the circumstances are still fresh in the memory of the people concerned.

Articles 13 and X(27) give the Subcontractor representation in arbitration proceedings involving his rights. This is implicit in an earlier clause, but it is stated specifically for clarity and to assure that the arbitrator admits the sub as an interested party. An arbitration proceeding between contractor and owner may be inadequate to cover the subcontract matters; when possible, the arbitration should also name the subcontractor as a party and make a separate finding regarding the subcontract. Otherwise, a later separate proceeding may differ on the claim, leaving the contractor either paying or receiving money when he is not directly involved. Both subcontract forms, in differing later clauses, attempt to clarify this situation.

Articles 13 and X(29) establish arbitration as the method of adjudicating claims between the Contractor and the Subcontractor. This article does not, however, allow for arbitration except where the general contract provides, and the quasi-judicial board of review hearing U.S. government contract disputes will not give an opinion between the general and the sub.

Clauses in the AGC contract not provided in the AIA contract include a provision that the subcontract is solely for the benefit of the signatories. This is to avoid the implications of the hold-harmless clause, which allows claims by employees or the public against a party of the contract. Since injured employees can collect only a limited amount under workmen's compensation laws, they seek to collect larger amounts from other parties, and the contract or subcontract is often used as a basis for such suits.

The AGC subcontract provides clauses relating to temporary site facilities furnished by the parties and reiterates in Article VIII, Insurance, that the subcontractor shall not be responsible for the acts of others (not only omitting the hold-harmless clause, but definitely stating that it shall not apply to the sub, even though it may apply to the general!). The subcontract provides that *Contractor's equipment shall be available to the sub only under specified conditions.* The lifting equipment in particular, either a tower crane or elevator, is used for subs. Some generals charge for such use, some offer it free within the availability for work other than the general contract work, and some make a definite obligation to furnish services to the sub.

Both standard subcontracts are in multiple-page format; that is, they are written on several pages, each of which must be initialed by the parties to authenticate them. The contractor who designs his own subcontract may reduce the print size and line spacing to fit a four or even eight-page folding sheet. This makes the various articles, at least in four pages, easier to compare with each other and simplifies locating any item. It also has the advantage of appearing simpler; attorneys rarely recognize that the legal-size paper and double-spaced lines of the standard legal form cause a "legal reaction" in most businessmen, causing them to look for their own lawyer. A stock printed form is much more likely to be accepted by a subcontractor. The Lenox Contracting Co. contract which appears in Appendix D was printed this way and was used for years with almost no complaints by subcontractors; this form contains some much more rigid requirements of the sub than are found in the standard requirements.

5.19 OWNER-BUILDER CONTRACT

The Lenox Contracting Co. contract in Appendix D illustrates a form used on a particular job. It has been adapted to an organization in which the construction firm was an affiliate of the owner, appropriate clauses were inserted to comply with state laws, and provisions were made for the type of project—a shopping center.

Owner-builders are owners who do their own general construction, contracting directly with specialty contractors, who are therefore prime contractors. However, these specialty contractors are usually called subcontractors since they do only part of the work. The Lenox agreement is termed a contract, although the parties are referred to as "contractor" and "subcontractor."

Article II. The provisions regarding *bound by any and all parts of said plans* referred to the nature of the project—a large one, with different contractors at different times on different parts. Connections were required from one part to another, and the subcontractor had to refer to drawings other than his own to clarify these details.

Article III. The Project Manager has the authority which would normally be exercised by the architect. Although he was also the owner's representative, none of the subcontractors mentioned losing any protection by the failure to have an architect. The Subcontractor was made responsible for missing details; that is, he could propose his own detail if he did not agree with the Project Manager's decision.

Article VI. The Subcontractor is committed, not to complete the work at a specified time, but to pursue it with a specified number of workmen. This provision is unusual, since the general then assumes the responsibility not only for determining the number of the subcontractor's men necessary, but also the responsibility for providing the work ahead of the subcontractor. This implies a degree of control of the work not usually assumed by the general.

The disadvantages are that there is an immediate yardstick to measure performance. A subcontractor's chief problem is stretching his manpower to cover his obligations. The subcontractor, unlike the general, confines himself to a limited work force, in order to attain greater efficiency and minimum turnover of workmen, which results in lower costs. For this reason, his services are in demand; otherwise, the general could do the work directly. But the subcontractor cannot plan his work when the demands made on him by the contractors (usually several at one time) he is working for vary. Usually, he does not have records which permit accurate forecasting of manpower needs for specified delivery dates, and the generals rarely can offer him continuous work during the course of a project. If a definite work force is required of him, he can plan his own work further in advance.

On the other hand, the contractor can never be sure that a sub will meet specified delivery dates if he cannot control the sub's work force. The sub is not in violation of the contract until the completion date is not met; it is then too late. Furthermore, the general has no effective control of the sub regarding which part of the subcontract is completed first, and some parts may be necessary for other work to proceed. Theoretically, the contractor is entitled to specify the order of work, but practically he has no way to prove noncompliance and therefore any action he takes against the sub might be considered a breach of contract. When the sub's work force is specified, his failure to meet the obligations imposed on him is immediately apparent; the general can take measures justified by the sub's breach of contract.

Article VII. This article relating to the sub's failure to maintain progress contains teeth not found in most contracts. The general is not obliged to terminate the sub's contract for lack of progress; as a practical matter, terminating the contract does not improve progress. The subcontractor is protected against the general's terminating his contract, not because of legal barriers, but because the general has nothing to gain by starting over with another subcontractor. Under Article VII, the general has another option; he may take over *part* of the work, leaving the sub to continue to the extent of his ability. The shoe is then on the other foot; the sub can quit the job, which makes the *subcontractor* the party who definitely terminates the contract, and leaves the sub in the breach without the defenses normally available to him; or he can continue to cooperate. If he continues, he can expect that his costs (which include the amount the general spends on labor or material) will go up, but this is minor and he still has a contract. He, too, often has subcontractors who are looking to him for payment. If he quits, his subs can make a new contract with the general to finish the work; because of his breach, he will then be responsible for the cost of the new contracts as well as amounts due from his old contracts.

The effect is to give the general considerable power to correct situations which lead to slow work. If the sub refuses to buy more expensive materials in order to get prompt delivery, the general may do so; and the general may order overtime.

Article XIV. None but labor compatible with other labor on the job shall be employed on work covered by this contract. This is usually interpreted by the sub as requiring union workmen. It requires the sub to bring union men on the job when the job is union, but it may equally well require him to use open-shop labor if other workmen want an open shop and his own men are union and refuse to work. He is not guaranteed that the work will be either open-shop or union. But if the job changes from union to open-shop during the construction period, he must change with it; he cannot maintain that the general is at fault if his own men refuse to work. The union has no status, in effect, under this contract; the men who support their employer are those who will work. The concept of changing from union to open-shop work in the middle of a job is not far-fetched; on one job, when pickets were placed on the job on Friday, an electrical contractor agreed to be on the job full strength by the following Wednesday with open-shop men (or perhaps the same men, having dropped their union obligation). However, the trade striking did not have the support of the workmen in other trades, and this word got to the union leaders very quickly. A labor union which pickets a job, when its own members are merely on strike and are not being displaced, gets very little support from other trades.

Article XVI. This article, Checking Drawings, represents an alternative method of avoiding charges (to the general contractor) which would occur because of discrepancies in the drawings. Clause 11.14 of the AIA contract seeks to avoid this situation by requiring "participation" of the sub in the preparation of "coordinated" drawings. Unfortunately, one sub often has his work fabricated or even installed before the interference is discovered. Article XVI requires each sub to check all other subs' drawings *available* to him. Note that there is no restriction that the drawings for which the sub is responsible need be contract drawings. This clause is much more restrictive on the sub than is the standard AIA clause.

For example, let us assume that a store ceiling is to have recessed electrical ceiling fixtures, sprinkler-system piping, plumbing drains from the floor above, and air-conditioning ductwork. In addition, there may be structural interferences, where the ceiling is close enough to the steel joists or beams that the space is restricted. The electrical conduit, fortunately, is usually of a size that may be worked into any leftover space. In this situation, each of the contractors concerned furnishes shop drawings. The electrician furnishes catalog cuts of the fixtures showing the recessed distance; the location is usually detailed on the contract drawings. The electrician, however, might not furnish details of junction boxes or the direction from which he brings out his conductors from the fixture; other detail he might omit occurs when he needs supports from the structure, rather than from the ceiling itself, because of the weight of fixtures.

The plumber provides few or no shop drawings, as he uses the contract drawings. The contract drawings are schematic, however; they will lack the size of elbows and roughins (distances from the floor hole to the wall). Because of these variations, and because the slope is usually determined by the specifications or code requirements rather than being shown on the drawings, considerable study may be necessary to determine the possible variations of the lines.

The sprinkler layout is furnished by the sprinkler contractor, but it may not include elevations of his lines and the size of fittings. He usually has considerable freedom to change his piping in the plan, but he must slope to drain. He will defend his layout as being notice to the general where his piping occurs, and he will cut his piping in the shop for this layout. However, the layout may not be the plans from which the piping is cut; the plans more likely used would be more detailed drawings. If the general has not received *all* the shop drawings, as well as shop drawings showing only general layout, the sprinkler contractor is taking his chances that the piping he has cut will fit the work of others without interference.

The ductwork man has the greatest freedom. Although the size of ducts and their location in the plan is shown on contract drawings, they may be varied even to changing the sizes and the cross-sectional area. He provides a shop drawing, but it is usually vague; the ductwork, which may be done in the shop as well as in the field, can readily be altered to fit changed conditions as the work progresses. Yet because his work is bulky, it seems visually inflexible, and the ductwork man is often encouraged to proceed before the pipe trades.

In the case described above, Article XVI could be used against *any* of the contractors—but only to the extent of his *own* work. If his work is placed in the only location possible, the others must work around him. As a practical matter, each sub is placed on notice that he must investigate the situation before he proceeds; it is a stoplight for the work. The sub will then demand clear instructions before proceeding, and the project manager or the superintendent must give him definite instructions about how to proceed. The clause deprives the sub of the defense that his work, when it must be changed or removed, was installed according to the contract drawings. Such a clause can be fatal to the progress of the work if combined with a timid or inexperienced project manager who is loath to assume the responsibility of working out these interferences on the job and to be responsible for the mistakes which will invariably occur.

However, the sub is *not* giving up any claims to which he would be entitled under the contract drawings. If he must put in adapters or elbows not shown on the contract drawings, he still is paid for this work; they must have been proven to be *necessary*. This clause is directed at the mechanical trades; few subs in the trowel trades, or structural steel detailers, are qualified to interpret mechanical drawings and quite reasonably could refuse a contract with such a clause.

Article XVIII. Invoices are often checked at the job, while other legal notices are sent to the main office of the contractor. Article XVIII states the address to which invoices are to be sent.

Article XIX. Acceptance of final payment of this contract, by Subcontractor, shall constitute a release of all other claims of Subcontractor arising out of this contract. This is a stiff requirement and can be used to force the sub to settle outstanding claims in order to get badly needed money. It is a "sleeper" in that the sub has often forgotten this clause when the end of the job comes around, and he inadvertently loses rights to extras. In court tests of this clause, it has been held that the sub cannot preserve his outstanding claims by a qualified release. The sub's claims, which have been accepted by the general, may be recognized on the ground that they were omitted in error, but this defense is not dependable and may be disregarded entirely by a public body. A sub should not accept this clause, therefore, with any assurance that he can preserve claims after he accepts the final payment.

To the general, the advantage of this clause is that the sub is obliged to present and justify his claims before final payment. Subcontractors are often lax in making claims for extras, and do not present them properly (that is, do not provide details and justification for the claims). This clause also obliges the sub to present claims which have been authorized in the field and assists in protecting the general from poor judgment in authorizations by his employees. A clause that the sub be required to make claims for extra work with his monthly billing is also helpful; unfortunately, a sub may have a large extra authorized and have ordered materials for it months before a billing for the work is made.

Article XXIII. Should any provision of this contract be held invalid by any Court, the remaining provisions hereof shall not be affected thereby. This is the same type of *saving clause* which has been used in labor contracts for many years.

5.20 WITHHOLDING SUBCONTRACT PAYMENTS

Many contractors attempt to secure performance from subs by withholding payments. It is the practice most resented by subs, who are nearly helpless against it. They may be forced to concede a disputed extra, for example, in this way. Such withholding is illegal, but the sub can usually collect only interest, which is much less than his real loss. The subcontract terms attempt to protect the subcontractor against this abuse.

If the general is able to pay the sub before the owner's payment is received, and ordinarily does so, this continuing practice becomes an ethical, if not a legal, obligation on the part of the general. Some contractors take responsibility for paying the subs on time, as a matter of principle; it would be expected that these generals would get a better price from the subs, but a bid depository system, which exposes the bids to the other bidders, tends to prevent the better price, and therefore gives little incentive for prompt payment. Some subs are unable to meet their payrolls if they do not get paid on time and therefore will not enter a contract in which their payment depends on the owner's payment. A contractor may require anything from a sub in consideration for an early payment which was *not* promised to the sub; but the effect is the same as withholding a payment when it is regularly due—in both cases payment is delayed to force the sub to make a concession unrelated to the payment. Any delay in payment to a sub raises such an emotional reaction that normal relationships are disturbed, and subs should therefore be paid as early as funds permit. If the supervision of the work, or the workmanship, is unacceptable, it is preferable to terminate a contract (and keep *all* outstanding funds) than to delay payment; an enemy off the job is preferable to a lukewarm friend on it.

In one arbitration case, a subcontractor otherwise entirely at fault was not properly paid, and for this reason alone he received a substantial settlement.

5.21 SUBCONTRACTOR DAMAGES FOR DELAY

The subcontractor should expect to be responsible for delays he causes, but the standard contracts are not at all clear on this point. At best it is difficult to determine or prove such responsibility, but the author recommends the following:

Damages for delay. If the Contractor incurs a liability to the Owner because of delay in completion of the work, and the work of this Subcontractor is incomplete, the Subcontractor shall be liable to the Contractor for all such liability, whether liquidated damages or damages due to alleged breach of contract. The Subcontractor may not use as a defense against such claims that other work on the job was incomplete, unless the Subcontractor is delayed by other such work. When payment is withheld from the Contractor by the Owner because of delay, regardless of whether the matter has yet to be determined by arbitration or court action, the Contractor may withhold the same amount from the Subcontractor's payment, pending final determination of the claim.

The Contractor may not collect or withhold from all Subcontrators on one job a sum larger than his actual damages or withheld amounts, and the Contractor may not agree to a settlement with the Owner which will deprive the Subcontractor of the process or arbitration or court action, ws may be allowed under the prime contract. The Contractor may withhold payment from the Subcontractor on any invoices due the Subcontractor, regardless of whether the invoices originated under this contract or another. The Contractor's right of action against other Subcontractors who may contribute to the delay shall be granted to the Subcontractor who pays the cost of delay or suffers delay in payment.

The *damages for delay* paragraph shall not be used to justify claims against the Subcontractor if the cause of the delay is delay in completion of preceding work by another subcontractor or the General Contractor. However, if the delay in the Subcontractor's work does not cause a liability to the Owner by the Contractor, but does cause a delay in payment to the Contractor, the Subcontractor shall be liable for 2 percent per month of such delayed payment. Collection of such damages shall not relieve the Subcontractor of any other damages which may arise under this Subcontract, and the Contractor may take other action to complete the contract under paragraph 11.10. The Subcontractor may not use as a defense against the collection of this percentage charge the fact that the payments delayed may not cause actual damage to the Contractor.

Punch list items. When the Subcontractor has neglected an item of work requiring no more than 1 worker-day to complete, or has been unable to complete it for any reason (including delay in work of the Contractor), he is obliged to complete the item within 7 calendar days of becoming aware that the work is delaying payment on the job or delaying the work of others. If he does not complete the item in the specified time, the Contractor at his own option and without affecting any other right the Contractor may have under the contract, may order the work to be completed by others at the Subcontractor's expense. If the Subcontractor has material needed for the work, he is obliged to deliver it to the Contractor in this 7-day period, or the Subcontractor will become liable for material orders placed for the material by the Contractor, whether the material ordered by the Contractor is actually used or not.

Likewise, after 7 days the Subcontractor becomes liable for any expense incurred by the Contractor, including cost of trips to the job by others, even though the other do no actual work. Lack of action under this paragraph by the Contractor will not abridge the Contractor's right of recovery for delay as expressed elsewhere in this subcontract. This paragraph shall not limit the right of the Contractor under paragraph 11.10.

5.22 PERFORMANCE BONDS

On public work the engineer is given the responsibility of determining if the low bidder is responsible financially and is capable of performing the contract. As a practical matter the engineer is inclined to accept the low bidder, relying on the surety (bonding company) to have made any necessary investigation of the financial and technical ability of the firm. The surety, for a fee of about 1 percent of the contract price, undertakes to guarantee that the contract will be performed and that debts incurred in ther performance of the contract will be paid. Public agencies operate under statutes which require that contractors furnish a performance and payment bond when the contract is signed.

Contractors vary widely in their attitude toward bonding subcontractors. Some firms require bonds of all subs over a defined amount, both to share the risk with a surety, and to increase their own bonding capacity. Sureties and bankers look with favor on the practice of bonding subs, and will increase their commitments to a general who bonds subcontractors.

Other generals do not normally require them on subcontracts. Bonding companies are primarily operated by lawyers, who are naturally inclined to improve their situation by court action or threat of court action. There are too many instances in which sureties have refused to make full payment of their obligations to general contractors who are in no position to tolerate delayed collection by continuing with protracted court suits. In short, it is a case of a small company against a big one. The general can exercise considerably more pressure against a sub than he can against a surety, and if the general takes over the sub's contract to complete bonded work, he may find himself in a protracted suit with the surety as to how the work was done.

In addition, the surety is not a party to the contract and therefore may not be subject to the arbitration procedure. The general may win an arbitration case with a sub, only to find that the sub is not solvent and that the decision is not enforceable against the surety. If the surety assumes the contract rights, he presumably will assume the arbitration obligations. A general contractor may do better to increase his overhead by the amount the bond for subs would cost, and use the expenditure for closer supervision both of the work and of the payments to subs.

It has been proposed that to save money, the bonding requirements for U.S. government contracts should be dropped. The situation here is entirely different; if no bond were required, both the losses of bankrupt contractors and the cost of investigation would be placed on the government, who would also be in a monopoly position in determining qualifications of bidders. Few people believe that the government could do this with a saving of money, and if the bonding requirement were dropped, there would be no effective way to determine whether money was actually saved.

In a report to Congress in June 1972, the General Accounting Office suggested that the U.S. government discontinue requirements for performance bonds. The National Association of Surety Bond Producers pointed out that the government would then have to default contractors who are short of money, whereas sureties often make them loans. NASBP estimated that the sureties' average loss ratio on U.S. government work was about 51 percent of the fees charged.

In a 1968 Montana case, a subcontractor was required to furnish a performance bond to the contractor. The bond furnished stated that in the event of any sub's default, the surety shall be informed within 10 days; and furthermore that the surety had no responsibility to anyone except the general contractor. On this basis, the surety refused to pay a material bill, but was instructed to pay, on the grounds that the intent of the agreements (including the contract and subcontract) was that all suppliers would be paid, and that *substance, not form, should control the surety's responsibility.*[5] This is of some assurance to contractors who, like most, cannot understand the language on the bonds they purchase.

5.23   COST-PLUS SUBCONTRACTS

Cost-plus subcontracts have the same advantages and disadvantages of cost-plus prime contracts. The general is usually in a better position to protect himself from loss on cost-plus subcontracts than is the owner, since the general is better represented on the site and is more familiar with the sub's work. Cost-plus subcontracts may be used when the general needs the sub's license—for example, in electrical or plumbing work, which the general cannot do with the direct labor even if he hires licensed men, since the firm as well as the individuals must be licensed. Since the cost-plus subcontractor is for all practical purposes the employee of the contractor, there must be a good reason for using the contract arrangement rather than direct employment. Some trade superintendents are able to work consistently at cost-plus subcontracts with several contractors, none of whom could hire his crew regularly. If the subcontractor is an architect, an engineer, or a land surveyor, payment may be on the basis of payroll but is called a *professional fee contract.*

5.24   BANKRUPTCY

Bankruptcy is a federal court proceeding under which a court-appointed person (the *receiver*) receives the assets of a persone or firm and pays off the creditors proportionately. The bankrupt firm is relieved of its indebtedness and may start over again. The firm is not relieved of certain obligations which are not normal business debts, such as claims for personal damages and money withheld from employees' wages for social security and income tax payments.

A bankruptcy proceeding may be initiated by the firm in debt or by its creditors. The firm would seek relief from the creditors when its owners feel that there is no chance of paying them from operations. The creditors would seek bankruptcy proceedings to prevent owners of the firm from taking out funds or to prevent the firm from paying some creditors in preference to others. One reason why a contractor incorporates his firm, aside from raising capital funds, is to protect himself and other owners from claims against the company.

---

[5]*Weissman v. St. Paul Insurance Co.,* 448 P2d 740 (1968), from *Building Design and Construction,* November 1975, p. 21.

A contractor may owe more than he owns and may be operating at a loss, but as long as he can raise enough payroll to operate, creditors will usually sustain him. This is because the amount which may be obtained from the bankrupt estate is so small that a creditor will continue extending credit, for only in this way has he any reasonable hope of payment. If the cause of bankruptcy is a single mistake, the chances are good that a company will get back on its feet. It is essential that such a firm deal honestly with its creditors, of course. In such a situation a sort of partnership comes into being; the creditors cannot cut off credit for fear of losing what is already owed them, and the firm must continue buying from them, rather than on the open market. Two factors tend to bar the firm from the open market in this situation: It may be refused credit (depending on how widespread is the knowledge of the firm's position), and failure to buy from dealer or subcontractor who is already a creditor may cause that creditor to bring action to foreclose on part of the contractor's assets. This action triggers similar action by other creditors who do not want to be left out—and eventually brings bankruptcy proceedings by the creditors. For this reason a contractor may say, "I owe too many people to go bankrupt."

5.25   RECOGNITION OF THE INSOLVENT FIRM

As pointed out elsewhere, the balance sheet of a contractor may not give much information on his net worth, but certain factors on the balance sheet or in the accounts are indicators of solvency—or insolvency.

1. A high current indebtedness in proportion to assets, as in any firm, but if the assets are largely in jobs which are in early stages, the appraisal is doubtful. If uncollected changes are listed as assets, the possibility that they are contested and may be uncollectible should be checked. The accounts receivable account will show when and from whom they were due. Evaluating this information requires some knowledge of local firms. An overdue bill from a firm which normally pays promptly, for example, may indicate a disputed item or a premature billing.

2. The age of equipment, as shown by depreciated value in relation to new cost, should be checked. New equipment often cannot be sold for its book value, whereas depreciated equipment can bring its book value or more, since depreciation rules usually allow equipment to be depreciated more quickly on the books than it actually drops in value. If the equipment is bought used and an equipment mortgage nearly covers it, such an asset may be worthless. On the other hand, a low depreciation rate, which is not normally desirable because it results in early payment of income taxes, increases the apparent profit of the firm. Such a rate may indicate that the books are being used to maintain the firm's credit, rather than to maximize profits after taxes.

3. The salaries of the officers, and any connection of the firm's business with other firms with the same owners, are suspect because it is the easiest way to siphon off assets from an ailing firm.

4. The gross indebtedness and the gross assets are not so important as the amount owing and receivable on individual jobs under way, particularly on the most recent ones. If a contractor has received payment for subcontractors' work from the owner but has not paid the subs, this is prima facie evidence that he is insolvent.

But how is the outsider, with no access to the contractor's books, to recognize an insolvent firm? Usually, it is not recognizable until the contractor falls into the spiral of a losing contractor trying to get out—paying subs late, increasing work contracted, and pushing jobs which are in an early stage to get funds from them, to the detriment of jobs which, though nearer completion, have expenditures that are exceeding income.

Any contractor should make his greatest effort to finish jobs that are nearly done, in order to get the retained percentage. If he does not, it may be for reasons which indicate that he is in trouble, such as these:

1. There may be no retained percentage because the owner has paid in full.
2. The contractor may have obtained payment for his own work, and for some of the subs—and the retained percentage is being used as a promise to pay creditors. That is, he may have promised more payments out or the retained percentage than there is money to pay them with, and so long as he does not get the retention, this will not be discovered.

A losing contractor may be forced into low bidding (which, of course, may add to his losses) in order to obtain jobs on which he can get advance payments. This will be the next characteristic of a failing firm, after failure to meet supplier invoices. Advance payments are not so labeled, but the contractor can often draw more on a contract than he has spent on it. Such payments are in advance so far as the contractor's condition is concerned.

The most definite indicator, as mentioned before, is the contractor's refusal to pay subcontractors out of current payments for their work. At times, the contractor may be able to conceal from the subs the amount of money he has requested.

A contractor may let old unpaid bills drag on for months; and when it becomes evident that he is dealing with a limited number of suppliers, in contrast to normal dealing, this may mean that he is restricted to suppliers who are creditors. He may cut down on inventory and patronize small suppliers on a hand-to-mouth basis. These are all merely indicators, and only a combination of them is significant.

5.26   AVOIDING BANKRUPTCY

If you should be in a position where you are trying to pull a firm out of trouble, you of course will attempt to avoid the preceding indicators. What you want to do must be tempered by what is possible. If you are new to the firm, you may find this to be an advantage; creditors will have obtained worthless promises from everyone else in the company and will grasp at any straw which offers a hope of getting their money.

These steps are recommended in the order shown:

1. To the disregard of all existing promises to the contrary, pay subcontractors and material suppliers their share of current receipts on each particular job. That is, when you collect for materials or work they have furnished, pay them for it. This may force the company to cut down on new work, which should be arranged with creditors.

2. Make a complete and detailed estimate (accountants would call it an audit) of the situation of the company. Do not hesitate to do this because the answer may be minus; if insolvent, find it out. Make two statements: one showing the liquidation value of the company as it would be under bankruptcy proceedings (taking into account legal costs) and another showing the amount of money which would have to be invested to acquire such a company by starting in business, bidding work under way, and purchasing the equipment owned.

3. On the basis of these statements, decide whether you can make out with the company as it is, or if reorganization is needed. If the work under contract at that time does not have the profit which allows a reasonable regular payment on bad debts, don't make a decision to continue based on the profits from work not yet contracted. This attitude may be modified if the company has contracts under negotiation which offer a good chance of being closed in the near future. If the owners have property which is not subject to the corporate debts (if the work is bonded, they may be pledged to that work, and therefore personally liable for corporate debts), see how much they will put into the company for it to survive. If they have a false optimism, it may be very difficult for anyone to rescue the company.

4. If the statements do not warrant requesting (with a chance of success) a bank loan (and banks are the first to get their money out of a failing firm), try the surety for any work under bond. The surety fulfills all the qualifications for a lender—if the company goes down, it will lose the money anyway, and the surety will keep the situation quiet. The fewer people who know the situation, the less likely are rumors to spread which would bring claims from creditors who realize the condition of the company.

5. If it appears probable that these sources of credit may get you out of trouble eventually, make an agreement with all creditors of unsecured obligations, not covered (that is, by bond), to pay a small amount frequently, including a reasonable interest rate. If they will accept a company note for the amount, so much the better. "Unsecured obligations" are not a mortgage against equipment or other assets, but are debts for materials for which the company has been paid by the owner but no longer has the money to pay the creditor, and fixed costs such as rent and telephone. You may make agreements with creditors to reduce your overhead expenses (such as cutting out some telephone service) in return for terms on the back debts.

   Loans on equipment or other real assets may be renegotiated if the equipment mortgage is below the resale value. Equipment returns should be made before a creditor is pushed to foreclosure, since foreclosures are publicized.

6. The last-resort step is a reorganization of the firm. The creditors are told exactly what the situation is, particularly the difference between the value of the firm as a going business and its value in liquidation. They are then invited to settle their claims by accepting common stock, preferred stock, or bonds. At this point, the owner may save his reputation but very little else; if the firm survives, the owner may be working on a salary for a long time to come. The owner may resign any position with the firm, if it appears that the creditors do not consider him equal to the task of continuing its direction. However, this offers him a chance to come out with his credit reputation relatively untarnished, and with a chance, after all debts are paid, of regaining control of his company.

Any involvement in a bankruptcy may seriously affect an owner for the rest of his life, since creditors are always doubtful that the bankruptcy was caused by an error, and may believe it was rigged to relieve the owner of responsibility while he was taking funds out of the company. These doubts exist because it is not unusual for a man to escape personal losses by abandoning a company which has little cash or by keeping property in his wife's name.

The name of a corporation should be carefully watched by creditors, since a minor change may indicate there are two companies, as "Smith Brothers Company," and "Smith Brothers, Inc." A creditor may find he is holding an obligation of one company while another has the credit rating.

Reorganization of firms unable to meet their obligations are often made by intervention of credit and financial management associations, composed of credit managers of local firms. These associations plan out-of-court arrangements, as well as providing credit information on firms generally, help make collections, and provide courses in credit management. Some associations are the Chicago-Midwest Credit Management Association, the New York Credit and Financial Management Association, and the San Francisco Board of Trade. Creditors' committees have obtained bank loans for firms in trouble by agreeing to subordinate their claims to the bank; that is, the creditors agree that they will allow the bank to claim the insolvent firms assets before the other creditors are paid.[6]

5.27 QUESTIONS FOR DISCUSSION

1. Your boss allows you to purchase materials and subcontracts if they are within budget amounts, but you must get his approval for purchases that exceed this amount. What are the advantages and disadvantages of this policy, and how do you think he should change it as you become more experienced?

2. What problems may arise from using purchases orders in lieu of subcontracts? Prepare the clauses to be added to a purchase order for the following contracts:
   a. Concrete block labor only
   b. Blown-in insulation
   c. Tar and gravel roofing
   d. Electrical work

3. Which of the contracts above are least, and which are best, adapted to a purchase order form?

4. For what type of construction is a liquidated damages clause preferable, and for what type is a *Time is of the essence* clause preferred, and why?

5. If a subcontractor is not bonded, what additional work is required of the project manager, as compared to the situation with a bonded sub?

6. A firm which gives every indication of insolvency requests credit from you, a supplier, for a job with a performance and payment bond protecting suppliers. Would you grant him credit? What factors would influence your decision? What information would you seek and where? Would you be influenced by whether the firm already owed you unsecured bills?[7]

[6] *Wall Street Journal,* June 17, 1975, p. 44.
[7] Bills may be secured by collateral, by a bond, or by current lien rights.

CHAPTER 6

ACCOUNTING

It is often said that a contractor cannot be an accountant. Many contractors assume that it would be too complicated to exercise supervision over their accountants, whether the accountants are independent firms or part of their companies. But as accounting is essential to the business, so is adequate supervision by the manager. Accountants are staff professionals, like engineers, doctors or lawyers, and have in common with other professionals the ability to proceed along accepted and established lines in any number of directions. If an accountant receives no proper instructions or information from the manager, the accountant must fill in his own blanks—and in many cases is obliged to make management decisions which the manager does not recognize as such. An accountant serves the organization; the more ways in which the accountant can adapt to the organization's needs, the better he serves the firm. In one organization eight days were required from the end of a payroll period until the men could receive their pay; any other timing was just "not in accordance with their methods." A compromise was finally worked out whereby pay was reported several days in advance in order to make payrolls in accordance with construction practice—yet only a 40-man payroll was involved, which could have been done in an hour by any payroll clerk! Unfortunately, in this case, as in many others, the accountant did not realize a choice existed. He was so firmly attached to punching buttons for a computer that he had forgotten, or had never learned, the basic methods he was supposed to supervise.

6.1 BOOKKEEPING AND ACCOUNTING

Bookkeeping refers to the mechanical process of making entries in a set of books, and deriving the totals from them. The bookkeeper is expected to receive direct instructions as to how the work is to be done: what books are to be used, the number and name of accounts, how numbers are to be transferred from one place to another. He may get instructions from an accountant or from the owner, and these instructions may come infrequently. A competent bookkeeper may receive instructions once a month or less often. He is expected, however, to follow predetermined methods and not to interpret policies in terms of setting up appropriate accounts. A bookkeeper usually has some formal training in this field, but a high school graduate may learn his work on the job. There is no legal definition of a bookkeeper's qualifications, as he does not serve the general public except for preparation of income tax returns (for which no special license is required).

An *accountant* is a professional, often licensed by the state as a *Certified Public Accountant* (CPA). Formal education and knowledge is required. An accountant is expected to be able to originate a set of books and specify how they are to be kept, in accordance with good practice and the intent of the employer. He has learned an established framework within which a number of accounting methods are possible. However, any books set up by one accountant should be comprehensible to another, and an accountant should be able to recognize accounting methods in use in a firm, and to continue or interpret them.

The public license entitles the accountant to advertise himself as CPA, but there is no legal requirement that a contractor must use an accountant's services. Only when the public is affected is a CPA required by law for examining records, as in estates, banks, and corporations. A contractor who has no stockholders to account to or who does not offer his stock on the market is not obliged to have an accountant. The Internal Revenue Service (IRS) does not require that an accountant keep a firm's books or even examine them.

An accountant may be a regular full-time employee (most corporations have CPAs as corporate officers controlling funds), or he may be an independent consultant who examines the books and instructs the bookkeeper at intervals.

A manager does not supervise his accountant in a technical or detailed way. A manager does not tell an engineer *how* to design a building, but the manager is certainly interested in having the right building designed. The manager should have the accountant give him the answers he wants but need not specify how they are to be obtained.

An *audit* is an examination of the records of a firm, or of part of them, usually by an accountant. The audit of a construction firm's payments is too important to trust to an accountant and is normally done by a corporate officer familiar with the purchases and contracts for work under way. If the payments of a firm are to be safely entrusted to anyone but the manager, with assurance that these payments are being properly checked against authorizations, not only the accounting department but also the project manager's organization must have a reliable procedure for all clerical work. On the other hand, a manager often retains invoice approval as the only way to keep an inefficient or informal organization (usually an informal organization is inefficient in its handing of paper work) from becoming an unprofitable one. Audits may be made by the Internal Revenue Service, either because they have reason to suspect the methods used or because the firm has been selected at random for examination. A contractor with a simple bookkeeping system may have his books audited by a CPA to enable the contractor to obtain surety bonds, or to check on the ability of his bookkeeper.

6.2 INCOME TAXES

There is a popular notice that there are ways money can be spent to save taxes, but there is no way that a construction business can be conducted so that money can be wasted without reducing profits. A popular conception is that firms can save taxes by spending a smaller amount on contributions or unprofitable operations. There have been a few such loopholes in the past, usually involving charitable contributions, but even these required a real loss on the part of the taxpayer.

Tax evasion, the illegal manipulation of accounts to avoid taxes, is a crime. *Tax avoidance* is the legal arrangement of one's affairs to avoid the payment of taxes. The simplest way to avoid taxes is not to make a profit at all. Another method—and one considered more desirable by most people—is to avoid payment of taxes on the increase in value of one's estate by spending the money on investment rather than on personal expenses. That is, by leaving the money in the business, you have *unrealized* profits; you have made a profit, but you have no money from it. This is also called *paper profits.*

When this avoidance is planned, financial operations are usually designed to reduce taxes by eventually paying the tax rate based on *capital gains* rather than on *ordinary income.* The difference in rate was once almost two to one, but it has been reduced by recent legislation.

A really determined businessman may find ways to avoid payment of any taxes at all on unrealized profits (but eventually he must either die or give the property away); the majority of loopholes in tax laws are concerned with these methods of tax avoidance. The practical methods of tax avoidance for a contractor are confined to delaying his income tax payments for a year or so, so that he may have the use of the money in the meantime. He may also arrange his affairs so that he does not have to pay income tax on profits he has not received.

A builder of an apartment house, or a contractor with heavy equipment, must make payments on the principal of his loan, or *amortization,* which comes out of his profits. He plans to receive enough cash to make thess payments, as part of *cash planning.* The cash paid on amortization is part of profits, so he must also have cash to pay taxes on these profits. *Tax planning* must be part of his planning for cash requirements.

Ther are two kinds of expenditures: *expense* and *investment.* Since *expenses* are deductible from profits, the government may be paying half of these payments, as income taxes are reduced accordingly. But *investment* funds are drawn from profits, not deducted from them. Whereas rental of equipment counts as an expense deductible from profits, purchase of equipment is considered an investment and cannot be deducted from profits. If the firm buys the machine, it may be sold and the return is not subject to taxes, but this is little consolation if the firm needs the machine indefinitely. For this reason, many firms rent machines when economics would otherwise show they should buy them.

A similar situation arises when a new corporation is formed. If you have $100 to invest in a construction company, you may buy $100 worth of common stock. All the profits from the stock will then be taxable. But if you buy $10 worth of common stock and lend the firm $90 for corporate bonds, the firm has the same amount of working capital and the first $90 you receive will be considered return of capital and therefore tax-free. The firm pays income taxes on the $90, but it is tax-free to the bondholder, as it is held in the firm as undistributed profits. The $10 stock then become worth $100 when the bonds are paid off, and the IRS will collect taxes on the $90 when the stock is sold.

Alternatively, interest may continue to be paid on the $90 bonds rather than paying them off; this payment is not taxable to the corporation, but it is taxable to the bondholder. A number of schemes involve reduction of corporate debt, for which interest is paid to nonprofit entities which themselves need pay no taxes on the interest received. A manufacturer may sell his plant to a university, for example, taking a long-term lease, in order to gain capital for further investment. *Tax loopholes* are the way *other* people get out of paying taxes, and therefore are not susceptible to definition except by reference to the purpose of the taxes—a political, not an accounting, problem.

6.3 NECESSITY FOR ACCOUNTS

Some contractors' prices or business decisions may not depend on whether they are making or losing money. If a contractor is selling an item at a standard price (as a cement finisher, who places slabs at a unit price), he cannot change his business or his price; he may only change his location or work for someone else. His indicator of profit or loss is whether he has enough money to pay his living expenses. For such a contractor, accounts are of little use.

However, the federal government has a more detailed interest; it wants to know, basically, what his living expenses are, since that is his profit. Any system that provides this information is adequate for the government's purpose. Its method of forcing a record to be kept is simple; it assumes that any receipts of the business are profits unless the contractor can prove otherwise, and he is required to keep a record of income. If a contractor has no such record, the IRS assumes that all bank deposits represent income. The contractor's simplest method is to deposit all income in a bank account and to keep all invoices in a box. At the end of the year, the invoices are added up (payroll paid out must be represented by a notation of some kind, usually a check to a payroll account or individual checks to the workmen) and subtracted from the total deposits; the difference is the profit for the year. If the contractor does not trust banks, he can add to the box a copy of the bills he sends out, or a copy of the receipts he has signed. There are now two kinds of paper—income and outgo. This is basically a *voucher system* of bookkeeping, and the addition at the end of the year is the *profit and loss statement.*

A general contractor, however, has some control over the kind of job he does, and the price he bids it for. He is consequently interested in knowing the profit or loss on each job. Unlike most businessmen, he initially cannot determine his profit for a week or a month; this knowledge comes much later. *His initial reason for starting a bookkeeping system is to determine the cost of each job.* He therefore does not emphasize detailed costs, as do manufacturers and merchants; his needs confuse accountants who are accustomed to *profit and loss* bookkeeping. Profit and loss bookkeeping for a contractor's entire business is so complicated that he doesn't attempt it until he is well along in volume, and then he must give his full attention to it. Accounting and line management can never get so far apart, in interest and communication, in a contracting business as in other kinds of businesses. A liquor store, a department store, or an automobile plant may establish definitions and accounting practices which require very little change from month to month or year to year. In construction, changes occur rapidly and transactions may be different from one supplier to another. As with banks, real estate brokers, and stockbrokers, large sums of money pass through the office with very little staying very long, and construction operations are even more variable. An accountant may therefore set up a department with extensive books and computer-programmed sheets, which give no more practical information than does the one-man contractor's box of bills. Construction accounting depends on a daily flow of information into the accounting department and a clear understanding of how reliable this information may be. Contractors with heavy equipment costs may be able to determine their profit more frequently than the time they keep their equipment—several years at least.

Construction accounting is primarily cost accounting, and as the accounting department grows, it requires more complete information. Some needs of the manager are these:

1. *Prompt cost information on jobs under way,* particularly on equipment and labor costs. As detailed in earlier chapters, these records are not necessarily kept by the accounting department.

2. *Current information which will assist in forecasting cash needs.* The cost information is required for accuracy, but an estimate can be made without it. Usually the forecast requires a comparision of costs of the various parts of the job with prospective income from each part.

3. *Protection against embezzlement or fraud* on the part of the contractor's employees. *Fraud* designates embezzlement, theft, or any other method by which an employee profits at the expense of his employer. Some methods, such as a rakeoff from purchase or commercial bribery, are dishonest but perhaps not criminal.

4. *Profit and loss estimates.* These are always based on some arbitrary assumptions; the shorter the period for which profits are estimated, the more approximate are the assumptions. A firm must have either a profit *or* a loss, but the accountants' term is "profit *and* loss" or "operating statement."

5. *Net worth of the business, from balance sheets.* This information must often be submitted to owners to demonstrate qualifications to bid, with bids to show the ability to complete the job, and to sureties to secure performance bonds.

6.4   ACCOUNTING TERMINOLOGY

Accounting methods are occasionally referred to as a *voucher system* or a *book system*. In the preceding example of the contractor with a box of invoices, he was using a *voucher system*—each transaction was represented by a piece of paper. If he recorded each invoice in a book, he would have a *book system*. Either system results in a set of books; the voucher system allows a number of invoices to be collected in an envelope (the cover of which may bear a register of the items within it), which may be represented by a single entry in the books. Virtually all modern accounts are kept by a variation of the voucher system, if a large number of transactions is involved, to reduce the labor and volume of maintaining accounting books.

The old-fashioned full manual entries in cost accounts continue to provide physical control of material if issues are made from a central warehouse to scattered work where physical control within a job site area is not possible. The alternative is for a supervisor familiar with requirements to review all issue tickets, which detail quantities and material issued to the work—a laborious procedure.

Accounting may be said to be *single-* or *double-entry*. A list of payments from cash would be a single-entry book. Single-entry books provide no check on possible error of any entry and are therefore not used, but many bookkeepers work with only one side of the transactions and are therefore not familiar with double-entry books. That is, they get into a small part of the system and cannot recognize other parts of it. The larger the firm, the more easily this can occur, and a bookkeeper for a large office may know very little about a complete set of books for a small contractor. With *double-entry* accounting, each transaction is recorded in two places; if the totals correspond, the books are *balanced* and the assumption is that there are no errors. That is, the books are balanced unless there are two errors which balance each other or unless the bookkeeper puts in a number to balance the books. For example, a payment is recorded both in the cash book and in an account describing what is bought or paid for with the money.

Under double-entry bookkeeping, each entry must be made in such a way that the books balance both before and after the entry. This requires two sides, which are rather arbitarily called *credit* and *debit*. Unfortunately, the layman associates these with plus and minus, or good and bad. There is no such association; it is the *difference* between them that is important.

We are all acquainted with a "credit" balance, but the popular connection with "good credit" is rather tenuous. A person with an account at a store normally has a *debit* balance in the store's books; payments are *credited.* These may be considered merely as *left* and *right* (debit and credit), the sides on which they are usually entered. On a balance sheet they appear as *assets* and *liabilities*—which sounds clear until one notes that the net worth of the business is listed on the liabilities side. Actually, *net worth* is placed on the credit side to make the two sides balance.

6.5 JOURNALS AND LEDGERS

In order to keep individual accounts—such as for each job, for equipment, or for equipment costs—there must be some way to charge twice, and to verify that both charges were made. This is done by entering charges in a separate kind of book, the journal. It is not necessary to keep a journal, but if one is not kept, errors are hard to find. Check stubs can serve as a journal, if they indicate to what accounts the payments are to be charged. There may be a variety of journals, but they have in common that entries are made in them in such a way that they are recorded and then charged out again to different accounts. A journal is merely a record of an entry having been made; once that has been made, it is no longer referred to (unless, of course, the books do not balance). If you keep track of your expenses in a permanent account, but write checks for them, your stub book is a *cash journal*—the source of entries both for your cash balance and for the things you bought. The permanent account is a *ledger,* which contains the records of how money was spent by classification. Your account record at a store is a ledger page, as is a page listing the expenditures for a construction job.

It is common to keep all expenditures for a construction job on a single page. This page may have several columns, for materials, labor, job overhead, subcontractors, and the like. The cost breakdown of the job is then limited to the number of columns on the page. If detailed costs are wanted, several ledger pages may be used, or a separate cost ledger may be kept, with a page for each cost account. The cost ledger is checked for totals against the ledger page for the entire job; the cost ledger need not be a part of the accounting books at all and may be kept in the project manager's office. If there is an error in the cost ledger, it does not affect the rest of the system; if the error is small, it may be neglected. An accountant, however, is accustomed to keeping all his books as part of the same system and will object to carrying, for example, two hundred pages where he could otherwise have but one.

Although a voucher system is used for the accounting books to reduce the number of entries, it may be preferable to record every item on the cost account. This cost account page can then be reproduced (in many cases the accountant's books are on sheets too large for photocopy reproduction) and used for the following purposes:

1. The *foreman or superintendents* in charge of the work are given a copy of the page each week. This enables them to spot mischarges or material which may have been picked up by their men and not delivered to the job.
2. The cost clerk receives a copy for the weekly cost report; he may report on all accounts or on those accounts with a minimum charge, for the *project manager.* For example, the project manager may get a copy of all pages over $500 weekly to review; and he may get a weekly report of the status of the 50 items with the largest charges for the week, with status over or under estimate.

3. If the item represents an *extra charge* to the owner, the item page is used as the billing, merely by making another copy. If a firm, such as home repair or plumbing repair firm, has many jobs for which charges are made on a time-and-material basis, the voucher system of bookkeeping may not be practical, since the customer wants a detailed list of materials and labor charges for his own job.

6.6 ASSETS ACCOUNTS

Accounts are of two types: those representing a permanent situation and those which are a temporary record, changed to a permanent account (closed) at the end of an accounting period or at the completion of a job. In construction, *accounting periods* (at the end of which a profit and loss statement is made) are quite irregular and may be as long as a year. Even then, many accounts are not closed (since the job is not finished) but are given a label such as "Cost of Work in Process." *Asset accounts* include: equipment, work paid for in advance (as insurance premiums), office equipment and cost of work under way (since it represents a debt by an owner to the contractor), and accounts receivable. The corresponding *liability accounts* include accounts payable and, to make the accounts balance, the net worth of the company

6.7 EQUIPMENT VALUATION

One of the main reasons for the unreliability of contractors' balance sheets is the unpredictability of the value of heavy equipment. Equipment is usually shown in the asset accounts at original cost with a deduction for depreciation, to show whether the value is overstated or understated in the balance sheet. Three valuations can be used: value in use, value which the equipment would bring at a forced sale, and normal market value. These valuations are used for determining the net worth of a firm, but the book value, based on a depreciation method, will only accidentally be identical with any of them.

A contractor has a considerable choice of depreciation rates. If he uses the highest possible rate, his reported profit will be low and his net worth low. This is of some disadvantage in obtaining bonds. But on the principle "*You can't pay bills with tired iron,*" most firms depreciate their equipment as fast as possible. This means that they report less than true profits in early years and greater than true profits in later years. From a tax standpoint, this defers taxes into later years, so the cash may be used in the meantime; this may be essential, as the firm may not have the cash. That is, the firm may have made a profit on the books, but the profits actually went into payments on equipment, which is an increase in capital, but taxable nevertheless. The IRS is not limited to collecting as income taxes only the money one may have.

The policy of taking maximum depreciation can result in book losses; the profit of the company may be represented by the difference between actual and book depreciation. For example, suppose that a contractor has $100,000 worth of equipment. He depreciates it at the rate of $35,000 per year, but the actual depreciation is only $20,000. His profit is reported as $15,000 less than it actually is; he shows a loss on his books although he really had a profit.

Overdepreciation to the point of causing a book loss occurs more often in real estate, since buildings have a long life and often actually increase in value over a part of their life, so the land may increase in value faster than the improvements are depreciated. The resulting book loss may be more than desired, as it can continue over a period of years and cannot be credited against later profits. Consequently, one may later pay income tax on profits which were never actually made, but which were used up by earlier losses which may no longer be deducted from profits.

6.8   THE BALANCE SHEET

A sample balance sheet, on the form recommended by the AIA and the AGC for the prequalification statement furnished to owners and A/E's is shown as Figure 6-1. The statement shows a net worth of $260,645 for the firm (the sum of capital stock and surplus, although the AGC form considers capital stock not part of net worth) with an accounts receivable/payable ratio of 363,951/203,026, or 1.71. This ratio is a key check of current ability to pay debts, as it represents the money available this month to pay current debts (sometimes a longer period is used, approximating the length of the average contract outstanding). The firm illustrated could readily reduce the payables, by using reserve securities. Some of the items are discussed below.

Cash. The cash has been allowed to fall quite low, with possibly just enough to pay the first January payroll, before some billings are received. Since there are a number of billings, one of them would probably be sufficient for the payroll; the existence of securities indicates that bank funds could be obtained, or some of the securities sold. *Listed securities* means that the securities are listed on one of the major stock exchanges and can therefore be cashed immediately. Unlisted securities may be equally good, but the price is not known daily, and the market is *thin*—that is, fewer shares are traded and several days or more may be needed to cash them. They are therefore not as valuable for cash requirements as are listed securities.

Accounts receivable from completed contracts. This information is more reliable than the sums due from work in progress, as the firm need do no further work to collect them; they are firm, not *maybe,* receivables. Other payments, especially retainages, may be delayed for minor items and cannot be relied on for payroll money.

Accounts receivable on uncompleted contracts. This is the important source of income for the company. The difference of this $323,072 from the amounts due to suppliers and the subs' $173,026, or $150,046, appears to be a company asset represented by working capital. That is $150,046 of the billing is for work done and already paid for by the firm.

However, this value cannot be confirmed without consideration of *all* outstanding contracts of the firm. It may be simply current overbillings—that is, billings in excess of work done. From the statement, the monthly amount of work done by this firm cannot be determined; but the current receivables should be for the current month, so *at least* $323,072 in volume is done each month. But since many firms collect at times other than at the end of the month, the volume could be much greater. This firm could have as much as $3 million worth of work under contract and be overbilled (that is, may have collected in excess of work done) 10 percent of this amount, or $300,000. Since this is more than the firm's net worth, the firm *could* conceivably be insolvent.

. 13. Give condensed current financial statement:

Condition at close of business December 31 19			

ASSETS — Dollars / Cts.

ASSETS	Dollars	Cts.
1. **Cash:** (a) On hand $ 200 , (b) In bank $2.523 , (c) Elsewhere $ ---	2,723	
2. **Notes receivable** (a) Due within 90 days	---	
(b) Due after 90 days	---	
(c) Past due	---	
3. **Accounts receivable from completed contracts, exclusive of claims not approved for payment**	1,540.3	
4. **Sums earned on uncompleted contracts as shown by Engineer's or Architect's estimate**		
(a) Amount receivable after deducting retainage	3,230.72	
(b) Retainage to date, due upon completion of contracts	2,500.0	
5. **Accounts receivable from sources other than construction contracts**	476	
6. **Deposits for bids or other guarantees:** (a) Recoverable within 90 days	---	
(b) Recoverable after 90 days	---	
7. **Interest accrued on loans, securities, etc.**	742	
8. **Real estate:** (a) Used for business purposes	4,201.6	
(b) Not used for business purposes	3,500.0	
9. **Stocks and bonds:** (a) Listed — present market value	1,572.9	
(b) Unlisted — present value		
10. **Materials in stock not included in Item 4** (a) For uncompleted contracts (pres. value)	1,232	
(b) Other materials (present value)	746	
11. **Equipment,** book value	2,507.6	
12. **Furniture and fixtures,** book value	1,732	
13. **Other assets**		
Total assets	4,889.47	

LIABILITIES

LIABILITIES	Dollars	Cts.
1. **Notes payable:** (a) To banks regular	3,000.0	
(b) To banks for certified checks	---	
(c) To others for equipment obligations	3,200	
(d) To others exclusive of equipment obligations	---	
2.*Accounts Payable: (a) Not past due	1,730.26	
(b) Past due	---	
3. **Real estate encumbrances**	2,207.6	
4. **Other liabilities**	---	
5. **Reserves**	---	
6. **Capital stock paid up:** (a) Common	1,500.00	
(b) Common	---	
(c) Preferred	---	
(d) Preferred	---	
7. **Surplus** (net worth) Earned $ 110,645 Unearned $	110,645	
Total liabilities	4,889.47	

CONTINGENT LIABILITIES

CONTINGENT LIABILITIES	Dollars	Cts.
1. Liability on notes receivable, discounted or sold	NONE	
2. Liability on accounts receivable, pledged, assigned or sold		
3. Liability as bondsman		
4. Liability as guarantor on contracts or on accounts of others		
5. Other contingent liabilities		
Total contingent liabilities		

* Include all amounts owing subcontractors for all work in place and accepted on completed and uncompleted contracts, including retainage.

14. Will you, upon request, fill out an approved form of detailed financial statement and an additional form of Job Plan and Equipment Questionnaire?   Yes

5

Figure 6-1
PREQUALIFICATION STATEMENT
(*Courtesy of The Associated General Contractors of America, 1957 E Street, N.W., Washington, D.C. 20006.*)

Real estate and stocks and bonds. These indicate that there has been a profitable history; the firm did not have to buy these items to do business and therefore at one time had surplus cash. The low amount due on equipment—$3,200 on a book value of $25,076—also indicates stability. The amount due on equipment *can* exceed the book value; a tractor on a five-year life basis may be depreciated 40 percent the first year, but if it is being paid for on a three-year contract, the amount paid the first year on the principal is much less. However, this stability may be due to a small volume of work in the past (or to an accountant who made the items come out this way, by the expenditure of relatively small amounts but large percentages of the items, on equipment balances due) so the relative volume of work undertaken by the firm is an important consideration.

The fact that there is a bank obligation ($30,000) even though the firm owns securities ($15,729) may mean merely that the owner wants to continue borrowing to assure availability of bank loans in the future, or it may mean that he needs the money only a short time and is looking to long-time appreciation of the securities.

6.9   EXPENSE AND RECEIPTS ACCOUNTS

The one-man house builder who puts his invoices in a box and pays cash used the box as his *expense accounts.* His receipts for bank deposits, or copies of receipts he gave to customers, were put in another box, his *receipts account.* At the end of the month, he adds them both up and calculates the difference. He is *closing his books* for the end of the period. This difference he must record for his tax report; it is his *income account.* If he adds up his tickets only once a year, the tax return is his income account. In either case, he may forget the boxes (except that he must store them for examination by the tax collector); they are *closed accounts.* They were temporary accounts, used only for that period. If he makes up a balance sheet (by counting his tools and adding bills for which he has not yet collected), he won't need to use these old boxes; they will not appear on his balance sheet. If he is collecting by jobs, he may keep his boxes by jobs; the boxes for unfinished jobs minus his receipts for these jobs will be a current asset and will appear on the balance sheet as the item shown in Figure 6-1 as *Sums earned on uncompleted contracts.* The tax collector takes little interest in a balance sheet, but it is an added detail for the contractor who wants to prove that he has a business with adequate capital or liquidation value.

At the end of the period, the expense and receipts accounts are transferred (closed) to the income account and to the net worth of the company (or paid out as dividends) by listing them on a sheet called the *Profit and Loss Statement.*

6.10   CHOICE OF TAX REPORTING METHODS

A contractor may choose from four methods of computing income for tax purposes: cash, accrual, completed-contract, or percentage. All these methods give the same results for contracts completed and paid for the same year in which started, and for overall calculation of profit when the firm is liquidated. However, the choice is important when a substantial proportion of contracts are in progress at the end of the year, and doubly so if the firm is expanding. The objectives of the tax accounting methods are:

1. To delay reporting income in the earlier year in order to use the deferred tax payment for another year, so that in the meantime it can be invested or used to reduce borrowing costs, and

2. To avoid paying taxes on money not yet received, which payment could seriously reduce operations, by reducing working capital.

The *cash method* of reporting income is the simplest, and also the safest method to avoid a capital shortage due to tax payments. Calculating income by the cash method is simple and probably more beneficial to the contractor than are the other methods. Accountants generally prefer other methods, since the cash method gives very little information on the state of the business. However, separate systems must usually be used for tax reporting and profit accounting in any case.

Under the cash method, money paid out is expenses or cost for the year, and money received. The difference is income. If the cash balance does not increase during the year, taxes will be paid on the cash taken out of the business. Since taxes are due on the net cash received during the year, there will be cash available to pay taxes. An expanding firm, which is using its profits to finance new jobs, will have no taxes to pay on these reinvested profits.

Equipment purchased, however, is not included as cash expenditures, but depreciation is expenses even though not paid out. Profits reinvested in equipment will be subject to taxes; this affects heavy contractors most seriously. Most of a building contractor's working capital is tied up in receivables (unreimbursed job costs), which are allowable expenses.

One disadvantage of the cash system is the lack of control available to the contractor. Its receipts are under the control of others. A slow year, with past work completed and new work started near the end of the year, may be reported as an unrealistic loss which in some cases may be undesirable; the contractor may have such losses for so long a period that he cannot use them all against future profits.

Under the *accrual method,* very similar to the percentage-of-completion method, income includes all amounts the contractor is entitled to bill, which is virtually all costs incurred even if not paid, plus his markup. Costs are all costs incurred, for which the owner can be billed.

The key advantage of the cash method, that no tax is paid on the profit included in the final payment until it is received, is retained in the accrual system. However, many contractors have included their retention in current income, and the method may not be changed without IRS (Internal Revenue Service) permission. The accrual method assumes that the contractor's end-of-year payment requests will be paid as submitted; if not, he will pay tax on it anyway.

The *completed-contract method* requires tax payment for each job in the year completed. The chief difficulty is that the IRS requires that a job be considered complete in the year in which it is finally accepted. The final payment is often not received for some time. The IRS considers that contractors "play games" with the completion date; it is certainly expensive to consider a job completed when the time and amount of payment is still doubtful. The IRS no longer allows a portion of the contract to be counted as cost and reserved for warranty work; in some cases, the contractor obtains final payment with the stipulation that some delayed work will be "warranty work"; the IRS would therefore consider the job completed before all work has actually been done. Of course, the only evidence of the final acceptance may be one letter, or none at all, in the contractor's files; to determine the time of completion, the IRS must examine both financial and contract files.

The IRS also no longer allows a contract to be counted as incomplete because there are claims or disputes pending. The completed-contract method is therefore less attractive for tax reporting than it has been in the past.

The *percentage-of-completion method* allows the contractor to place a percentage for income due to him on each contract at the end of the year. The method used to determine the percentage may differ from one contract to another, so the contractor has some discretion in determining his receivables. The percentage does not have to be the same as the billings to the owner.

The *accrual method,* when retention is counted as billings, is essentially the same as the percentage-of-completion method. The percentage-of-completion or accrual method can be used with billings as income and payables as expenses, and this is the simplest method for the accountant, followed by many contractors. It has the advantage that year-end billings may be high or low as fits the tax situation.

It is meaningless to say that the contractor's billings and even purchases and conduct of the work should not be affected by tax reporting. These operations vary widely for many other reasons, and the people involved should know that operations at year end should not be allowed to *increase* taxes; it is of advantage for the firm if the project manager is informed of the tax situation. For years, U.S. government personnel in comparable positions have been informed that they are to spend as much money each June as practical, so the payments will not be reported in the following fiscal year.

For example, let us suppose that a major shipment will arrive about December 31. If it is otherwise immaterial to construction, should the project manager arrange for the invoice to arrive early, to increase taxes under one method of accounting and decrease them under another? Even one week's acceleration or delay in an invoice may make a material difference in taxes, especially on the cash system. Suppose that a contractor has a $75,000 prefabricated steel building due. If he has a $25,000 line of credit at his bank, he can arrange a letter of credit or loan on the building itself for the $75,000. But the difference between the two methods of payment in taxes may be $75,000 in profit. A letter of credit is merely a guarantee of indebtedness, so he still owes the vendor. But a bank loan enables him to pay the vendor, and he owes the bank; he has paid out an additional $75,000 cash in the tax year. These kinds of transactions must be planned in advance, rather than completed without regard to tax effect and the record delivered to the tax preparer at the end of the year.

6.11 THE PROFIT AND LOSS STATEMENT (OPERATING STATEMENT)

This statement shows the expenditures and receipts of the firm for a certain period, usually a month or a year. Since contractors' statements for short periods are inaccurate and require considerable clerical work, contractors usually make a formal statement only once a year, with estimates at other periods.

There is no reason—in theory—that a contractor cannot calculate his work done and its cost for any period, just as a merchant does. However, the items carried by a merchant may be as few as a thousand kinds for a small store and ten thousand for even a fairly large one; the items a contractor handles are almost unlimited. Each piece of material for which the contractor pays material and labor is a different item. Items are standardized as far as possible, but each *standardization*[1] increases the inaccuracies. How does one determine, for example, the

---

[1]*Standardizaion* means to classify different things in the same category, whether they are purchases, forms, or kinds of material or labor. This cuts down clerical labor and makes it possible for managers to make decisions readily which would otherwise require investigation, and it transfers responsibility to the man who does the standardizing. Without standardization, we would not have automobiles or college degrees.

cost of the *last* piece of mortar or dirt to be cleaned off the floor? Or the *last* hole to be filled and retamped, after it has settled? Or the time the superintendent must spend to get the *last* subcontractor to repair the *last* damaged faucet? It is the cost of these items at the end of a job which are hard to determine, and for which the general contractor's cost is most variable. Estimates made during the job must therefore include a generous margin for finishing work, as well as an attempt to compute the cost of work partially done for which the available unit prices are for complete items. Estimates of jobs under way may be made, but the cost is never certain until the job is done.

Income may be reported for tax purposes by cash inflow and outflow (with a few exceptions, particularly the purchase of assets such as equipment), by the completed-contract method, or by the percentage-of-completion method of accounting. By the completed-contract method, all jobs are carried as work in progress, and no profit is recognized until a job is complete. Under the percentage-of-completion method, the percentage of completion is estimated at the end of the year, as is the amount to complete, and a portion of the expected profit is reported for that year.

The chief advantage of the completed-contract method is that there is no income tax on incomplete jobs; income tax is delayed as long as possible. The percentage-of-completion method is more flexible, however, and allows profit to be reported in a year convenient for the contractor. Keeping current costs on jobs under way, which is essentially a percentage-of-completion method of bookkeeping, does not prevent the contractor from using the completed-contract method for income tax purposes. Percentage-of-completion accounting can be quite inaccurate, especially when first begun; there is no reason to pay taxes on *possible* profits.

Sample profit and loss statement. An end-of-year statement (which may be established by the contractor on any date, but which may be changed only by permission of the IRS) for a completed-contract method of accounting may be as follows:

1.	Gross income for jobs completed during year	$1,300,000
2.	Cost of jobs completed during year	1,100,000
3.	Cost of overhead and undistributed charges during year	50,000
4.	Net profit before taxes	$ 150,000

Each of the items above could be detailed as desired. Most contractors keep costs detailed only by jobs, not by vendors or material, so they cannot state the amount of purchases from any vendor or the amount of any particular kind of material used. The vendor information is very useful at times—for example, if the contractor is deciding whether to set up a subsidiary operation as a specialty corporation or lumberyard. If machine accounting is used, each entry may be keyed to this information, and results may be obtained for any item for which the entries have been keyed. Whether machine accounting is to be used depends on the situation and the detail to which results are desired.

If a percentage-of-completion statement is to be prepared, it may be as simple as the following:

1.	Gross income for work done		
	Clark Job	37% × $250,000	$ 92,500
	Federal Building	100% × 500,000	500,000

Seamon County Hangar	70% × 300,000	210,000	
Oppenheimer Building	20% × 500,000	100,000	$902,500
2. Direct cost of work done		800,000	
3. Overhead and distributed costs		60,000	860,000
4. Net profit before taxes			$ 42,500

In the same manner as with the completed-contract method, many subsidiary accounts may be included in the statement above, but only the total is necessary to determine the profit. The architect's estimate may be used to determine the amount of work done, and *all* material costs may then be combined in Item 2, Direct Cost. The percentage-of-completion method therefore may be made simpler than the completed-contract method; it is essentially the same as cash accounting. Combining all direct costs in this manner, however, deprives the firm of the entire planning–direction–comparison cycle, which is the essence of the construction management process.

6.12   THE MONTHLY INCOME STATEMENT

A monthly income statement may be made in a firm with a cost-reporting system, provided certain assumptions or inaccuracies are allowed.

1. Any distribution of overhead accounts is approximate. It is particularly doubtful to consider estimating expense a general overhead item chargeable to current production or current profits. It is just as logical to charge estimating expense for the past unsuccessful jobs to the new job when it is obtained, or to put a flat charge on all jobs, independent of what estimating expense actually was while the job was being completed.
2. The estimates of current work done by the general contractor during a period as short as a month may be subject to large errors on long jobs; this applies both to materials and labor. In the long run, this will even out, but an error of even 5 percent of the job, when it shows up in the work done during a month, will be quite significant, compared to the profit for the month.
3. Large swings in monthly profits will occur due to one-time happenings, such as errors or favorable purchases. Closing out jobs and obtaining new work may produce disproportionate changes in anticipated profits.

6.13   OVERHEAD, PROFIT, AND DISTRIBUTED EXPENSES

Overhead costs have the attributes that:

1. It is inconvenient to charge them to individual jobs.
2. They are paid for a considerable period in advance.

3. They are purchased in minimum quantities greater than are needed for any one job.

4. They must be paid, or contracted to be paid, in advance of securing a contract, regardless of whether a contract is secured.

Overhead costs can readily be charged as a percentage of direct costs, as is conventional, without disruption of the cost system, as long as the gross contracts do not greatly vary and jobs are the same type and size. The contractor should always consider overhead cost changes. Construction planning is from job to job, but overhead costs are usually committed a year in advance.

A simple example of this is Jones Construction Company, which has been going along with $2 million a year gross with a 6 percent overhead cost, or $120,000 a year. One year Jones obtains a number of large jobs and doubles his gross. If he holds his overhead down, he will have greater profits and greater income tax. So he decides, during the course of the year, to expand his estimating department, management staff, and other facilities so that he may continue the higher rate of business sales. As he goes into the next year, he has a $240,000-a-year overhead rate, and is faced with a possible loss when his gross income reverts to $2 million. His logical decision at this point, which few contractors will do, is to cut back his cost of obtaining new business as fast as it becomes evident that the new business is not being obtained. His expansion should have been of this sort. His expanded overhead was mostly out of profit originally, and should be so recognized; by cutting back he is accepting his loss.

Overhead cost planning is therefore based on current conditions. If expansion is planned, it requires greater cost, which in other businesses would be called sales cost. But overhead consists of a fixed short-term cost based on need. A contractor should not believe that his kind or size of job is limited by his overhead cost; if he has qualified people, variations in job size are merely variations in cost. Methods of reporting overhead cost vary greatly, but the fixed-cost portion, which cannot be cut, is a lesser percentage for a larger firm. The least flexible contractor is the popular "office in his truck" operator, who must cut his standard of living to reduce his overhead, and increase gross income by increasing already long working hours. These increased hours are often at the expense of efficiency, and either way the small contractor is the first to discontinue his business when conditions change.

Overhead is usually considered expense which cannot logically or easily be charged directly to a job; sometimes expenses of delivery and payroll taxes and insurance (social security, workmen's compensation, unemployment and liability insurance) are charged to overhead, even though they may be broken down by jobs. A contractor who rarely loses money on a job, and who has a formula for determining bids which is basically a method of determining his competitor's bid, may not need a precise method of determining his costs. A firm bidding large jobs under stiff competition cannot afford to put in a percentage for overhead without a determination of its actual overhead costs. Interest on the firm's capital is usually ignored as a cost, but interest on borrowed money may be charged as a general overhead item. In this way the cost of capital requirements for work is ignored.

Distribution of overhead is recommended to the extent that costs are due to a particular job, including the following:

1. All direct costs associated with payroll, including payroll checks and clerks' time, are to be charged to that payroll.

2. All the estimator's time associated with the job are to be charged accordingly. The original estimate is usually not a large item, but estimators also handle management matters during the course of a job.

3. The office executives' time and expenses are to be charged to the job, to the extent that the cost can be identified. This reduces their time to two charges—new business and management of old jobs.

4. Overhead costs are to be charged not only to the job, but also, so far as possible, directly to the trade involved—particularly when a cost is associated with a subcontract item as opposed to an item of work done by the contractor.

One advantage of obtaining this information is that the actual overhead costs can be estimated much more accurately. This is important not only to avoid high-overhead jobs (like hospitals) and to gain low-overhead jobs (like warehouses) at the same markup, but also to determine when the existing office force will be inadequate. If the estimate calls for overhead expenses beyond the expenditures of your office, it is a signal that the office will be snowed under—an occurrence which often makes contractors wonder if they are in the right business. A job force can be inefficient by itself, but the office force's failure to accomplish their work is reflected by higher costs on *all* jobs.

If a general contractor sublets 70 percent of a job, his direct labor may be as little as 10 percent and his overhead as high as 5 percent; yet the overhead may be thrown in as a percentage and the direct labor estimated in great detail.

Another advantage of the overhead distribution recommended here is to make possible certain comparisons, to decide whether work is to be sublet rather than done with the contractor's own forces. It is easy for a contractor to reason that his labor and material is less than he would pay a sub, and he has his overhead "anyway." A total estimate of cost will expose not only the cost of supervision of contractor-force work, but the cost of estimating and of the officers as well. It may be more economical to hire consultants or office services for some of the "overhead" functions.

6.14 SAMPLE MONTHLY STATEMENT

If overhead is fully distributed for work done for the jobs, the remaining office expense will be the cost of obtaining new work, which is on the monthly statement as an expense. The simplest way of doing this is by applying it as a current cost in proportion to the work in place, for example:

Futuristic Construction Company
Statement for
July 1973

Work in place for month	$230,000
Overhead cost for month	24,000
Total cost of work done for month	$254,000
Estimated value of work done for month	263,000
Estimated profit for month	$ 9,000

The preceding items are summaries of extensive calculations. The work in place, for example, must include payments to subs, materials on site, and retained percentages. A number of jobs could be at a loss, but the losses are concealed in the total. For checking the efficiency of various managers, it may be desirable to show the totals broken down by jobs, or even by materials and labor.

6.15 CASH REQUIREMENTS

As a contractor, you must operate with your available cash. It is important that the cash requirements of each job be known before the job is bid; if you have a known capacity for additional work, no detailed computation is required. But once you have a contract, a forecast should be made to determine how much money will remain for *another* job.

This determination is made by a calculation of the money to be received as progress payments during the job, and how much must be paid out during the same period, considering each month separately. The following pages illustrate how this may be done.

Cash forecasting requires "making up" accounts and expenditures and may be quite detailed. An example is illustrated in Tables 6-1 through 6-5. The first step (Table 6-1) is to make an A/E's payment breakdown from the estimator's bid estimate. The *estimated cost* items are taken from the summary sheet of the bid estimate. The amounts are adjusted in order to reflect items such as overhead and profit which are not accepted as separate items in the payment breakdown, and to collect as much money early in the job as possible. In the example, the first four items—clearing, foundations, masonry, and concrete—have been increased $51,000, or 20 percent over the estimate, although the profit and overhead are only 6 percent. The amounts for *pay items* are adjusted down the line and are checked against the total. However, these amounts are not collected in full as progress payments; and A/E retains part of the amount, here assumed to be 10 percent. This 10 percentage is deducted (and shown as a final payment at the end) to give a realistic method of computing the cash received.

Figure 6-2 illustrates the process of payment and the various creditors on a construction project.

The first payments originate with the construction loan, usually made by a commercial banker who places a mortgage on the property. If the loan is comparatively small and the owner is substantial, this may be an unsecured loan, which has no claim on the property.

Before starting the job, the contractor may obtain a line of credit, usually unsecured, or secured by the current billing from its banker. This loan is usually on a portion of the contractor's working capital. This money is used to pay contractor's employees and materials purchases. The subcontractors obtain loans in the same manner as the general; their bank loan is secured by the sub's lien rights, which usually are superior to those of the general contractor. The suppliers and laborers are paid from these funds.

The owner draws monthly on the construction loan for construction during the month, and pays the contractor. The contractor in turn pays the bank for the monthly loan, and pays the subs their share. The subs in turn pay their bank. The monthly payment will not pass through those banks who make their loans for 90 days or longer; such loans are commonly extended repeatedly until the end of the job. However, since bank examiners frown on too many such extended loans, the contractor may "roll over" these loan by paying them off, then signing a new note and promptly borrowing the money back. If the contractor uses money due the subs to do this, it would be technically the same offense as diverting the subs' payment to other purposes, although it could be done while the sub's checks were in transit.

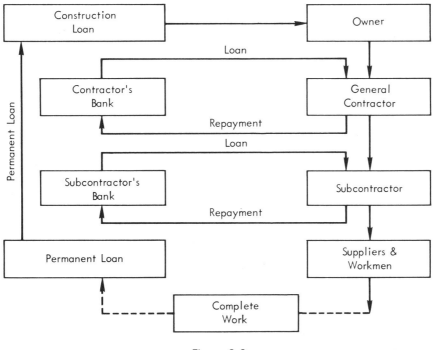

Figure 6-2
FLOW OF PAYMENTS

The permanent loan may pass through several hands in a day; checks are even written ahead and mailed with the payment is received.

Upon completion of the work and sometimes after the building starts to return a previously agreed income, the permanent loan proceeds pass to the construction lender through the owner. Since the permanent lender is last in line for claims during construction, he will take adequate legal steps to protect his rights, including an assignment of the original construction mortgage to predate his claim of record and place himself at the head of the claims chain.

The principal lending risks are by the construction lender and the respective banks. The construction lender may not have adequate security if the work is unfinished or permanent financing is not complete, and the banker relies on the borrower's equity in the business.

Each of the borrowers in this chain usually, but not always, adds some capital of his own, and has additional capital invested in other work. It has happened that many of the contractors went out of business, and all lenders except the one granting the permanent loan lost money, because the job was never completed.

6.16 PAYMENT METHOD BREAKDOWN (CASH FLOW)

Payments by contractors generally are one of three types—those which *must* be made on the due date if the firm is to remain solvent (usually only payroll is considered to be in this category), those which *should* be made on the due date, in order to receive a discount and maintain one's credit rating (materials and those subcontracts for which definite promises have been made), and those for which the contractor is responsible only to pass the payment on to the subcontractor when it is received from the owner. Some firms consider all their debts to be in the first category (that is, to be paid when due), even if there is no legal obligation as in the case of subcontract payments. These firms do less business with the same amount of money, but presumably they get better prices and show a higher profit.

In this example, the *cash outflow weekly* includes payroll payments; it may also often include items which are quoted with a discount in seven or ten days instead of a discount if paid by the tenth of the following month.

It is presumed that the contractor will pay his monthly bills from the owner's payments, or at least will keep these separately in the breakdown to see how much cash the bills will require. These are designated *cash outflow monthly*. Subcontractors are separated as a third category; even if the contractor has a policy of meeting his bills on the tenth without regard to his receipts, it is desirable to keep these monthly and subcontract payments separate; the manager can then assess the probability that cash will be required on this job for that purpose, and consider the requirements of other jobs at the same time. That is, he would assume that *not all* his receipts will run late at the same time.

Payment inflow by time shows that time that various items of work will be done. It is the same as that used to prepare the Gantt chart and is prepared by the project manager in accordance with scheduled job progress. (See Chapter 7, Figure 7-1, for a Gantt chart.)

6.17 INACCURACIES IN ESTIMATES

The calculations of cash requirements are subject to error in size of items and in time of completion. An inaccuracy in the estimate will make a corresponding change in the final cash requirement. A contractor usually assumes that the subcontract amounts and timing (which under the usual arrangement requires no cash on the part of the general contractor) are firm, and his own labor costs as subject to error. Either may be in error, and the contractor should have a cash reserve to cover both high labor costs and the possibility that a subcontractor may default and the work will have to be relet at a higher price. This contingency may be as great as the profit, requiring that a corresponding amount of cash be invested throughout the job.

6.18 SUBCONTRACTORS' RETENTION

The *payment method breakdown* portion of Table 6-1 shows which contractors are to receive their retention on completion of their work, which may be much sooner than the end of the job. In the illustration, the masonry, roofing, and tile contractors are shown as paid currently, while all other contractors are given their retention (denoted by the figure "R" in parentheses) as their work is completed. The decision whether to pay subs currently depends on the individual sub—particularly how much lower a price he will give if paid on completion of his part of the work. The expression often used is that subs *carry the general,* that is, that the subs, but not the general, have an investment in the job; but the fact that the general has no investment in the job (as, in this case, is true for one month) does not mean that the subs put up more than their normal monthly payment and retention. In the example, the general is paying the masonry

Table 6-1. Cash Forecast
James County Jail

	Pay Item Breakdown			Payment Method Breakdown			Payment Inflow by Time — Month					
Item	Estimated Cost	Amount for Pay Item	Cash Inflow	Cash Outflow Weekly	Cash Outflow (Matl) Monthly	Subs Cash Outflow	1	2	3	4	5	6
1. Clearing	$ 2,000	$ 10,000	$ 9,000	$ 1,000	$ 1,000		$ 4,500	$ 4,500				
2. Foundations	10,000	20,000	18,000	4,000	6,000		6,000	12,000				
3. Masonry	127,000	140,000	126,000			$127,000		30,000	$ 45,000	$ 45,000	$ 6,000	
4. Concrete	110,000	130,000	117,000	70,000	40,000			30,000	57,000	30,000		
5. Plastering	40,000	42,000	37,800			37,800 (R.2,200)				20,000	17,800	
6. Metal Work	50,000	60,000	54,000	10,000	40,000			27,000	27,000			
7. Doors, Windows	10,000	12,000	10,800	2,000	8,000				5,400	5,400		
8. Roofing	10,000	11,000	9,900			10,000				9,900		
9. Tile	5,000	5,500	4,950			5,000					4,950	
10. Electrical	100,000	100,000	90,000			90,000 (R.10,000)		10,000	10,000	40,000	20,000	$ 10,000
11. Air Conditioning	100,000	100,000	90,000			90,000 (R.10,000)			10,000	50,000	30,000	
12. Plumbing	40,000	40,000	36,000			36,000 (R.4,000)		10,000	10,000	16,000		
13. Painting	15,000	13,000	11,700			13,500 (R.1,500)					6,000	5,700
14. Site Work	10,000	9,000	8,100	4,000	6,000		2,000	2,000	2,000	2,100		
15. Overhead	40,000			10,000	30,000							
16. Profit	23,500											
17. Retention		69,250	69,250			27,700						69,250
18. Total	$692,500	$692,500	$692,500	$101,000	$131,000	$437,000	$ 12,500	$125,500	$166,400	$218,400	$ 84,750	$ 84,950

$101,000
131,000
437,000
23,500 (Profit)
—————
$692,500 Check

$ 12,500
125,500
166,400
218,400
84,750
84,950
—————
$692,500 Check

contractor on completion of the work by marking up the masonry work amount, payment for which is received in the early part of the job. But before the work is over, the general must put up this markup himself. Two other subs—roofing and tile—are being paid in full on completion. The larger contractors—electrical work and air conditioning—are being paid directly the amount received for their work. These contractors, whose work extends over the life of the job, may themselves collect early in the work; subs, like the general, try to collect early also but their billing does not usually affect the cash position of the general contractor one way or the other.

6.19   WEEKLY COST ITEMS BY MONTHS

Table 6-2 shows the anticipated weekly cost items. The largest item is for concrete work. If progress is 50 percent greater during the third or fourth month, cash requirements increase by $15,000 during this period, illustrating that more frequent payments are necessary for a cash-short contractor to increase his progress. Cash requirements are determined by volume/period, not just volume/month. If the volume rises and the period shortens, cash requirements may be the same. When payments are being made by a mortgage company rather than by an owner, the billing is often based on the stage of the work rather than done on a monthly basis. In this case, the contractor may *reduce* his cash requirements by greater progress. This is because the payment actually covers the last week's work, which is an increasing part of the total, as the frequency of progress payments increases.

Monthly Cost Items, Table 6-3, and *Subcontractor Payments,* Table 6-4, are next separated. Those subs who are shown in the breakdown at their actual value and are paid the amounts received for their work—such as electrical, air conditioning, and plumbing subs—may be omitted from all these schedules, since they do not affect the general contractor's cash requirements.

Table 6-2. Weekly Cost Items

Item	Total Cost	Costs by Months 1	2	3	4	5	6
1. Clearing	$ 1,000	$ 500	$ 500				
2. Foundations	4,000	2,000	2,000				
4. Concrete	70,000		10,000	$30,000	$30,000		
6. Metal work	10,000				5,000	$5,000	
7. Doors, windows	2,000				1,000	1,000	
14. Site work	4,000	1,000	1,000			2,000	
15. Overhead	10,000	1,000	2,000	2,000	2,000	2,000	$1,000
TOTALS	$101,000	$4,500	$15,500	$32,000	$38,000	$10,000	$1,000

Table 6-3. Monthly Cost Items

Item	Total Cost	Costs by Months					
		1	2	3	4	5	6
1. Clearing	$ 1,000	$ 500	$ 500				
2. Foundations	6,000	3,000	3,000				
4. Concrete	40,000		10,000	$10,000	$20,000		
6. Metal work	40,000		10,000	30,000			
7. Doors, windows	8,000			4,000	4,000		
14. Site work	6,000	1,000	1,000		4,000		
15. Overhead	30,000	5,000	5,000	5,000	5,000	$5,000	$5,000
TOTALS	$131,000	$9,500	$29,500	$49,000	$33,000	$5,000	$5,000

Table 6-4. Subcontractor Payments

Cost Item	Total Cost	Payments by Months					
		1	2	3	4	5	6
3. Masonry	$ 127,000		$30,000	$45,000	$ 45,000	$ 6,000	$ 1,000
5. Plastering	40,000				18,000	18,000	4,000
8. Roofing	10,000				9,000		1,000
9. Tile	5,000					4,500	500
10. Electrical	100,000		10,000	10,000	40,000	30,000	10,000
11. Air conditioning	100,000			10,000	50,000	30,000	10,000
12. Plumbing	40,000		10,000	10,000	16,000		4,000
13. Painting	15,000					6,000	9,000
17. Retention*							
TOTALS	$437,000		$50,000	$75,000	$178,000	$94,000	$39,500

*Included in 6th period payments.

6.20   SUMMARY OF CASH FLOW BY MONTHS

The inflow and payments are shown together here to calculate the overall cash requirements. Weekly payments are advanced a month, since unlike other payments, they must be paid before an owner's payment is received, even though the owner's payment is prompt. During the first month of construction (month "O"), current bills are paid to the amount of $4,500. Sometimes a pay item is set up for *mobilization*, upon which the contractor may draw without having completed any particular work; in such a case, even this first $4,500 may be collected before it is actually expended. In the example, it is assumed that this is not the case; the contractor must put up the first $4,500.

During the next month (month 1), the payrolls begin at the rate of $12,500 for the month, and debts for monthly payment are incurred amounting to $9,500 for the month. After the end of the first month, $12,500 is received, which will pay the bills and leave $3,000 toward payroll. At the end of the "O" month, some payment would have been received, but this is ignored, as is the fact that the second month's payroll is not covered at the time incurred. If the payments and receipts are compared during the middle of the "O" month, the balance would be different. It is assumed that the payment received at the end of the "O" month is sufficient to pay the payroll of the month 1. This is one of many approximations which reduce the accuracy of the calculation. Another approximation is the payment received at the end of one month, which will cover those payments which are actually made a week to 10 days after delivery. The delay between men working and being paid, usually three to four working days, is also ignored.

The cash flow for month (Table 6-5) indicates the net payout or receipts for the month. For the first two months, payments exceed receipts as the expenditures grow faster than the enlarged pay items can cover them. By month 2 this situation has leveled off, and the contractor need put no *more* money into the job until month 4, when the retention exceeds the amount the contractor has allowed for overhead, profit, and enlarged early pay items. This must happen sooner or later, since an enlarged pay item at the beginning *must* reduce a pay item somewhere else. The maximum demand comes at month 5, just before final payment is received—here,

Table 6-5. Summary of Cash Flow

	Cash Flow by Months						
	0	*1*	*2*	*3*	*4*	*5*	*6*
Cash inflow*		$ 12,500	$125,500	$166,400	$218,400	$ 84,750	$ 84,950
Paid to subs			50,000	75,000	178,000	94,500	39,500
Paid monthly		9,500	29,500	49,000	33,000	5,000	5,000
Paid weekly	$ 4,500	15,500	32,000	38,000	10,000	1,000	
Cash flow for month	-4,500	-12,500	+14,000	+4,400	-2,600	-15,750	+40,450
Accum. total (Cash in job)	$-4,500	$-17,000	$ -3,000	$ +1,400	$ -1,200	$-16,950	$+23,500

Average cash requirement $8,250.

Note: Minus (−) indicates investment; plus (+) indicates that money is being taken out of the job.

For work paid for before the 10th of the following month.

230

$16,950. If the work is completed, but the retention not paid, the contractor has even more invested; the payment ($69,250) less the subs' share ($27,700) and his anticipated profit ($23,500), or $18,050. If the profit does not materialize, he also has that amount of his own money invested.

Note that of the cash requirement just before the end of the job, $11,000 occurred because the masonry subcontractor was paid in full during the course of his work (10 percent of his contract). One may say that this was covered by the increase in the pay amount at the first stage; but the contractor could still have increased this amount and not have paid the sub his retention. There is not necessarily any correlation between the two.

6.21 PAYING MONTHLY BILLS

Suppose that the contractor had elected to pay his material bills on time, regardless of when the owner's payment was received. The maximum requirement for these bills would have come in the third month, when the bills were $49,000. This is a one-time payment, and the cash requirement does not continue. It would therefore have been of advantage to the contractor to borrow for the few days the owner's payment was late. It also establishes a definite cost incurred by the contractor, if he later can make a claim on the owner for the interest because the owner delayed payment. If the contractor also held himself responsible to the subs, he could have paid them with another $75,000, making the total cash requirements at that time $165,000—the amount of the late payment, less the $1,400 the contractor was ahead (had received cash in advance) after the payment was made. So the policy of paying the subs on time could have raised the *maximum* cash requirements of the job from $17,000 to $163,400! Since the contractor may have obtained a bond for the job with as little as $40,000 in the till, it is unlikely that any but the strongest contractor would pay all bills on the tenth. As a practical matter, whether bills are paid promptly depends on the relationship with each supplier or subcontractor and on the contractor's total obligations at that time. By requesting a few larger suppliers to wait a few days, the smaller suppliers can be paid promptly.

The *average* cash requirement is the algebraic sum (deducting the surplus of some months from the shortage in other months) of the cash in the job, divided by the number of months; in the example, the average cash requirement is $8,250. If all jobs were scheduled so that surplus funds from one shifted to others perfectly; this average would be the maximum cash requirement. The surplus funds so shifted must be the contractor's, not payments received in advance for the subs. Diverting funds other than contractor's capital and profit, as pointed out previously, may result in criminal charges and revocation of license. Managing the cash requirement so that the requirement is constant would be a rarity. The general cannot control his own rate of progress, because of supply and subcontractor difficulties; consequently, he cannot control his cash requirements.

6.22 DEPRECIATION

The actual cost of items produced, as complete construction projects, services for them, or parts of jobs, can be computed by any convenient method which satisfies the manager and owners of a firm. Some of these costs are necessarily approximations. The overall profit can be calculated by the difference between income and payments, with one exception: *expenditures for capital goods,* which is not allowable as an expense when paid out. The manager, to decide whether to buy or lease equipment, may use any method of calculations he wishes; but for income tax purposes, the allowable methods are restricted. The depreciation method used determines *when* the cost of capital goods (which is taken here to mean equipment) is to be charged as a cost of doing business—that is, how much a construction job may be charged for the use of equipment.

The contractor normally wants to charge his equipment off as expense as soon as possible to reduce his current taxes (which will increase taxes in a later year), using this money as capital for a time. Some contractors, however, are not interested in this *accelerated depreciation* because the manager wants to show as large a profit as possible for the owner-stockholders. Stockholders are aware that profits may be greater than reported for tax purposes, and many annual statements show estimated profits separately from tax profits. The income tax which has been deferred is a liability to be paid in later years.

By charging off as large a part of the equipment cost as possible, the contractor may sell the equipment and show the profit on sale of equipment, not on his business. This profit is treated differently from ordinary business profits; at one time it was treated as capital gains, so the income tax rate was half as great as on ordinary income. Presently, with some ifs and buts (which apply to all statements in this brief consideration of complicated tax laws) this profit is tax-free as long as it remains in the business, but eventually it must be shown as profits.

Depreciation is the allowance for the cost of exhaustion, wear, minor damage, and obsolescence of equipment. It may be calculated by a variety of ways:

1. *Straight-line depreciation,* or a constant dollar amount based on the original cost. For example, a $1,000 machine which has an estimated 4-year life is depreciated by 25 percent of $1,000, or $250 per year. The advantage of this method is its simplicity. In an accounting sense, it is not conservative in that it usually overstates profits when equipment is new. Few machines can be sold for their book value determined in this way during the first portion of their useful lives, and on older machines the reported depreciation is too great.

2. *Declining-balance depreciation,* by which a constant percentage of the book value of the machine is charged each year. If 25 percent were being used, the first year's depreciation on the machine above would be 25 percent of $750, or $187.50, and so on. There would always be a residual value unless a salvage value were set as the bottom book value, after which no depreciation would be charged. Since if the same percentage rate were charged, as in straight-line depreciation, this method would result in a much longer *useful life* (that is, the time necessary to charge off the value of the machine), a larger percentage is normally used than would be used in straight-line depreciation. For a 4-year life, tax rules allow this to be double the straight-line rate, or 50 percent in this case. So the allowable first year's depreciation is 50 percent of $1,000, or $500; the second, 50 percent of $500, or $250.

    The declining-balance method is more realistic in determining the sales value of equipment than is the straight-line method, and it provides a faster allowable write-off. It is also more realistic when applied to used equipment, which can be depreciated at the same rate as new equipment without consideration of its age. The straight-line method requires determination of the new value and age, if used equipment is to be depreciated on the same basis. Tax authorities do not clearly state the allowable life for depreciating used equipment, which is a shorter period than for new equipment.

3. *The sum-of-the-years'-digits method,* which is similar to declining-balance depreciation, except that the rate during the later years is high enough to reduce the balance to zero. With this method, the annual rate is obtained by dividing the years of life remaining by the sum of the numbers to that point; for a machine with a 4-year life, the first year's rate would be $4/1 + 2 + 3 + 4$, or 40 percent; the second year $3/1 + 2 + 3$, or 50 percent; the third $2/1 + 2$ or 67 percent and the fourth $1/1,100$ percent. This method offers an initial rate nearly as high as the declining-balance method and a higher rate in later years. It is generally more satisfactory, but calculation is more complicated since the annual rate varies.

4. *Unit-of-production method,* by which depreciation may be charged on a unit price per hour or per cubic yard. This is a more accurate method for internal bookkeeping and is used by most firms internally but not for tax reports. That is, the annual depreciation is divided by the expected hours of work, to arrive at a figure which represents the cost of that part of the operating expense, for internal bookkeeping. The chief disadvantage for tax purposes is that when the equipment is idle, it may not be depreciated, resulting in higher cash requirements for income tax payments at the worst time, when work is slack. If usage is quite high, however, there is always the possibility that a machine may be depreciated to zero in a year—which is quite possible, if a conservative number of hours, such as 1,500 hours or so per year, has been used for determining the hourly rate.

   A similar hourly or per-cubic-yard year rate may be used profitably for machines built for a particular job of known size. A number of years ago, the largest truck in the world was built to haul on one job, where the haul was downhill. It could not be used elsewhere because it was powered for the downhill haul and could not pull its rated load on a level road! Such a machine can be depreciated by yardage, since the yardage (although not the time to complete the work) is known when the machine is purchased.

Technically, the IRS does not prescribe depreciation rates but merely examines and approves (or disapproves) the rate used by contractors. It publishes allowable standards, but it does not automatically disallow other methods. The present minimum approved rate for general contractor's equipment is 4 years, with 8 years for office equipment and 2 ½ years for automobiles. The highest depreciation rate for automobiles is obtained by using a 3-year life, however; if less than 3 years is used, the declining-balance method may not be double the straight-line method (that is, the *double declining-balance* method may not be used).

6.23   TERMINOLOGY IN TAX ACCOUNTING

Certain terms peculiar to tax accounting, though seldom encountered by the construction manager, are basic to understanding allowable tax depreciation rates. A *vintage* account contains the value of similar machines bought the same year and subject to the same depreciation rate. By combining several machines, both bookkeeping and checking by the IRS is simpler. Some of these combinations are quite illogical; a power chain saw, with a life of less than a year in steady use, may be grouped with a generator that will last twenty years or even more; both, however, are items of construction equipment. A *guideline class* is a group of machines which the IRS considers identical for tax purposes. Automobiles, trucks, construction equipment and office equipment are the guideline classes with which a contractor is concerned. A vintage account contains the records of the equipment in the same guideline class purchased in the same year. *Half-year* and modified *half-year conventions* refer to the date on which depreciation of equipment begins; the person buying equipment should be aware of the company's advantage in this respect. It may be that a $100,000 bulldozer, bought at the end of the year, may be depreciated beginning six months earlier. With a 50 percent rate for 6 months, this $25,000 depreciation may delay income taxes of as much as $12,500, which will not be entirely paid for several years. This is a considerable help to some firms, in financing the purchase; the government may be making your down payment! *Depreciation reserve* is an account which shows the reduction in value of equipment as it is depreciated. A $1,000 machine, for example, is always a $1,000 machine; but when it is depreciated fully, the depreciation reserve balances the $1,000, so the balance sheet value is zero. Viewed another way, it is a way to charge cash income without showing a profit. If you rent a machine for $4,000 bare (no gas, operator, or repairs) and receive $4,000 cash, it *may* be charged to depreciation reserve instead of to profits. You have the money, but you have no profit to pay taxes on.

If a machine included in a vintage account (combined with other machines, that is) is sold, the money received may be charged to the depreciation reserve. What happens to it eventually is rather complicated, but you have the money, as in the preceding renting example, and you do not have to pay income tax on it for some time. The manager should be able to call on his accountant to determine when such situations may exist.

A *repair allowance* has been given by the IRS as its idea of the maximum amount which should be spent on a machine for repairs each year. The IRS rate is now 12.5 percent for construction equipment and 16.5 percent for automobiles. If you spend *less* than this amount, the IRS will accept it; if you spend more, they are inclined to believe you are rebuilding a machine, or engaging in such extensive work that it should be capitalized rather than charged as a current expense. The IRS 12.5 percent allowance is approximately the same as the AGC *equipment ownership expense*, an estimate of what such repairs should be. You should end the tax year with this amount spent or the equipment in good condition; delays which require higher expenses may require justification. Improvements to a machine which raise its production as much as 25 percent are not chargeable to repairs in any case.

6.24   HISTORY OF DEPRECIATION ALLOWANCES

Since many firms use tax interpretations which no longer need to be followed, it is useful to know the development of present rules. After the introduction of income taxes in 1913, there were no formal rules until 1933, when a 25 percent reduction of depreciation lives (with the effect of raising taxes) was proposed in the Congress; this was not passed, but the Treasury Department undertook to accomplish the same effect by requiring taxpayers to prove the justification for their depreciation rates. The *IRS Bulletin F,* which has been the authority for many years, first appeared with suggested depreciation lives in 1931 and was revised several times in later years. In 1948 the declining-balance method of depreciation was first authorized, but the double declining-balance was not allowed until 1954. In 1962, *Revenue Procedure 62-21,* the so-called *Guidelines,* was established and the *Reserve ratio test* instituted. This restricted the percentage of cost which could be charged off and prevented equipment from being fully depreciated. The guideline lives were about 20 percent shorter than the *Bulletin F* lives, increasing allowable depreciation rates. In June 1971, the *Asset Depreciation Range* rules were adopted, making it possible for equipment classes to be fully depreciated, and again decreasing allowable lives (and increasing allowable depreciation rates). From 1942 to 1962, the proceeds of sale of equipment above its book value was taxable at the capital gains rate, which gave a tax reduction (not merely a tax delay, as is the case with adjusting depreciation rates) to firms with high depreciation rates. This was changed so that such profits became ordinary income in 1962, but the 1971 rules permit charging this income to the depreciation reserve—in effect, delaying the income tax until after the entire group of equipment has been depreciated to zero value, or sold. Income tax rules are constantly being changed, both by legislation and by court decisions. The manager should know the rules used in his own firm; *if he does not use them to his advantage, they will work to his disadvantage.* The manager should plan the method to be used and check the effect of further purchases and sales before they are made.

6.25 INFLUENCE OF DEPRECIATION RATES ON OPERATIONS

Generally, the depreciation rate to be used is entirely a tax matter, in that equipment should be bought and sold according to operation considerations, except as cash availability and interest cost considerations may affect a purchase or disposal. However, there are situations in which the depreciation rate has more important influence on the conduct of the business.

When a low depreciation rate is being used, as a straight-line rather than a declining-balance method, this affects the actions of department heads whose earnings are calculated on the basis of the earnings of their departments (which is the case, even though there is no formal "calculation" or the earnings). If a manager must choose between buying a new machine and some other alternative—obtaining an old company machine, buying an old one, or renting—he will decide on the basis of the method used for depreciation by the company. If the company depreciates a new bulldozer at 20 percent per year, based on the new value, a purchase becomes very attractive. If the manager is being charged the same dollars-per-year rate for a 5-year-old machine as for a new one and the new machine carries a one-year guarantee for parts and labor, he certainly will be inclined to purchase the new one. In a large company, I made every effort to obtain the new machines and those just depreciated to $1.00 (at which time there was no further depreciation). One would look for a machine in its sixth year, with a 5-year life, rather than a 14-year-old machine. Since the firm did not include overhauls in the charge for the machines (but it was charged against the manager unlucky enough or conscientious enough to have the machine when it was overhauled), the advantage of the new machine was even greater. *Regardless of how the equipment may be depreciated for tax purposes, internal charges should reflect true costs.*

The decision to retain old machines and the charges made for old machines affects the supervisor's decision to retain them. A machine becomes valueless when the time lost for repair creates a cost higher than the total costs of depreciation, interest, and maintenance. But if the value is zero, there is no depreciation or interest, and the comparison is between cost of lost time and maintenance cost. A machine can normally give 85 to 90 percent availability and becomes uneconomical for most uses at less than 75 percent. But for emergency or occasional work, it may be useful with a much lower availability ratio; that is, if you need a machine 10 percent of the time, it is repaired during unneeded time, and the availability during the time it is needed becomes very high. Such machines may be worth more per hour, while working, than are regular machines, since they work only during emergency periods. The old bulldozer, capable of working a day or two at a time, can be valuable to keep on a job for emergency drainage work or moving stuck vehicles.

If machines are sold at a higher price than the book value when vintage accounts are used, the remaining machines acquire a lower book value; when the last machine in the account is sold, this profit becomes taxable. Selling an old machine may then cause an income tax liability on the account received for the machine and give only a small increase in working capital.

Some managers take the view that if a machine is available it will be used, in spite of high cost and detriment to the work. Automobiles are usually used as long as they will roll; yet a $75 truck may be quite efficient if it is needed primarily as a storage platform, to reduce hand hauling and rehandling of materials.

6.26   EQUIPMENT ACCOUNTS AND REPLACEMENT

A vintage account, which offers tax advantages, is a very crude measure to determine the operating costs of equipment, to make decisions on what kind to purchase, and to decide when equipment should be replaced. More detailed accounts are kept for internal cost accounting. Complicated formulas presented by the engineering academic fraternity to determine obsolescence and equipment costs are practically useless because the necessary data to use these formulas is not available.

The accountant and manager decide the number of machines to be combined into a cost account; the manager determines the criteria, which are often changed on the advice of the accountant who points out the cost of detailed accounts. An account may be for one machine, when the firm has 30 identical models, or all possible machines may be combined into a single account. If the manager assumes that costs are different for each unit and that the costs obtained by individual records will be useful, an account must be kept for each machine. The assumption that one machine is different from another, when both are the same model and age, is a rather mystical intuition rather than a belief founded in the methods of assembly-line production. If a machine has had a breakdown in the past, it would logically have fewer in the future. Some machines are known as "lemons," because of their constantly high repair rate, but such a machine may suddenly become the one with the lowest costs for years afterward.

There is an advantage in differentiating between machines under different supervision, but this supervision may continue for a year or more before the cost data are significant. It is recommended that all units of the same model be considered identical; under the same conditions, breakdowns occur on a random basis and the difference between machines proves nothing. Likewise, the repair rate on machines does not rise with age; an older machine has a higher repair rate than a new one, but after 2 or 3 years the difference is lost. The age of equipment, as determined by the year it was manufactured, is not important when repairs are no longer being made on original equipment.

Assuming that machines of the same model are identical, they may be listed in one cost account, which will include no other machines. The measure of efficiency then becomes a comparison of one manufacturer or model against another. This method is reliable only if there are a fairly large number of machines in each account; a comparison of two or three machines against each other may prove nothing, as the statistical sample is too small. Still, these data are better than none.

The decision to buy a certain kind of equipment or to replace a machine is not one which can be made on the basis of the cost account alone. The account does not reflect the frequency of breakdown or the cost of work delays caused by breakdowns, but only the cost of repairs. The shop repair load and supervision should also be considered; if for a period complete repairs are not made, repair costs may materially increase. Gasoline-powered equipment has a high frequency of breakdown but a low cost of repair as compared to diesels. The cost of operation is only one factor to be considered in judging a machine; the other factors (such as availability) may be reported, or carried only in the memory of the equipment supervisor. It is possible to make equipment decisions on the basis of field reports alone, but few contractors receive reports sufficiently detailed to do this.

6.27 VALUE OF USED EQUIPMENT

The relative cost of work done by various kinds of equipment depends on the use and circumstances on the particular job as follows:

1. *Output.* Generally, well-cared-for equipment is capable of the same power and working speed as when it was new. Any used machine that does not produce efficiently should have appropriate repairs or overhaul. Equipment is being constantly improved, however, and later models often have improvements which increase their efficiency. Machines available are often not the ideal size for the job, so there will be design inefficiency.

2. *Reliability.* It is generally assumed that new equipment is more reliable than is old. Since the *downtime* (time unavailable because of repairs) of certain machines—a loader serving a number of trucks, for example—is very expensive, reliability is much more important for some kinds and uses of machines.

   The principal cause of job breakdown in an old machine is poor maintenance or repair; in a new one, poor design or workmanship. A knowledge of a used machine's past history is helpful; high repair costs may mean either that the machine is wearing out or that everything has been repaired. What was the nature of past repairs? If there were repeated similar damages, a design fault or a failure to repair the real damage may have been the trouble. A faulty relief valve in a hydraulic system may cause damaged hoses, connections, a broken pump shaft, pump, and driving motor damage—one by one. If a tractor is repeatedly down for track work, is it because the running gear is being replaced one part at a time?

3. *Price.* New equipment is priced between manufacturer's cost and what the contractor who needs it immediately will pay for it. This is more or less stable. The price of used equipment must only be less than that of new equipment and varies widely. The price of each item is what the contractor who needs it most can afford to pay for it.

The person who is a mechanic, operator, and contractor can derive his income from his work continuously in all three occupations. He may be able to outbid a general contractor for equipment, and still work for him on a contract basis for less than the contractor can do the work himself. There is nothing contradictory in this. If the machine breaks down, the operator becomes a mechanic, purchasing agent or whatever is necessary, but he lacks capital for a new machine. As he expands, the costs of breakdowns rise and justifies more reliable equipment.

The value of used equipment is therefore greatest for the person who has the operation and skill best suited to use it efficiently, and he will be the highest bidder. If you, as a manager for a contractor, can use the machine as efficiently as a prospective buyer, it is worth as much to you as to him, and there is no reason to sell it. The decision to sell or keep a machine therefore depends on the value of the machine to you as compared with the value to someone else. By attempting to establish the sale point by a cost accounting method, you try to determine this point more accurately than can the prospective buyer; if you can do that, the cost account is justified.

6.28   FRAUD

To control fraud by job personnel, you need to consider all possible schemes which are open to them. The common methods by which job personnel may steal, embezzle, or divert funds from the contractor include the following:

1. *Theft of money from petty cash.* This may be done either by reporting it as an outside theft or not reporting a theft at all, leaving the shortage to be discovered at the end of the job. This loss may be minimized by a small amount of cash on the job and frequent auditing. Few contractors have enough cash on the job to make the loss important.

2. *Theft and sale of materials and equipment.* This is the most difficult of all to prevent. A watchman should not allow materials to leave the job without a permit from the superintendent, and the timekeeper should have a card or brass responsiblity system for small tools (each man has a brass tag which he leaves when he checks out hand tools). Proper calculation of materials needed, with proper identification of the use of materials on purchase orders, will help disclose theft and lead to protective measures while there is still something left to protect. An adequate key control system or list of men with access to various areas is essential, and it is especially important to obtain keys from men who are discharged. Keys should be changed occasionally on a long job. If the work justifies, locks with a proprietary keyway control of copies can be maintained. A proprietary keyway is a lock system using keys for which blanks are not commercially available; there is a one-time charge of about $2,500 for this keyway by the lock manufacturer. Even with such systems, keys can be handmade, and in any system a key may be made from another, or if enough different keys are available, a master key may be made from them.

3. *Theft of vehicles and movable equipment.* This type of theft is becoming more serious as professional criminals become interested. All equipment should have reliable locks (an ignition lock which may be shorted out is not reliable), and equipment should be kept in a fenced enclosure or other location where it may be guarded rather than left on the work, particularly on weekends. Remember that any driver can obtain a copy of a key in a few minutes.

4. *Kickbacks from paychecks.* This possibility occurs when union wages have become higher than the market and there is considerable unemployment. The only adequate way to avoid kickbacks is for the superintendent to be familiar with the employees, encouraging the confidence of the better ones and discharging supervisors who accept kickbacks.

5. *Payroll padding.* This is a real danger. There should be a check made independently of the foremen of the number of employees on the job, and payoff checks should not be delivered by the foreman. In such a case, he may pay off in cash when a man is discharged, carry the workman another few days on the payroll, and put the check in his

pocket when it comes through. Endorsements of checks offer no protection, since endorsements are not checked and considerable time would be needed to do so. Most people cannot identify a forgery even of their own signatures, much less that of a stranger.

6. *Contract rakeoff.* Contract rakeoff from letting contracts is the most dangerous in amount, is the most difficult to prove, and may not be illegal if proven. If a superintendent takes bids, then gets a payment from the *low* contractor before sending them to the office for the award, only the successful contractor and the superintendent know of the payment. If discovered, it is not a provable loss to the general contractor, so how is a crime to be proven? A contractor who keeps direct personal relationships open with subs has the best opportunity of avoiding this situation; also, opening of sub bids in public rules out the possibility of contract rakeoff.

7. *Kickback on short shipments.* This may occur when the superintendent accepts short or damaged shipments without complaint and receives a payment from the supplier. This appears in any records as theft, but of course job protection fails to stop it. Here again, the manager should be as close to the job, and to the suppliers, as possible, and material should be received by a person other than the superintendent.

8. *Retention of payments.* Payments to the subs are often sent to out-of-town jobs so the superintendent may obtain releases when the checks are picked up. The superintendent may cash these checks, sign the releases himself, and keep the money. This may be done for only a short time, but a considerable amount of money may be embezzled, and the superintendent will be very difficult to locate. Do not assume that checks may be cashed only by payees, or that forgeries may be detected (by anyone but an expert, by which time it is usually too late).

   If a superintendent intended to cash a check, he would certainly plan in advance. He might incorporate a firm of a similar name or in another state, locate a person of the same name, or by various ways open an account in the name of a payee. Although the possible methods are not known to many superintendents and are therefore not repeated here, it can readily be done. This would eventually show up (or an irate creditor would show up), and the possible amount can be very large.

9. *False invoices.* This is the greatest danger, as it is difficult for the accounting department *ever* to discover the fraud. Purchases are made, or subcontracts signed, with fictitious firms (or real firms, owned by the superintendent) for items not needed for the job. The receiving reports, delivery slips and invoices are all in order. Equipment may even be purchased, and sold at the end of a job, but without ever having existed. Because of the occurrence of this kind of fraud on a very large scale a number of years ago (when non-existent warehouses were filled with nonexisting goods), accounting methods were modified to concentrate on verifying inventories; but on a construction job, verification of the items bought is very difficult. To find the items the accountant must go back to the same person who may be the thief—and then does not know if the item is the correct one without checking all *other* contracts and purchases for a duplication. On a complicated job, no one person may know all the details; the accountant can look only for paper correlation of documents, not for physical correlations.

If purchasing and contracting procedures are sloppy, it may be impossible to discover at what level fraud occurred, even if discovered. A fake firm may make a proposal in competition with real firms, but the purchase is duplicated under different descriptions. Such fraud requires both elaborate fake documents and a knowledge of the payment system at least equal to that of anyone else in the firm, but it does not necessarily require a dishonest person in a position of responsibility, nor does it require, in many cases, forged signatures. An honest person, unacquainted with the duties he has accepted, can unwittingly make the fraud possible for others to perpetrate. College-educated managers are particularly susceptible to this type of fraud, being propelled into responsible positions on their decision-making ability without knowledge of the details behind the operation of the business.

The best protection against these frauds is to require several persons to be familiar with the work. The superintendent, being most familiar with the entire work, usually must be a part of any fraudulent scheme involving his work, but if the superintendent is not given full information on contracts, purchases, and payment for his job, or if he is relieved from checking the work of his timekeeper or materials clerk, there may be extensive fraud without his knowledge. Overcentralized handling of funds, supervisors stretched beyond their capacity by attempts to cut "overhead," and disinclination to trust supervisors and clerks with information because they have no "need to know" are all helpful to internal fraud.

6.29 PAYROLL ACCOUNTING

Statutory requirements state that certain payments and reports, including the following, must be provided to federal and state authorities:

1. *Social security payments* (more properly termed F.I.C.A. payments) deducted from workmen's pay for transmittal to the federal government.
2. *Income tax withholding amounts* deducted from workmen's pay and transmitted to the federal government, and to some states.
3. Payment of the *employer's portion* of *social security* payments.
4. Payment of *unemployment compensation insurance* to the states and the federal government.
5. Payment of *workmen's compensation insurance,* based on the percentage of wages and type of work done by each man. In some states, these payments are made to the state, whereas in some other states the payments are made to private insurance companies.
6. Payment of *public liability insurance* to private insurers.

Other reports may be required by the job specifications or by state or local agencies. On union jobs, payments are made to pension, vacation, health and welfare funds, apprenticeship, and other *fringe benefits*[2] according to the local contract. Public contracts may require a total minimum wage; for union members, fringe payments are made to a trust fund, but for non-union workers, an equivalent amount is paid directly to the worker. Typically, labor costs per hour might be as shown here:

Basic hourly wage		$10.00
Social security	6.5%	0.65
State unemployment	2.7%	0.27

[2]To an employer, *fringes* are all payments other than the hourly rate; to a labor union, *fringes* include only payments not required by law.

Federal unemployment	0.5%	0.05
Workmen's compensation	3.2%	0.32
Public liability	0.5%	0.05
Vacation		0.30
Health and welfare		0.30
Apprentice training		0.02
Total		$11.96

These items are employer-paid fringe benefits, not deductions from workmen's pay. An amount equal to the employer payment is deducted from the workmen's pay for social security, and the employer sends both portions to the Treasury. The total costs on payroll, above, amount to about 20 percent, but in some cases they may be as high as 50 percent. It is therefore important that the contractor clearly specify in cost-plus contracts that this amount is to be a charge not included in the overhead markup. Some contractors charge only for direct pay, making the markup large enough to cover payroll taxes and fringes.

Since the method of computing these charges varies, records must show that the charges were properly made. *Social security and unemployment charges* are each paid only up to an annual maximum for each employee. The maximum varies almost annually and differs for the two kinds of charges. In construction, the employer often pays more on a workman's social security account than does the workman, although the percentage is the same for both. This occurs when a workman with several employers pays more than the specified maximum social security and gets the excess payment returned at the end of the year; but his combined employers, none of whom individually pay more than the maximum, have paid together over the maximum for the individual, and none of them can claim the surplus.

Worker's compensation payments vary with the trade and with the work done in that trade, as illustrated in Table 6-6, which shows rates for several states.

The *code number* for occupational classification (as termed; but not, as noted below, is the rate an occupational one) is standard throughout the nation, published in a guide which may be obtained by contacting your state insurance commission or insurance agent. The codes shown are but a few of the total. The guide is a small book with detailed descriptions of all classifications, not limited to construction work. There may be advantages in some types of work in using codes which are not usually considered construction codes; there are often two or more possible codes which may be used for a particular work item. The contractor initially pays a deposit based on his expected payroll and classifications, when he obtains compensation insurance. The agent who computes this deposit may not give the contractor details of the codes; he may use "general" or "N.O.C." (not otherwise classified), particularly if the contractor's representative is primarily interested in the least possible paperwork. Since no one in the contractor's office is aware of the difference between codes, the firm may overpay indefinitely. This is a consequence of separation of accounting and construction supervision, to the extent that the accountant may never actually see a construction job. The contractor calculates his own payments, but using the codes and rates furnished him, based on the previous report.

For example, a contractor in the state of Washington who builds one- and two-family residences will pay 5.11 percent for general (unclassified) carpentry, but only 3.87 percent if properly classified. This 1.24 percent overpayment is both a reduction in his working capital and a continued expense, since premiums are paid in advance. If he has a $10,000 weekly payroll and pays insurance quarterly, the added investment which is lost each quarter is $1,162.

Payroll classification is neglected because it is not automatically provided in either profit or cost accounting and is not required of the contractor. The usual classification is calculation by trades, as carpenters, bar setters, laborers, and plumbers. Where large jobs have separate

Table 6-6. Typical Worker's Compensation Insurance Rates
as Percentage of Gross Payroll

Effective July 1, 1979

Classification of Work	Code	Ala.	Alas.	Cal.	Conn.	D.C.	Ill.	N.J.	N.Y.	Ore.
Carpentry—1-, 2-family	5645	3.76	7.09	8.43	10.09	14.41	5.51	6.63	7.50	16.11
Carpentry—3 stories or less	5651	4.05	8.62	8.43	10.09	14.41	7.02	5.61	10.01	11.34
Carpentry—interior cabinet work	5437	2.58	6.19	8.43	6.40	10.99	3.42	5.19	NA	9.17
Carpentry—General	5403	4.28	8.90	10.47	12.50	18.37	8.27	6.94	11.33	25.12
Chimney—brick construction	5000	8.13	24.52	17.02	44.17	78.09	21.36	26.82	34.59	36.60
Concrete work—bridges, culverts	5222	5.87	11.08	17.02	14.51	25.50	8.50	6.66	9.22	13.10
Concrete work—dwelling, 1-, 2-family	5215	2.51	8.26	5.17	8.91	23.69	3.76	6.66	NA	8.58
Concrete work N.O.C.	5213	4.85	9.14	8.56	8.52	24.93	9.67	6.10	12.03	15.09
Concrete or cement work—floors, sidewalks	5221	3.68	6.70	4.12	7.67	12.94	3.64	5.89	6.81	9.21
Electrical wiring—inside	5190	2.29	5.41	4.28	3.59	10.67	3.32	3.24	4.70	5.62
Excavation—earth, N.O.C.	6217	5.06	7.89	6.14	8.27	17.01	5.31	5.98	8.03	15.06
Excavation—rock	1605	4.95	8.56	10.34	11.08	48.25	28.17	6.10	7.43	14.03
Glaziers	5462	5.09	8.74	8.20	9.87	20.81	8.34	5.98	10.35	7.32
Insulation work	5479	4.62	8.07	7.52	6.81	13.86	7.24	8.12	5.22	11.69
Lathing	5443	2.28	5.46	5.28	5.82	11.12	2.68	3.98	6.95	15.01
Masonry	5022	3.18	10.44	8.75	11.01	23.95	6.24	6.91	10.58	15.11
Painting and decorating	5474	3.36	8.09	7.98	9.73	14.86	6.31	7.38	12.14	11.34
Pile driving	6003	11.83	24.50	19.35	8.42	67.37	14.56	9.17	17.11	31.60

Table 6-6. (Continued)

Effective July 1, 1979

Classification of Work	Code	Ala.	Alas.	Cal.	Conn.	D.C.	Ill.	N.J.	N.Y.	Ore.
Plastering	5480	3.74	7.78	9.27	6.84	12.82	4.38	3.98	13.01	11.39
Plumbing	5183	3.10	6.90	5.17	4.24	12.05	4.18	3.38	7.26	7.71
Roofing	5551	7.88	20.00	16.42	18.13	33.80	14.40	14.90	NA	30.25
Sheet metal work—erection and install	5538	3.20	6.28	5.62	7.71	13.11	4.46	4.21	10.95	8.96
Steel erection—doors and sash	5102	3.37	8.69	6.32	6.06	21.50	5.28	NA	5.64	16.41
Steel erection—interior ornamental	5102	3.37	8.69	6.32	6.06	21.50	5.28	4.94	5.64	16.41
Steel erection—structure	5040	6.53	36.15	11.93	27.98	59.37	22.90	18.67	21.01	30.18
Steel erection—dwelling, 2-story	5060	10.23	25.79	17.93	19.17	59.19	25.13	11.52	12.39	23.25
Steel erection—N.O.C.	5057	9.85	22.18	16.23	26.67	73.47	23.68	8.85	20.08	28.41
Tile work—interior	5348	2.63	2.99	4.97	4.61	33.94	2.52	5.33	6.84	7.28
Timekeepers and watchmen	5610	3.16	8.02	6.99	7.72	21.56	5.91	3.40	5.00	18.31
Waterproofing (brush)— interior	5474	3.36	8.09	7.98	9.73	14.86	6.31	7.38	12.14	11.34
Waterproofing (trowel)— interior	5480	3.74	7.78	9.27	6.84	12.82	4.38	3.98	13.01	11.39
Waterproofing (trowel)— exterior	5022	3.18	10.44	8.75	11.01	23.95	6.24	6.91	10.58	15.11
Waterproofing (pressure gun)	5213	4.85	9.14	8.56	8.52	24.93	8.67	6.10	12.03	15.09
Wrecking	5701	15.63	30.72	NA	54.19	51.55	21.94	36.11	NA	47.81

Note: All insurance is subject to published rates and rules in effect at the time of coverage, and may change at any time. This table is furnished for illustrative use only and should not be used for estimating.

Source: Herbert L. Jamison & Co., New York, N.Y.

243

payrolls, a rate may be used which appears to be applicable to the job as a whole. A four-story building may all be calculated as concrete work over three stories in such a case; the discrimination, far from being useful, may result in higher rates. Laborers are difficult to classify. They may start a job as wreckers (at 65 percent rate, in Arizona) and at times work as lather helpers (at 7 percent in the same state).

Throughout the nation, rates may vary from 78 percent, for chimney constructors in the District of Columbia, to 1.1 percent, for lathers in Indiana. Because of these variations, *cost accounting is necessary for labor classification as well as for cost control.* If cost items also discriminate between codes, as is usually the case, the labor costs can be used for the base of insurance. *For example,* one cost item for rock excavation and another for earth excavation gives required information. For compensation insurance, the insuror does not require payroll or discrimination between skilled and unskilled trades. The method of computation need not be exact; *for example,* suppose that you have $850 for rock excavation and $125 for earth, but the correct payroll total is $1,000. You then use the proportions 850:125 and report totals of $871.79 and $128.21. You have made a reasonable allocation of payroll, and this is all that the insurer requires.

Although the contractor computes his own premiums, these are regularly audited by the insurer's accountant, in the contractor's office.

Liability insurance is also based on payroll, but there are only a few rates for one contractor. The differences are great, as between 1 percent and 30 percent, and the contractor must be careful to properly classify labor cost for the more expensive categories, as underground. Remember, rates are not determined by the overall job or by the individual worker, but *by the actual labor cost for the work performed.* This applies to both liability and worker's compensation insurance.

State variations in rates occur since there is no federal control on compensation insurance. There is no consistency in rates, as they are determined by the administration of the system as well as by the benefits required by state legislatures. It is essential that an estimator know both rates in all states in which he bids and the base for the rates, which is not necessarily total payroll. A state may act as insurer, collecting premiums rather than requiring commercial insurance. Wyoming does not classify labor at all, but charges a flat 5 percent for all work, dropping to 1 ¾ percent after the first 12 monthly payrolls. Adjustments are usually made to the overall rate in all states, based on the cost of accidents due to the individual contractor's operations. Florida, until recently, applied the insurance rate only to the first $100 per week for each worker.

Rates have risen rapidly in recent years, in advance of other costs. Since the rates are based on wages, the actual cost is rising with wages when the rate is uniform; but rising benefits, medical costs, more liberal administration, and higher insurer profits and costs have driven up rates radically. In Florida in 1972, for a carpenter who made $280 per week, the general rate was 7.69 percent or $7.69 a week. In 1979, the rate of 14.53 percent (of the full wage) on $400 per week required payments of $58.12 per week. The rate on the same classification in Washington, D.C., rose from 2.84 percent in 1972 to 18.37 percent in 1979. When bidding long jobs, or even to keep hourly rates for estimating up to date, the estimator must consider this "percentage on the percentage" of inflationary increases.

Deductions from workers' pay are shown on the *weekly* payroll, also called the *payroll journal* since it serves as a journal sheet to accumulate and record deductions and job charges. Each week, the total earnings, hours, and deductions must be shown in three locations; on a statement and check for the worker, on the payroll, and on the individual's record. In addition, some owners require that a payroll copy be submitted to them weekly, and various government agencies require the payrolls to be available for inspection. There are a number of methods to

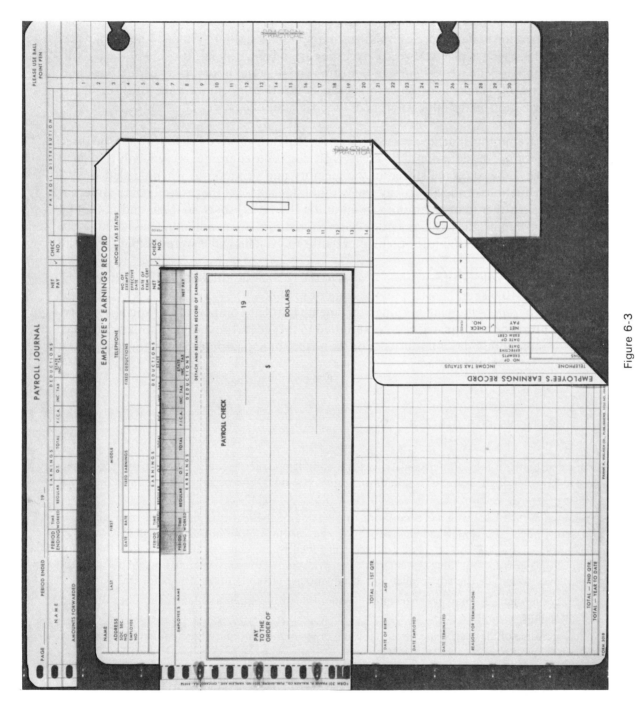

Figure 6-3
A THREE-IN-ONE PAYROLL SYSTEM
(*Courtesy of Frank R. Walker Company.*)

make these entries without writing or typing them individually: either manually, by an accounting machine which prints several places at one time, or by computers which put the information on tape. Figure 6-3 is a *three-in-one payroll system,* since three entries are made manually at one time by carbons; this is also called a *peg system,* since the sheets must be aligned with pegs on the side of a board. Several firms manufacture similar systems. The employee's statement on his check is placed directly over his earnings record and the line of the payroll where he appears for the week. The same information—name, hours, rate, deductions, and pay—are written only once and the carbons reproduce the information on all three items. From the information on the upper margin of the check, the check proper is then typed or written separately.

When an accounting machine is used, the three items are inserted in a long roller and three sets of type reproduce the data simultaneously. If an electronic computer is used, only the check need be printed; other data are stored in the memory and printed out when needed.

Payroll deductions of social security and withholding taxes are the property of the Internal Revenue Service rather than an amount owed to the IRS. It is money in trust, and cannot be avoided by bankruptcy: the IRS has a prior claim to creditors. If an individual has failed to pay withheld taxes, he remains liable for them after bankruptcy, if his assets were insufficient to pay these taxes.

If a person has failed to pay withheld taxes to the IRS, he is guilty of a federal felony, subject to a fine of $10,000 and costs, plust 5 years of prison.

A *quarterly report* showing earnings for unemployment compensation, social security, and income tax records of the workers is submitted to the federal government; this report may be copies from monthly payroll journals, or if an electronic computer is used, a copy of the memory tape may be substituted.

6.30 QUESTIONS FOR DISCUSSION

1. For what kinds of jobs and circumstances is a complete ledger account preferable to a voucher system? If you want to keep cost accounts on the job, how does this affect the accountant at the home office?

2. Paid invoices are usually filed in accounting by date or alphabetically by payee. As a manager, how does this affect you and what change would you ask for? How is this affected by computer accounting? How are you affected if invoices are filed by purchase order number? Is there any connection between your filing system for delivery tickets awaiting invoice and filing of the invoices themselves? What are the advantages and disadvantages of the Uniform Index for this purpose?

3. List ways in which (a) a balance sheet and (b) a monthly profit and loss statement may be in error or be misleading.

4. What are some of the decisions a contractor may make that reduce liquidity (available cash)? That increase liquidity?

5. How would you determine the overhead logically chargeable to a job when it is completed?

6. "Unsuccessful sales efforts, such as advertising and estimating, should be a deduction from profit rather than a cost." Explain this statement.

7. Should you pay all suppliers and subcontractors equally promptly? Why or why not?

8. Under what circumstances would each of the depreciation methods be preferable?

9. How may unrealistic depreciation rates cause real (not paper or unrealized) losses in a firm? How may the method of charging equipment costs cause losses?

10. In what ways are all depreciation methods unrealistic?

11. Should the value of equipment on a firm's books be different from its market value? Why or why not?

12. Explain "You can't pay bills with tired iron" and the experience indicated by a man making this statement.

13. Some construction firms are said to be equipment-poor because ownership of equipment increases operating costs. Explain.

14. You are a contractor's employee, assigned to verify a rumor that a project manager is "on the take" (taking bribes) on his job. How would you go about checking this?

15. You are a project manager of a firm suffering from a cash shortage. A U.S. government inspector offers to increase your progress payment $200,000 for $200. What would you do?

CHAPTER 7

PLANNING AND SCHEDULING

Planning is the separation of a construction job into small parts, estimating the time and manpower to complete each part, and arranging them in sequence in conformance with space, manpower, and equipment requirements. *Scheduling* is the assignment of dates to the parts, and rearranging or changing them to complete the work in the most economical manner. *Planning* may also designate the method for each portion of the work, which includes a time-and-motion-study routine.

Estimating is an essential part of planning and the starting point for all further plans. For many contractors, a job estimate is the only plan necessary. Customarily, estimates have been made in terms of money and unit prices, rather than in labor-hours, as the labor-hour estimate must then be converted to dollars. Most estimators have a keen sense of the labor-hours represented by the unit prices they use, and adjust rates constantly by new calculations of labor-hours. The academic fraternity is inclined to make the labor-hour recording and estimating system basic, on the assumption that labor-hours per unit do not vary with wage rates. If you do not have to furnish a new estimate each week, or twice a week, you can afford to favor more precise but slower methods. There are many cases where labor-hours per unit vary with wage rates, when different labor markets are compared. That is, better workers tend to move to high-wage areas and crowd out lesser skilled workers, who return to low-wage areas.

From the job estimate and his past experience, a contractor can judge the approximate number of workers to use on a project. Suppose that he has a $500,000 project with a general construction payroll of $100,000. If he estimates his maximum work force at 40, not including the subs, at $100 per day each, or $4,000 per day, a rule of thumb is to halve this for the average, or $2,000 per day for an average throughout the job. The time for completion is then $100/$2,000 per day, or 50 days. To this should be added start-up time—say 20 days—and final punchout time—allow another 20 days. This gives us 90 days for the job, or 18 weeks plus holidays, or 19 weeks. The minimum cash requirement is about three weeks' payroll, 15 times $4,000, or $60,000—which he can reduce, if he is lucky, by drawing ahead on the job.

This computation has two weak points: the estimate of maximum work force and the possibility of extensive subcontract work for which the general contractor has no labor estimate.

The general contractor has experience in the type of job he is doing and will look suspiciously at any more complicated method of scheduling which gives more optimistic results.

The job may be planned and scheduled when the estimate is made, but this would be wasted money, if the contractor does not get the job. With an estimate cost nearly as great as his net profit, a contractor is not inclined to expand the functions of his estimator and increase estimating cost.

7.1 PURPOSES AND TYPES OF JOB PLANS

The planning for a job may be done by the project manager, or he may have help from other departments.

Plans include the following:

1. The *estimate*
2. The *budget*, as previously described
3. The *time schedule* and sequence for completion of each part of the work, made by CPM or other methods
4. The *cash flow budget,* as detailed in Chapter 6
5. *Manpower planning,* to include the contractor's manpower requirements for workmen and supervisors, and requirements of the subcontractors
6. *Plant,* which includes office and office procedures, job organization and individuals who may be available, equipment which may be rented, transferred from other jobs or purchased, and temporary construction.

7.2 TIME SCHEDULES

The basic elements used in preparing time schedules are the estimate items on which the cost of the project was based. These items are combined, as a rule, although if large items have been used—such as including several floors of columns rather than one floor—portions may be separated to form items suited to the schedule. These portions must be further divided if a *lag time by trades* is to be calculated. This lag time is the time one trade must have worked on a schedule item before another trade can begin. Some of the summaries used in the schedules are work schedules, bar charts, labor graphs by trades, and the CPM (critical-path method) chart.

A work schedule is a list showing the time at which various items of work may begin and end. If a CPM system calculated by computer is used, a work schedule is the machine printout. If no CPM chart is used, the schedule is prepared manually. For a house, for example, the schedule will call for the time each crew, subcontractor, and trade is to start and when materials are to be delivered. The data made up for CPM may be processed manually if they are not extensive, and the schedule made up accordingly. The schedule is the useful information reaching the job superintendent or foreman. Some managers make up very approximate schedules, or schedule a few days ahead; some are very detailed. Before this information was shown on CPM charts, it was visualized on the building itself. Good managers can plan from experience on previous similar jobs alone without breakdowns into items.

There is an old story about how to weigh a pig without a scale. One balances a pig against a stone, then weighs the stone. Much present planning is of this type—by making the analysis indirect, details are supposed to provide some information which one cannot obtain directly. If you must ask a superintendent how long each item of work on a building takes, why not ask him how long the building takes, and put all your guesses in one place? Many managers can estimate large items more accurately than small ones, chiefly because they have more opportunity to discover their errors on overall estimates than on the pieces.

7.3   THE BAR CHART

The bar chart usually used in the form of the modified Gantt chart illustrated in Figure 7-1, gives a summarized overall view of progress on a job, in terms of overall work done and portions of the work in which delays have occurred. An advantage is its simplicity, in that it may be prepared in a day and kept up to date with an hour's work a week. Of course, it may be made as complete as a CPM diagram, with comparable work, in which case it becomes a bar chart for each CPM activity.

The bar chart is readily understood and shows the situation at a glance. It illustrates the amount of money in the job and can readily be used to forecast cash requirements for the owner, as well as income to the contractor. This chart does not provide cash requirements of the contractor, since contractors' payments to others are not shown. The progress report data from a Gantt chart may be summarized with other jobs, which is important for federal agencies, who must keep track of spending for budgetary purposes. The Gantt chart lacks detail, particularly in that it fails to show critical items which, although of small cost, will delay the entire job. This fault does not show up as long as other work is ahead, so the value of the work is up to schedule.

The modified Gantt chart, in Figure 7-1, includes the percentage on the bars and the overall completion curve, to give the following:

1. *The status of the work* at any time for each item, by crosshatching the lower half of each bar to show the amount of work done at the scale on the upper part of the bar itself.

2. *The schedule for the work to be done,* which shows in the upper half of each item bar. The percentage completed is shown at the beginning of each month. The crosshatching of the upper part of the bar merely shows the date of the report, not that work has been completed to the percentage marked on the bar. The upper crosshatching shows the passage of time on the horizontal scale and will therefore be the same for all bars at all times. However, the scale of work done is irregular; there is no implication that the work for any item is to be done at a uniform rate. Each upper half of the bar could be plotted as work done against time and would appear as an S-curve.

3. A percentage scale on the right, combined with the horizontal time scale, serves to plot the *total work schedule.* For the first of each month, the amount of work scheduled to that time is added and plotted as a curve. This curve, which may be on a separate sheet, is the traditional S-curve, so shaped because of low output during startup and completion.

4. The overall status of work to date is shown as a dashed line, plotted on the same scale as the S-curve of the overall work schedule. The chart shown has been brought up to date for July.

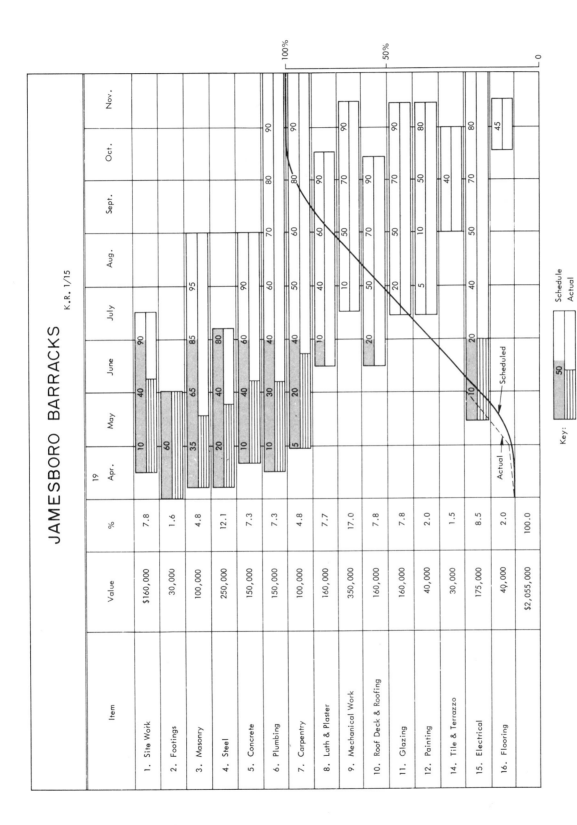

Figure 7-1

A MODIFIED GANTT CHART

251

To read the schedule, the two bars for each item are compared. If the lower is further advanced to the right, the work is ahead of schedule. The percentage of the total that the work is ahead or behind schedule is interpolated between the first and the last of the month's percentages for that particular bar. Site work, for example, should be 90 percent complete, but is between 40 and 90, interpolated as 60 percent. It is therefore behind by 30 percent of the total.

The overall status is read from the two curves at the bottom, marked *actual* and *schedule*. If the actual is above the schedule, work is ahead by the percentage read off the scale to the right; in this case, the actual work is 29 percent and the scheduled work is 25 percent.

When a CPM system is used, the computer may also print out a bar chart showing time during which each CPM item is under way, without the percentage of completion at any particular time and without a comparison of the schedule to date. With a Gantt chart to show overall job progress and an experienced project manager to spot omissions, the detail of CPM planning can be achieved without the cost.

7.4   PREPARING THE GANTT CHART

Calculations for the Gantt chart may be made with an office calculator without any computation sheets. Printed blanks for the chart are available commercially.

The procedure is as follows:

1. Fill in the items of work and the values from the estimator's breakdown.
2. Calculate the percentage each item is of the total (this requires only one entry for the total and one for each item, in a desk calculator).
3. From your own judgment or that of others, schedule each item by percentages each month.
4. Using the *percentages,* strike a cumulative product *total* for the first of each month on a calculator. (For May 1, for example, $7.8 \times 10\%$, $1.6 \times 60\%$, and so forth); the sum will be the percentage of work scheduled at that date and may be plotted on the overall schedule curve as the work scheduled. On the first of each month, enter the reported percentage completed, and add up product totals on a calculator as before; the result is the actual percentage of completion at that time.

   If copies of the chart are to be distributed, the original is kept on tracing paper and prints are distributed. This, with present technology, makes the use of colored copies impractical in most offices, although in some firms they may be colored by hand.

7.5   CRITICAL-PATH METHOD

The critical-path method (CPM) is a standardized system and nomenclature for arranging items of work for a construction project, so that the sequence of work is readily apparent and the time schedule can be easily prepared. The overall completion time may then be estimated, and items which may delay the work may be located. By combining the critical-path method with job cost control, both progress and cost may be monitored with less effort than if separate systems are followed. At the present time, this combination is not common; in fact, there are few contractors who maintain both a time and cost control system.

It has been claimed that the critical-path method makes possible monitoring which has not been done before, and that it is an essential tool of management. Its effectiveness depends on who uses it and for what purpose, and its necessity depends on the contractor's or owner's skills in other methods. Since the clerical work involved becomes lengthy on a job of any size, the use of a computer for calculation is common. When the computer is used, the critical-path chart, which gave the system its name and with which the name is chiefly allied, becomes unnecessary except as a planning tool to the programmer; and it becomes so large that it may be read only if it is placed on a floor or on a wall, where a rolling ladder is necessary to read it. Variations such as PERT are of importance only to a student of the topic.

In this chapter, the principles and nomenclature of the method will be presented with a very simple example; this will enable you to use the products derived from the method. Considerable study and practice are required to plot a chart, and preparation of the data for a chart required a major effort of highly experienced people.

7.6   PHASES OF THE CRITICAL-PATH METHOD

Application of the critical-path method consists of three phases: *planning, scheduling,* and *control-monitoring.*

Planning consists of breaking down the work into *activities;* that is, parts which are complete in themselves. By definition, an activity may be completed without waiting for other action such as an approval, delivery of materials, or transfer of men or equipment. Activities are then arranged in order of dependence on each other. A time for completion (time necessary to do the work) is essential, and if uses other than time scheduling are to be made of the data, then material cost, labor cost, labor-hours of each trade involved, and responsibility (superintendent or subcontractor) for each activity may also be included. The time at which the work must be paid for may be added, although the AGC method does not include this refinement. In Figure 7-2, for example some of the activities are: *Prepare submission of A/C equipment, Approve submission, Fabricate A/C equipment.* The data for each activity specify which activity must precede it and which one must follow; in this way the order of activities is designated.

Planning is the major part of the work of making a CPM chart and must be done by the contractor's and subcontractor's personnel. Most faulty charts are the result of poor planning. A number of writers consider this phase the most valuable part of the process because persons responsible for the work are obliged to plan it in detail and list the work to be done. For the contractor, it may be the only benefit of using a CPM system.

Scheduling consists of adding the activities end to end and adding up the total time to see how long the job will take. The line of activities which add up to the longest time form the *critical path;* identifying this line of activities, which limit the completion, is the objective of using the system. Once you have established the critical path, you know what work will delay the job and can concentrate on speeding up that work. Scheduling can be done by a consultant in this work, and if the results are to be determined by computer, the schedule must be in exactly the right form to fit the program.

The *program* is the set of instructions given the machine, telling it how to handle the data. If you are told to "add up all the figures I give you," you are being programmed. Computer programs can be designed to suit your problem, but the problem must first be solved by hand, converted to computer language, and then tested by working a problem both manually and by computer. If you have less than 30 or 40 identical problems, forget the computer—it is too expensive. So far computers have shown little talent for programming themselves, which would be necessary to solve a *new* problem.

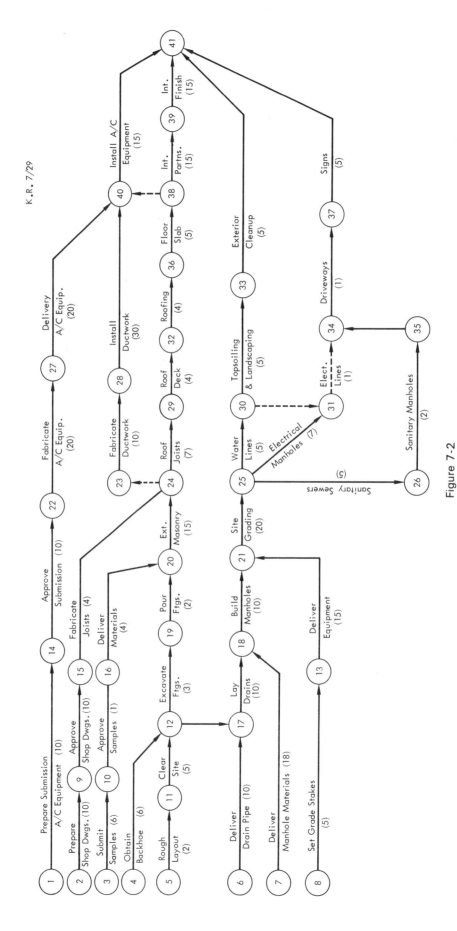

K.R. 7/29

Figure 7-2
A CPM DIAGRAM

254

The writer considers a job of less than $10 million too small for CPM planning to be useful, but this depends on the experience of the supervisors.

Control-monitoring consists of giving the computer corrections based on changes, or on work which is completed ahead of or behind time. If a cost system is being used in conjunction with CPM, cost corrections may also be made. In this phase, the computer is of great value. Since the late completion of one activity may affect hundreds of others, the clerical work in tracing out such changes would be prohibitive. Because of the human tendency to be optimistic about making up past losses, schedules are rarely revised regularly. To do so would require the manager to admit that he does not think he will make up the lag. Besides, if the delay is known to be on the critical path, no amount of analysis will change the completion date; physical action is required.

The control phase may also include controlling costs and cash requirements, using methods basically the same as outlined for manual operations in preceding chapters.

7.7 CPM NOMENCLATURE

The terms used above are fairly well standardized. There is almost no limit to the number of terms used in this system (which are being added to every day, in academic CPM books), but the terms below are those used in the AGC manual.

The *arrow diagram* is the CPM chart variation used by the AGC and illustrated in Figure 7-2. Each arrow represents an *activity,* or independent item of work as described previously, and is labeled on the chart. The beginning and end of activities are shown by circled numbers representing *events.* An *event* is nothing more than the beginning of one activity and the end of another (except, of course, for the event used to start the whole job and the one used to end it). For each of these activites, the *input*, or information needed, may be these:

1. *Definition* of the activity. The work order shown in Chapter 2 may be used as a definition. (The work order was used long before CPM was first used.) Although the work order is for a smaller unit than a usual CPM activity, work orders and CPM activities should coincide in order that only one system of progress and labor cost control is necessary.

2. *Duration* of the activity, for time planning. Only *definition* and *duration* are essential for a critical path diagram as it is usually made.

3. *Cost estimate,* which ties the activity into the cost accounting system. The activity may have separate labor and material estimates. There may be several cost items per activity, several activities per cost item, or they may coincide. They must be correlated; that is, one cost item must not include two activities, if one of the activities is also included in *another* cost item.

4. *Resource estimate*—that is, the number of men or machines, whichever appears to be important, or both, for each activity. Apparently computer programs at this stage are designed to add up the requirements for each day, but not to correct previous schedules from resource estimate calculated, or to respond with general schedule corrections to resource data given to the machine. This is an important disadvantage of the CPM

schedule. Resource leveling is scheduling activities to reduce the maximum demand for men or equipment. CPM does not tell you how to assign workers.

5. *Trade indicator,* or a designation on each activity as to who is responsible for it. CPM printouts are very complex and lengthy; each person is interested only in his part of the schedule. By adding a numeral designator for the subcontractor or foreman, you can get a printout for just his portion of the work.

7.8 CPM SCHEDULING

In the *scheduling phase,* a diagram is made as shown in Figure 7-2. Each circle, representing an *event,* shows where one activity must be completed before another is started. For example, Activity 5-11 is the designation for *Rough layout* (an activity), and the time (2) below, indicates that this activity will take 2 days. Activity 11-12, *Clear site,* cannot start until 5-11, *Rough layout,* is completed. The sequence must be set up during the planning phase; the chart is made merely to accommodate planning data and make it easier to understand.

Laying drains, 17-18, cannot begin until after the drain pipe is delivered, 6-17, but it also cannot begin until drains are laid out, 12-17, which occurs after other operations. Laying of drains can start 10 days after the job is begun, according to 6-17, but according to the sequence 5-11-12-17, 11 days of work precede the laying of the drains. The 11-day period therefore controls.

If you follow through all the various possible paths from the beginning of the job (events 1 through 8) to the end at 41, the longest will be found to be 5-11-12-19-20-24-29-32-36-38-39-41, and the total elapsed time will be found to add up to 77 days, which is greater than any other path. This chain of operations is the *longest,* the number of days to complete it is the *greatest,* and it is therefore the *critical path* of events. If any event in this chain may be shortened, the overall completion of the job is shortened. The chain of events along the top, from 1 to 41, shows the work to be done by the A/C contractor. This path adds up to 75 days, so he will finish two days earlier than required. He is said to have 2 days' *float time,* or time to spare in a chain of operations. If the critical path is reduced, by better planning or changing the method of operation, to 75 days, then the A/C path becomes critical. To reduce the overall time further, there must be reduction in both paths which have now become critical. *There may be any number of critical paths at the same time.*

One of the objectives of the critical-path method is to reduce the overall completion time at least cost. It is assumed that the original scheduling is done at such a rate that the work is completed at least cost, and any increase in men or equipment, or improved delivery, will actually result in increased cost of the item.

Suppose that we want to complete this job more quickly. We take each critical operation (operation on the critical path) and study it. Can it be done more quickly by another method? In this instance, let us assume that we may complete masonry faster by using a subcontractor for work we intended to do with our forces, continuing to do as much work as possible ourselves (Activity 20-24). We make a graph of the proposed changes and the added cost, as in Figure 7-3, the time-cost chart. On this chart, it is assumed that we can calculate the cost for three points: for the normal time, for a time saving several days, and for an absolute point beyond which the saving in time is impractical, the *crash cost.*

In this instance, the masonry completion at Event 24 is 27 days from the beginning of the job (adding 5-11-12-19-20-24). We may cut this time, by added cost to the masonry, to 24 or

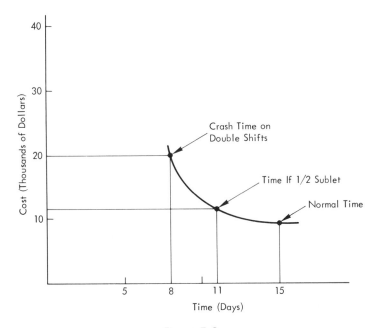

Figure 7-3
A TIME-COST CHART FOR ONE ACTIVITY

even 20 days. But after the masonry is completed, we can do nothing until the roof joists are delivered; according to our chart, these will be delivered 24 days after the beginning of the job (2–9–15–24). Consequently, we must also shorten the joist time if the masonry is to be a crash program, and the air-conditioning machinery installation must be shortened still further. If time-cost charts are made for each activity which becomes critical, and if the calculation is carried through the various ramifications, we can derive a cost for any time of completion up to the point where all activities on the critical path (which, of course, may now be a new path) are on a crash program. This may be done by computer, with the various cost curves assumed either to be a straight line (the saving in time is directly proportional to the added cost), or a curve as we had here (the daily cost of time saved becomes greater with each additional day saved), or a two-point diagram. With a two-point (discontinuous curve) diagram, there are only two alternative ways of performing an activity, each with its own cost and time.

The objective, of course, is to shorten the time for those activities that yield the greatest time saving for each dollar. Such a complete analysis is seldom performed, as the data is difficult to obtain and usually unreliable. However, you may demonstrate that changes in a few activities can result in a substantial saving of time at little cost.

During the control-monitoring phase, reports on time of completions and starts are made, and completions are recalculated to stay current. Although the CPM method gives little useful information after the original chart if it is not kept up to date, few contractors maintain the monitoring phase. As mentioned before, one reason for this is the disinclination of contractors (and owners) to admit to themselves that lost time will not be made up, and to make a new schedule based on current events.

When a CPM chart is required, the contractor has very little time to prepare it. He is often not allowed to change it once submitted; it becomes an addition to the contract and may be used against him if the job is delayed. Under such circumstances, the schedule is no longer useful as a practical tool and the contractor must furnish a defensive schedule. Some attributes of such a schedule are:

1. Float is eliminated and all activities are made critical. In this way, a delay of any activity may be used to justify a time extension for the contract. Since a good schedule, which reflects labor conditions, will have no float, the elimination of float does not mean the schedule has been tampered with for this reason.
2. Placing activities which require hard-to-get materials early in the schedule then justifies time extensions when the items are late. The schedule is then followed, and the preceding work is completed, and the schedule becomes self-fulfilling; a delay actually occurs when it has been forecast to occur.

A similar distorted valuation of activities is made when this valuation is used for progress payments. The contractor may be obliged to give breakdowns of subcontracted work before the contractor can obtain them from the subs.

The solution, obviously, is to use the CPM chart and progress payment breakdown as an aid to administration of the contract, not as part of the contract itself, and ample time is required for its preparation. A/E's realize this, but governmental agencies are not so knowledgeable and cooperative.

Contractors are often discouraged by CPM planning as it is used by owners to monitor construction activities. Since such a schedule enables an inexperienced person to better understand construction activities, owners demand even more detailed schedules to determine if the contractor is progressing satisfactorily. The contractor is incapable of using such detailed planning, and the result is that increasingly detailed and nearly worthless data are given the owner, who becomes more confused. There is no substitute for construction experience to supervise and judge construction work. One can make a beautiful schedule for a 3-year job and follow it beautifully for 2½ years before it becomes apparent the work is a year behind schedule. A patient undergoing a heart operation does not ask a doctor for an explanation of what is to be done, and then hire a first-year medical student to supervise the doctor during the operation. Yet an owner puts such faith in a system of scheduling that he relies on persons with little or no construction experience to supervise the work. An owner cannot let a contract on a lowest-price basis, and then try to change the contract by asserting his own cost-value system. If the owner wants a job built to minimize his own costs, the contract should be designed accordingly. If he wants to monitor progress, he must hire persons qualified to do so—and these persons are rarely design professionals. The owner should not specify or impose a scheduling system if he does not understand the construction process.

7.9 CPM AND MANPOWER SCHEDULING

Stability of employment is an important requirement for labor efficiency. To supervise, instruct, or reward a workman, you must employ him for a reasonable length of time. This truism is repeated because CPM scheduling, as usually presented, considers stability as of minor importance. CPM writers point out the importance of office work, such as giving proper priority to shop drawings and deliveries, from the overall view of the project. But at the job level, what can be done? The superintendent is told what jobs are important, but he can do only two things in response: hire more men or assign the men he has in a priority to stay up with the schedule.

After all office submissions and orders, subcontracts, and approvals are properly obtained, building construction requires the right kind of men at the right place. The delaying factor is then lack of workmen in the *critical trade*—that is, the trade working on activities which are critical at the time. The CPM chart may show float time, based on an assumed manpower availability, which actually does not exist. *No path has float time unless it is known that the men are available.* Availability on paper is not sufficient; they must be properly supervised and have reasonably stable employment. If the project manager does his job by obtaining materials and preserving the free assignment of men by the superintendent, *all paths requiring workers of the critical trade are critical.* Other paths do not become critical. Since one of the chief objectives of planning is to avoid situations where men in one trade are scarce, the planner tries to make men of each trade as fully employed as men of the other trades. It is irrelevant to speak of sufficiency of labor on a building project; except on very confined projects—and even then only in the early stages—supervisory and organizational problems do not allow employment of the maximum number of men who can work efficiently. A foreman may say he has all the men who may efficiently work on a job, but a superintendent can point out many locations where another man can be used. Construction management is the organization of material and supervision which will make possible the largest employment of men; *supervision is a managed labor shortage during the course of a job.* If by proper CPM planning, the manager adjusts the scheduling of activities to accommodate available men (resource leveling), *all activities of critical trades are on critical paths and no float time exists.* This is the objective of labor planning. If there were float time, men would be removed from other activities to speed the critical items.

7.10 EXAMPLE OF RESOURCE LEVELING

A sample section of a CPM diagram for a single trade is shown as Figure 7-4. This diagram is as it would be made without consideration of variations in the work force. Path 1 (1-5-7) is critical, with float times for the other paths as shown. In addition to the completion time necessary for each work item, the number of men necessary is shown.

The *manpower requirement* of this diagram is plotted in Figure 7-5, the vertical measurement representing the number of workers on each activity, and the horizontal dimension representing the number of days they must spend to complete the work. The diagram represents a 16-day operation, limited by critical items 1-5 and 5-7. The beginning of each job on paths 2 and 3 is shown to occur on the earliest starting date. That is, if the superintendent had all the workers he needed, he might follow the schedule shown. In this way, 24 workers would be required at the peak of the work. This schedule may be modified without changing the time required for each item. There is no need, for example, to start item 2-4 before 3-4 is finished, using up some of the available float time after activity 6-7. In this way, the maximum number of workers could be reduced to 20, but the size of the crews would have to be changed.

Going a little further, assume that the number of workers used may vary freely, although this changes the assumption that the number of workers used for each operation has been fixed at the minimum cost. Figure 7-6 shows the assignment of workers, plotted in the same manner as in Figure 7-5, but with the new assumption. Figure 7-6 illustrates the manpower assignment that would require the least transfer of workers between jobs, as well as the minimum overall labor force. Initially, the labor force of 15 workers is determined by computing the number of

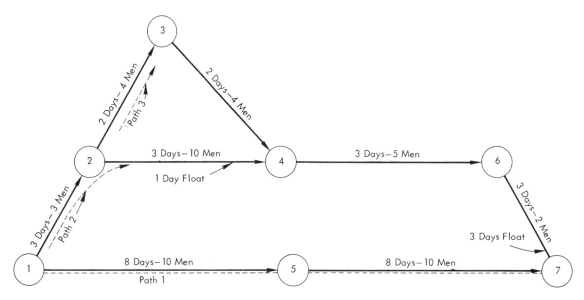

Figure 7-4
PORTION OF A CPM CHART

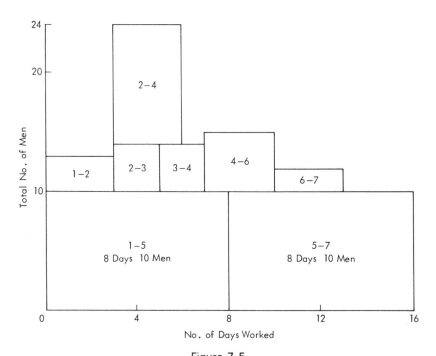

Figure 7-5
MANPOWER DIAGRAM FOR FIGURE 7-4

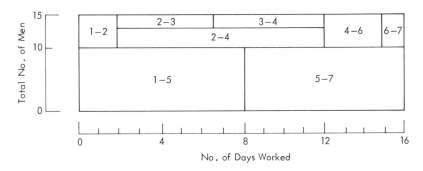

Figure 7-6
ADJUSTED MANPOWER DIAGRAM

worker-days of this trade (the entire chart segment shown) and dividing this total by the 16 days allotted for the work (or by dividing the total estimated payroll cost by the number of working days, for daily payroll).

7.11 SUPERINTENDENT'S ASSIGNMENTS

With 15 workers, the superintendent's initial decision comes at the completion of item 1–2, when these workers are released. How many workers are to be assigned to 2–3 and how many to 2–4? In order to keep the same workers on 1–5 throughout and to keep this crew together at the completion of 1–5, he assigns 10 workers to this path, or series of operations; the others are assigned proportionately to the other two jobs. In some cases, fractional workers are scheduled; this means that a worker must be transferred during the period at which activity is under way.

If the labor cost estimate rather than worker-days is used throughout, the labor cost reports become progress reports for incomplete activities. In turn, reports of complete items become checkpoints for labor costs, and worker-days are not important. That is, time is measured by money expended rather than by days, and such costs as overtime hours, holidays, weather delays, and differences in rates of pay are included in the figure for labor costs.

A concrete pouring schedule is often prepared weekly, two weeks in advance, by the superintendent. The foreman, subs, and suppliers can then plan their work by days for a week in advance, and for another week within a day or two. The job foremen need this information for planning their work force, and are little interested in projections of weeks or months in advance. The concrete schedule may be printed from a transparency of the structural plans, and may require weekly information from the controlling trade foremen.

7.12 VARIATION IN COMPLETION TIME

In the example, 15 workers are required for 16 days for the entire operation. If an earlier completion time is desired, the labor force may be increased, but the diagram is similar. Completion times are advanced proportionately, since all activities affect the overall completion date of the work.

In summary, the use of the CPM diagram will be to determine the proportionate assignment of workers during critical times. The work force adjustment is made in Figure 7-6 to meet a completion date for one trade. But there are other choices; do the carpenters do everything possible to get ready for the electricians, for example, or for the bricklayers? The CPM diagram will show the time allotted to each of these trades, but it is often necessary to make adjustments; who can increase their work force most easily, electricians or bricklayers? The carpenters' effort is diverted to the trade which can move up its own schedule, or the effort is lost—we have "float" again. The overall CPM schedule must be made for all trades. It is quite practical to plan materials deliveries so that the remaining variable is the availability of manpower; even if an ample supply of workers is available, a limit must be placed on the work force if work is to progress efficiently and reduce the time delays between trades. *Consequently, the ultimate purpose of CPM planning is to furnish the superintendent with the information that enables him to assign men so that all paths are critical.* This departs from other writers' interpretations of CPM planning in that emphasis is placed on a single variable—labor availability—rather than assuming the labor force to be adequate.

7.13 SCHEDULES FOR REPETITIVE WORK

The construction time required for a number of identical jobs is the construction time for one plus the time for the longest common operation. This is illustrated in Figure 7-7 for eleven houses. If properly (or even reasonably) scheduled, each operation must take the same number of working days, here 21. If an operation is faster early in the construction, a time occurs during which no work on the last house is being done until the slower operation begins there. Each operation, if too fast, will have to slow down for operations preceding it, or if the preceding operations are

Operation	HOUSE NUMBER										
	①	②	③	④	⑤	⑥	⑦	⑧	⑨	⑩	⑪
Clearing	1	3	5	7	9	11	13	15	17	19	21
Foundations	4	6	8	10	12	14	16	18	20	22	24
Masonry	6	8	10	12	14	16	18	20	22	24	26
Framing	10	12	14	16	18	20	22	24	26	28	30
Roofing	13	15	17	19	21	23	25	27	29	31	33
Siding	17	19	21	23	25	27	29	31	33	35	37
Plaster	25	27	29	31	33	35	37	39	41	43	45
Flooring	31	32	34	36	38	40	42	44	46	48	50
Cabinets	32	34	36	38	40	42	44	46	48	50	52
Painting	35	37	39	41	43	45	47	49	51	53	55
Cleanup	38	40	42	44	46	48	50	52	54	56	58

Figure 7-7
SCHEDULE FOR REPETITIVE WORK IN WORKING DAYS FROM
BEGINNING OF PROJECT

also staying ahead, an idle time occurs until slower operations catch up. If some subs take houses by groups, in order to make a reasonable amount of work at one time, the time for completion of their work on *one* house may be, for scheduling purposes, the actual time they spend on three houses, depending on whether they waited until three houses were ready for work to start before they began the first of the three. Large sub crews are therefore usually a cause of delay in the work since they will not start their work on a house as soon as it is ready for their operation. *Crews that are too large are as often a cause of construction delay as those that are too small;* such oversize crews come late, leave the job, and delay in getting back again.

In Figure 7-7, one house may be completed in 38 days, and one crew may complete any operation in 20 days. The entire project will then take, if properly scheduled, 58 working days to complete. The original schedule is assumed to be such that each operation begins with an optimum crew as soon as preceding interferences are disposed of.

The dark lines show variations of how schedules may be made, which correspond to paths of a critical-path diagram. The clearing crew completes all work in 22 days; the last house then faces the same operations as the first. Or if House 1 is followed to the siding stage, then the siding crew followed to House 7, then the time for operations on House 7 to the cleanup crew, then the cleanup crew to the end of the job, one arrives at the same time for completion. Viewed as a CPM diagram, each square represents an activity, dependent on the activity (square) above it for availability of working space and depending on the activity to the left for availablity of manpower. This is an example of how a CPM diagram can be constructed to take availability of resources (workers) into account. This simple diagram represents 121 activities and nearly twice as many events. For a nonuniform job, where workers work as individuals rather than uniform crews, the number of dependent situations could be as high as the product of the number of workers and activities. The various paths possible in Figure 7-7 will be equal if work proceeds with a uniform work force. Any delay delays all houses beyond that point. In CPM parlance, each path is critical.

7.14   A MANPOWER CHART

The manpower chart is a result of, as well as an aid to, CPM or other scheduling methods. Figure 7-8 shows the projections of manpower for a project. This projection may be made from the labor estimate and checked against sequence of activities. It may also be made by computer from the CPM activities, although it will not appear flat—no change in the number of men working—unless the activities are very long. Assuming that manpower is limited and that a constant work force will be used at some stage, the critical trades requirement illustrated in Figure 7-8 will serve as the basis for shifting activities, changes in design, or other methods of improving the time needed for work.

An example of a manpower chart generated from the CPM schedule for the Bath County Pumped Storage Project is shown as Figure 7-9. The scale on the left is for numbers of workers and also for thousands of cumulative labor-hours. It is evident that the work force cannot increase as radically as shown, such as hiring hundreds of workers simultaneously. The CPM schedule may be adjusted to change this situation, or because this is a very complicated operation, the job superintendent may be trusted to work it out the best he can. In such situations, the inaccuracies and approximations of the system are understood, and like the builders of the pyramids, we revert to the intuition of the experienced manager.

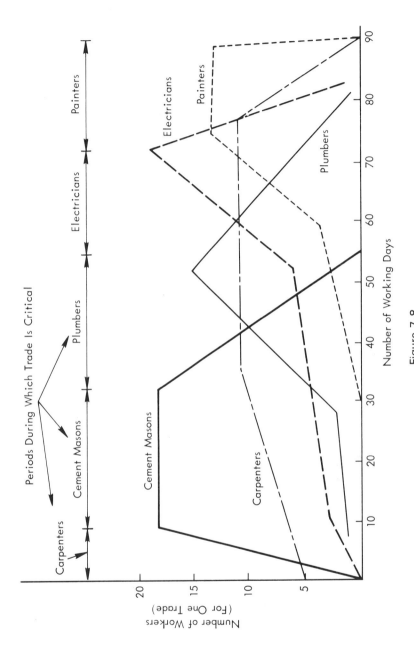

Figure 7-8

MANPOWER FOR CRITICAL TRADES

264

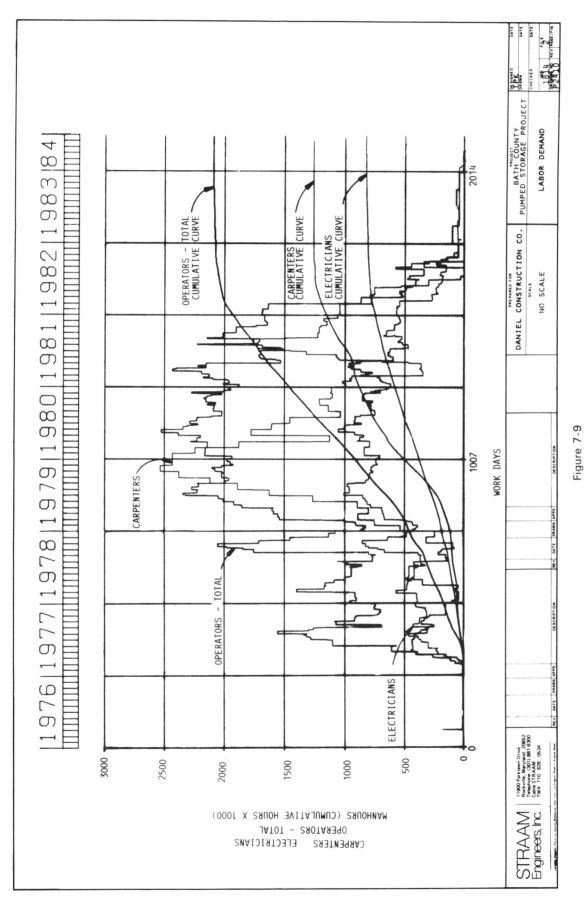

Figure 7-9
LABOR DEMAND
(Courtesy of Stramm Engineers, Inc.)

7.15 LAG BETWEEN TRADES

Delays may be caused by poor planning (as assumed by the CPM approach), by shortage of men (as assumed in the critical manpower shortage approach) or by delays in organization such as obtaining men or materials. Delays do not mean that the job is poorly planned; planning well all the activities and manpower still does not eliminate the very important delay involved in transferring work from one trade to another. Conventional CPM planning requires that the defined activities be so small that the next operation cannot begin until the preceding activity is *completely* done. There are nearly always areas where the next trade may begin although these areas are too small to plan. More important, it is inefficient to move a trade into an area unless the workers can keep working; that is, in order to work closely behind a trade, the next trade must have its rate of advance clearly established. Delays between trades often take longer than does the actual construction, and planning methods to date do not take this into account. It is just too hard to figure. But a superintendent on the job is efficient only if he can do this kind of scheduling.

On repetitive work, this lag can be worked out in the course of a few units. When the units are somewhat variable but basically similar, as in stores of a shopping center, the work can be planned on the job. Where the work of each trade is highly variable in different areas, as for a hospital, planning is more difficult. Area planning is insufficient; it is necessary for the several mechanical trades, who work on a linear basis (running long lines through several areas) to be able to work in the same area at the same time. If the estimates of manpower are worked out and plotted as labor against time, as in Figure 7-10, the horizontal difference between the resulting lines represents the time lag between trades and the slope represents the rate at which each trade is working (which varies with the number of workers). In the example, bearing wall

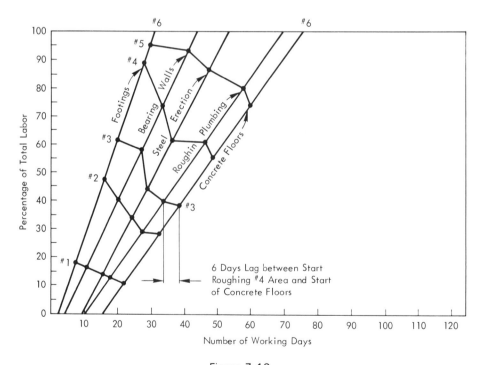

Figure 7-10
LABOR GRAPH BY TRADES

266

construction may begin immediately after footings, as footings go faster and the walls will not catch up. Steel erection follows at the same rate, nearly, as walls, so there must be a "unit" difference between succeeding trades that is greater. In the example, no. 1, no. 2, no. 3, and so forth, represent stores of a shopping center. The points at which a trade will complete a store is calculated from the estimate, and the fact that the lines connecting no. 1, no. 2, and so on, from one trade line to another are variable in length and direction indicates that the amount of work the trades have to do in a store is variable. In CPM planning, each unit may be a separated activity, or the planner may just plan activities to go on at the same time, leaving details of planning to the superintendent. Again when planning gets tough, the superintendent has to do it anyway.

As an example of closely following trades, a store was built in which at one end the tile work, plastering, and finish ceiling was under way while at the other end, the concrete floor—on which all other operations depended—had not yet been poured. One such store took 3 months to finish, while another identical store took only half the time, due solely to planning of the trades within the building. You may refuse to accept conventional planning—that is, you may change the normal sequence of trades—when you have the knowledge to back it up and are willing to revise the planning when it does not work. This is another way of saying the planning must be done by a thoroughly knowledgeable manager, and the supervisor on the job must have full authority to replan it as necessary.

7.16 DELAYS

Either the contractor or the owner or both must assume the cost of delay in the delivery of a completed job. Since the contractor directs the work, he is logically responsible for the cost of delay; but since, in practice, he does not assume the responsibility, he is loath to take it. The owner, on the other hand, is already taking a risk in building the structure and is more inclined to take this risk than to pay the amount which would be necessary to persuade the contractor to guarantee the completion date. The principal costs of delays are the interest cost during construction and the income lost because of the delay.

The owner is likely, therefore to assume the direction of the work when the cost of delay is high. He may engage a contractor on a cost-plus arrangement, in which the owner is practically the contractor, having assumed all the risks, all the direction, and the right to change his employee-contractor when he wishes. This arrangement also allows the owner to start work as plans are completed for various parts of the project, without negotiating new prices. Alternatively, he may make the contractor a partner in the enterprise. The contractor is then given the entire responsibility for cost and a share in the cost of delay, collecting a corresponding share of the profits in the enterprise. These arrangements often result in the contractor becoming a turnkey operator—that is, furnishing his own design—or a builder rather than a contractor.

When time is critical, there is a constant conflict between the contractor who has one economical rate of completion, and the owner, who has another. Ordinarily, the owner has an important stake in early completion, and the contractor wants flexibility to meet situations, including delay in the work, as they arise. The saving the contractor makes on overhead cost by early completion may be quite small. High-value personnel are seldom kept on a construction job when they are not working near their capacity; and they may not work at capacity because of the danger of their being "snowed under" if the job speeds up or more time is required on some portions. Key personnel must be able to work in advance, pushing vendors and subs for shop drawings and material approvals, pushing the architect for changes which will become necessary; in no other way will they have the spare time necessary to assure immediate action on day-to-day problems.

On the other hand, the contractor may sometimes be more interested in early completion than is the owner or architect. The architect has problems of his own and is often the bottleneck in plans, approvals, and early ordering of extras. Since the architect also decides if the contractor is to blame for delays, the architect is in effect judging himself; and of all the people involved, the A/E alone does not pay in some way for delays he causes in the work. Legally, the A/E is responsible for delays, but practically, collection of damages is both difficult and unwise.

Some examples of critical times which must be met are lease dates for stores, in which the tenant must meet certain dates for seasonal sales; production dates for industrial plants, where the output is already sold or is necessary in order to meet completion dates of other plants; and expiration dates on permanent mortgage commitments. Unfortunately, the costs of delays and the saving or loss to the contractor due to advancing completion dates are not well enough known to put a price tag on negotiations between two firms, and there are few contractors who attempt to estimate these costs with the accuracy with which they estimate the least accurate labor cost of an item.

7.17   LOWEST COST—FAST OR SLOW?

In most contracts the fastest job is the cheapest. It depends on whether the contractor is short of workers or of supervisors. If there are few qualified workers, they can be kept on the job longer, and be best used, by putting a small force on for a longer time rather than moving in a large force and moving it out again. But if supervisors are lacking, it is advantageous to use as many workers per supervisor as possible and to keep them in one place. Larger gangs of workers are therefore more efficient, and this large gang will complete a smaller item of work more rapidly. Most subcontractors prefer to use as large a gang of workers as possible, for better supervision, but rarely is a job staffed to the greatest number of workers who can work on the job efficiently. Either a shortage of workers or of supervisors will make a maximum work force impractical. Of course, workers can work overtime, in either case; but a 10-hour day is usually enough to bring in workers, so the extended overtime which would have any significance in the work is rarely necessary. The increase expected in the work force by increasing hours 25 percent (to a 10-hour day) and increasing pay by 37 percent is much greater than the increased work done in these two hours. However, competent supervisors are more difficult to find and do not move to follow overtime jobs so promptly. Once foremen are obtained, they must be watched closely long enough to determine their competence, and this also takes supervisory time. The most efficient way to stay on schedule is to keep an adequate supply of supervisors, more than necessary, and to preserve the alternative of increasing the work force on short notice.

7.18   PROGRESS MEETINGS

Once the overall plan has been made, it must be followed up with the superintendent and subcontractors on the job. Many contractors hold weekly progress meetings to exchange reports on delays; sometimes such meetings are held by the A/E or the *owner's representative. Owner's representative* refers to the construction supervisor of the owning corporation, who supervises the A/E for his firm. The term also indicates the A/E, when the owner does not deal directly with the contractor.

The contractor may prepare a weekly schedule of work to be done; if concrete is being poured, the schedule may be dates marked on a print and distributed to each subcontractor involved. These meetings may be very useful, but the project manager should always compare the time lost by persons not involved in the matter under discussion with the time needed to hold meetings separately with the subcontractor affected on each point. A strong chairman is necessary to prevent the meetings from degenerating into bull sessions which convince the subcontractors only that the manager does not know what he wants.

7.19 SUMMARY

Planning and scheduling is an essential part of construction management and is infinitely more complicated when one tries to reduce it to paper than is implied by trade and professional literature. The basic skill needed is ability to plan items by interference between trades, costs, and labor-hours required; elements of the accuracy necessary for the more sophisticated CPM procedures just are not available. Likewise, CPM procedures can be handled by manual methods, but jobs so simple that manual methods are sufficient can usually be scheduled by the superintendent while he drives to work in the morning.

A number of useful tools are shown here; not mentioned is the source of information, which must be compatible with the kind of units you are using. Computer people call this lack of information the GIGO principle, for "garbage in, garbage out," and in the case of CPM output, this description covers most available data. It is up to the individual manager to plan his work in accordance with the tools and knowledge at hand.

7.20 QUESTIONS FOR DISCUSSION

1. You are a new college graduate in engineering and are offered a position in CPM scheduling at the home office of a large construction firm. What are the advantages and disadvantages of such a position, and what would you want to know about it?
2. A construction job can often be completed faster with a reduced work force. Explain.
3. The critical-path chart has served its purpose before it is drawn. Explain.
4. With computer-planned CPM, the chart is not necessary. Do you agree or disagree? In what way is the chart required?
5. Resource leveling is an auxiliary function of CPM and does not materially affect the planning. Do you agree or disagree, and why?
6. CPM is useful in single-family housing projects, as in any other job. Discuss.
7. Can one trade be critical throughout an entire project? If so, under what circumstances?
8. The contractor will not accept the cost of late completion of large jobs when the completion date is critical. True or not, under what circumstances, and why?

CHAPTER 8

LABOR UNIONS

8.1 BASIC POLICIES TOWARD UNIONS

Any decision regarding policy or action toward a labor union requires experience and judgment.[1] The writer does not encourage any particular attitude toward unions but tries to explain the available alternatives. A construction firm has some or all of the following basic policies, depending on the circumstances.

Union contract. This is the common method by which a contractor "goes union" or is known as a "union contractor." The firm signs a contract with the unions who claim jurisdiction over the firm's work in one of the following ways:

1. The firm, as an *individual,* may contract directly with the local chapter of a union, or *local.* The unions prefer this form of contract, as they may strike contractors one at a time, keeping the men working for other contractors; this causes losses for the contractors, but with comparatively small loss to the union. For this reason, local contracts are rare.

2. The contractors, as an *association,* may contract with the local unions. If the union strikes for greater pay, it then must strike all contractors. The association may be formed for the purpose of negotiating union contracts, or it may be a number of contractors from a local association. Other contractors in the same association may choose to remain outside the contract.

3. If the contractor works in many areas, he may sign an international agreement. The contract is between the main office (international office) of the union in Washington

[1]For details of daily operation and jurisdictional disputes, see the author's *Desk Book for Construction Superintendents,* Chapters 7 and 10 (listed in the Bibliography).

and the contractor. The contractor agrees to conform to the local association's contract wherever the work may be, and to use union subcontractors. This is an important concession, since the general contractor often agrees to use union labor only of the trade with which he signs an agreement and is not responsible for his subcontractors. In return for the international agreement, the contractor need not contract with each local where he works, and the contractor will not be struck because of issues involving the local contractors' association. The most important such issue is wages; the out-of-town contractor may continue to work during a local strike, agreeing to pay retroactively any wage raise which may later be agreed upon by the local association. Such conduct is considered unethical by local contractors, since the union is able to use the out-of-town job to continue to work and to reduce the pressure from its members to end the strike. The contractor with an international contract may choose to support the local agreement, however, and close down during a strike.

4. Where the area covered by the contract is larger than the jurisdiction of any union local, the agreement is a *master agreement*. The construction unions in southern California have such an agreement.

5. On very large jobs, such as the Florida Rocket Base at Cape Canaveral, Disney World, and the BART Transportation System in San Francisco, a *project agreement* may be made between the general contractor and the union. Such an agreement is essentially an agreement with the local unions, but it is signed by the international (the national leaders of the union) as well as by the local unions. The contractor then requires his subcontractors to comply with the agreement for their respective trades. The chief purpose of such an agreement is to avoid disputes between the unions and the subcontractors, to assure that wages will remain stable (when the length of the job permits this), to avoid travel time which may otherwise be required by local unions some distance from the project, and to secure prompt handling of jurisdictional disputes by the international office of the union. The agreement also avoids entanglements by the contractor in local wage disputes, as does an international contract, except when the project agreement expires and must be renegotiated.

No union contract. If the contractor chooses not to sign a contract with the union, he may operate under either union or open-shop conditions.

If a contractor operates under union conditions, he hires only workers who are members of the union or who apply for such membership, and he conforms to union pay and working conditions. He may hire his men from the union hiring hall or he may not, but he must make the health and welfare and other fringe benefit payments as required by the local association's contract. The unions are often more concerned about these payments than about a contract, since they have found that collecting money is often more difficult than collecting signatures on a contract. The advantage of this arrangement for the contractor is that he may change to open-shop work at any time, should it be worthwhile to do so. However, it may be held by the National Labor Relations Board that an implied contract exists in such situations. It is of advantage to the union, since it can keep members on such a job if it is on strike against the contractors' association. If faced with a strike against one of his union subcontractors, a contractor in such a position may take over the sub's workers and supervisors, continue to work, and still pay the subcontractor the amount of his subcontract at the end. The general contractor has the advantage of an international contract, while retaining his freedom of action. The local members of the contractors' association will naturally resent such a situation but may not know of it.

The *open-shop contractor* is one who does not call on the union for workers and who usually does not conform to union working rules and wage rates. The most common saving through open-shop work is that the contractor can use unskilled labor for work which would require skilled labor on a union job. A large part of carpentry work consists of driving nails, holding pieces, or even cutting—work any person can learn in a short time and can do well, provided that he has immediate supervision and is confined to the work he has just learned. Characteristically, the open-shop contractor pays workers in accordance with their skill, which means a variety of wage rates for each occupation; hence the open-shop is also called the *merit shop,* in that a man is paid in accordance with his merits. An open-shop contractor may employ both union and open-shop subcontractors. Since it is more difficult to find skilled workers in the mechanical trades and the work cannot be divided up so readily between unskilled and skilled workers, the electrical and plumbing contractors are often union even on a job where the general contractor uses open-shop labor.

8.2 PURPOSES OF UNION ORGANIZATION

Contractors may usually reduce labor costs by paying workers in proportion to work done, either by piecework payments or by an hourly rate based on production. They also may reduce costs by hiring workers who may be used for several different kinds of work. Such practices and resulting low labor costs in the mobile home industry, as well as lower construction standards for mobile homes, has led to virtual elimination of conventional homes from the low-cost market. When workers are paid by work done, the total payment received by the workers is less than otherwise, and the more skilled workers receive a larger wage, which reduces the pay of less skilled workers who cannot produce as efficiently. Piecework has been used first as an incentive. However, when new workers have become available at lower rates, the piecework rates have been lowered; similarly, hourly rates based on production have been lowered when cheaper labor has become available.

Such actions by contractors are the basic reason for the existence of construction trade unions; they try to protect hourly earnings from being reduced and to equalize pay among workers, so the qualified worker with the lowest degree of skill can earn as much as the best worker for the same time worked. To some extent they try to equalize annual earnings. Annual earnings are shared to some degree by rotating job referrals; that is, by equalizing job opportunities, sending workers to work first who were first laid off. However, they do not consider the length of previous jobs, so better workers normally will hold jobs longer and will therefore be employed a higher percentage of the time. When average employment in a trade is 90 percent, one may find individuals who work less than half the time, either for lack of skill or just bad luck.

Of the other issues between contractors and union, the most important is the wage rate. Once a union has the entire labor force in its membership, it can obtain higher wages. In most of the country, unions may not raise their own demands much higher than the prevailing open-shop labor cost, or their employers will be placed at a disadvantage in competitive bidding.

Labor unions are basically *trusts,* or agreements to restrain trade, and as such were placed under legal restraints at various times prior to the passage of the antitrust laws of the United States in the 1890s; but they were then specifically exempted from the antitrust laws. I oppose any restraints of trade, but this view is not popular; the architectural and engineering organizations, for example, relinquished agreements (many of which were embodied in state laws) not to bid in competition only after attacked by the antitrust division of the U.S. Attorney General's

Office; like other occupations and businesses, they favor free competition for businesses other than the one they are themselves engaged in.

Trade vs. industrial unions. Traditionally in the United States, construction workers have organized by trades, since such a union need organize only a fraction of the workers in order to be effective; their high degree of skill makes substitution of other workers for those on strike very difficult. The construction firms are also organized along trade lines, so that the employees of one firm are confined to one or a few trades. Until the 1930s, trade organization was the dominant form of labor union in the United States. At that time, John L. Lewis organized the automobile workers, who, with others, formed the CIO (Congress of Industrial Organizations) and separated from the AFL (American Federation of Labor). After a number of years, the two organizations combined to form the present AFL-CIO. The Teamsters are an industrial union but are closely allied on the local level with the construction trades (and were a party to the Disney World Agreement). Like other industrial unions, the Teamsters are comparatively indifferent to trade jurisdiction among their own members and organize all workers they are able to claim. Consequently, they become involved in disputes with nearly every union which drives equipment or handles material. The Teamsters generally drive over-the-road equipment and load and unload material at warehouses. However, they operate equipment mounted on their trucks for loading and unloading them, such as concrete mixers and cranes, which leads to disputes with the union of Operating Engineers.

The CIO and AFL unions often dispute plant construction maintenance, the CIO being inclined to fight any invasion of construction workers into work which may be done by their maintenance workers, such as maintenance and remodeling. These disputes have at times led to plant riots.

8.3 CONTRACTORS' TACTICS AGAINST UNIONS

Among methods used by contractors in the past to delay or prevent union organization and end strikes are these:

1. *Secret informants,* usually private detectives, have been used to locate union organizers and to end strikes. In other industries, these people have assaulted or killed organizers. There is no longer any purpose in using secret informants, as there is little reason on the part of the organizers to remain unknown.

2. *Discharging union members,* and *blacklisting* them by distributing the names to other contractors has been done so union members may not obtain work anywhere in the industry. This practice is now illegal.

3. *Lockouts* have been used, whereby if one employer's men, or part of them, go on strike, he increases pressure on the union to end the strike by cooperating with other employers (or other portions of his own organization) by closing down operations. Technically, lockouts are now illegal, but as it is legal for members of an employers' association to close down if one is struck, the effect is the same as a lockout. When the employers in an area attempt to force the unions into a single contract for a larger organization, they may refuse to renew the separate contracts, which becomes a strike when the unions refuse to work without a contract.

4. *Boycotts* of union contractors have been done by suppliers refusing to service union jobs. This is illegal and has not been practical for many years anyway; the more common boycott is by the union, which is also illegal.

5. *Strikebreaking* has been used, in which the contractor hires men to replace men on strike. This is legal, providing the original strike by the union is not over illegal acts of the employer. Strikebreaking is rarely used by contractors in recent years, as it often leads to violence and sabotage. The immediate result is abandonment of the job by trades not engaged in the original strike.

6. *Injunction,* or court order against the union for illegal acts can be used against strikes. At one time, the strike itself was illegal; in recent times, use of the injunction is limited to terminating the picketing of contractors not a party to the strike.

    Under the National Labor Relations Act, unions may obtain a supervised election of the workmen to choose whether they want union representation. This has very little significance in the construction industry, as the employees seldom work for one employer long enough to set up elections. Unions therefore represent the workmen almost entirely because of contracts with employers made without elections.

7. *Refusal of credit* to union contractors and to owners who employ them has been used against unions by influencing banks and other lending institutions.

8. *Payments* have been made to owners and contractors who may not be controlled by other means, indemnifying the contractor or owner against the difference between union and nonunion bids, where the union bid is lower.

9. *Lawsuits against unions for damages due to illegal union actions.* The international headquarters of unions and the locals refrain from accumulating funds which might be taken to satisfy court judgments. Particularly in jurisdictional disputes, the International attempts to prevent strikes.

    During 1970, the Carpenters' officers supported a jurisdictional strike against Kaiser Engineers in Missouri;[2] as a result, the owner (as it was a cost-plus job) sued both the International and the local, and was awarded a judgment of $914,000.

    In this case, an agreement between the millwright and pipefitter locals regarding erection of gates and material handling facilities had been made at a pre-job conference, but the International later supported job picketing even though Kaiser had suspended work on the disputed portion.

10. *Double-breasted operations.* It is quite common for an owner or group to own two firms, one of which uses union labor and the other which has no contract with the union and does not hire from the union hall. The purpose is to operate in separate markets, either geographically or by type of construction. There have been rulings by the NLRB and the courts that under certain circumstances, the labor contract applies to both firms and the open-shop employees must be paid back union wages, with fringes and benefits. This would usually result in employees not skilled in the trades, but using tools, to be classified as mechanics.

    These circumstances may occur when an existing union firm establishes a new firm to work in the same market, and maintains links with the new firm, such as sharing officers, supervisory personnel, and offices. The criteria in determining if the union contract is applicable are common ownership, management, control of labor relations, and interrelation of operations. If the firm is open-shop and establishes a union

[2]*Norada Aluminum, Inc.* v. *Carpenters,* 85 LRRM 2147, U.S. District Court, Eastern District of Missouri, No. 72-C-33(1), December 28, 1973.

firm, there seems to be less trouble. Also, a union firm which buys an open-shop firm is much less likely to have trouble.

The contractor planning such an operation should be informed on such matters. It is desirable that the firm establishing the new open-shop operation should follow as many of the following rules as possible, compatible with the circumstances.

a. Keep ownership separate.

b. Offices should be in separate locations, with separate employees and officers.

c. Separate banking and credit operations of the two firms.

d. Hire separate employees initially, and do not interchange them between corporations.

e. Do not refer inquiries from one firm to another, or interchange equipment.

f. The firms should bid different sizes and types of work, or in different areas.
   If some of these rules do not fit the situation, then any joint facilities by the two companies should be accounted for by intercompany charges.

11. *Separate job entrances for union and open-shop contractors.* A gate reserved for one contractor has been defined as one in which the *materials essential to the contractor's normal operations enter,* by the National Labor Relations Board panel in a recent case, by a 2-1 decision. An electrical contractor who was being picketed, J.F.Hoff Electric Company, installed a separate gate to prevent picketing of other contractors' gate. The picketing spread to the main gate after electrical fixtures were delivered there. The fixtures were purchased by the owner and stored in the owner's warehouse, for future installation by Hoff. Although the majority of the panel held that the immunity of the main gate to picketing was lost by using it for materials essential to Hoff's operations, the dissenting member said the Supreme Court decision outlawing site picketing was violated.[3]

Most of these methods are entirely or to some extent illegal, depending on the circumstances. They should not be considered merely historical, however, simply because they have not lately appeared in the newpapers or court records. Many of them are by their nature very difficult to prove or even detect; a number are methods of industrial warfare, used more often against other businesses than against unions.

8.4 UNION STRATEGIES FOR CONTROLLING WORK FORCES

In attempting to control the work force, some of the basic union strategies are:

1. Depriving the employer of a work force and materials by strikes and picketing

2. Gaining control of the existing work force by organization

3. Furnishing work to employees through a union hiring hall

---

[3]*Engineering News-Record,* Vol. 202, No. 18, May 3, 1979, p. 54.

4. Creating skilled workers by training

5. Preventing the entry of open-shop employers into the industry

Some of the tactics used in carrying out this strategy are discussed in the following paragraphs.

Strikes. A strike is the most obvious and most frequently used tactic. A construction strike is now far from the noisy—and often bloody—event of the first years of union organization. The workmen more often do not use the work *strike,* but say *stop work, pull out, go home,* or *go out.* Since the employer nearly always waits for economic pressures to force the strikers into a more conciliatory position, pickets are unnecessary, and the trades not on strike continue working as before, even for the same employer.

To some extent, a contractor may continue with construction by using only trades not on strike. The work of the trade on strike, however, will not be done by other trades, and eventually the work must stop. The unions allow work to continue because they anticipate going back to work and know they will eventually be paid to do the work of the trade on strike, making full employment (where normally there would be partial unemployment) or overtime. The workers on strike will lose wages during the strike period but can look forward to the same overall earnings for the year. Likewise, they would cause their fellow workers to suffer, as well as the employers, if they used sabotage or violence. Mature labor organizations, dealing with the same employers for many years, adopt a much less extreme attitude than do their fellows engaged in early stages of organization.

Picketing. Picketing consists of displaying a sign at entrances to the job, requesting other workers not to go to work; that is, not to *cross the line.* In other industries, *mass picketing* is used; that is, using so many pickets that the effect is physically to prevent workers from crossing the line, rather than requesting them not to cross. Mass picketing is illegal, and court injunctions often specify the number and spacing of pickets that may be placed on a job. Usually, picketing is more a symbolic act in construction disputes; two pickets are the largest number normally assigned to one gate. Pickets sometimes prop their signs against the gate and sit down under a tree or sleep in their cars.

American workers, including many who are not union members, consistently refuse to cross picket lines. This makes picketing effective. The courts have recognized the peculiar nature of this form of free speech as an effective, positive act preventing the employer from engaging in business and have therefore restricted its use. These restrictions and the statutes under which they are enforced require that picketing be confined to the employer with whom the workers have a dispute. This may be done by showing, on the sign, the trade or union with the dispute, and the employer to whom the action is directed. Since pickets are usually respected regardless of what may be on the sign, employees not picketing may continue work only by using a separate entrance. For example, the mechanical trades may have a separate gate; if the general contractor has a strike, the pickets may not picket the gate which is not used by the general contractor. This has proven to be an effective way to continue work when one trade is on strike and picketing the employer. Since the law varies with each year's decisions and statutes, the manager must stay abreast of the changes. This text does not pretend to be authoritative on the latest modifications. The unions can usually circumvent laws which they feel restrict them unduly, since they can accomplish their purpose before court action has become effective.

8.5 ORGANIZING

Labor unions are created by organizing workers. In modern times, the dominance of some trades in urban areas and the resulting high standards of entry into the union have given the impression to the public that union membership is a privilege difficult to achieve. But few unions dominate their trades completely, and they are constantly engaged in drawing open-shop and independent workers (who work directly for owners, or as subcontractors, and whose earnings are not on an hourly basis) into the union.

An important decision made by the union is the minimum skill required of new members. Many union members are not now qualified in all skills of their trade. The basic requirement of these new members is that they be able to hold a job at the trade. The new man, even though his experience is limited and his wage below a third- or fourth-year apprentice, will not want to accept apprentice status. The unions must therefore modify their standards to gain these men, and in this way many union members are workers who have never served an apprenticeship. Much of the widespread criticism of the ability of union workers is of such members. The union would like to maintain admission standards to limit membership to those workers with a complete knowledge of the trade, but it must compromise to avoid the undercutting of the wage rate by workers who are skilled in only one part of the work. Therefore, when a superintendent calls for carpenters, he needs to specify form carpenters, door hangers, framing carpenters, or flooring layers. If he needs a carpenter who can cut rafter framing, he is wise to say so. A call for bricklayers in Miami may be difficult to fill, as blocklayers are in demand and therefore there are many more of them. (The cheapest kind of work is the type done in the locality; specialists are attracted and compete with each other.) Likewise, a call for blocklayers in the northern states may bring a response, "Well, we have some people who *can* lay block." Manhole builders, who work without guides for either line or elevation, have different workmanship standards from wall bricklayers, and many workers cannot do both efficiently. The heavy walls on which the men selected by Frank Gilbreth for time studies were working in the early 1900s have all but vanished, and with them the specialists on "throwing bricks in the wall."

A letter from the Bricklayers' president to the membership, reprinted as Figure 8-1, indicates the continuing importance of organization to unions.

The original and still most widely used method of organizing is to convince each worker that he can get better wages by joint action. The difficulty encountered is that the worker, who is often personally acquainted with his employer, fears that he will merely be replaced. The individual knows that all employers must compete and therefore raises must be demanded simultaneously of all employers. The union must make its wage rate competitive with the prevailing wage in the industry. Homebuilders, in particular, enjoy a lower union wage rate than do industrial and commercial contractors. When union work is available, the organizer has a simpler job of recruiting men to join the union, but many members leave the union when only open-shop jobs are available.

A more popular and efficient method is to organize the employer, then to accept his entire work force into membership in accordance with their ability as journeymen, fourth-year and third-year apprentices, and so forth. Employers have been forced into the union by picketing of their jobs, but the effect is sometimes to close down all work *except* that of the employer being

AS IT LOOKS FROM HERE

By President Thomas F. Murphy

Our Primary Function Is Organizing

I have always thought and still do that the primary function of a labor union was to organize all of those who work at any branch of their trade or occupation. For this reason we resist Right-to-Work laws because, while it is a catchy phrase, it in no way provides employment for anybody and gives no one the right to work unless the employer accepts him.

For the same reason, it seems to me that we are somewhat naive in our ideas that we can not accept into membership those who apply unless we can provide a job for them. Providing jobs is certainly the thing all of us would like to do, but there are times when people are unemployed through no one's fault or for no reason other than there is just no work.

It is surprising how many times when we turn down an applicant for membership for the reason we have no work available, the rascal then finds a job at our trade. Therefore, it seems to me that it is a simple bit of business on our part to accept into membership those who apply, and I am sure anybody who applies for work in our organization is reasonably qualified at some branch of our trade. We are not all "Fancy Dans" with the trowel (I, of course, am one of those good Bricklayers you hear about) but there is a practical ability about the Bricklayers, Tile Layers, Marble Masons, or whatever, that does not apply somehow in other trades. Without downgrading my friends in the Building Trades, driving a nail or sawing a board is relatively simple. As for painting, ask any wife who has undertaken to freshen up a kitchen or bathroom. Leaking faucet, anyone?

Some of our members going from one jurisdiction to another with good cards seem to have a difficult time securing a job. We call ourselves an International Union which means that our card ought to be good anywhere in the United States or Canada, and so long as there are reasonable local employment opportunities, our brother-members ought to be accorded certain privileges as a result of their membership. To exclude those who work at any branch of our trade because we have the idea somehow that keeping people out we keep all the work for ourselves, is wrong.

I just read a poem about the life of a union business agent which in a satirical manner indicated that if he tries to talk on a subject, he is trying to run things. If he doesn't, it is because he is too dumb. If he goes on a job, what's he doing there? If he doesn't, he's neglecting his duties, and so forth.

It appears to me that whenever a job starts anywhere, it is a function of the Local Union to appoint a competent Shop Steward to look after things and then leave the job alone and devote our time to doing something constructive by organizing the unorganized and making it easy for reasonably qualified people to join our organization in order to combat the menace of the non-union worker.

Thomas F. Murphy

Figure 8-1
BRICKLAYERS' LETTER
(Reprinted by permission of the International Union of Bricklayers and Allied Craftsman.)

picketed. In the 1950s at Dover Air Force Base, pickets were not allowed in the base and remained at the front gate; *all* work ceased for weeks *except* that of the open-shop employers being picketed!

More commonly at this time, a subcontractor becomes a union employer (in union terms, a *fair* employer; any nonunion employer is *unfair*) in order to get work with union general contractors, particularly those with national union contracts who may not employ open-shop contractors.

A number of states prohibit union as well as closed-shop contracts, maintaining the right of any person to work anywhere. These *right-to-work* laws tend to make unions easier to deal with, as the employer has the option of hiring new workers and discharging any union members who refuse to work. These laws make union contracts unenforceable in important respects, but the union shop is based on custom rather than on contract and union employers in the construction industry may not, as a practical matter, take advantage of such laws.

Both the Taft-Hartley Act and the right-to-work laws are restraints on the right to contract of both employer and union; the nonunion worker, not the employer, is to be protected. The employer always has had the option to refuse to sign restrictive clauses in union contracts. Although the union is deprived of legal means of enforcing the contract, it still has the weapons which induced the employer to sign a union contract. It is unrealistic to believe that the employer will challenge the union later if he would not challenge it before the contract was signed.

Consequently, only the nonunion worker, and the public who eventually pays the higher labor costs, have an interest in enforcement of these laws. They are enforced only when a public prosecutor is unusually active, or when there is an opportunity for an employer or a nonunion worker to get money damages. The unions are careful to keep their considerable assets out of the power of locals who are the active offenders and therefore subject to lawsuits. The union funds are property of the national headquarters or in independent trust funds, and national officers are careful to avoid illegal actions.

8.6   UNION HIRING HALL

The local union, as well as its office, is referred to as the *hall*. When the union is acting as a personnel employment agency, it is referred to as a *hiring hall*. Technically, workers are not hired but are referred to an employer, in the same way an employment agency would refer them; as a practical matter, the men are hired on arrival at the job and may be entitled to payment for two hours reporting time if they are refused work.

The continued control by a union over its members is supported by its control over hiring. Customarily, a superintendent calls the union office at the end of the day to request workers to start work the next day. Sometimes calls may be made and filled during the day. The superintendent may fire workers with as little as an hour's notice and order new ones to replace them. When construction workers are out of work, they are expected to be available for work at the union office or at a telephone. If the BA (business agent) cannot reach the worker in time, he sends someone else; according to one rule, if a member cannot be reached by telephone within two hours, he goes back to the bottom of the list and will not be called again until all other idle members have been referred to jobs.

If no workers are available, the superintendent's request is held until workers are available; one of the BA's duties is to keep informed of anticipated layoffs, so he can estimate when workers will be available. If a labor shortage continues, the BA will recruit workers from other cities or will recruit new members. Workers brought from other locals usually receive no travel pay or wages other than what local men are paid. There may be substitutions of one craft for another in the same union, as structural for reinforcing ironworkers or fitters for plumbers; or the BA may make other concessions, such as allowing laborers to do part of the work. These concessions are more often not discussed; the BA merely ignores the fact that someone else is doing work he claims.

The Taft-Hartley Act requires that hiring halls refer both union and nonunion workers if the hall is the employer's sole source of workers. The Laborers Union follows this rule in many cities, collecting the initiation fee over a period of weeks or even months after a worker is placed. But as jobs are short and the turnover rate is high, workers may return for work once or twice a year for several years, working at other jobs meanwhile. Workers start payments on their initiation fee (which may be as little as $150, a fraction of the fee for other unions) several times, and the hall is operated like an open-shop employment agency.

Virtually all other unions confine their referrals to union members, although they may agree to place former members and to accept new members when their own people are fully employed. Since the hiring hall prequalifies or screens workers and responds to calls in a matter of hours, it is more efficient than any agency available to open-shop contractors and provides a valuable service both to its members and to the contractors it serves. Usually, idle workers are sent out for work in the order of their registration *on the list*. This priority is modified by a number of circumstances, including the following:

1. Request of an employer for workers by name; in some cases a worker has priority on calls from his former employer, even if he is not requested by name.

2. Request for foremen, who are usually on the list for journeymen but may get priority on foreman jobs. Designation of foremen is a prerogative of the superintendent; but when the superintendent does not know the local workers, he may request one whom the BA recommends.

3. Request for specific skills, so several other workers may be passed over for a specialist in the work requested.

4. Apprentices and older workers, both of whom receive less than the standard wage; the union may insist that a proportion of these workers be hired.

5. Workers who find their own jobs, without waiting for the BA to send them out. This practice is discouraged by the union, as it may make it necessary for *all* idle workers to make rounds of the jobs looking for work. Many locals prohibit this practice.

8.7 TRAINING

Unions do not generally favor the teaching of construction trades in public schools as a full-time course, since such schools would be outside their influence and would be likely to develop a qualified open-shop labor force. Since vocational training is expensive and construction work lacks prestige, public agencies are seldom interested in establishing schools for the construction trades. The public schools cooperate in apprentice-training programs to the extent of providing classrooms not otherwise used and may make a nominal payment to part-time teachers.

Training programs are usually initiated and paid for by employer-union associations, or they are not offered at all. The public school system, with its white-collar orientation and limited funds, does not provide a significant number of workers trained for construction.

The usual apprentice program provides evening instruction for one or two evenings a week for three to five years. The apprentice works full-time, or as nearly so as possible, and the BA tries to assign him to varied kinds of work in accordance with a standard schedule. A record is kept of the time spent at each of various skills of the trade. In the basic (general contractor) trades, unions are as much interested in increasing the number of apprentices as are employers, or even more interested. Contractors often prefer not to hire apprentices, especially in the early years, as they believe the labor cost to be higher.

The United Association (plumbers and fitters) has unusually complete texts and teaching aids, developed and used only by their union-industry apprentice programs and not publicly available. The Operating Engineers, who are trying to organize surveyors on engineering construction, have a training program for them and a national recruiting program so they may provide surveyors to employers needing them.

However, too few construction workers receive apprenticeship training for such training to furnish a primary source of workers, except for the electrical trade, or to be important in union control. In 1972 less than 20 percent of journeymen entering the construction trades did so through apprentice training. In that year, 63,106 new apprentices were registered, but 30,159 dropped out. Among the reasons for this are low pay (usually less than laborers'), lack of available work because of lack of contractor acceptance, lack of work due to union restrictive regulations, boredom with poor training, and impatience with the length of the training program. Both the unions and the federal government have resisted changes proposed by contractors. Although the federal government is not necessarily to blame for the failure of the program, it does prevent adaptation of the program to one which would produce more trained workers, and has resisted shorter programs or extending training of one apprentice to more than one trade.

8.8 LICENSING

Municipalities have a virtually unlimited right to determine who may be contractors in their jurisdiction, and to approve or disapprove the work done. The popular charges that inspectors (in the pipe and electrical trades, in particular) will not accept nonunion work and that nonunion contractors cannot be licensed in large cities are too persistent to be disregarded. In any situation where the municipal authorities are known to have strong views, the possibilities of illegal action by them must be considered. The arbitrary nature of the authority they exercise makes proof difficult or impossible; more often it is exercised to resist a development using a type of construction they do not approve, such as modular homes, or tenants they do not approve, as in government-subsidized housing.

Governing bodies or contractor licensing boards can also restrict licenses by requiring a minimum capital, without reference to the size of the work the contractor is licensed to do.

8.9 SABOTAGE

Sabotage of construction work is a very real danger in areas where a solid union position is being attacked by a single contractor. This may occur either during working hours or at night and usually takes the form of cutting electrical cables or in other ways destroying materials.

8.10 ADMISSION OF CONTRACTORS

In union areas, an employer must have a union contract; the union may choose to protect the existing contractors and maintain construction costs. If competition is too strong in an area, the pressure is greater on each contractor to keep down wages, require more production, or to shift to open-shop operations. This protection from out-of-city competition is not so commonly provided to general contractors as to subcontractors, particularly labor subcontractors, who are poorly financed and not well skilled in estimating. Unions discourage or prohibit subcontractors from subbing labor only, that is, hiring out a work crew rather than providing materials, equipment, and labor.

The mechanical trades may refuse to provide workers to general contractors; in most cases, this is unimportant since municipal regulations require licensed specialists for mechanical work. The mechanical trades have difficulty in enforcing labor jurisdiction when they claim work not in the mechanical contracts or classified as specialized work by the municipality, such as switching temporary circuits, watching heaters, and unloading materials. The general contractor can simply refuse to give this work to a sub, and the union cannot legally picket the job. The mechanical contractor may therefore have a problem, so he usually includes such work in his bid or does it without extra payment.

Unions do not necessarily prefer to work for local contractors, or even for contractors in preference to owner-builders. They want an employer who can afford to lose money on one trade, and who can rely on the average of the job to make it up elsewhere. They naturally prefer employers of proven financial responsibility and for that reason may prefer to work directly for the owner rather than for a contractor. Unions may prefer working for a new contractor in an area; although the new contractor does not have the established contacts that the local people do, he also does not inherit local animosities.

8.11 REQUIREMENTS ON U.S. GOVERNMENT CONSTRUCTION

In recent years the federal government has insisted on employment of minority groups, particularly blacks and women, in certain proportions, and considerable friction has arisen over the refusal of some locals to comply. National union leaders, who have cooperated with the government in this program, have even approved the use of nonunion labor and the crossing of picket lines in instances where locals have refused to admit black members.

A number of unions have admitted blacks ever since the unions were formed; some black locals are charter members of the national unions, such as those of lathers, plasterers, cement masons, bricklayers, carpenters, and laborers. There have been mixed locals with black business agents, usually in the South, because that is where the blacks were in the majority. More often, black and white locals were separate; in 1958, a contractor could choose black or white union bricklayers in Atlanta, but the crews were not mixed on the wall. Some all-black unions in the South have remained segregated by choice. Since locals are independent in all such matters, many locals with all-white membership, particularly in the North, refused to accept black union members from other areas.

Friction continues, but on the same job nonunion blacks may be tolerated by white locals who would walk off the job if a white nonunion worker came on the job. On a Michigan job in 1956, a nonunion black electrical contractor did part of the work, protected by a tough state law; the union members just pretended he was not there. The local officers were able to enforce their decision on the members in this case, but often they cannot.

Other federal government requirements specify a minimum wage (not connected with the general minimum wage law for all industry). Under the *Davis-Bacon Act of 1931,* the Department of Labor sends circulars to contractors to determine the prevailing wage and then sets the wage rates for each trade on each government-owned construction project. The law does not intend that only union rates be established, but there have been widespread complaints for many years that the Department of Labor has requested rates only from union contractors, so that the resulting rates were not really prevailing rates. The result is that contractors who use less skilled workers at lower pay rates cannot compete for the work. Since black construction workers at the present time are, on the average, less skilled and less well paid than white construction workers, the Davis-Bacon Act tends to bar blacks from government construction projects, while other departments of the government are attempting to get more blacks employed. There are usually similar laws on state-owned work by the state involved.

The procedure for determination of *prevailing wage rates* by the Department of Labor requires that the workers in the area be counted according to the wage rate they receive. If a greater number receive the same rate, that is the prevailing rate; but if they are paid at different rates, the prevailing rate is the rate at which the greatest number (not necessarily the majority) are paid. If as many as 30 percent are paid the same rate, and there are no similar number receiving another rate, then the 30 percent may establish the prevailing rate. If fewer than 30 percent receive the same rate, rates are averaged. Since open-shop contractors pay varying rates but union contractors do not, this is virtually an order that union rates shall be used if as many as 30 percent of the workers are union—even though these 30 percent work much less than 30 percent of the hours worked. To take an extreme case, if 30 percent of the contractors are union (actually, the count is by contractors, not workers, which makes disparity still greater; many union men work open-shop when union work is unavailable) and pay $8 per hour, and if the other 70 percent pay varying scales from $2 to $4, the prevailing wage is $8. According to the U.S. Comptroller-General (an independent auditing agency of the U.S. Congress), the Davis-Bacon Act increases the cost of federal construction, as compared to private work, 5 percent to 15 percent.

Federal laws require that overtime be paid on government work for time worked in excess of eight hours a day. The Wages and Hours Law requires overtime payments only on time worked in excess of 40 hours a week. The difference may be considerable when long days are being worked but bad weather prevents full weeks of work. The extra cost of the overtime paid for time over eight hours a day, as compared with overtime paid for work in excess of 40 hours a week, can be 10 percent when 10-hour days are being worked. This is assuming that the job is open-shop and that the time worked for the week totals only 40 hours because of bad weather.

Construction employment of workers under 18 is also restricted by the Wages and Hours Law.

8.12 FEATHERBEDDING

Featherbedding is any requirement by a labor union that a worker who does no useful work must be hired and paid. Unions successfully enforce contract working rules which require minimum size steel erection crews, oilers on equipment which has an operator, and operators on automatic stationary equipment such as air compressors, conveyors, and welding engine-generators. They are also usually able to install nonworking job stewards.

In 1974, the National Labor Relations Board determined that a demand for a nonworking steward was featherbedding. In this instance J.R. Stevenson of Hempstead, N.Y., was to hire a job steward for the Teamsters, and provide him with a heated, telephone-equipped trailer. The NLRB ruled that the failure of the employee to perform *relevant service* constituted featherbedding.[4]

Except for the instance mentioned, labor unions are usually willing to trade away any rights for unnecessary workers, particularly in an area where open-shop competition is severe.

Some labor leaders are quite outspoken in their support of labor efficiency as organized labor's hope to combat open-shop work. Martin J. Ward, of the plumbers, believes in only one work rule: A man shows up on time, works the full time required, and does the job efficiently. *If he does not, he gets fired.* This sounds more like the proverbial nineteenth century capitalist than a modern labor leader, and emphasizes the extent that labor leaders, especially in the construction trades, are misunderstood. It also reflects a difference in attitude between many workers and their national leaders. The national leaders have little to do with making or enforcing working rules, so they can afford to have an attitude more conducive to good public relations.

8.13 UNION SHOPS VERSUS CLOSED SHOPS

Under a *union-shop* agreement, the employer may *employ* only union members, but he may *hire* anyone, provided that the new employee joins the union or pays union dues. That is, the new employee becomes a union member *after* he is hired. A *closed-shop* agreement requires that the employer hire only union members.

The Taft-Hartley Act (Labor-Management Relations Act) of 1947 affirmed the right of any employer to hire any man he pleased, regardless of his having signed a contract to the contrary. The closed shop, but not the union shop, was forbidden as against public policy.

The right-to-work state laws prohibit both the closed and the union shop. Under the Taft-Hartley law, a union may require the worker to pay dues, but may not require him to join the union. The Taft-Hartley law does not require a union shop but permits the employer to agree to it. The right-to-work laws do not require any kind of agreement, but they do not permit a union shop. The Taft-Hartley Act permits the states to pass their own right-to-work laws, if they choose.

Theoretically, under the Taft-Hartley Act, a worker is not required to join the union, may not be fined, and is not subject to union discipline. However, the courts have not recognized the practical necessity of joining the union in a union shop and have upheld the right of unions to discipline their members, even to the extent of collecting fines through court action from reluctant members who have been practically (but not legally) forced into the union by an employer's contract.

[4]*AGC Newsletter,* Vol. 26, No. 34, September 4, 1974.

The Taft-Hartley Act also regulates the charges unions may make for initiation fees and dues. Although these fees appear large (the initiation fee for the Operating Engineers may run into thousands of dollars) for the privilege of working, fees are less than private employment services charge for the same services.

In larger cities and in areas where unions are strong, the closed-shop operation predominates. In areas where the union's hold is weak and much work is done in open shops, a business agent may readily forgo a closed-shop for a union-shop operation. The right-to-work law can never be acceptable to a union, since it not only allows nonunion people to work alongside union members, but it also enables workmen to refuse to pay union dues even after the majority have elected the union as bargaining agent. The attitude of the union is that nonunion men derive union benefits without paying union dues. Union men will work alongside nonunion workmen only during the organizational phase of the formation of a union—which may, however, last many years.

8.14   SUPPORT BETWEEN TRADES

There is no legal obligation of one trade union to support another at the local level. Nonunion workers of one trade often work alongside union workers of another trade. One trade union will normally support another by observing its picket line, whether the pickets are legal or not. However, each union must judge the possible harm to its own members by lost work, against the possible benefits to another union which is not yet organized; if the striking union is too weak, the stronger may refuse to support it. That is, the established union must be satisfied that the objective of the strike is one that may reasonably be attained. Rarely does a local advise its members to cross a picket line; usually a notice is given to the picketing union that its pickets will not be recognized, and the pickets are withdrawn.

A plumbers' local union picketed a jobsite in Dallas, even though no plumbers were employed there, *to force the contractor to sign a subcontractor clause* which would have restricted the contractor to use only union subcontractors. The National Labor Board's General Counsel refused to consider the case, and the lower court held the union activity to be legal. However, the U.S. Supreme Court in 1975 held that the *union was subject to the antitrust laws* when engaged in such activity, and the picketing was therefore illegal. The decision was five justices against four, which means it could be reversed after a new justice is appointed.

A federal district court also ruled in 1974 that the Oregon operating engineers' union violated a federal antitrust law when it attempted to force a contractor to sign a subcontractor clause which prohibited him from subcontracting work to anyone who had not signed the labor agreement.[5]

8.15   ELEVATOR CONSTRUCTORS

The Elevator Constructors are unique among construction trade unions, in that they retain an employer- or industrial-type organization. Their jurisdiction is the elevator shaft, and they claim all work in it—whether work is done elsewhere by the carpenters, electricians, or plumbers. They will do steel and even concrete work necessary for support of their equipment, bore the

---

[5]*AGC Newsletter,* Vol. 26, No. 43, November 6, 1974.

hole for hydraulic elevators, and lay flooring tile in the cab. As they must deal with relatively few employers compared to other trades, the union is nationally oriented and has been able to develop uniform national working rules and arbitration procedures. Wages are not negotiated between the union and employers but are set as the average of the highest five of the seven basic trades specified in the wage agreement. It is said that the union has had only three strikes since 1904; workmen have spent their working lives in the trade without even hearing about a strike. The locals cannot authorize strikes; this must be done by the national headquarters. Practically every factor characteristic of trade unions has been modified by the Elevator Constructors. Its organization is similar to industrial rather than to other construction unions.

8.16 HISTORY OF LABOR UNIONS

The development of construction labor unions in the United States has paralleled the development of specialized methods requiring skilled trades people. Before 1900, numerous inventions had been introduced, both to provide improved services in buildings and to accelerate construction. The new materials and equipment had not become well enough accepted, however, to greatly affect the industry. In 1885 the Bessemer process of steelmaking made possible economical rolled steel beams, and the development of structural steel frames and of reinforcing steel for concrete structures soon followed. In 1889 the first steel-frame building was erected and the electric elevator was introduced, and in 1890 the Waldorf-Astoria Hotel in New York became the first building with both of these improvements. Increasing use of steel resulted in the organization of ironworkers, and a shortage of structural steel in 1897–1898 gave a big push to the use of reinforced concrete. Brickmaking plants developed from local ovens to large-scale commercial enterprises.

By 1880 American schools began to teach architecture. The power shovel came into general use, first for railroad work. Increasing changes in contractor organizations by 1890 made evident the deficiencies of the old apprentice system, with personal indenture to the contractor, and changes were begun. In 1890 the electric motor was used for hoist power on construction jobs, although the internal combustion engine was not introduced for another 15 years. The pneumatic riveter and the cement gun were introduced in 1898.

Local unions, led by the bricklayers, plasterers, and carpenters in New York, began to establish the present type of closed shop about 1880. Active union organization, particularly in the basic trades, was not matched by employer organization, and opposition to the closed shop did not become effective until after 1900. Construction employers rarely engaged in organized opposition to the operations of union contracts, and even the much later open-shop campaigns of the 1920s were financed and often directed by firms outside the construction industry.

After 1900, technical advances in both design and construction methods were rapidly adopted by the growing industry. Immigration of over 1 million persons a year led to a high rate of urban development, and in many areas some trades were nearly monopolized by immigrant groups of one nationality. Extensive rebuilding of San Francisco was required after the 1906 earthquake and fire. Our cities were largely built during 1900-1914, and vast areas of the major cities were covered by row houses, forcing high-rise construction in the central areas. Improvements in machines made many skilled workers obsolete; half of the skilled stonecutters, for example, were replaced by unskilled workers. The new steel-frame buildings, equipped with extensive electrical and mechanical systems, required specialized contractors, and corresponding trade unions multiplied; at one time there were *50 different craft unions* in New York City alone, as compared to 17 today.

By 1905 closed-shop agreements were made in the major cities, although organization of either unions or employers was not active at the national level. The large number of craft unions led to continual jurisdictional disputes over the new methods being introduced, and the national labor organization was not strong enough to enforce its rulings. Often work was disputed not only between different crafts, but between different unions of the same craft as well. It is reported that 95 percent of all strikes during this period were caused by jurisdictional disputes.

In New York in 1903, one hundred thousand workers were idled by a dispute between rival carpenter unions, which later involved other trades. Although the National Building Trades Council was formed by independent locals in 1897 and a Building Trades Department established in the AFL in 1907, these agencies were ineffective in settling disputes; according to a statement by the secretary of the Bricklayers in 1922, not one union accepted a decision against itself before 1918. Introduction of metal trim and its award to the Sheet Metal Workers in 1906 led to a 20-year dispute between the Carpenters Union and other trades; the Carpenters were expelled by the AFL because of their refusal to accept this decision, but they eventually gained their point by refusing to join a national organization unless this work was restored to them. Since regaining it, they have retained it to the present. Continued disputes between the plumbers and pipefitters were occurring by 1900, finally ending when the crafts were forced by the AFL National Convention in 1911 to amalgamate.

In a 10-year period, 1898 to 1907, nine national trade unions were formed, of which six survive today. The growth of subcontracting led to a proliferation of crafts and to formation of small unions opposed by the large, established ones. Most of these smaller unions were finally combined into larger ones. The Bricklayers absorbed the tilesetters, stone masons, terazzo workers, and some of the plasterers, but they failed to gain jurisdiction over the greater part of the plasterers and cement masons, who formed their own combined union. The plumbers, after combining with the pipefitters, joined the sprinkler fitters to form what is today the United Association.

Only the Ironworkers (who at that time had a *1 percent annual fatality rate* from job accidents) faced strong resistance from employer organizations, but by 1912 they had organized 55 percent of the workmen in this trade. The number of electricians more than tripled during the period 1900–1910, although the national organization was not founded until 1908. The Painters had their first national convention in 1901, and their membership doubled during the following decade. The Elevator Constructors were founded in 1901 and three years later had organized workers in all major cities. Otis Elevator alone employed 90 percent of the elevator workers.

During the 1920s, there were union-busting efforts in some cities, marked by blacklisting of union members by employers, and refusal of suppliers to sell to nonunion contractors. In the 1950s and later, the first legislation to curb the power of unions was enacted by Congress in the Taft-Hartley Act, and right-to-work laws were passed by many states; and antiunion activity in the form it had taken in the 1920s had been abandoned. The unions, in turn, have accepted legal methods of organizing workers.

Jurisdictional labor disputes can be as violent as disputes with contractors. In 1966 an estimated 150 electrical workers rioted over conduit under a Miami expressway; 14 were arrested and six hospitalized, one by a stab wound.[6]

[6]*St. Petersburg* (Florida) *Times,* March 1, 1966.

8.17 UNION ORGANIZATION

Practically all union construction workers are members of a *local* organization which in turn is a member of the *national trade union* (since these unions include Canada, they are called *international unions*), which in turn makes up the American Federation of Labor. The Teamsters, at present, are not members of the AFL. At the local level, the various trade locals are members of a district trades council for coordinating strike efforts and for public relations. The local may also join with its employers to form an organization for the improvement of common aims—usually for the advertising of the product, such as plastering or masonry. Both AFL and CIO unions form a council in each state for coordination, public relations, and state political action. The construction manager has no contact with the state group in his work.

The local, however, is the basic unit and wields the power of an independent organization (except in the Teamsters Union, which is associated with many District Councils but is not usually considered a construction trade). The national headquarters may exercise influence but not give orders; it may revoke the charter of the local, and it controls some trust funds, but this does not require the local to conform to its direction. If the charter is revoked, a new local may be established by the national headquarters, and the members must be persuaded to leave the old organization for the new one. If a worker does not do so, he may be unable to work in areas not within the jurisdiction of his local. This is extreme action, however, and is almost never done. The local members elect a business agent (BA), usually a full-time, paid executive, with a president and other part-time officers as needed. Normally only the BA is a paid officer and receives the same salary as a journeyman in the trade.

The local's operations are based on a charter from the international, which includes the constitution and bylaws of the local and the *working rules*. The working rules may be printed as a booklet, either with the bylaws or with the local contract. The contract may or may not include the working rules. The working rules state the method of computing overtime and the tools to be used, and may describe the jurisdiction of the union, although this is usually in the constitution as well. There are no uniform working rules in use by all locals of a trade, although for new locals the international may furnish typical rules. The working rules for different locals of the same union are similar.

The *job steward* is a journeyman designated as the union representative for each job. He may be elected by the workers, may be merely the first person to arrive, or may be appointed by the business agent. He checks the incoming workers for their books, showing whether their union membership is currently paid up, and their *referral slips* from the business agent, directed to the superintendent, for employment at that job and defends his trade jurisdiction. Normally, union matters are settled between the job steward and the foreman of the trade; business agents and superintendents are called upon only if there is an unresolved complaint.

The *international representative* is the field representative employed by the international headquarters. His duties include these:

1. Settlement of jurisdictional disputes with other unions at the local level.
2. Assistance in manpower planning of large projects, with the locals involved and the owner or contractor.
3. Mediation of labor disputes between local unions and contractors. He is obviously not impartial, but he often can assist settlements which are delayed because of personal conflicts, and can help to standardize agreements in order to make working conditions and jurisdiction similar throughout the country.

4. Advice to local unions, in respect to legal procedures and information on what other locals are doing. When international contracts are involved, he takes a more direct interest, since the international may have made obligations to the contractor, which the local must follow through on.

5. Inspection of locals to check compliance with international rules and payments and compliance with labor laws, and generally to provide information to the president of his union.

An international representative is usually a former BA.

8.18 DISNEY WORLD AGREEMENT

As an example of union practice, the Disney World Project Agreement is selected because it was drawn in a state where the unions do not control the labor force as in the northeast United States and in many urban areas, because the location was one in which many of the trades did not have men working much of the time, and because the international leaders of each union signed the agreement. The full agreement is with seventeen unions and has over 100 pages.

General agreement portion of the Disney World Agreement. The general agreement of the Disney World Agreement is between the Allen Contracting Company, a California firm, and the international presidents of 17 unions and consists of only nine pages. The general agreements sets forth the general promises of the internationals but makes them subject to agreement made with the various locals, who have individual agreements comprising the balance of the contract. Some of the key clauses are these:

1. The provisions apply to the construction of Walt Disney World, *notwithstanding provisions of local or national union agreements.* This is essential to the project contract; otherwise, it would not be a project contract, but only a collection of local contracts. *This agreement shall not apply to work performed under a legitimate manufacturer's warranty.* Equipment often must be repaired on the site because of faulty manufacture. The manufacturer usually refuses to do this with construction help, which is more expensive than his shop help and less skilled with the equipment. The use of the manufacturer's workers, even though it may be of another AFL local, is a frequent source of friction.

2. *Nothing in this agreement shall limit the right of the Contractor to subcontract or select his Subcontractors.* Presumably this is to permit local subcontractors who were not previously union, but who must become union, to work on the job under union conditions and also to permit out-of-state subs. Since the agreement is by job, not by employer, it would presumably be possible for subcontractors to work other jobs with open-shop labor. This is not uncommon, although the subcontractor usually has a separate corporation, although with his same supervisory personnel, for such open-shop work. In some areas, unions will not permit the formation of additional subcontractors.

3. *No rules, customs, or practices shall be permitted or observed which limit or restrict production or limit or restrict the joint or individual working efforts of employees.* Such a clause is common but is often not observed. In this contract, for example, the Operating Engineers still have inserted their rule that an operator may not work on more than two machines in one day.

4. *The contractor may utilize any method or techniques of construction and there shall be no . . . restriction . . . of the use of prefabricated materials . . . tools, or other labor-saving devices. . . . The contractor shall determine when overtime will be worked. . . .* Such clauses are common, but these are more definite than usual. Certain restrictions have been retained; for example, brushes for painters may not be wider than 6 inches. The contract does allow a wide variety of special shop-built items required for this particular job.

5. *. . . it is recognized that in rare cases personnel having special talents or qualifications not employed under this Agreement may participate in the installation or setup of unusual items.* Normally, this clause would be significant because of the use of "rare," which is rather indefinite. In this case, with an arbitrator to determine the definition, and because of the unusual nature of the work, it was appropriate. Unions often refuse to accept that their workers cannot do special work and require a "watcher," who is a union member, to be present.

6. *. . . the steward shall have the right to receive, but not to solicit, complaints or grievances. . .* This says that the steward may do only what he should be doing anyway. A steward is not expected to *solicit* grievances; perhaps the contractor's experience in California had been unfortunate.

7. *The steward shall, in addition to his work as a journeyman, be permitted to perform during working hours such of his normal duties as cannot be performed at other times.* On large jobs, it is not uncommon to require a nonworking job steward. In this case, with hundreds of workmen of each trade, it is doubtful that the steward would do any useful work on the job itself.

8. *The selection of applicants . . . to jobs shall be on a nondiscriminatory basis and shall not be based on, or in any way affected by, union membership. . . .* The Taft-Hartley law does not allow an employer to hire only from a union hiring hall. The unions, to circumvent this rule, have set up "nondiscriminatory" hiring halls, in which they register nonunion workmen. This is of advantage to the union because such men must join the union (except in right-to-work states) after being hired; but it is doubtful that many nonunion workmen find jobs in this way. Later in this contract, this restriction is circumvented by giving preference to men who have previously worked for union employers.

9. *The contractor may request and the system will refer employees who have previously been employed by the Contractor within the last twelve (12) months, not to exceed 40 percent of the men by craft requested periodically.* Since the BA's control of the men is by spreading work equally, he prefers to send men in his order, rather than as requested by the contractor. The contractor, on the other hand, wants to get back men whom he knows to be good workers. The 40 percent compromise gives the contractor substantially what he wants, since he would rarely want more than this proportion of former employees.

10. *In the event the union is unable to fill any requisitions for applicants within forty-eight (48) hours the contractor may employ applicants from any other available source.* This is

a standard clause in many contracts. It is seldom profitable for the contractor to invoke this clause, as he would require several days or weeks to locate other men. It does enable the contractor to bypass the hiring hall; the contractor could take men off other jobs or from other cities, but he would get union workers in most cases.

11. *There shall be no strikes, work stoppages, picketing or slowdowns . . . the contractor may discharge any employee violating this provision.* This is the important part of the agreement; it is good, of course, only for the period of the agreement at the agreed wage, which in this case was for four years. Disputes under this clause are to be settled within 48 hours by a specified arbitrator; however, the arbitrator has very limited powers and is not an arbitrator as defined by state law.

   A grievance procedure is included, but it does not include jurisdictional disputes, work stoppages, or lockouts. Other grievances are subject to the board of two men selected by each party to the dispute and one selected by these two. This is the type of arbitration once favored by the AIA clause in construction contracts, which has been replaced by the standard AAA clause. The Disney World contract, however, does include the AAA clause as a backup if the four men selected fail to agree on a fifth arbitrator. If a workman is discharged for engaging in a strike, he may appeal to the grievance machinery, but the contract clause is too involved and vague to be enforceable.

12. *Contractor shall provide a convenient and sanitary supply of drinking water . . . sanitary toilet facilities . . . a safe place for storage of tools, and facilities for changing clothes.* Water and toilet facilities are required by state law, and nearly all union contracts provide for storage and change rooms, although they are often not provided on smaller jobs.

Attached to the general agreement, each trade union made its own contract as part of the agreement. The Carpenters, of greatest concern to general contractors, took only three pages, while the Operating Engineers' contract was twice as long as the general agreement (the portion signed by all trades jointly).

The plasterers' union, usually the Operative Plasterers and Cement Masons, did not sign this agreement. The cause evidently was that the Bricklayers, who include plasterers and cement masons in areas where there are not enough workmen to form a local of their own, included the plasterers and cement masons in the area. The plasterers, although part of the same local, have separate working conditions and a separate contract, signed by the same business agent for both crafts.

The Disney World contracts with the locals are substantially the same as individual contracts with locals on an ordinary project (or any other project), although details vary.

Apprentices under the Disney World Agreement. The Disney World Agreement provides for wage rates for apprentices based on a percentage of a journeyman's wages. A new apprentice receives from 40 percent to 65 percent of a journeyman's wages. The low figure is the electricians', who, because of the high wages and security of their trade, can obtain apprentices at that rate. Several unions provide for a payment to an apprentice-training program for each hour's wage to the journeyman. Like payments for industry promotion, these are considered fringe payments, although the employer derives as much (or more) benefit from them as does the union. A number of agreements specify the number of apprentices per journeyman; the lathers designate one apprentice to five journeymen, the glaziers one apprentice to one journeyman; others do not specify. The lathers' proportion is probably insufficient to meet demands, even if the proportion stated is maintained. The lathers' rule is stated as a requirement, specifying it neither as a minimum or a maximum.

The apprentice periods are three or four years, varying with the trade. Occasionally apprentices are paid more than the minimum; the union discourages this but does not strongly complain. The business agent may feel that for one apprentice to be paid more than another leads to disruption of the program. Such surveys as have been made indicate that apprentice programs, for the basic trades at least, train foremen rather than journeymen, since most apprentices become foremen shortly after completing their apprenticeships.

Overtime under the Disney World Agreement. Only the Carpenters and Bricklayers agreed with the general agreement stipulation of time and a half for overtime, although the glaziers allowed that rate for up to sixty hours per week, and the cement masons for one hour of overtime per day. The International Union of Operating Engineers, commonly referred to as "Operating Engineers" or "Operators," allow time and a half for heavy construction (all work other than buildings) but not for building construction. A common rule for Operating Engineers is time and a half for overtime, but double time when they are working with other trades who receive double time. The purpose is to encourage overtime, but to get double time in situations where the time-and-a-half rule is not the cause of overtime.

Since the Federal Wages and Hours Act is generally applicable to construction, time and a half is required for time worked over 40 hours a week in any event. Double-time requirements are most often favored by the union because the increased cost pressures the contractor into hiring more men instead of relying on overtime, thus enabling the union to spread available work among as many men as possible. In some cases the union favors double-time requirements because the men do not want to work long hours.

A desire to work overtime is often reflected in lower rates for overtime. Some locals prohibit overtime because workmen may work for straight-time payment for overtime and not report it to the union. At a local union meeting when the members discussed enforcement of an existing rule not allowing any overtime to be worked, it was assumed that the contractor was paying straight time, but the younger members with families, who wanted overtime, opposed enforcement of rules which would stop it. The BA was instructed to enforce the rule, but it was obvious that he was not going to take the matter seriously.

Foreman under the Disney World Agreement. In the Disney World Agreement, the Carpenters required that a carpenter directing fewer than five workers be paid at foreman rates. If more than five carpenters are employed, one must be designated a foreman, but no further requirements that foremen be appointed or that there be nonworking foremen were made. Usually, they require, as do nearly all trades, that a foreman be appointed when a certain number, usually about six, are supervised by him. With a specified number of foremen, usually about two or three, a general foreman must be designated. The Bricklayers, noted for their independence, do not make such a requirement, but they require only that a worker serving as foreman be paid an extra hour a day. Although the Bricklayers have a number of trade customs they enforce, they do not necessarily appear in the contracts. This reflects their confidence that they do not need to depend on contracts; a bricklayer is more likely just to leave the job if he does not like it rather than complain to the steward. The Ironworkers, on the other hand, not only require that a foreman be appointed as soon as a second man arrives on the job but also require that a man working alone receive extra pay. The glaziers not only do not require that a foreman be appointed, but agree that their journeymen will accept instructions from foremen who are not glaziers.

The foreman's pay rate is established, except in the Bricklayers, by a differential over the journeyman rate. This differential is small and has not risen as fast as wages have. The requirement to pay the foreman's rate to a worker is therefore not as expensive proportionately as in past years.

The foreman is a representative of management and is exempt from responsibility for enforcing union rules. The United Association rule states:

> *The selection . . . of foremen is the responsibility of the Contractor. In keeping with this agreement, he shall act as agent of the Contractor only and shall not apply or attempt to apply any regulations, rule, bylaw or provisions of the Union Constitution in any respect, or any obligation of union membership.*

This removes the foreman from threat of fine by the union for carrying out instructions given him by the contractor.

The United Association includes plumbers, pipefitters, and others. In this instance, the local is a combined one and a single contract is used for all crafts. The UA may also have separate locals for different crafts in the same union.

Fringe Payments under the Disney World Agreement. In this area of the Disney Agreement, the Carpenters have the fewest benefits; a health and welfare fund, from which benefits such as health insurance and life insurance are paid. Other contracts may include pension benefits, vacation pay, and advancement funds for the industry. The Electricians have a national employee benefit fund in addition to their local health and welfare fund. Benefits such as pensions and vacations must be paid into a reserve, since an individual usually does not remain on the payroll of one contractor long enough to receive benefits from an employer's fund.

Tools and vehicles under the Disney World Agreement. Many journeymen acquire quite extensive and costly sets of tools and make use of their automobiles or light trucks for their employer's benefit. Since an employer would hire these workers in preference to those not so well equipped, all journeymen would eventually be forced to do the same. The contracts therefore list the tools that are required in a number of trades and specify that the worker is to furnish no other tools. Usually, the employer is required to furnish tools which are used up rapidly, such as welding gloves. He must also furnish tools that are expensive, such as power equipment, and those which cannot be readily carried in a car, such as long levels. Some rules require that the worker furnish files but that the contractor replace them when they are worn.

Carpenters, who require more time to sharpen tools than do other trades, are expected to report to work with their tools sharp, and have been refused by foremen for any employment unless they *are* sharp. They are entitled to leave with them sharp, that is, to sharpen them on the contractor's time and if laid off, to be given time to sharpen tools. The Carpenter's contract under the Disney World Agreement includes the requirement. To save cost, contractors on large jobs either send saws offsite for machine sharpening or assign a worker to this work. Some carpenters object to having saws sharpened by machine, but in the Disney Agreement, it is asserted as a *right* of journeymen to have the saws sharpened by the contractor. Saw sharpening is part of the carpenter's trade, but many carpenters cannot do a satisfactory job and hand sharpening is more time-consuming than machine sharpening.

Surveyors under the Disney World Agreements. The Operating Engineers have been attempting to get jurisdiction of surveyors and have made special efforts to train workers for larger jobs. *Surveyors* here refers to those who do construction layout and occasional measurements of quantity. It does not refer to *land surveyors,* who establish land boundaries, or *engineering surveyors,* who do mapping and quantity measurements for consulting engineers. In the Disney World contract, the Operating Engineers provide all surveyors who work for the contractor or subs, but not part-time workers such as supervisors and not workers employed by consulting engineers. The clause follows the California practice as Disney World and the contractors were both from California; however, it is not a national practice. The surveying necessary for building construction is not considered by the trade unions to be a craft but a skill; that is, several kinds of craftsmen are expected to be able to do it in connection with their own work. The prescribed apprentice course for carpenters requires that apprentice carpenters receive over twice as much practice with line and grade instruments as does a civil engineer in his college courses.

Extra-pay work under the Disney World Agreement. In addition to overtime, the Disney Agreement requires that workers receive extra pay for a variety of jobs. Shift work usually requires 8 hours' pay for fewer hours' work on the night shifts. There are small differentials for working on a scaffold, by height. Working with dangerous or uncomfortable materials, such as creosote or fiberglass, requires extra pay in most trades. In many cases, extra pay is for work requiring greater skill, such as for painters who hang wallpaper. In the Disney World contract, electricians are granted double pay for working on scaffolding over 40 feet high and for working in tunnels or under water, and they are granted time and a half for working in the rain.

In general, construction workers work in all weather for regular pay. When weather conditions are bad, they are not paid when they do not work; usually, there is no conflict between supervisors and workers, since neither are interested in work under inefficient and uncomfortable conditions. There are many variations of this custom in local contracts.

Special requirements under the Disney World Agreement. Other clauses specify that the Painters may not use a brush over 6 inches wide, the Insulators may not use larger than 14-quart buckets, and the Plasterers, 12-quart buckets. A number of trades require a ten-minute coffee break morning and evening, and others get it by custom. The Lathers specify that they will not be required to *mouth* their nails, although this is the common method of nailing rocklath and the same contract requires nails to be blued and sanitary. The Painters' contract provides that workers may be discharged for reporting to work drunk, and the Electricians' contract states that a *journeyman may be required to work on his own time to correct his mistakes.* The Insulators' contract states that the foreman *and* the steward are to receive reports of inferior work; this implies that insulators are obliged to report inferior work by other journeymen of their trade.

Hiring procedures under the Disney World Agreement. Several unions did not accept the general part of the Disney World Agreement but assigned priorities to job applicants by past union work or by residence. The intent in most cases appeared to be to assure work to local residents in preference to those living outside the area, which included a large part of the state, as much as to assure work only for union members. Some of the union hiring clauses were probably illegal, as they required union membership indirectly by specifying that the workers must have previously worked under a collective agreement of the union.

Safety under the Disney World Agreement. There are a number of safety requirements, but unions are not, generally speaking, greatly concerned with safety in their contracts. Only the Electricians' contract mentioned the stud gun, for example, a firearm that shoots nails into concrete or steel and is capable of firing with the same force and range as a .22 or .32 pistol. The Electricians' contract required only that apprentices not be allowed to use the stud gun during their first three years of work; in other areas, unions have required that it be used only on overtime (when fewer men are on the job) or have outlawed it entirely. In the Disney World contract, the Electricians also required that energized circuits of 440 volts or over be worked by workers in pairs. In one area, the Bricklayers specify for safety reasons the grade of lumber to be used for scaffolding.

Wage payments under the Disney World Agreement. At one time, union rules required payment in cash; none of the contracts under the Disney World Agreement state a requirement, but the business agent has authority to see that reasonable check-cashing facilities are provided. Unions object to the use of out-of-state banks, as the checks are often hard to cash. The standard clause and practice is that workers who are laid off must be paid at the time they leave; but if they quit, they may be required to wait until the next regular payday for their wages. The number of days lapsing between the end of the week worked and payday is specified; this was once no more than three days, but with the advent of computer bookkeeping, pay often is delayed longer. The Disney World contracts allow as much as a week; the glaziers need not be paid until the next payday even if laid off but if "discharged," must be paid at once.

Working contractors under the Disney World Agreement. The unions usually allow a contractor or a partner in a firm to work with his tools, but practice differs. In the Disney World contracts, the United Association allows one member of the firm to work with his tools; the Electricians allow the owner to work if he employs one journeyman but not more than two. The rules are intended to prevent individuals who might sell their own labor at cheaper hourly rates than the union rate (*scale*) from underbidding union contractors.

8.19   CRAFTS AND TRADES

A *craft* is a kind of work at which all journeymen are considered skilled if they are of that craft. *Trade* and *craft* are roughly synonymous, and neither are precise terms, although trade often designates a union or general term, as carpenters, and craft the skill of an individual, as pile driver. Any bricklayer, for example, can lay concrete block and can usually do floor tile work. But within one union local, one may differentiate blocklayers, bricklayers, and stone setters. The Carpenters' Union includes pile drivers, millwrights (who do not necessarily work in wood; their work is setting machinery), floor layers (of asphalt tile), drywall men (who nail, or *hang,* sheetrock but do not tape or smooth the joints), and finish or rough carpenters. Often workers specialize in form work, finish work, cabinet work, or framing and may be requested from the BA for such skills. Such specialists may have their own locals where there are enough workers. In 1972 the Organization of Architectural Employees, a professional union with three San Francisco architectural firms, joined the carpenters!

Some skills are common to a number of trades. Welding, for example, is a part of the training of carpenters, ironworkers, bricklayers (for masonry ties), boilermakers, operating engineers, pipefitters, and plumbers. In these trades, one will find workers who can weld and some who cannot.

8.20   RECENT GROWTH OF OPEN-SHOP WORK

Increases in construction labor rates are often blamed on labor union contracts, and open-shop work has increased at the expense of union work, when the union wage has been far above the market rate. The National Constructors Association, the largest group of union contractors in the industrial field, reports that in 1972, 32 percent of all industrial construction—double the proportion of three years ago—went to nonunion builders. The Associated General Contractors, generally considered an organization of union contractors, reports that it has 1,000 more nonunion members than it had 3 years ago; 35 percent of its 9,000 members are now nonunion. In Baltimore 75 percent of all construction was reported as open-shop.

Studies by various sources state home building was 57 percent organized in 1936, 68 percent in 1957, but was down to 20 percent in 1978. Various 1978 studies estimated that construction work as a whole was 40 percent to 80 percent union. In residential work, open-shop contractors have dominated most geographical areas, but open-shop contractors are now making major advances into industrial construction. The president of the United Association, Martin J. Ward, reported that in a single week $1 billion of industrial construction was let to nonunion contractors, including a $600 million nuclear plant and a $350 million generating plant, and warned that the National Constructors Association, whose 30 members have been averaging $10 billion worth of construction a year, may be lost entirely to union craftsmen if union productivity is not increased. Labor leaders have been outspoken in attempting to inform their members of the importance of the situation. Robert Georgine, secretary-treasurer of the AFL-CIO Building Trades Department, is quoted as blaming both unions and union employees for "a serious problem of featherbedding, where two guys are assigned to a piece of equipment that needs only one."

8.21   QUESTIONS FOR DISCUSSION

1. What are the differences between a master agreement, a project agreement, and an international agreement? Under what circumstances does each provide advantages for the contractor? For the unions?

2. In a union area, only the very small and the very large contractors can work under open-shop conditions. Explain.

3. A contractor's control over his men depends as much on the stability of their employment as on whether they are union members. Explain.

4. In what ways does Thomas F. Murphy's letter (Figure 8-1) differ from stories you have heard about other unions? Why are they different?

5. Why are the unions interested in promoting apprenticeship programs?

6. An open-shop contractor is at a disadvantage when he bids on public work with a specified minimum wage higher than he is paying; but if the minimum is below his accustomed rate, he may still be competitive. Why or why not?

7. The building trades unions became organized before the labor union movement elsewhere in the country was established, and where they were defeated were able to spring up again in a short time. Why or why not?

CHAPTER *9*

BIBLIOGRAPHY

This bibliography is an introduction to the academic study of construction management, an appraisal of material for college courses and for supplementary reading, and a guide to people in the industry who want a review of what is available.

The construction industry has suffered from periodic economic depressions even more severely than has the rest of the country. Each depression and each war has driven managers and skilled workers from the industry, and years were required for recovery. The literature of management shows this trend. The period ending in 1914 had seen continued expansion for nearly 50 years, culminating in the construction revolution of new materials and equipment. There has been no comparable period in this century, although the stability since 1950 has again led to substantial improvements in management.

The demand for literature on construction management in the early 1900s paralleled the development of more efficient methods pioneered in other industries by Gilbreth and Taylor. As the dividing line between engineers and constructors was than rather indefinite, many of the engineers were well versed in construction.

This period was marked in 1908 by the publishing of the first works of Fred B. Gilbreth, and in 1909 by Halbert P. Gillette's book on cost accounting, followed in 1910 by his *Cost Data Handbook*. In 1911 D. J. Hauer wrote his first volume of *Economics of Contracting,* in which he pointed out the demand for professional construction education. In 1912 Taylor's *Concrete Costs* was published; in 1915 Hauer's second volume and Frank R. Walker's *Building Estimator's Reference Book* were published. During this same period, William Arthur wrote a book on estimating data and the Audel carpentry trade series first came into print.

These volumes attempted to systematize both construction, by means of motion study, and improved direction and estimating, by organized data and time study. The writers of the time recognized the size of the task and attempted to put into writing the knowledge of the tradesmen, previously handed down for generations by journeyman to apprentice. With the advent of World War I, this effort slowed down, and not until the 1960s was there repeated such an attempt at writing on construction and improving apprentice teaching aids.

In the years following 1900, Taylor and Gilbreth were consultants to the manufacturing industries; Gillette edited the magazine which later became the *Engineering News-Record*. After World War I, economic crises nearly wiped out the construction industry entirely, and it was then curtailed by World War II. Since then, a long stable period has encouraged the reappearance of literature on all phases of construction.

The first part of this bibliography provides alternative texts and readings for undergraduate courses. The second part is a guide to some of the books which may be found in libraries, for historians, writers, and graduate students. History in construction has particular value, as new situations require methods not known to modern supervisors, as the methods have not been often used in recent years. Except for such essential permanent pamphlets as the *Green Book* of labor jurisdication, pamphlets and magazines are not generally included in the list.

BOOKS IN PRINT

These publications were available in 1979. The books included are those considered particularly useful and outstanding, and of course only a part of the books now in print.

Addresses of publishers are:

American Institute of Architects (AIA), Publishing Department, 1785 Massachusetts Avenue, N.W., Washington, D.C. 20036. A complete publication list and list of dealers with prices is available on request. The AIA no longer provides publications from the Washington office.

Associated General Contractors of America (AGC), 1957 E Street, N.W., Washington, D.C. 20006. Publications and price list on request.

McGraw-Hill Book Company (McGraw-Hill), Box 402, Hightstown, New Jersey 08520.

Prentice-Hall, Inc. (Prentice-Hall), Englewood Cliffs, New Jersey 07632.

Frank R. Walker Company (Walker), 5030 North Harlem Avenue, Chicago, Ill. 60656.

John Wiley & Sons, Inc. (Wiley), One Wiley Drive, Somerset, New Jersey 08873.

Commercial publishers will ship books on approval (without prior payment or even without purchase order), even on a telephone call.

American Arbitration Association. *An Outline of the Law and Practice of Arbitration under the Construction Industry Arbitration Rules of the American Arbitration Association.* New York: 1978. A 33-page typewritten pamphlet free from the Association at 140 West 51st Street, New York, New York 10020.

Every contractor should have a copy of this pamphlet; sooner or later he will need it. Citations to cases on arbitration procedure, with brief statement of decision. This list is intended for attorneys primarily, but is valuable for contractors as well, as a guide to case law regarding arbitration and its application.

American Insurance Association. *Summary of State Regulations and Taxes affecting General Contractors.* New York: 1978. 123 pages. A summary of licensing, prequalification, corporate permits, and taxes affecting contractors in all states. The contractor who operates in new states needs this to first determine if he can bid at all, and later to comply with state laws. Periodically updated, available from the Association at 85 John Street, New York, New York.

Associated General Contractors of America publishes a large number of pamphlets and books.

These are available to nonmembers as well as members. The publications include, among others:

1. *The Use of CPM in Construction,* a small hardbound explanation of CPM, well written and explaining the Association's recommendations.
2. Standards for AGC pumps. These are the small self-priming gasoline pumps seen on every site.
3. *Contractors Equipment Manual,* a guide to calculation of equipment costs. This is not at all like the *Associated Equipment Manual,* with which it is sometimes confused. The AGC manual is the cost of owned equipment to the contractor; the AED manual commercial rental rates.
4. Recommended bidding procedures, foremen's pamphlets on theft and safety, forms for contracts, and others.
5. *National Jurisdictional Agreements,* a paperbound reprint of trade jurisdictional agreements not included in the *Green Book.* It is sometimes referred to as the *Gray Book.* A well-studied copy is helpful for everyone concerned with trade assignments of work.

Audits of Construction Contracts. New York: American Institute of Certified Public Accountants, Inc., 1965. 102 pages.

This book is a guide for public accountants making independent audits for contractors' books, and for making proper charges by construction personnel. It is not a simple book, but it should be read by project managers or construction managers who are concerned with the overall profit situation, particularly if their compensation is based on net profit. It is essential for a contractor's accountant. It does not deal with details of charges as does Coombs, but with overall standards, particularly as to how company profits are reported.

For students, it will not normally be suitable except for those with previous auditing course material. It is useful for accountants who teach accounting courses for construction students.

The book is available from the American Institute of Certified Public Accountants, 666 Fifth Avenue, New York, New York 10019.

Bureau of National Affairs. *Construction Craft Jurisdiction Agreements.* Washington, D.C.: Bureau of National Affairs, 1971. 200 pages.

This book duplicates the jurisdicational agreements found in the AGC *National Jurisdicational Agreements* above (trade jurisdicational agreements not included in the *Green*

Book). The BNA publications have been brought up to date on each odd year since 1957, except for 1961, and will probably continue to be more up to date than the AGC version. The BNA version is larger and easier to read. It is available from BNA, 1231 25th Street, N.W., Washington, D.C. 20037.

Clough, Richard H. *Construction Contracting.* 2d ed. New York: Wiley, 1969. 392 pages.

Clough takes for granted that a lump-sum contract, with an established contractor, is the only proper way to build a project. He has an opinion that a contractor is "entitled" to profit and that ethics include a commitment to make a profit. The bid is evidently assumed to cover public contracts ("with a few exceptions, the proposal must be accompanied by some form of security . . .") and an owner's point of view. The author states that the book is written to apply to small and medium-size contractors, but the organization shown includes six departments— with an accompanying chart which shows a purchase order, for example, passing through five people (not counting stenographers) before going to the subcontractor. Clough presents a separate contract department and an engineering department which excludes purchasing, estimate, and construction supervision. The structure, in general, has the appearance of a government bureau.

Clough has made a special effort to tie in the theoretical course of business administration through a chapter bibliography and by the language used. This is the best-selling construction management textbook on the market; it is complete, well-written for college students and construction managers, and up to this time, the best book available for the professional. The treatment does not profess to show variations or opposing points of view, or to explore details.

Clough. *Construction Project Management.* New York: Wiley-Interscience, 1972. 266 pages.

Clough here describes the administration required of a project manager. He implies a separation of the project manager's and the superintendent's duties by the home office, even on jobs large enough to have a full-time project manager; the two act as a "partnership." About 70 percent of the book is devoted to CPM applications, and the balance to cost accounting and financial planning. There is an implication that the project manager's role is to devote himself to these topics, and at times "management" is considered to be officers of the corporation who do no detailed work themselves. The view of the project manager having only a semiclerical role—collecting and summarizing data rather than managing in terms of estimating, purchasing, and directing the job personnel and subcontractors—is not unusual among contractors and is typical of centralized organizations. Clough implies that the centralized organization is the optimum, since he considers that delegation of responsibility should occur only when absolutely necessary. The presentation of cost accounting is conventional and brief. There is a short chapter on contract administration, primarily concerned with general contractor payments from the owner.

The majority of the topics in this book are most useful to an owner's representative, the cost chapters being of particular application on a cost-plus job.

Construction Cost Control. American Society of Civil Engineers, Construction Division. Originally published 1951, reprinted in 1955 and 1979. 88 pages.

A paperbound primer for construction cost, well written and illustrated, with numerous examples and forms. Its value is illustrated by the continued demand for three decades. Available from the Society at 1707 Street, N.W., Washington, D.C. 20036.

Coombs, William E. *Construction Accounting and Financial Management.* New York: McGraw-Hill, 1958. 491 pages.

This book is exceptional both in content, which is wide and practical, and in its writing, which is straightforward and simple. It is intended for accountants with no construction experience and for that alone should be furnished to all new accountants. But it is also written in a way that is adequate for contractors, superintendents, or project managers. Coombs' application is almost entirely on construction, and he covers related subjects as well as accounting, particularly purchasing and construction organization. He says of cost accounting that if operations people are as conscious of accounting requirements as they should be, and vice versa, it is immaterial who keeps cost accounts. He is familiar with labor cost systems which give useful results the following day, but this seems to apply only when manual systems are used. Computer systems usually slow down operations in a small firm. The book is a fairly complete guide, not only to accounting but to personnel and purchasing procedures as well. Coombs' assumption, apparently based on his experience, is that the accountant should be able to step into any office function when it is found that there are no better qualified people available. Since purchasing and personnel departments are the last to be established in a contracting organization, this is a reasonable assumption. Coombs treats the things to be done, and why; he does not pretend to standardize procedures or to describe the duties of persons. Consequently, the text is as applicable to a contractor with five employees as to one with five hundred. He treats the project manager as the key person, who should have as much authority as he is qualified to assume, and presents the wide ranges of alternates available with excellent arguments for and against centralized accounting.

He does not present basic accounting; that is, how the data are put into the books once gathered, except in a general way. This is no impediment for the nonaccountant reader, however. Computer accounting is not presented, as it was not common in the industry at the time the book was written (1958), but this also does not detract from the systems presented, as it affects only the final step of entering the data in the books. Coombs has presented details without confining himself to them, so that a reader can set up his own specialized system, or understand a variety of systems, from this presentation. It is the single most important book for contractors, accountants, and project managers available today.

Deatherage, George E. *Construction Company Organization and Management.* New York: McGraw-Hill, 1964. 328 pages.

The four Deatherage volumes are a modification of a correspondence course for project managers offered by Deatherage for many years. The books were published at the same time but under different titles; they are here reviewed in the order in which they were written. They were written for students who had no access to other literature, and consequently include considerable material (probably 75 percent or more of the total) which are forms available from other sources, such as AIA or AGC. The series was evidently not well received, as the first printing has been unsold for 20 years.

Construction Company Organization and Management reviews organizational charts for several kinds of operations and offers some advice on organizing a company, with particular emphasis on specifying exactly how various items will be handled and by whom, in the organization. Deatherage gives the full text of the AIA contract forms, and a number of other forms. The book is source material, with inadequate explanation for student use except in connection with other texts. In some respects, the series repeats fundamental information which is required by freshman drawing students, but not by project managers with no formal college education, for which the course and book were intended. It is hard reading for construction personnel.

Deatherage. *Construction Office Administration.* New York: McGraw-Hill, 1964. 316 pages.

In this volume Deatherage reviews insurance, social security, compensation, lien laws, bonds, and bidding procedures in the same manner used in the previous book—that is, by reprinting extensive portions of standard forms and literature. The section on office engineering, considered to be design or preparation of shop details from the overall designs, is very useful to supervisors of process piping work but is much more than is expected of construction firms on nearly all other work. The section on cost controls, cost engineering, bookkeeping and clerical functions represents his own experience and explains his own ideas; this section alone is worth the price of the book. Deatherage received the preceding day's labor reports by 10 A.M. each day on a force of 1,800 men—which most managers consider impractical. This book also stands alone in the management writing field in concentrating on the difficulty of getting good cost data from the field, and emphasizing the effect that gathering the data has on job efficiency.

Deatherage. *Construction Estimating and Job Preplanning.* New York: McGraw-Hill, 1965. 316 pages.

In this volume Deatherage explores construction methods, defined in the Preface "as that combination of men and machines selected to perform each specific part of the work to be done." The discussion is consistent with that definition, concentrating on gang work rather than on motion study. He refrains from calling his discussion "industrial engineering," preferring *methods engineering.* The influence of Frederick W. Taylor is quite apparent, even to some of the symbols (which, of course, are common to time- and motion-study texts). Preplanning is his term for the planning of construction methods which is done at the time the estimate is prepared, or shortly thereafter. *Scientific management* is used by Deatherage in the original, or classic, sense, with no compromise with the later school of management which considers management to be a manipulation of ideas and a making of decisions rather than the direction of men. *Supervision* and *Administration* are to Deatherage the management phases performed on the job and in the office, respectively. The difference in rhetoric and emphasis appears important to the casual reader, but does not affect his basic principles.

Deatherage has quite long check lists and extensive cost coding listing, which are useful for reference; his writing is, in many respects, a source book rather than a text for study. He emphasizes the importance of planning the labor force in detail before starting work.

Deatherage. *Construction Scheduling and Control.* New York: McGraw-Hill, 1965. 322 pages.

Deatherage includes CPM scheduling as a portion of the content of this book; to the modern student and some professors, the CPM presentation requires an entire book. In the Introduction, Deatherage points out:

> *The creation of a CPM network is, once all the facts are at hand, a simple mathematical process. Securing the facts comprises fully 90 percent of this process, and this is substantially increased where the work is extremely complex and subject to almost daily changes, either in design or in purchase and delivery of materials.*

He defines scheduling as a mechanical process for formalizing the planning functions. His neglect of CPM lies in the above; that is, CPM programming is not, in the modern ungrammar, *that difficult.* His discussion of purchasing, subcontracts, and materials expediting shows that he was familiar with work on a large scale. Deatherage was for many years a superintendent for Union Carbide both in the United States and abroad. His presentation of project management is short since most items have been covered in previous chapters, and goes into some detail about company manuals and detailed instructions for handling correspondence and other office details. Deatherage, unlike many present-day writers, was accustomed to setting up large jobs in considerable detail, being responsible for the entire operation; it was, therefore, essential that he know the importance of proper procedures in saving the manager's time.

The Deatherage books, as a whole, suffer from the inclusion of forms and typical contracts, laws, and other reference data in the body of the book rather than in the appendix. As a result, the reader gets lost in the stiff language of these references, confusing them with the writer's own language. Deatherage, however, was an excellent writer who expressed his principles logically; his books suffer from the lack of his own observations rather than from too much of them. The introductions themselves should serve as a guide for future writers in construction management and can stand on their own as magazine articles. His work is in the tradition of Gillette, Taylor, and Gilbreth, with whom modern writers have little in common. Literally hundreds of constructions problems the project manager will encounter are mentioned in Deatherage but in no other book.

Mr. Deatherage died soon after completion of this series.

Dunham, Clarence W., and Robert D. Young. *Contracts, Specifications, and Law for Engineers.* 2d ed. New York: McGraw-Hill, 1971. 540 pages.

This is a widely ranging exposition of elementary law, contract general conditions, and some interpretations of technical specifications. The writing is simple and well organized, without omitting parts of items covered. It is suitable for engineer training and is not intended for contractors. Dunham, the engineer half of the authors, takes a rather paternalistic view of specifications writing, reserving the engineer powers beyond those in the standard contracts but at the same time urging the engineer to give the contractor every opportunity to make a profit. He considers unbalanced bids proper—both for early payment and for the contractor to take advantage of his own investigation of the work. Dunham considers that bidder's alternates should be rejected as improper, but he does not mention the gain from them. He implies that a hold-harmless clause as broad as possible should be used and desires to give the engineer wide powers that are not subject to appeal. He does not favor payment for materials stored on the site, although it is apparently the wording of this phrase, not the intent, to which he objects. He believes that general contractors should not obtain competitive quotations from several subcontractors.

Designers openly refuse to bid in competition and they to some extent carry this attitude into their discussion of contracting operations. They are inclined to allow—or even require—limiting competition between contractors in order to obtain better quality of construction, and they believe, lower costs. Dunham does not lean so far in this direction as do others, particularly William R. Park.

Dunham proposes that the right of contractors to withdraw bids *before* bid time be restricted or prohibited. He considers "builder" and "contractor" as synonymous. The portion of the book devoted to general law, of which construction law is only a part, is extensive, although the authors in the preface explain that they are not trying to teach law. Because of its extensive treatment, the text is suited to a curriculum in which the students take no other law course.

Engineered Performance Standards

These are estimating data prepared by the Naval Facilities Command (Bureau of Yards and Docks to many of us), U.S. Navy. These are over 30 publications prepared by the Navy from 1959 to the present, and continue to be updated, for the use of Naval shore facilities in estimating maintenance work; most of them are applicable to construction, and are particularly useful for small renovation work. There are several general publications explaining the use of the standards and the methods by which they were derived. The methods are not only intended for estimators with little experience, but the attitude is that returned costs from actual work is undesirable. Nevertheless, both the studies, which are as detailed as the Taylor *Concrete Costs* and infinitely more comprehensive, are valuable to the estimator both as a guide to the methods used and for the data given. If the government language is penetrated and the differences in attitude ignored, the series will be valuable to every estimator. A price list is available from the National Technical Information Service, 5285 Port Royal Road, Springfield, Virginia 22151. The set includes over 4,000 pages and sells (1979) for about $200 complete. Most contractors will want only a few of the standards; the set includes nearly all the building trades.

The set is used also by the Army and Air Force. There are several index numbers for each standard, and they seem to be available from other sources. The numbers used below are the ones assigned by the Navy.

The general publications are necessary as a key to the tables; the trade standards list below are incomplete by themselves. The listings below are samples. The series is suitable for an estimating course, and in fact both instructor's and student's manuals are available.

Engineered Performance Standard P-700.2. Planners and Estimator's Workbook, about 150 pages, 1974.

Intended for use in a course for new estimators in Navy yards. Explains use of the tables, with examples and many problems. The writing is vague and is probably the result of review by too many people; it is not recommended for inexperienced estimators without help. Most of it refers to naval paperwork procedure, which is inapplicable to a private contractor; but the differences are readily understood by an estimator with even a little experience.

Engineered Performance Standard P-702.0. Carpentry Handbook, 334 pages, 1978.

Contains time studies of 590 tasks, as well as tables and other data on labor estimates on carpentry construction, excluding concrete formwork.

Engineered Performance Standard P-711.0. Pipefitting, Plumbing Handbook, 183 pages, 1966.

A detailed listing of worker-hours required to perform about 473 tasks in pipefitting and insulation removal, construction, and maintenance; also 32 tabular sheets corresponding to specific labor factors.

Gilbreth, Frank B. *Bricklaying System.* Easton, Pa.: Hive, 1974. 321 pages.

In this book originally published in 1909, Gilbreth (the father of the 12 children portrayed in the book and movie *Cheaper by the Dozen)* explains with numerous examples and photographs how construction in general and bricklaying in particular was done, and how most of it is done today. The Lowell Laboratory at MIT is illustrated, with explanation of how this 44,000-square-foot masonry and steel building on piling was built in two months and 10 days. Gilbreth was the earliest motion-study consultant, working into other industries with what he had learned on his own construction work. The books of the 1890–1910 era by Gilbreth, as well as of H.L. Gantt, Lillian B. Gilbreth (the lady with the dozen children), C. B. Thompson, and the writings of Frederick W. Taylor and others, are now available from Hive Publishing Company, Alpha Building—Box 1004, Easton, Pennsylvania 18042. Presumably, modern writers have improved on the theory, but the original books are better written and more adaptable by contractors.

Gilbreth, Frank B., and Lillian M. *Fatigue Study,* Easton, Pa.: Hive, 1973.

Original publication date not stated, this book was jointly written by Frank and Lillian Gilbreth, after she earned her Ph.D.—Frank never went to college. This book emphasizes that efficient production must not be tiring, which appears to show Lillian's strong influence. The elimination of fatigue had already been an essential element of motion study; this book, written after Frank Gilbreth's industrial experience, emphasizes individual chairs for each worker, and reduction of both mental and physical fatigue, standards not yet attained in industry.

Gilbreth, Frank B., and Lillian M. *Fatigue Study,* Easton, Pa.: Hive, 1973.

This is a pocket-size edition of Gilbreth's instructions to his job superintendents, first published in 1907 and probably written largely some time before that. It includes hiring, deliveries, and accounting, prescribing job planning, methods of construction, and handling of materials. It is written in the common language of the superintendent, most of which has not otherwise appeared in print.

For the construction manager writing a policy manual for his own firm's job superintendents, this remains the best guide available: much of it is obsolete, many methods will differ from his own, but he gives a great many ideas and points to be covered in such a manual. The modern manager can use this booklet as a check list and guide for improvement. Until 1973, this book had been practically unobtainable for 60 years; even Lillian Gilbreth had not copy, receiving a copy from Deatherage in 1965. It will become obsolete only when a better one is written.

Gilbreth, Frank B. *Motion Study.* Easton.: Hive, 1972.

A facsimile reprint of the 1911 edition. The primer in motion study, with particular emphasis on bricklaying, as the author felt that this was an operation which everyone was familiar with.

Gilbreth, Lillian M. *The Quest of the One Best Way.* Easton, Pa.: Hive, 1973.

Probably published in 1954. This is a brief biography of Frank by his wife Lillian, which has no bearing on construction management except for Gilbreth fans. Lillian was 76 when this was published, and continued to lecture on industrial engineering in her nineties; she died in 1972, at 94. Frank died in 1926. This booklet covers at least 44 years in 64 page. To do this, she will devote three lines to the birth of as many children, drops names like Rice, Gantt, and Frederick Taylor. It is as much of herself as of Frank, although she refers to herself in the third person.

Green Book. Common Name of the *Plan for Settling Jurisdictional Disputes Nationally and Locally,* National Board of Jurisdictional Awards, and *Agreements and Decisions Rendered Affecting the Building Industry,* 1970. 152 pages.

This paperbound book is available from various trade unions, including *The Lather,* 6530 New Hampshire Avenue, Takoma Park, Maryland 20012. The *Green Book* is published in the "union size," 3 ½ by 5 ½ inches, in which union working rules are usually printed. The National Board is strictly a union organization, unrelated to the National Labor Relations Board (which sometimes finds itself in the same business) and founded for the purposes of settling interunion disputes. This booklet is essential for the superintendent or any person reviewing the superintendent's determination of jurisdictional awards. It is probably the booklet best known to journeymen, other than their own working rules. Decisions in the book are not classified other than by date, although it has an index; it is not easy reading, since it uses language of the trade which may not be known to some people of the trade or in some areas. This booklet merits detailed study, especially since there are agreements which do not bind unions not signatory to an agreement.

Havers, John A., and Frank W. Stubbs, Jr., editors. *Handbook of Heavy Construction,* 2d ed. New York: McGraw-Hill, 1959.

A series of chapters each written by an expert in his field. The first edition was edited by Mr. Stubbs, now deceased, and Mr. Havers has expanded it considerably. This is an essential part of every medium and heavy contractor's library, covering many phases of construction, unduplicated in other publications.

Lincoln, James F. *Incentive Management.* Cleveland: Lincoln Electric Company, 1951. ($1) 286 pages.

Lincoln. *A New Approach to Industrial Economics.* New York: Devin-Adair, 1961. ($1) 165 pages.

These books are available from Lincoln Electric Company, 22801 St. Clair Avenue, Cleveland, Ohio 44117. They are not written with special application to construction, but the principles Lincoln proposes and the results he has obtained are probably more applicable to construction today than to any other area. The two books are virtually the same, one having been written ten years later than the other. The earlier book has more detailed (but not so recent) sales and labor cost data than the first.

Lincoln presents the Protestant work ethic and application of the Golden Rule with some repetition and apparent irrelevance to the question he presents, which is labor efficiency and how to achieve it. The outstanding difference between this presentation and that presented by any rugged individualist of the nineteenth century is that he has figures to show that from 1934 to 1960 his labor costs and selling prices went down or remained firm in the face of rising prices of raw materials and labor earnings. A man who pays workers an average of $6.99 an hour for factory labor when the competitive rates are $2.54 an hour, while restricting his output to avoid antitrust charges should he run his competitors out of business, should not be ignored merely because his explanation is more in religious than in industrial engineering terminology.

Lincoln has constantly raised the bonus to his employees and has kept labor turnover to the neighborhood of ½ percent monthly, by increase in efficiency. The heart of his argument is that if you enlist the help of the workers and pay them for their improvements, no labor union or supervision by the owner in the usual sense is necessary. During World War II, he was fined by the government for paying too high wages to his workers; Lincoln pointed out that had he paid the same amount of money to twice the number of employees, depriving the armed services of 2,500 men as his competitors did, he would not be punished! Each book has its own appendix, tracing the development of the Lincoln payment system as compared with his competitors. Lincoln, while using idealistic clichés, was an unorthodox disturbance to labor unions, government, and management experts alike by his insistence that money to express appreciation for efficiency was the strongest form of motivation.

Nichols, Herbert L. *Moving the Earth*, 2d ed. Greenwich, Conn.: North Castle, 1962. About 1,350 pages.

This is the incomparable reference book in the heavy construction field—well written, with readable type, and by far the most complete book published. The book is primarily on construction methods, but I recommend Chapter 11 on business methods, accounting, and equipment cost accounting as valuable additional reading in this topic. After rejection by the large publishers, it was privately published in ten printings totaling seventy thousand copies to date. It is probably the best-selling book published in the field of construction since 1915. It is available from North Castle Books, 212 Bedford Road, Greenwich, Connecticut.

O'Brien, James J., and Robert G. Zilly. (eds.). *Contractor's Management Handbook.* New York: McGraw-Hill, 1971. About 800 pages.

This is a series of articles by a number of writers and provides a certain amount of repetition and many omissions. It is not as inclusive and does not treat subjects as deeply as does the *Handbook of Heavy Construction;* writers on management are few, although for the *Heavy Construction Handbook* a more formal background of knowledge exists to draw from. This management handbook is useful for added readings in construction management courses. For the contractor, it is useful as a reference for matters on which he is also getting other information. It is not difficult reading, either for construction managers or for students, and like other McGraw-Hill Handbooks, is written by practicing professionals.

Oppenheimer, Samuel P. *Directing Construction for Profit; Business Aspects of Contracting.* New York: McGraw-Hill, 1971. 286 pages.

Oppenheimer covers the field of construction management as laid out by other writers, with particular emphasis on examples rather than general discussion. Several chapters are devoted to the preparation of the job by the superintendent and detailed treatment of his duties, particularly in regard to the items he must check. A section on sales is well presented, from a standpoint both of making contacts and of negotiating price after the initial contact is made. His presentation of the psychology of arriving at a contract price is a subject about which contractors are aware, but which is dismissed by other writers with a few words, to sidestep the fact that we just don't have the extensive experience in negotiations to make any generalizations. The book throughout emphasizes careful attention to detail, particularly careful field supervision to reduce labor costs, and is written in a straightforward and direct manner. It is useful either as a text for a construction course or as additional reading.

The author's point of view is that of the small contractor and subcontractor (the subcontractor, particularly) has special problems not considered elsewhere in management literature.

Parker, Henry W., and Clarkson H. Oglesby. *Methods Improvement for Construction Managers.* New York: McGraw-Hill, 1972. 311 pages.

Parker has simplified the book beyond the point which he considers necessary, as he has not had the hoped-for response from contractors in courses offered to them. The simplification, combined with comments on the failure of contractors to cooperate with him, gives one the impression that the book is written for college students who, when they graduate, will have an overrated idea of their knowledge. The material is good on work simplification procedures that Parker advocates. A great deal of the book has been borrowed from other disciplines, particularly psychology, to prove the author's point. Students should not be exposed to adverse comments regarding contractors, however, as it is only too likely to reinforce the superior attitude they easily develop.

Reiner, Laurence C. *Handbook of Construction Management.* Englewood Cliffs, N.J.: Prentice-Hall, 1972. 348 pages.

The most important omission is the lack of an index for the appendix, which consists of 114 pages. There are a number of controversial statements, such as: "Construction deals with the basic problem of shelter, the architect's services must include not only his own services but those of any engineers that may be required" (evidently stated as fact, not an opinion), and the unconscious humor in "contract which the owner and the architect-engineer have decided will be most advantageous to all parties concerned" (presumably including not only the contractor but also the subcontractors). The description of the construction process is simple and straightforward, although heavily emphasized in regard to the architect's viewpoint. Evidently, the book is intended for a first-year junior college course, or an introduction for architectural students. The book, to this reviewer, lacks depth—in his effort to refrain from overburdening the reader, the definition of the chapters and the purpose of the operations do not seem to have been clearly explained.

Richardson Engineering Services

This group publishes Estimating Standards which are outstanding in that all trades are included, rather than just general construction as in Walker, both for material and labor. All items are illustrated. The series includes light and commercial building construction, and process piping. It is revised annually; information is available from Richardson Engineering Services, P.O. Box 370, Solana Beach, California.

Royer, King. *Applied Field Surveying.* New York: Wiley, 1971.

A field manual for new construction survey rodmen and instrument men, which will cover the basic surveying operations. It lacks laser beam information, and is out of date in that practice in calculation shown is logarithmic, no longer necessary in the day of electronic calculators. It is used widely in community colleges, but lacks theoretical explanations desired for engineers. It is intended as a laboratory manual for engineering surveyors, or as the sole book for vocational schools.

Royer, King. *Desk Book for Construction Superintendents.* 2d ed. Englewood Cliffs, N.J.: Prentice-Hall, 1980.

A new edition of a book with wide sales in the trade and some acceptance in community colleges. It is written as a guide for construction foremen who are to be on their own as a job superintendents, in the public relations and administrative aspects of the job. The chapter on labor relations and labor jurisdiction is unique in the literature.

Surety Association of America. *Contract Bonds.* New York, 1978. 60 pages.

A pamphlet explaining the services of sureties with a summary of surety's action and loss, if any, in 75 actual cases. This booklet gives even an experienced contractor information on what sureties do that is logical to cut their losses, rather than taking adversary position to the contractor or owner, and contains a list of several hundred surety firms. Available on request from the Association at 125 Maiden Lane, New York, New York 10038.

Sweet's File

These sets are distributed through the local agent of J.R. Dodge Corporation (which also publishes the Dodge Reports). There are several of these sets: building, light construction, and engineering construction. The building set is the largest and the only set of value. The sets are distributed only to designers who specify products; therefore, contractors and small home builders have difficulty in obtaining them. Since the Dodge reporter is anxious to get information from general contractors regarding subcontract work to be let and information on lettings, the construction manager can often persuade him to obtain an older set for the manager. Architects are given new sets but must return old ones; the old copies are then distributed to firms further down the priority list.

These volumes consist of manufacturers' advertisements, bound into volumes about 4 inches thick. There are about 12 volumes in the set, which is valuable for purchasing, obtaining addresses, and information on brand name products. The new set is delivered about January of each year.

Thomas Register of American Manufacturers. New York; Thomas Publishing Company. 2,208 pages, published annually, about 11 volumes.

Thomas Register is the most comprehensive list of manufacturers published in the United States. The address, approximate total sales, type of goods, trademarks, and names are listed, with a cross index on trademarks and company names. It is not concentrated on construction materials but includes the principal manufacturers and thousands of minor ones. An important aid to the purchasing agent, it is available from the publisher, 461 8th Avenue, New York, New York 10001.

Uniform Construction Index: A System of Formats for Specifications, Data Filing, Cost Analysis and Project Filing. Washington, D.C.: Construction Specifications Institute. About 100 pages.

This index is published by the AIA, AGC, Construction Specifications Institute and other professional organizations, and is primarily a system of indexing specifications and advertising literature; a cost accounting system is also incorporated into it. The system represents an agreement of the American Institute of Architects, Associated General Contractors, Construction Specifications Institute, Council of Mechanical Specialty Industries, and National Society of Professional Engineers, with other contributors. If a uniform system is to be adopted by a contractor to cover CPM schedule items, estimate and cost records, and cost accounting, this is the obvious one to use; however, it suffers from the same disadvantage as all predesignated schedules in that each contractor will find he can more evenly distribute his costs using a system to fit his own type of work.

Walker, Frank R. *Practical Accounting and Costkeeping.* Chicago: Walker, 1968. 250 pages.

This book is periodically revised by unstated authors, as is the *Estimator's Reference Book*. It explains the bookkeeping system from the point of view of the forms used, as Walker is primarily interested in selling forms. In a general way the book includes daily and weekly construction reports and estimating; it is for reference or study, not merely for reading, and is the equivalent of a first course in accounting. Walker gives a complete presentation of the accounting which may be necessary for contractors with varying volumes of work, using illustrations of actual entries for many different kinds of accounts.

This book is useful for students in construction courses who are not expected to be versatile (as are the graduates of most 4-year university courses). I believe there are simpler methods of cost breakdowns than are used here, but there is no better book on the market on this topic.

Walker. *The Building Estimator's Reference Book,* 19th ed. Chicago: Walker, 1977. 1,254 pages.

Of a number of estimating handbooks published in the early 1900s, only Frank R. Walker's 1915 publication has survived. Walker periodically renewed this book, last in 1947, which was about the tenth edition. After his death, the company was sold and McClurg-Shoemaker of Chicago was engaged to keep the publication up to date. Walker's introduction was deleted from the 17th edition, and the personal author touch has been entirely removed; no one now puts his name on it. Walker has long been the standby of the building industry, and even estimators who considered it distorted preserved it on their shelves. It is so far above any other estimating book on the market in its field that it is recommended for an estimator's library. It is also recommended for estimating courses, although it is not a simple book for that purpose.

The *Reference Book* covers virtually all items which might occur in building construction work from the viewpoint of the general contractor, describing and illustrating all work, as well as giving costs in worker-hours with space for substituting one's own labor costs. The explanation of bid preparation and overhead costs, while brief, has been recently expanded and a few pages added on CPM. This book is published in a new edition every few years, and most estimators have an earlier edition in use. The later editions have omitted the rates for workmen's compensation insurance, which in view of insurers' disinclination to give data on rates, is difficult to obtain elsewhere.

The book is greatly improved in the 19th edition by expansion to the standard book size, 5 ½ by 8 ½ inches, and the use of heavier paper.

OUT-OF-PRINT BOOKS

Books listed are outstanding in some particular. A few books on estimating or methods are included because they appear to be the first of their kind or are outstanding in other respects. The Gilbreth books have been removed from this list as they have now been republished by Hive Publishing Company.

For those interested in obtaining out-of-print books, there are a number of firms specializing in this business, one of which is International Bookfinders, Inc., Box 1, Pacific Palisades, California 90272.

Arthur, William. *The New Building Estimator.* New York: Scientific Book Corporation, 1930. 1,023 pages.

This is a late edition of a reference guide originally published much earlier. It is similar to Walker but is generally more clearly written and more detailed in regard to the mechanical trades.

Beach, Wilfred W. *The Supervision of Construction.* New York: Charles Scribner's Sons, 1937. 488 pages.

This was originally published as a series of articles and consists of a case study of the duties of an architect's "superintendent" on a Chicago school job. It is well written in narrative form, intended as instruction for the young architect regarding his duties as a construction supervisor. The use of "superintendent" to designate the architect's inspector to the exclusion of the term "job superintendent" makes the text rather confusing. It is notable for the approach to professional instruction by a specific job description rather than a general discussion.

Betram and Mailel. *Industrial Relations in the Construction Industry: The Northern California Experience.* Berkeley: University of California Press, 1955. 70 pages.

The history of the labor movement with an evaluation of the effects of working rules, restrictions on output, and so on, is written without excessive use of political history but with reference to the overall effect of union contracts or the lack of them. Since the same pattern has been followed throughout the country, this pamphlet is useful for union organization background, collective bargaining patterns, and working data for course material. Historical data on labor unions, in detail, is difficult to find.

Rules for Estimating. Boston: Building Trades Employers' Association, 1918. 44 pages.

This is not an estimating text but a professional's guide to standardizing local estimating methods for payment by quantities. It includes excavation, masonry, concrete, wood framing, screens, shingles, and trim. The constitution and membership of the Association are included. Apparently no later attempts were made to republish this manual, which was an early American attempt to follow British methods in standardization of quantity surveying.

Carpenters' and Builders' Association of Chicago: Official Directory. Chicago: Carpenters' and Builders' Association, 1909. 280 pages.

This book provides interesting period information, including their contractors' working rules, such as refusing to pay for "ordinary" plaster damage, no cutting for other crafts unless specified, no paying for building permits unless they have a general contract, no responsibility for settlement or damage. This book contains also union working rules (such as forbidding piecework and allowing for arbitration), provides annual wages for apprentices (beginning at $312 a year), and includes the 1908 Chicago Building Code, which is the greater part of the book. It includes the mechanics' lien law, the membership of the Chicago Estimators' Club, and the gas fitters' code.

Connor, Frank L. *Labor Costs of Construction.* Chicago: Gillette, 1928. 195 pages.

A well-written and explicit discussion of costs and production, written for the trade with very little extraneous language. Most of the book is in terms of dollar costs, but labor rates are also given. The wage rate used was 50 cents per hour for common labor and $0.80-$1.00 for skilled workers. The author gives his opinions in pronounced terms, such as "most experienced contracts refuse to bid on federal work at all" (because of rigid inspection) and "it is not un- common to see designs where from 10-20 percent of the owners' money is wasted by require- ments calling for expensive or useless material and unnecessary labor."

Dingham, C.F. *Construction Job Management.* New York: McGraw-Hill, 1928. 220 pages.

This pocket-size book is directed to the job superintendent, with specific notes on man- agement and lists specific pieces of equipment for 68 of its pages. The balance of the book reviews general construction methods for the superintendent. Although many methods (such as job-slaked lime) are rarely used now, the main part of this account is still current. There is a two-page bibliography, with heavy emphasis on general construction methods. Some mention is made of the use (or nonuse) of big men as foremen to impress subordinates, especially for supervising common labor. There is also a comparison of races (the darker, the slower but more painstaking), which would not be included in a modern publication. Bricklayers, for ex- ample, should be big blond men, and carpenters smaller and darker. Such segregation is as logical as the methods by which many large modern corporations pick their managers.

Dingham. *Estimating Building Costs.* New York McGraw-Hill, 1944. 401 pages.

This narrative book could easily be a summary of Walker's *Reference Book.* A good chapter, "Short Cut Methods of Estimating," shows development of unit structural costs, a very useful concept seldom found in full detail. This method prices walls, floors, roofs, and so on separately, and may be used to readily determine the relative cost of changes or of alter- native layouts of a building.

Educational Survey of General Building Contractors in the Philadelphia Area. Philadelphia: General Building Contractors Association, Inc., and Pennsylvania State University Continuing Education Ser- vices, 1961. 92 pages.

This study of Philadelphia contractors is useful far beyond the subject indicated by the title, since it provides information on the organization, pay structures, and job descriptions of a large number of contractors and represents an effort to put into firm numbers the organiza- tional types of various contractors, with detailed information on their size and type of work. Although the sample is too small for general use, it is one of the few surveys of its type that have been made. The large variations in wage data are not adequately explained; probably the duties of various men vary widely, even though the organization chart is similar.

This is the only complete study found on education of present construction supervisors and their description for further education. It is very limited in scope, being confined to contractors in the city of Philadelphia, but the methods are thorough and point the way to state or national studies of the same type. The survey includes 190 contractors and 114 managerial employees. Not only the small numbers but also the limitation to one city prevent conclusions on a national level. Based on this survey, and undoubtedly on subjective judgment, a program of university study was designed for a four-year program. The study was evidently paid for by the local building contractors' association and was doubtless affected by the fact that McQuade, then chairman of the National AGC Education Committee, is a Philadelphia contractor.

Questions relating to job security and satisfaction were asked of a number of employees; the difference in tone between this publication, and those financed by the government or independent sources (such as Parker's studies at Stanford) is apparent. The academic studies commonly picture the contractor as an uninformed manager who needs guidance; the contractor approach is to find out what are prevailing practices throughout the industry before venturing any conclusions.

Gillette, Halbert. *Handbook of Cost Data.* Chicago: Myron C. Clark, 1910. 1,854 pages.

This is a monumental work by a very productive writer. The publisher was a pioneer in the field of construction engineering and published the Gilbreth books. This book has quite complete and specific cost data for heavy construction; its 1,854 pages repeat very little of the cost data found in Walker. The section on railways is particularly complete and now rare. The book includes cost keeping, excavation, roads, stone masonry, reinforced concrete, water works, sewers, bridges and culverts, and other heavy construction work. Because of its scope and detail, this is still a valuable reference, if only to show what should be included in a cost book. The book sold 11,000 copies of the first edition in 1908, its first year of publication; the 1910 edition had four times the pages of the first edition. Only a handful of modern books on construction have done as well.

Gillette, Halbert, and Richard T. Dana. *Cost Keeping and Management Engineering.* Chicago: Myron C. Clark, 1909. 344 pages.

The approach of this detailed cost accounting text is typified by this example: "Bookkeeping was first devised and subsequently developed by merchants. Cost keeping was devised and developed by engineers. The merchant is a student of profits; the engineer is a student of costs."

This attitude of early (pre-1945) advocates of formal construction management was later dubbed by Taylor "scientific management." This very complete cost account book for construction has many examples and detailed forms. It is the earliest book covering this topic.

Haber, William. *Industrial Relations in the Building Industry.* Cambridge, Mass.: Harvard University Press, 1930.

This is an excellent summary of both labor union history and industrial conditions, nearly all of which is as current now as when published. The author confined his study to the closed-shop cities of his time—New York, Chicago, and San Francisco—avoiding the open-shop cities of Los Angeles and Detroit. His observations on labor efficiency and working conditions are therefore in comparison with other industries and do not refer to union as compared with open-shop conditions; it may be well-tempered with the more objective study by Haber and Levinson in this respect. His factual information is broad, and a number of observations which support unions must be taken at face value. He points out that when construction was at a peak, there was a lack of apprentices in all fields, documenting, the cause of high wages for construction workers as the refusal of men to enter or work at a trade. This was not the case in the nineteenth century, when there were more applicants for the trades than could be absorbed.

This book is the only such work discovered, and in spite of its age, it is therefore indispensable for serious study of the trade union organization. There are detailed histories of the labor struggles in Chicago and San Francisco and the success of the union-breaking plan in San Francisco. In that city, after the unions' refusal to accept an arbitrated wage cut, some of the employers combined to form the "American Plan"—essentially a right-to-work program. They raised $1,250,000 and imported several thousand out-of-town men, giving them a six-month work guarantee. Contractors who contracted with the union were refused materials and credit; a union contractor who underbid others might find that the employers' organization would pay the owner the bid difference to deprive the low bidder of the contract. If a union subcontractor appeared on a job the entire job was deprived of materials; and in one case, a general contractor was paid to cancel his contract with a union subcontractor. It is valuable for both employers and unions to know the history of this struggle (as well as the history of the Chicago strikes, where the employers were unsuccessful) as an example of what may occur and for the unions to know what is meant by "union-busting tactics"—a term often used in any dispute today, however small. In the San Francisco affair, the chief antagonists of the nonunion forces were the union contractors, not the workmen.

Haber, William, and Harold Levinson. *Labor Relations and Productivity in the Building Trades.* Ann Arbor, Mich.: University of Michigan, 1956. 266 pages.

In this objective study of the influence of labor unions on productivity, the data are derived from many contacts with contractors and are as informative as one can be on the subject in which firm data are hard to find. This is a primary reader for anyone who is concerned with labor productivity.

Hauer, D.J. *Economics of Contracting,* 2 vols. Chicago: Baumgartner, 1915. 269 pages, 334 pages.

Hauer stated his belief in the Introduction as follows:

> *Contracting is often spoken of simply as a business. . . . but conditions have changed in the last twenty-five years [1885-1910], so that today it is considered a profession . . . it is a disputed question, not only among educators, but among engineers and contractors as to whether or not contracting can be taught in our colleges.*

Hauer wrote a very detailed and competent book to serve as instruction material in such a course. Some portions are now only of historical interest, but a great deal is still useful.

The first volume is a guide to the business and an elementary explanation of the contracting system and contains much information on labor relations, particularly with immigrants. Trade directions are not detailed, but the entire work is intended for the job superintendent as well as for the general manager. It has a 10-page annotated bibliography—the earliest, and almost the only, example in construction books.

The second volume expands many of the previous topics, with some repetition. Added topics are estimating and bidding, making contracts and obtaining bonds, finance, legal aspects, handling men, office filing systems, corporate organization, and lines of specialization.

Hauer was associate editor of *Engineering-Contracting,* the predecessor of *Engineering News-Record,* under Halbert F. Gillette as editor.

Hauer. *Modern Management Applied to Construction.* New York: McGraw-Hill, 1918. 187 pages.

A quote from Benjamin Franklin: "In American business life there has arisen almost a hue-and-cry for methods of higher efficiency."

Hauer notes that the contractors who are among the first to apply scientific management to their work will reap the greatest rewards. In my copy an unknown hand has added, "*Frank Gilbreth was the first; he died broke.*" This is a plainly well-written discussion of scientific management, general planning, and time-and-motion-study.

Moder, Joseph J., and Cecil R. Phillips. *Project Management with CPM and PERT.* New York: Reinhold, 1964. 283 pages.

The first few chapters serve as a reasonably simple presentation of CPM methods; the later chapters are progressively more difficult, beyond my ability (my mathematics goes no further than calculus). It appears to be a good text for those who intend to become professional CPM consultants or to write programs for this use. The book treats resource allocation by a check made after CPM planning is completed, and the writers consider cost records as a more complicated outgrowth of CPM. They give a well-illustrated outline of a computer estimating and cost control system, but they do not consider the manual work necessary to carry it out, other than as a computer feed-in process.

National Building Studies: Incentives in the Building Industry. Special Report 28, Department of Scientific and Industrial Research. London: Department of Scientific and Industrial Research 1958. 43 pages.

The orientation of the British is toward a practical and detailed approach to research and writing in construction. This kind of research, to the extent that it exists at all in the United States, is too expensive and is intended for a scattering of "systems" builders, or it is done by scattered government departments who have no construction people assisting them.

This study is a survey and report on existing pay systems, not including piece rates, in effect on repetitive construction in England and includes varieties of incentive schemes, their performance and incentive in relation to organization and administration, collection of information, recording of bonus data, and method of analysis. This thorough report on the success of incentive payments in relation to cost of basic wages does not analyze actual production increases. Evidently incentive payments were necessary to raise wages so men might be retained, so extra payment has been made as incentives rather than increasing hourly pay rate. This report is valuable for a study of pay raises in construction; even when men are available at the basic wage, the problem of attracting superior workmen remains.

National Electrical Contractors' Association. *Manual of Labor Units.* Washington, D.C., NECA. 115 pages, regularly revised.

An estimate of worker-hours for installation of all types of electrical materials, for differing types of construction. As a standard reference, it is useful for substantiating extra claims and is generally generous in the values given. It is written by the Research Department of the National Electrical Contractors' Association (NECA) and is distributed only to members, who must be licensed electrical contractors.

A loose-leaf manual, new pages are sent to members from time to time, but changes do not often occur. Electrical contractors, even those who do not use the manual, follow the method of worker-hours per piece; few other trades consistently do this. It is listed under books not in print since it is not available from the publisher by the public.

Schobinger, George, and Alexander Lackey. *Business Methods in the Building Field.* New York: McGraw-Hill, 1940. 350 pages.

Well-written, but not profound, this is a useful text for basic understanding of the business. Authors point out that they introduce no new concepts; improvements, including improvement in design, were pointed out by the Roman Vitruvius and many others. Here are some of the quotations:

> *When we mean to build*
> *We first survey the plot, then draw the model*
> *And they we see the figure of the house*
> *Then we must rate the cost of the erection;*
> *Which if we find outweighs ability*
> *What do we do then but draw anew the model*
> *In fewer offices, or at least desist*
> *To build at all?*

> *Henry IV, Part 2,* Act 1, Scene 3

Or, in modern language, back to the old drawing board!

> *"Mechanical power and mechanical knowledge have advanced more in thirty years than they advanced in three-thousand years before"—and concludes that emphasis on false appearances and rapidly decaying materials have made us a Society of Plasterers rather than of architects.*

> *Specification for Practical Architecture*
> London, Bartholomew, 1840

Skillings, D.N., and D.B. Flint. *Catalogue of Portable Sectional Buildings,* 1862. 53 pages.

This is a catalog of prefabricated wood buildings for a variety of uses: railroad sheds, army barracks, summer cottages, schoolhouses, churches, hospitals, special tropical designs. Erection details to show construction. This is recommended for those who believe prefabricated buildings to be a new development. Prefabrication of buildings is also reported by other sources to be well known in the early 1800s, and prefabrication was used in Dutch shipyards at least 100 years earlier.

Taylor, Frederick Winslow. *Scientific Management.* New York: Harper & Bros., 1947. 600 pages.

This book is a collection of Taylor's *Shop Management, The Principles of Scientific Management,* and testimony before a Special House Committee, with a foreword by Harlow S. Person. The material was originally published in 1903 to 1912. To some extent, these items repeat each other, having been written at different times and not intended for combination.

Taylor, Frederick Winslow, and Cecil Thompson. *Concrete Costs.* New York: Wiley, 1912. 709 pages.

Frederick W. Taylor was a machinist in the late 1800s who, by better organizing the work under his charge and then later through work as a consultant, first established the techniques of time study and *scientific management.* "Scientific management" is not a popular term, since it is still identified with Taylor and unfortunately with some of his imitators who made hurried and incomplete time studies without taking into account the reaction of the workers. Taylor was a thorough man, who estimated 5 years as a reasonable time to establish scientific management methods in a plant and gained acceptance largely through a studied demonstration to the workers that increased output was in their interest.

Cecil Thompson worked for seventeen years with Taylor's support to gather costs of construction operations. Taylor and Thompson apparently did not complete the studies they contemplated, and this volume is the only one found resulting from the studies. The treatment is quite detailed and thorough, intended for use in both estimating and selecting economical designs. There are tables of previously reported costs, for operations, and for partial operations such as the separation of portions of the work involved in concrete formwork.

This book still serves as a basis for preparing cost data in a systematic way and includes not only cost data but descriptions of the effect of different concrete mixes and of the equipment used in concrete from the quarry to the finished structure. Some observations are that the wages of a man are assumed equal to that of a horse (and today the cost of riding horses stays close to this rate); with skilled labor at 50 cents per hour, three-coat plaster costs 35 to 40 cents per square yard—somewhat less efficient than today. Brick walls 8 inches thick are quoted at a subcontract labor cost of $13 to $18 per thousand bricks, or about 25 worker-hours per thousand bricks. This is comparable to today's costs, and far from the popular conception that the old-time bricklayers were several times as efficient as those today.

The important figures are on concrete formwork and steel placement, as the methods have not changed greatly since 1912. Taylor gives quite detailed tables and includes the variance for using wood-wedge or steel-clamp column forms, for cutting on the floor and for prefabricating. For a 24-inch concrete column 12 feet high, he gives 12.7 worker-hours, or 0.132 hour per square foot. Walker (1970) has 0.155 hour, with the advantage of plywood. Walker gives the fabrication as 29 percent of the total time; Taylor has 23 percent.

Taylor lists 18 operations, all figured separately, with respective times, to make up a concrete column form (not including setting). His estimates are different, not proportional, for variations in size of members and size of material used to form them. The use of 2-inch material for wall forms, for example, is only 3 percent more expensive for labor than the cost for 1-inch material. Other charts are similarly detailed, with times for carrying, applying pockets, and so on. Taylor's figures are the best published to this day, so far as the breakdown between the various operations is concerned. The best contractors today collect cost data on a square-foot basis, knowing well that the costs are not proportional to the area forms. The cost of beam forms, according to Taylor, is 50 percent less per square foot for a 16-by-32-inch beam than for an 8-by-16-inch beam. There were only 5,000 copies of Taylor's book in the first printing, which was probably the total number printed.

Tregold. *Carpentry.* Philadelphia: Carey and Hart, 1828. 280 pages plus drawings.

At the time this book was published, carpentry included principles of construction, and the book is what would now be described as a civil engineering text and includes descriptive geometry. Carpentry is defined as "the art of combining pieces of timber for the support of any considerable weight or pressure" and is a liberal art, the knowledge of which distinguishes the engineer from the workman. Quite modern description of materials, or at the least the basic materials available at that time, are included. This is primarily a design book with few framing details but with several large truss drawings. It is valuable for a historian of the period.

Engineering Extension of the University of California. *Conference on Construction Operations.* Los Angeles: University of California, 1956. 77 pages.

This valuable booklet is the complete presentation of the conference by construction executives of H. K. Ferguson, Bechtel, Morrison-Knudsen, and other companies, with details of their planning methods, and includes some information on operations research and linear programming.

Walker, Nathan, and Theodor K. Rohdenburg. *Legal Pitfalls in Architecture, Engineering, and Building Construction.* New York: McGraw-Hill, 1968. ($14) 297 pages.

This book provides a clear and well-indexed explanation of the legal principles involved in construction between design professionals and contractors. It is written from the point of view of the architect, even recommending the hold-harmless clause which has been the source of considerable contention, although not used in the AIA contract form. The authors uphold the powers of the architect as an arbitrator, but not giving the complete working and in my opinion, discuss the power or arbitration as applying to the architect without giving the circumstances. It is good that architects should study a contractor-based view (such as my text) and the contractor an architect-based view (such as the Walker and Rohdenburg text) to realize fully the divergence of opinion possible from the same set of cases.

This book mentions the three-way arbitration proceedings when two contractors are involved with the same architect and when a contractor and a subcontractor are involved with the architect, quoting typical contract clauses to cover such contingencies. Their clause gives the contractor the right to make the subcontractor a party to the arbitration proceedings by inserting a clause to that effect in the general contract. It appears to me that this is a right and a duty of the contractor in any event. The authors point out to the contractor the necessity of providing for arbitration in his subcontracts in a way compatible with the arbitration clause of the general contract.

The authors give a very good presentation of lien laws in the larger states and have provided a suitable text for engineers and architects, although prospective contractors, being concerned with many more transactions, need a more thorough exposure to law.

Ward, Jack W. *Construction Information Source and Reference Guide.* Phoenix, Ariz.: Construction Publications, 1966. 88 pages.

This publication is a bibliography of about a thousand books and magazines, intended primarily as a source material for courses in construction and including material manuals and similar material, as well as publications on methods and management. The price is changeable and the book is regularly revised.

Werbin, I. Vernon. *Legal Phases of Construction Contracts.* New York: McGraw-Hill, 1946. 273 pages.

Werbin. *Legal Cases for Contractors, Architects, and Engineers.* New York: McGraw-Hill, 1955. 497 pages.

Both these books are collections of key cases, indexed by the point considered, and are valuable for reference by the contractor or student who is able to find a copy. The decisions and to quote preceding cases. It is easy reading, although not organized as a textbook, and is more thorough and readable than more recent books in the field.

Wilson, J. *The Mechanic's and Builder's Price Book.* New York: Appleton, 1859. 175 pages.

Wilson shows in detail the prices of wood, brick, and stone work, painting and glazing, to which is added a dictionary of mechanical terms. This is a complete estimating handbook of the time, at a daily wage rate of $1.75 for carpenters and rates per 100 square feet for all kinds of work. In comparison with present standards, carpenter efficiency is now 50 to 100 percent higher. Material prices and stock dimensions are also shown. This is a valuable book for the historian, in an area where little information is available.

Wynn, A.E. *Estimating and Cost Keeping for Concrete Structures.* London: Concrete Publications, 1949. 221 pages.

This is a good text which includes steel takeoff and cost accounting, but the old English monetary system is used throughout.

APPENDIX A

AMERICAN INSTITUTE OF ARCHITECTS DOCUMENTS

THE AMERICAN INSTITUTE OF ARCHITECTS

AIA Document A701

Instructions to Bidders
1978 EDITION

Use only with the 1976 Edition of AIA Document A201, General Conditions of the Contract for Construction

TABLE OF ARTICLES

AIA DOCUMENT A701 • INSTRUCTIONS TO BIDDERS • THIRD EDITION • MAY 1978 • AIA® • ©1978
THE AMERICAN INSTITUTE OF ARCHITECTS, 1735 NEW YORK AVE., N.W., WASHINGTON, D. C. 20006

A701-1978   1

ARTICLE 1

DEFINITIONS

1.1 Bidding Documents include the Advertisement or Invitation to Bid, Instructions to Bidders, the bid form, other sample bidding and contract forms and the proposed Contract Documents including any Addenda issued prior to receipt of bids. The Contract Documents proposed for the Work consist of the Owner-Contractor Agreement, the Conditions of the Contract (General, Supplementary and other Conditions), the Drawings, the Specifications and all Addenda issued prior to and all Modifications issued after execution of the Contract.

1.2 All definitions set forth in the General Conditions of the Contract for Construction, AIA Document A201, or in other Contract Documents are applicable to the Bidding Documents.

1.3 Addenda are written or graphic instruments issued by the Architect prior to the execution of the Contract which modify or interpret the Bidding Documents by addition, deletions, clarifications or corrections.

1.4 A Bid is a complete and properly signed proposal to do the Work or designated portion thereof for the sums stipulated therein, submitted in accordance with the Bidding Documents.

1.5 The Base Bid is the sum stated in the Bid for which the Bidder offers to perform the Work described in the Bidding Documents as the base, to which work may be added or from which work may be deleted for sums stated in Alternate Bids.

1.6 An Alternate Bid (or Alternate) is an amount stated in the Bid to be added to or deducted from the amount of the Base Bid if the corresponding change in the Work, as described in the Bidding Documents, is accepted.

1.7 A Unit Price is an amount stated in the Bid as a price per unit of measurement for materials or services as described in the Bidding Documents or in the proposed Contract Documents.

1.8 A Bidder is a person or entity who submits a Bid.

1.9 A Sub-bidder is a person or entity who submits a bid to a Bidder for materials or labor for a portion of the Work.

ARTICLE 2

BIDDER'S REPRESENTATIONS

2.1 Each Bidder by making his Bid represents that:

2.1.1 He has read and understands the Bidding Documents and his Bid is made in accordance therewith.

2.1.2 He has visited the site, has familiarized himself with the local conditions under which the Work is to be performed and has correlated his observations with the requirements of the proposed Contract Documents.

2.1.3 His Bid is based upon the materials, systems and equipment required by the Bidding Documents without exception.

ARTICLE 3

BIDDING DOCUMENTS

3.1 COPIES

3.1.1 Bidders may obtain complete sets of the Bidding Documents from the issuing office designated in the Advertisement or Invitation to Bid in the number and for the deposit sum, if any, stated therein. The deposit will be refunded to Bidders who submit a bona fide Bid and return the Bidding Documents in good condition within ten days after receipt of Bids. The cost of replacement of any missing or damaged documents will be deducted from the deposit. A Bidder receiving a Contract award may retain the Bidding Documents and his deposit will be refunded.

3.1.2 Bidding Documents will not be issued directly to Sub-bidders or others unless specifically offered in the Advertisement or Invitation to Bid.

3.1.3 Bidders shall use complete sets of Bidding Documents in preparing Bids; neither the Owner nor the Architect assume any responsibility for errors or misinterpretations resulting from the use of incomplete sets of Bidding Documents.

3.1.4 The Owner or the Architect in making copies of the Bidding Documents available on the above terms do so only for the purpose of obtaining Bids on the Work and do not confer a license or grant for any other use.

3.2 INTERPRETATION OR CORRECTION OF BIDDING DOCUMENTS

3.2.1 Bidders and Sub-bidders shall promptly notify the Architect of any ambiguity, inconsistency or error which they may discover upon examination of the Bidding Documents or of the site and local conditions.

3.2.2 Bidders and Sub-bidders requiring clarification or interpretation of the Bidding Documents shall make a written request which shall reach the Architect at least seven days prior to the date for receipt of Bids.

3.2.3 Any interpretation, correction or change of the Bidding Documents will be made by Addendum. Interpretations, corrections or changes of the Bidding Documents made in any other manner will not be binding, and Bidders shall not rely upon such interpretations, corrections and changes.

3.3 SUBSTITUTIONS

3.3.1 The materials, products and equipment described in the Bidding Documents establish a standard of required function, dimension, appearance and quality to be met by any proposed substitution.

3.3.2 No substitution will be considered prior to receipt of Bids unless written request for approval has been received by the Architect at least ten days prior to the date for receipt of Bids. Each such request shall include the name of the material or equipment for which it is to be substituted and a complete description of the proposed substitute including drawings, cuts, performance and test data and any other information necessary for an evaluation. A statement setting forth any changes in other materials, equipment or other Work that incorporation of the substitute would require shall be included. The burden of proof of the merit of the proposed substitute is upon the proposer. The Architect's decision of approval or disapproval of a proposed substitution shall be final.

3.3.3 If the Architect approves any proposed substitution prior to receipt of Bids, such approval will be set forth in an Addendum. Bidders shall not rely upon approvals made in any other manner.

3.3.4 No substitutions will be considered after the Contract award unless specifically provided in the Contract Documents.

3.4 ADDENDA

3.4.1 Addenda will be mailed or delivered to all who are known by the Architect to have received a complete set of Bidding Documents.

3.4.2 Copies of Addenda will be made available for inspection wherever Bidding Documents are on file for that purpose.

3.4.3 No Addenda will be issued later than four days prior to the date for receipt of Bids except an Addendum withdrawing the request for Bids or one which includes postponement of the date for receipt of Bids.

3.4.4 Each Bidder shall ascertain prior to submitting his bid that he has received all Addenda issued, and he shall acknowledge their receipt in his Bid.

ARTICLE 4

BIDDING PROCEDURE

4.1 FORM AND STYLE OF BIDS

4.1.1 Bids shall be submitted on forms identical to the form included with the Bidding Documents, in the quantity required by Article 9.

4.1.2 All blanks on the bid form shall be filled in by typewriter or manually in ink.

4.1.3 Where so indicated by the makeup of the bid form, sums shall be expressed in both words and figures, and in case of discrepancy between the two, the amount written in words shall govern.

4.1.4 Any interlineation, alteration or erasure must be initialed by the signer of the Bid.

4.1.5 All requested Alternates shall be bid. If no change in the Base Bid is required, enter "No Change."

4.1.6 Where two or more Bids for designated portions of the Work have been requested, the Bidder may, without forfeiture of his bid security, state his refusal to accept award of less than the combination of Bids he so stipulates. The Bidder shall make no additional stipulations on the bid form nor qualify his Bid in any other manner.

4.1.7 Each copy of the Bid shall include the legal name of the Bidder and a statement that the Bidder is a sole proprietor, a partnership, a corporation, or some other legal entity. Each copy shall be signed by the person or persons legally authorized to bind the Bidder to a contract. A Bid by a corporation shall further give the state of incorporation and have the corporate seal affixed. A Bid submitted by an agent shall have a current power of attorney attached certifying the agent's authority to bind the Bidder.

4.2 BID SECURITY

4.2.1 If so stipulated in the Advertisement or Invitation to Bid, each Bid shall be accompanied by a bid security in the form and amount required by Article 9 pledging that the Bidder will enter into a contract with the Owner on the terms stated in his Bid and will, if required, furnish bonds as described hereunder in Article 7 covering the faithful performance of the Contract and the payment of all obligations arising thereunder. Should the Bidder refuse to enter into such Contract or fail to furnish such bonds if required, the amount of the bid security shall be forfeited to the Owner as liquidated damages, not as a penalty. The amount of the bid security shall not be forfeited to the Owner in the event the Owner fails to comply with Subparagraph 6.2.1.

4.2.2 If a surety bond is required it shall be written on AIA Document A310, Bid Bond, and the attorney-in-fact who executes the bond on behalf of the surety shall affix to the bond a certified and current copy of his power of attorney.

4.2.3 The Owner will have the right to retain the bid security of Bidders to whom an award is being considered until either (a) the Contract has been executed and bonds, if required, have been furnished or (b) the specified time has elapsed so that Bids may be withdrawn, or (c) all Bids have been rejected.

4.3 SUBMISSION OF BIDS

4.3.1 All copies of the Bid, the bid security, if any, and any other documents required to be submitted with the Bid shall be enclosed in a sealed opaque envelope. The envelope shall be addressed to the party receiving the Bids and shall be identified with the Project name, the Bidder's name and address and, if applicable, the designated portion of the Work for which the Bid is submitted. If the Bid is sent by mail the sealed envelope shall be enclosed in a separate mailing envelope with the notation "SEALED BID ENCLOSED" on the face thereof.

4.3.2 Bids shall be deposited at the designated location prior to the time and date for receipt of Bids indicated in the Advertisement or Invitation to Bid, or any extension thereof made by Addendum. Bids received after the time and date for receipt of Bids will be returned unopened.

4.3.3 The Bidder shall assume full responsibility for timely delivery at the location designated for receipt of Bids.

4.3.4 Oral, telephonic or telegraphic Bids are invalid and will not receive consideration.

4.4 MODIFICATION OR WITHDRAWAL OF BID

4.4.1 A Bid may not be modified, withdrawn or canceled by the Bidder during the stipulated time period following the time and date designated for the receipt of Bids, and each Bidder so agrees in submitting his Bid.

AIA DOCUMENT A701 • INSTRUCTIONS TO BIDDERS • THIRD EDITION • MAY 1978 • AIA® • ©1978
THE AMERICAN INSTITUTE OF ARCHITECTS, 1735 NEW YORK AVE., N.W., WASHINGTON, D. C. 20006

4.4.2 Prior to the time and date designated for receipt of Bids, any Bid submitted may be modified or withdrawn by notice to the party receiving Bids at the place designated for receipt of Bids. Such notice shall be in writing over the signature of the Bidder or by telegram; if by telegram, written confirmation over the signature of the Bidder shall be mailed and postmarked on or before the date and time set for receipt of Bids, and it shall be so worded as not to reveal the amount of the original Bid.

4.4.3 Withdrawn Bids may be resubmitted up to the time designated for the receipt of Bids provided that they are then fully in conformance with these Instructions to Bidders.

4.4.4 Bid security, if any is required, shall be in an amount sufficient for the Bid as modified or resubmitted.

ARTICLE 5

CONSIDERATION OF BIDS

5.1 OPENING OF BIDS

5.1.1 Unless stated otherwise in the Advertisement or Invitation to Bid, the properly identified Bids received on time will be opened publicly and will be read aloud. An abstract of the Base Bids and Alternate Bids, if any, will be made available to Bidders. When it has been stated that Bids will be opened privately, an abstract of the same information may, at the discretion of the Owner, be made available to the Bidders within a reasonable time.

5.2 REJECTION OF BIDS

5.2.1 The Owner shall have the right to reject any or all Bids and to reject a Bid not accompanied by any required bid security or by other data required by the Bidding Documents, or to reject a Bid which is in any way incomplete or irregular.

5.3 ACCEPTANCE OF BID (AWARD)

5.3.1 It is the intent of the Owner to award a Contract to the lowest responsible Bidder provided the Bid has been submitted in accordance with the requirements of the Bidding Documents and does not exceed the funds available. The Owner shall have the right to waive any informality or irregularity in any Bid or Bids received and to accept the Bid or Bids which, in his judgment, is in his own best interests.

5.3.2 The Owner shall have the right to accept Alternates in any order or combination, unless otherwise specifically provided in Article 9, and to determine the low Bidder on the basis of the sum of the Base Bid and the Alternates accepted.

ARTICLE 6

POST BID INFORMATION

6.1 CONTRACTOR'S QUALIFICATION STATEMENT

6.1.1 Bidders to whom award of a Contract is under consideration shall submit to the Architect, upon request, a properly executed AIA Document A305, Contractor's Qualification Statement, unless such a Statement has been previously required and submitted as a prerequisite to the issuance of Bidding Documents.

6.2 OWNER'S FINANCIAL CAPABILITY

6.2.1 The Owner shall, at the request of the Bidder to whom award of a Contract is under consideration and no later than seven days prior to the expiration of the time for withdrawal of Bids, furnish to the Bidder reasonable evidence that the Owner has made financial arrangements to fulfill the Contract obligations. Unless such reasonable evidence is furnished, the Bidder will not be required to execute the Owner-Contractor Agreement.

6.3 SUBMITTALS

6.3.1 The Bidder shall, within seven days of notification of selection for the award of a Contract for the Work, submit the following information to the Architect:

.1 a designation of the Work to be performed by the Bidder with his own forces;

.2 the proprietary names and the suppliers of principal items or systems of materials and equipment proposed for the Work;

.3 a list of names of the Subcontractors or other persons or entities (including those who are to furnish materials or equipment fabricated to a special design) proposed for the principal portions of the Work.

6.3.2 The Bidder will be required to establish to the satisfaction of the Architect and the Owner the reliability and responsibility of the persons or entities proposed to furnish and perform the Work described in the Bidding Documents.

6.3.3 Prior to the award of the Contract, the Architect will notify the Bidder in writing if either the Owner or the Architect, after due investigation, has reasonable objection to any such proposed person or entity. If the Owner or Architect has reasonable objection to any such proposed person or entity, the Bidder may, at his option, (1) withdraw his Bid, or (2) submit an acceptable substitute person or entity with an adjustment in his bid price to cover the difference in cost occasioned by such substitution. The Owner may, at his discretion, accept the adjusted bid price or he may disqualify the Bidder. In the event of either withdrawal or disqualification under this Subparagraph, bid security will not be forfeited, notwithstanding the provisions of Paragraph 4.4.1.

6.3.4 Persons and entities proposed by the Bidder and to whom the Owner and the Architect have made no reasonable objection under the provisions of Subparagraph 6.3.3 must be used on the Work for which they were proposed and shall not be changed except with the written consent of the Owner and the Architect.

ARTICLE 7

PERFORMANCE BOND AND LABOR AND MATERIAL PAYMENT BOND

7.1 BOND REQUIREMENTS

7.1.1 Prior to execution of the Contract, if required in Article 9 hereinafter, the Bidder shall furnish bonds covering the faithful performance of the Contract and the payment of all obligations arising thereunder in such form and amount as the Owner may prescribe. Bonds may be secured through the Bidder's usual sources. If the furnishing of such bonds is stipulated hereinafter in Article 9, the cost shall be included in the Bid.

7.1.2 If the Owner has reserved the right to require that bonds be furnished subsequent to the execution of the Contract, the cost shall be adjusted as provided in the Contract Documents.

7.1.3 If the Owner requires that bonds be obtained from other than the Bidder's usual source, any change in cost will be adjusted as provided in the Contract Documents.

7.2 TIME OF DELIVERY AND FORM OF BONDS

7.2.1 The Bidder shall deliver the required bonds to the Owner not later than the date of execution of the Contract, or if the Work is to be commenced prior thereto in response to a letter of intent, the Bidder shall, prior to commencement of the Work, submit evidence satisfactory to the Owner that such bonds will be furnished.

7.2.2 Unless otherwise required in Article 9, the bonds shall be written on AIA Document A311, Performance Bond and Labor and Material Payment Bond.

7.2.3 The Bidder shall require the attorney-in-fact who executes the required bonds on behalf of the surety to affix thereto a certified and current copy of his power of attorney.

ARTICLE 8
FORM OF AGREEMENT BETWEEN OWNER AND CONTRACTOR

8.1 FORM TO BE USED

8.1.1 Unless otherwise required in the Bidding Documents, the Agreement for the Work will be written on AIA Document A101, Standard Form of Agreement Between Owner and Contractor, where the basis of payment is a Stipulated Sum.

ARTICLE 9
SUPPLEMENTARY INSTRUCTIONS

THE AMERICAN INSTITUTE OF ARCHITECTS

AIA Document A201

General Conditions of the Contract for Construction

THIS DOCUMENT HAS IMPORTANT LEGAL CONSEQUENCES; CONSULTATION WITH AN ATTORNEY IS ENCOURAGED WITH RESPECT TO ITS MODIFICATION

1976 EDITION
TABLE OF ARTICLES

1. CONTRACT DOCUMENTS

2. ARCHITECT

3. OWNER

4. CONTRACTOR

5. SUBCONTRACTORS

6. WORK BY OWNER OR BY SEPARATE CONTRACTORS

7. MISCELLANEOUS PROVISIONS

8. TIME

9. PAYMENTS AND COMPLETION

10. PROTECTION OF PERSONS AND PROPERTY

11. INSURANCE

12. CHANGES IN THE WORK

13. UNCOVERING AND CORRECTION OF WORK

14. TERMINATION OF THE CONTRACT

This document has been approved and endorsed by The Associated General Contractors of America.

This document has been reproduced with the pemission of the American Insitute of Architects.

A201-1976   AIA DOCUMENT A201 • GENERAL CONDITIONS OF THE CONTRACT FOR CONSTRUCTION • THIRTEENTH EDITION • AUGUST 1976
AIA® • © 1976 • THE AMERICAN INSTITUTE OF ARCHITECTS, 1735 NEW YORK AVENUE, N.W., WASHINGTON, D.C. 20006

ARTICLE 1

CONTRACT DOCUMENTS

1.1 DEFINITIONS

1.1.1 THE CONTRACT DOCUMENTS

The Contract Documents consist of the Owner-Contractor Agreement, the Conditions of the Contract (General, Supplementary and other Conditions), the Drawings, the Specifications, and all Addenda issued prior to and all Modifications issued after execution of the Contract. A Modification is (1) a written amendment to the Contract signed by both parties, (2) a Change Order, (3) a written interpretation issued by the Architect pursuant to Subparagraph 2.2.8, or (4) a written order for a minor change in the Work issued by the Architect pursuant to Paragraph 12.4. The Contract Documents do not include Bidding Documents such as the Advertisement or Invitation to Bid, the Instructions to Bidders, sample forms, the Contractor's Bid or portions of Addenda relating to any of these, or any other documents, unless specifically enumerated in the Owner-Contractor Agreement.

1.1.2 THE CONTRACT

The Contract Documents form the Contract for Construction. This Contract represents the entire and integrated agreement between the parties hereto and supersedes all prior negotiations, representations, or agreements, either written or oral. The Contract may be amended or modified only by a Modification as defined in Subparagraph 1.1.1. The Contract Documents shall not be construed to create any contractual relationship of any kind between the Architect and the Contractor, but the Architect shall be entitled to performance of obligations intended for his benefit, and to enforcement thereof. Nothing contained in the Contract Documents shall create any contractual relationship between the Owner or the Architect and any Subcontractor or Sub-subcontractor.

1.1.3 THE WORK

The Work comprises the completed construction required by the Contract Documents and includes all labor necessary to produce such construction, and all materials and equipment incorporated or to be incorporated in such construction.

1.1.4 THE PROJECT

The Project is the total construction of which the Work performed under the Contract Documents may be the whole or a part.

1.2 EXECUTION, CORRELATION AND INTENT

1.2.1 The Contract Documents shall be signed in not less than triplicate by the Owner and Contractor. If either the Owner or the Contractor or both do not sign the Conditions of the Contract, Drawings, Specifications, or any of the other Contract Documents, the Architect shall identify such Documents.

1.2.2 By executing the Contract, the Contractor represents that he has visited the site, familiarized himself with the local conditions under which the Work is to be performed, and correlated his observations with the requirements of the Contract Documents.

1.2.3 The intent of the Contract Documents is to include all items necessary for the proper execution and completion of the Work. The Contract Documents are complementary, and what is required by any one shall be as binding as if required by all. Work not covered in the Contract Documents will not be required unless it is consistent therewith and is reasonably inferable therefrom as being necessary to produce the intended results. Words and abbreviations which have well-known technical or trade meanings are used in the Contract Documents in accordance with such recognized meanings.

1.2.4 The organization of the Specifications into divisions, sections and articles, and the arrangement of Drawings shall not control the Contractor in dividing the Work among Subcontractors or in establishing the extent of Work to be performed by any trade.

1.3 OWNERSHIP AND USE OF DOCUMENTS

1.3.1 All Drawings, Specifications and copies thereof furnished by the Architect are and shall remain his property. They are to be used only with respect to this Project and are not to be used on any other project. With the exception of one contract set for each party to the Contract, such documents are to be returned or suitably accounted for to the Architect on request at the completion of the Work. Submission or distribution to meet official regulatory requirements or for other purposes in connection with the Project is not to be construed as publication in derogation of the Architect's common law copyright or other reserved rights.

ARTICLE 2

ARCHITECT

2.1 DEFINITION

2.1.1 The Architect is the person lawfully licensed to practice architecture, or an entity lawfully practicing architecture identified as such in the Owner-Contractor Agreement, and is referred to throughout the Contract Documents as if singular in number and masculine in gender. The term Architect means the Architect or his authorized representative.

2.2 ADMINISTRATION OF THE CONTRACT

2.2.1 The Architect will provide administration of the Contract as hereinafter described.

2.2.2 The Architect will be the Owner's representative during construction and until final payment is due. The Architect will advise and consult with the Owner. The Owner's instructions to the Contractor shall be forwarded through the Architect. The Architect will have authority to act on behalf of the Owner only to the extent provided in the Contract Documents, unless otherwise modified by written instrument in accordance with Subparagraph 2.2.18.

2.2.3 The Architect will visit the site at intervals appropriate to the stage of construction to familiarize himself generally with the progress and quality of the Work and to determine in general if the Work is proceeding in accordance with the Contract Documents. However, the Architect will not be required to make exhaustive or continuous on-site inspections to check the quality or quantity of the Work. On the basis of his on-site observations as an architect, he will keep the Owner informed of the progress of the Work, and will endeavor to guard the Owner against defects and deficiencies in the Work of the Contractor.

2.2.4 The Architect will not be responsible for and will not have control or charge of construction means, methods, techniques, sequences or procedures, or for safety precautions and programs in connection with the Work, and he will not be responsible for the Contractor's failure to carry out the Work in accordance with the Contract Documents. The Architect will not be responsible for or have control or charge over the acts or omissions of the Contractor, Subcontractors, or any of their agents or employees, or any other persons performing any of the Work.

2.2.5 The Architect shall at all times have access to the Work wherever it is in preparation and progress. The Contractor shall provide facilities for such access so the Architect may perform his functions under the Contract Documents.

2.2.6 Based on the Architect's observations and an evaluation of the Contractor's Applications for Payment, the Architect will determine the amounts owing to the Contractor and will issue Certificates for Payment in such amounts, as provided in Paragraph 9.4.

2.2.7 The Architect will be the interpreter of the requirements of the Contract Documents and the judge of the performance thereunder by both the Owner and Contractor.

2.2.8 The Architect will render interpretations necessary for the proper execution or progress of the Work, with reasonable promptness and in accordance with any time limit agreed upon. Either party to the Contract may make written request to the Architect for such interpretations.

2.2.9 Claims, disputes and other matters in question between the Contractor and the Owner relating to the execution or progress of the Work or the interpretation of the Contract Documents shall be referred initially to the Architect for decision which he will render in writing within a reasonable time.

2.2.10 All interpretations and decisions of the Architect shall be consistent with the intent of and reasonably inferable from the Contract Documents and will be in writing or in the form of drawings. In his capacity as interpreter and judge, he will endeavor to secure faithful performance by both the Owner and the Contractor, will not

show partiality to either, and will not be liable for the result of any interpretation or decision rendered in good faith in such capacity.

2.2.11 The Architect's decisions in matters relating to artistic effect will be final if consistent with the intent of the Contract Documents.

2.2.12 Any claim, dispute or other matter in question between the Contractor and the Owner referred to the Architect, except those relating to artistic effect as provided in Subparagraph 2.2.11 and except those which have been waived by the making or acceptance of final payment as provided in Subparagraphs 9.9.4 and 9.9.5, shall be subject to arbitration upon the written demand of either party. However, no demand for arbitration of any such claim, dispute or other matter may be made until the earlier of (1) the date on which the Architect has rendered a written decision, or (2) the tenth day after the parties have presented their evidence to the Architect or have been given a reasonable opportunity to do so, if the Architect has not rendered his written decision by that date. When such a written decision of the Architect states (1) that the decision is final but subject to appeal, and (2) that any demand for arbitration of a claim, dispute or other matter covered by such decision must be made within thirty days after the date on which the party making the demand receives the written decision, failure to demand arbitration within said thirty days' period will result in the Architect's decision becoming final and binding upon the Owner and the Contractor. If the Architect renders a decision after arbitration proceedings have been initiated, such decision may be entered as evidence but will not supersede any arbitration proceedings unless the decision is acceptable to all parties concerned.

2.2.13 The Architect will have authority to reject Work which does not conform to the Contract Documents. Whenever, in his opinion, he considers it necessary or advisable for the implementation of the intent of the Contract Documents, he will have authority to require special inspection or testing of the Work in accordance with Subparagraph 7.7.2 whether or not such Work be then fabricated, installed or completed. However, neither the Architect's authority to act under this Subparagraph 2.2.13, nor any decision made by him in good faith either to exercise or not to exercise such authority, shall give rise to any duty or responsibility of the Architect to the Contractor, any Subcontractor, any of their agents or employees, or any other person performing any of the Work.

2.2.14 The Architect will review and approve or take other appropriate action upon Contractor's submittals such as Shop Drawings, Product Data and Samples, but only for conformance with the design concept of the Work and with the information given in the Contract Documents. Such action shall be taken with reasonable promptness so as to cause no delay. The Architect's approval of a specific item shall not indicate approval of an assembly of which the item is a component.

2.2.15 The Architect will prepare Change Orders in accordance with Article 12, and will have authority to order minor changes in the Work as provided in Subparagraph 12.4.1.

2.2.16 The Architect will conduct inspections to determine the dates of Substantial Completion and final completion, will receive and forward to the Owner for the Owner's review written warranties and related documents required by the Contract and assembled by the Contractor, and will issue a final Certificate for Payment upon compliance with the requirements of Paragraph 9.9.

2.2.17 If the Owner and Architect agree, the Architect will provide one or more Project Representatives to assist the Architect in carrying out his responsibilities at the site. The duties, responsibilities and limitations of authority of any such Project Representative shall be as set forth in an exhibit to be incorporated in the Contract Documents.

2.2.18 The duties, responsibilities and limitations of authority of the Architect as the Owner's representative during construction as set forth in the Contract Documents will not be modified or extended without written consent of the Owner, the Contractor and the Architect.

2.2.19 In case of the termination of the employment of the Architect, the Owner shall appoint an architect against whom the Contractor makes no reasonable objection whose status under the Contract Documents shall be that of the former architect. Any dispute in connection with such appointment shall be subject to arbitration.

ARTICLE 3

OWNER

3.1 DEFINITION

3.1.1 The Owner is the person or entity identified as such in the Owner-Contractor Agreement and is referred to throughout the Contract Documents as if singular in number and masculine in gender. The term Owner means the Owner or his authorized representative.

3.2 INFORMATION AND SERVICES REQUIRED OF THE OWNER

3.2.1 The Owner shall, at the request of the Contractor, at the time of execution of the Owner-Contractor Agreement, furnish to the Contractor reasonable evidence that he has made financial arrangements to fulfill his obligations under the Contract. Unless such reasonable evidence is furnished, the Contractor is not required to execute the Owner-Contractor Agreement or to commence the Work.

3.2.2 The Owner shall furnish all surveys describing the physical characteristics, legal limitations and utility locations for the site of the Project, and a legal description of the site.

3.2.3 Except as provided in Subparagraph 4.7.1, the Owner shall secure and pay for necessary approvals, easements, assessments and charges required for the construction, use or occupancy of permanent structures or for permanent changes in existing facilities.

3.2.4 Information or services under the Owner's control shall be furnished by the Owner with reasonable promptness to avoid delay in the orderly progress of the Work.

3.2.5 Unless otherwise provided in the Contract Documents, the Contractor will be furnished, free of charge, all copies of Drawings and Specifications reasonably necessary for the execution of the Work.

3.2.6 The Owner shall forward all instructions to the Contractor through the Architect.

3.2.7 The foregoing are in addition to other duties and responsibilities of the Owner enumerated herein and especially those in respect to Work by Owner or by Separate Contractors, Payments and Completion, and Insurance in Articles 6, 9 and 11 respectively.

3.3 OWNER'S RIGHT TO STOP THE WORK

3.3.1 If the Contractor fails to correct defective Work as required by Paragraph 13.2 or persistently fails to carry out the Work in accordance with the Contract Documents, the Owner, by a written order signed personally or by an agent specifically so empowered by the Owner in writing, may order the Contractor to stop the Work, or any portion thereof, until the cause for such order has been eliminated; however, this right of the Owner to stop the Work shall not give rise to any duty on the part of the Owner to exercise this right for the benefit of the Contractor or any other person or entity, except to the extent required by Subparagraph 6.1.3.

3.4 OWNER'S RIGHT TO CARRY OUT THE WORK

3.4.1 If the Contractor defaults or neglects to carry out the Work in accordance with the Contract Documents and fails within seven days after receipt of written notice from the Owner to commence and continue correction of such default or neglect with diligence and promptness, the Owner may, after seven days following receipt by the Contractor of an additional written notice and without prejudice to any other remedy he may have, make good such deficiencies. In such case an appropriate Change Order shall be issued deducting from the payments then or thereafter due the Contractor the cost of correcting such deficiencies, including compensation for the Architect's additional services made necessary by such default, neglect or failure. Such action by the Owner and the amount charged to the Contractor are both subject to the prior approval of the Architect. If the payments then or thereafter due the Contractor are not sufficient to cover such amount, the Contractor shall pay the difference to the Owner.

ARTICLE 4

CONTRACTOR

4.1 DEFINITION

4.1.1 The Contractor is the person or entity identified as such in the Owner-Contractor Agreement and is referred to throughout the Contract Documents as if singular in number and masculine in gender. The term Contractor means the Contractor or his authorized representative.

4.2 REVIEW OF CONTRACT DOCUMENTS

4.2.1 The Contractor shall carefully study and compare the Contract Documents and shall at once report to the Architect any error, inconsistency or omission he may discover. The Contractor shall not be liable to the Owner or the Architect for any damage resulting from any such errors, inconsistencies or omissions in the Contract Documents. The Contractor shall perform no portion of the Work at any time without Contract Documents or, where required, approved Shop Drawings, Product Data or Samples for such portion of the Work.

4.3 SUPERVISION AND CONSTRUCTION PROCEDURES

4.3.1 The Contractor shall supervise and direct the Work, using his best skill and attention. He shall be solely responsible for all construction means, methods, techniques, sequences and procedures and for coordinating all portions of the Work under the Contract.

4.3.2 The Contractor shall be responsible to the Owner for the acts and omissions of his employees, Subcontractors and their agents and employees, and other persons performing any of the Work under a contract with the Contractor.

4.3.3 The Contractor shall not be relieved from his obligations to perform the Work in accordance with the Contract Documents either by the activities or duties of the Architect in his administration of the Contract, or by inspections, tests or approvals required or performed under Paragraph 7.7 by persons other than the Contractor.

4.4 LABOR AND MATERIALS

4.4.1 Unless otherwise provided in the Contract Documents, the Contractor shall provide and pay for all labor, materials, equipment, tools, construction equipment and machinery, water, heat, utilities, transportation, and other facilities and services necessary for the proper execution and completion of the Work, whether temporary or permanent and whether or not incorporated or to be incorporated in the Work.

4.4.2 The Contractor shall at all times enforce strict discipline and good order among his employees and shall not employ on the Work any unfit person or anyone not skilled in the task assigned to him.

4.5 WARRANTY

4.5.1 The Contractor warrants to the Owner and the Architect that all materials and equipment furnished under this Contract will be new unless otherwise specified, and that all Work will be of good quality, free from faults and defects and in conformance with the Contract Documents. All Work not conforming to these requirements, including substitutions not properly approved and authorized, may be considered defective. If required by the Architect, the Contractor shall furnish satisfactory evidence as to the kind and quality of materials and equipment. This warranty is not limited by the provisions of Paragraph 13.2.

4.6 TAXES

4.6.1 The Contractor shall pay all sales, consumer, use and other similar taxes for the Work or portions thereof provided by the Contractor which are legally enacted at the time bids are received, whether or not yet effective.

4.7 PERMITS, FEES AND NOTICES

4.7.1 Unless otherwise provided in the Contract Documents, the Contractor shall secure and pay for the building permit and for all other permits and governmental fees, licenses and inspections necessary for the proper execution and completion of the Work which are customarily secured after execution of the Contract and which are legally required at the time the bids are received.

4.7.2 The Contractor shall give all notices and comply with all laws, ordinances, rules, regulations and lawful orders of any public authority bearing on the performance of the Work.

4.7.3 It is not the responsibility of the Contractor to make certain that the Contract Documents are in accordance with applicable laws, statutes, building codes and regulations. If the Contractor observes that any of the Contract Documents are at variance therewith in any respect, he shall promptly notify the Architect in writing, and any necessary changes shall be accomplished by appropriate Modification.

4.7.4 If the Contractor performs any Work knowing it to be contrary to such laws, ordinances, rules and regulations, and without such notice to the Architect, he shall assume full responsibility therefor and shall bear all costs attributable thereto.

4.8 ALLOWANCES

4.8.1 The Contractor shall include in the Contract Sum all allowances stated in the Contract Documents. Items covered by these allowances shall be supplied for such amounts and by such persons as the Owner may direct, but the Contractor will not be required to employ persons against whom he makes a reasonable objection.

4.8.2 Unless otherwise provided in the Contract Documents:

  .1 these allowances shall cover the cost to the Contractor, less any applicable trade discount, of the materials and equipment required by the allowance delivered at the site, and all applicable taxes;

  .2 the Contractor's costs for unloading and handling on the site, labor, installation costs, overhead, profit and other expenses contemplated for the original allowance shall be included in the Contract Sum and not in the allowance;

  .3 whenever the cost is more than or less than the allowance, the Contract Sum shall be adjusted accordingly by Change Order, the amount of which will recognize changes, if any, in handling costs on the site, labor, installation costs, overhead, profit and other expenses.

4.9 SUPERINTENDENT

4.9.1 The Contractor shall employ a competent superintendent and necessary assistants who shall be in attendance at the Project site during the progress of the Work. The superintendent shall represent the Contractor and all communications given to the superintendent shall be as binding as if given to the Contractor. Important communications shall be confirmed in writing. Other communications shall be so confirmed on written request in each case.

4.10 PROGRESS SCHEDULE

4.10.1 The Contractor, immediately after being awarded the Contract, shall prepare and submit for the Owner's and Architect's information an estimated progress schedule for the Work. The progress schedule shall be related to the entire Project to the extent required by the Contract Documents, and shall provide for expeditious and practicable execution of the Work.

4.11 DOCUMENTS AND SAMPLES AT THE SITE

4.11.1 The Contractor shall maintain at the site for the Owner one record copy of all Drawings, Specifications, Addenda, Change Orders and other Modifications, in good order and marked currently to record all changes made during construction, and approved Shop Drawings, Product Data and Samples. These shall be available to the Architect and shall be delivered to him for the Owner upon completion of the Work.

4.12 SHOP DRAWINGS, PRODUCT DATA AND SAMPLES

4.12.1 Shop Drawings are drawings, diagrams, schedules and other data specially prepared for the Work by the Contractor or any Subcontractor, manufacturer, supplier or distributor to illustrate some portion of the Work.

4.12.2 Product Data are illustrations, standard schedules, performance charts, instructions, brochures, diagrams and other information furnished by the Contractor to illustrate a material, product or system for some portion of the Work.

4.12.3 Samples are physical examples which illustrate materials, equipment or workmanship and establish standards by which the Work will be judged.

4.12.4 The Contractor shall review, approve and submit, with reasonable promptness and in such sequence as to cause no delay in the Work or in the work of the Owner or any separate contractor, all Shop Drawings, Product Data and Samples required by the Contract Documents.

4.12.5 By approving and submitting Shop Drawings, Product Data and Samples, the Contractor represents that he has determined and verified all materials, field measurements, and field construction criteria related thereto, or will do so, and that he has checked and coordinated the information contained within such submittals with the requirements of the Work and of the Contract Documents.

4.12.6 The Contractor shall not be relieved of responsibility for any deviation from the requirements of the Contract Documents by the Architect's approval of Shop Drawings, Product Data or Samples under Subparagraph 2.2.14 unless the Contractor has specifically informed the Architect in writing of such deviation at the time of submission and the Architect has given written approval to the specific deviation. The Contractor shall not be relieved from responsibility for errors or omissions in the Shop Drawings, Product Data or Samples by the Architect's approval thereof.

4.12.7 The Contractor shall direct specific attention, in writing or on resubmitted Shop Drawings, Product Data or Samples, to revisions other than those requested by the Architect on previous submittals.

4.12.8 No portion of the Work requiring submission of a Shop Drawing, Product Data or Sample shall be commenced until the submittal has been approved by the Architect as provided in Subparagraph 2.2.14. All such

portions of the Work shall be in accordance with approved submittals.

4.13 USE OF SITE

4.13.1 The Contractor shall confine operations at the site to areas permitted by law, ordinances, permits and the Contract Documents and shall not unreasonably encumber the site with any materials or equipment.

4.14 CUTTING AND PATCHING OF WORK

4.14.1 The Contractor shall be responsible for all cutting, fitting or patching that may be required to complete the Work or to make its several parts fit together properly.

4.14.2 The Contractor shall not damage or endanger any portion of the Work or the work of the Owner or any separate contractors by cutting, patching or otherwise altering any work, or by excavation. The Contractor shall not cut or otherwise alter the work of the Owner or any separate contractor except with the written consent of the Owner and of such separate contractor. The Contractor shall not unreasonably withhold from the Owner or any separate contractor his consent to cutting or otherwise altering the Work.

4.15 CLEANING UP

4.15.1 The Contractor at all times shall keep the premises free from accumulation of waste materials or rubbish caused by his operations. At the completion of the Work he shall remove all his waste materials and rubbish from and about the Project as well as all his tools, construction equipment, machinery and surplus materials.

4.15.2 If the Contractor fails to clean up at the completion of the Work, the Owner may do so as provided in Paragraph 3.4 and the cost thereof shall be charged to the Contractor.

4.16 COMMUNICATIONS

4.16.1 The Contractor shall forward all communications to the Owner through the Architect.

4.17 ROYALTIES AND PATENTS

4.17.1 The Contractor shall pay all royalties and license fees. He shall defend all suits or claims for infringement of any patent rights and shall save the Owner harmless from loss on account thereof, except that the Owner shall be responsible for all such loss when a particular design, process or the product of a particular manufacturer or manufacturers is specified, but if the Contractor has reason to believe that the design, process or product specified is an infringement of a patent, he shall be responsible for such loss unless he promptly gives such information to the Architect.

4.18 INDEMNIFICATION

4.18.1 To the fullest extent permitted by law, the Contractor shall indemnify and hold harmless the Owner and the Architect and their agents and employees from and against all claims, damages, losses and expenses, including but not limited to attorneys' fees, arising out of or resulting from the performance of the Work, provided that any such claim, damage, loss or expense (1) is attributable to bodily injury, sickness, disease or death, or to injury to or destruction of tangible property (other than the Work itself) including the loss of use resulting therefrom, and (2) is caused in whole or in part by any negligent act or omission of the Contractor, any Subcontractor, anyone directly or indirectly employed by any of them or anyone for whose acts any of them may be liable, regardless of whether or not it is caused in part by a party indemnified hereunder. Such obligation shall not be construed to negate, abridge, or otherwise reduce any other right or obligation of indemnity which would otherwise exist as to any party or person described in this Paragraph 4.18.

4.18.2 In any and all claims against the Owner or the Architect or any of their agents or employees by any employee of the Contractor, any Subcontractor, anyone directly or indirectly employed by any of them or anyone for whose acts any of them may be liable, the indemnification obligation under this Paragraph 4.18 shall not be limited in any way by any limitation on the amount or type of damages, compensation or benefits payable by or for the Contractor or any Subcontractor under workers' or workmen's compensation acts, disability benefit acts or other employee benefit acts.

4.18.3 The obligations of the Contractor under this Paragraph 4.18 shall not extend to the liability of the Architect, his agents or employees, arising out of (1) the preparation or approval of maps, drawings, opinions, reports, surveys, change orders, designs or specifications, or (2) the giving of or the failure to give directions or instructions by the Architect, his agents or employees provided such giving or failure to give is the primary cause of the injury or damage.

ARTICLE 5

SUBCONTRACTORS

5.1 DEFINITION

5.1.1 A Subcontractor is a person or entity who has a direct contract with the Contractor to perform any of the Work at the site. The term Subcontractor is referred to throughout the Contract Documents as if singular in number and masculine in gender and means a Subcontractor or his authorized representative. The term Subcontractor does not include any separate contractor or his subcontractors.

5.1.2 A Sub-subcontractor is a person or entity who has a direct or indirect contract with a Subcontractor to perform any of the Work at the site. The term Sub-subcontractor is referred to throughout the Contract Documents as if singular in number and masculine in gender and means a Sub-subcontractor or an authorized representative thereof.

5.2 AWARD OF SUBCONTRACTS AND OTHER CONTRACTS FOR PORTIONS OF THE WORK

5.2.1 Unless otherwise required by the Contract Documents or the Bidding Documents, the Contractor, as soon as practicable after the award of the Contract, shall furnish to the Owner and the Architect in writing the names of the persons or entities (including those who are to furnish materials or equipment fabricated to a special design) proposed for each of the principal portions of the Work. The Architect will promptly reply to the Contractor in writing stating whether or not the Owner or the Architect, after due investigation, has reasonable objection to any

such proposed person or entity. Failure of the Owner or Architect to reply promptly shall constitute notice of no reasonable objection.

5.2.2 The Contractor shall not contract with any such proposed person or entity to whom the Owner or the Architect has made reasonable objection under the provisions of Subparagraph 5.2.1. The Contractor shall not be required to contract with anyone to whom he has a reasonable objection.

5.2.3 If the Owner or the Architect has reasonable objection to any such proposed person or entity, the Contractor shall submit a substitute to whom the Owner or the Architect has no reasonable objection, and the Contract Sum shall be increased or decreased by the difference in cost occasioned by such substitution and an appropriate Change Order shall be issued; however, no increase in the Contract Sum shall be allowed for any such substitution unless the Contractor has acted promptly and responsively in submitting names as required by Subparagraph 5.2.1.

5.2.4 The Contractor shall make no substitution for any Subcontractor, person or entity previously selected if the Owner or Architect makes reasonable objection to such substitution.

5.3 SUBCONTRACTUAL RELATIONS

5.3.1 By an appropriate agreement, written where legally required for validity, the Contractor shall require each Subcontractor, to the extent of the Work to be performed by the Subcontractor, to be bound to the Contractor by the terms of the Contract Documents, and to assume toward the Contractor all the obligations and responsibilities which the Contractor, by these Documents, assumes toward the Owner and the Architect. Said agreement shall preserve and protect the rights of the Owner and the Architect under the Contract Documents with respect to the Work to be performed by the Subcontractor so that the subcontracting thereof will not prejudice such rights, and shall allow to the Subcontractor, unless specifically provided otherwise in the Contractor-Subcontractor agreement, the benefit of all rights, remedies and redress against the Contractor that the Contractor, by these Documents, has against the Owner. Where appropriate, the Contractor shall require each Subcontractor to enter into similar agreements with his Sub-subcontractors. The Contractor shall make available to each proposed Subcontractor, prior to the execution of the Subcontract, copies of the Contract Documents to which the Subcontractor will be bound by this Paragraph 5.3, and identify to the Subcontractor any terms and conditions of the proposed Subcontract which may be at variance with the Contract Documents. Each Subcontractor shall similarly make copies of such Documents available to his Sub-subcontractors.

ARTICLE 6

WORK BY OWNER OR BY SEPARATE CONTRACTORS

6.1 OWNER'S RIGHT TO PERFORM WORK AND TO AWARD SEPARATE CONTRACTS

6.1.1 The Owner reserves the right to perform work related to the Project with his own forces, and to award separate contracts in connection with other portions of the Project or other work on the site under these or similar Conditions of the Contract. If the Contractor claims that delay or additional cost is involved because of such action by the Owner, he shall make such claim as provided elsewhere in the Contract Documents.

6.1.2 When separate contracts are awarded for different portions of the Project or other work on the site, the term Contractor in the Contract Documents in each case shall mean the Contractor who executes each separate Owner-Contractor Agreement.

6.1.3 The Owner will provide for the coordination of the work of his own forces and of each separate contractor with the Work of the Contractor, who shall cooperate therewith as provided in Paragraph 6.2.

6.2 MUTUAL RESPONSIBILITY

6.2.1 The Contractor shall afford the Owner and separate contractors reasonable opportunity for the introduction and storage of their materials and equipment and the execution of their work, and shall connect and coordinate his Work with theirs as required by the Contract Documents.

6.2.2 If any part of the Contractor's Work depends for proper execution or results upon the work of the Owner or any separate contractor, the Contractor shall, prior to proceeding with the Work, promptly report to the Architect any apparent discrepancies or defects in such other work that render it unsuitable for such proper execution and results. Failure of the Contractor so to report shall constitute an acceptance of the Owner's or separate contractors' work as fit and proper to receive his Work, except as to defects which may subsequently become apparent in such work by others.

6.2.3 Any costs caused by defective or ill-timed work shall be borne by the party responsible therefor.

6.2.4 Should the Contractor wrongfully cause damage to the work or property of the Owner, or to other work on the site, the Contractor shall promptly remedy such damage as provided in Subparagraph 10.2.5.

6.2.5 Should the Contractor wrongfully cause damage to the work or property of any separate contractor, the Contractor shall upon due notice promptly attempt to settle with such other contractor by agreement, or otherwise to resolve the dispute. If such separate contractor sues or initiates an arbitration proceeding against the Owner on account of any damage alleged to have been caused by the Contractor, the Owner shall notify the Contractor who shall defend such proceedings at the Owner's expense, and if any judgment or award against the Owner arises therefrom the Contractor shall pay or satisfy it and shall reimburse the Owner for all attorneys' fees and court or arbitration costs which the Owner has incurred.

6.3 OWNER'S RIGHT TO CLEAN UP

6.3.1 If a dispute arises between the Contractor and separate contractors as to their responsibility for cleaning up as required by Paragraph 4.15, the Owner may clean up

and charge the cost thereof to the contractors responsible therefor as the Architect shall determine to be just.

ARTICLE 7

MISCELLANEOUS PROVISIONS

7.1 GOVERNING LAW

7.1.1 The Contract shall be governed by the law of the place where the Project is located.

7.2 SUCCESSORS AND ASSIGNS

7.2.1 The Owner and the Contractor each binds himself, his partners, successors, assigns and legal representatives to the other party hereto and to the partners, successors, assigns and legal representatives of such other party in respect to all covenants, agreements and obligations contained in the Contract Documents. Neither party to the Contract shall assign the Contract or sublet it as a whole without the written consent of the other, nor shall the Contractor assign any moneys due or to become due to him hereunder, without the previous written consent of the Owner.

7.3 WRITTEN NOTICE

7.3.1 Written notice shall be deemed to have been duly served if delivered in person to the individual or member of the firm or entity or to an officer of the corporation for whom it was intended, or if delivered at or sent by registered or certified mail to the last business address known to him who gives the notice.

7.4 CLAIMS FOR DAMAGES

7.4.1 Should either party to the Contract suffer injury or damage to person or property because of any act or omission of the other party or of any of his employees, agents or others for whose acts he is legally liable, claim shall be made in writing to such other party within a reasonable time after the first observance of such injury or damage.

7.5 PERFORMANCE BOND AND LABOR AND MATERIAL PAYMENT BOND

7.5.1 The Owner shall have the right to require the Contractor to furnish bonds covering the faithful performance of the Contract and the payment of all obligations arising thereunder if and as required in the Bidding Documents or in the Contract Documents.

7.6 RIGHTS AND REMEDIES

7.6.1 The duties and obligations imposed by the Contract Documents and the rights and remedies available thereunder shall be in addition to and not a limitation of any duties, obligations, rights and remedies otherwise imposed or available by law.

7.6.2 No action or failure to act by the Owner, Architect or Contractor shall constitute a waiver of any right or duty afforded any of them under the Contract, nor shall any such action or failure to act constitute an approval of or acquiescence in any breach thereunder, except as may be specifically agreed in writing.

.7 TESTS

.7.1 If the Contract Documents, laws, ordinances, rules, regulations or orders of any public authority having jurisdiction require any portion of the Work to be inspected, tested or approved, the Contractor shall give the Architect timely notice of its readiness so the Architect may observe such inspection, testing or approval. The Contractor shall bear all costs of such inspections, tests or approvals conducted by public authorities. Unless otherwise provided, the Owner shall bear all costs of other inspections, tests or approvals.

.7.2 If the Architect determines that any Work requires special inspection, testing, or approval which Subparagraph 7.7.1 does not include, he will, upon written authorization from the Owner, instruct the Contractor to order such special inspection, testing or approval, and the Contractor shall give notice as provided in Subparagraph .7.1. If such special inspection or testing reveals a failure of the Work to comply with the requirements of the Contract Documents, the Contractor shall bear all costs thereof, including compensation for the Architect's additional services made necessary by such failure; otherwise the Owner shall bear such costs, and an appropriate Change Order shall be issued.

.7.3 Required certificates of inspection, testing or approval shall be secured by the Contractor and promptly delivered by him to the Architect.

.7.4 If the Architect is to observe the inspections, tests or approvals required by the Contract Documents, he will do so promptly and, where practicable, at the source of supply.

.8 INTEREST

.8.1 Payments due and unpaid under the Contract Documents shall bear interest from the date payment is due at such rate as the parties may agree upon in writing or, in the absence thereof, at the legal rate prevailing at the place of the Project.

.9 ARBITRATION

.9.1 All claims, disputes and other matters in question between the Contractor and the Owner arising out of, or relating to, the Contract Documents or the breach thereof, except as provided in Subparagraph 2.2.11 with respect to the Architect's decisions on matters relating to artistic effect, and except for claims which have been waived by the making or acceptance of final payment as provided by Subparagraphs 9.9.4 and 9.9.5, shall be decided by arbitration in accordance with the Construction Industry Arbitration Rules of the American Arbitration Association then obtaining unless the parties mutually agree otherwise. No arbitration arising out of or relating to the Contract Documents shall include, by consolidation, joinder or in any other manner, the Architect, his employees or consultants except by written consent containing a specific reference to the Owner-Contractor Agreement and signed by the Architect, the Owner, the Contractor and any other person sought to be joined. No arbitration shall include by consolidation, joinder or in any other manner, parties other than the Owner, the Contractor and any other persons substantially involved in a common question of fact or law, whose presence is required if complete relief is to be accorded in the arbitration. No person other than the Owner or Contractor shall be included as an original third party or additional third party to an arbitration whose interest or responsibility is insubstantial. Any consent to arbitration involving an additional person or persons shall not constitute consent to arbitration of any dispute not described therein or with any person not named or described therein. The foregoing agreement to arbitrate and any other agreement to arbitrate with an additional person or persons duly consented to by the parties to the Owner-Contractor Agreement shall be specifically enforceable under the prevailing arbitration law. The award rendered by the arbitrators shall be final, and judgment may be entered upon it in accordance with applicable law in any court having jurisdiction thereof.

7.9.2 Notice of the demand for arbitration shall be filed in writing with the other party to the Owner-Contractor Agreement and with the American Arbitration Association, and a copy shall be filed with the Architect. The demand for arbitration shall be made within the time limits specified in Subparagraph 2.2.12 where applicable, and in all other cases within a reasonable time after the claim, dispute or other matter in question has arisen, and in no event shall it be made after the date when institution of legal or equitable proceedings based on such claim, dispute or other matter in question would be barred by the applicable statute of limitations.

7.9.3 Unless otherwise agreed in writing, the Contractor shall carry on the Work and maintain its progress during any arbitration proceedings, and the Owner shall continue to make payments to the Contractor in accordance with the Contract Documents.

ARTICLE 8

TIME

8.1 DEFINITIONS

8.1.1 Unless otherwise provided, the Contract Time is the period of time allotted in the Contract Documents for Substantial Completion of the Work as defined in Subparagraph 8.1.3, including authorized adjustments thereto.

8.1.2 The date of commencement of the Work is the date established in a notice to proceed. If there is no notice to proceed, it shall be the date of the Owner-Contractor Agreement or such other date as may be established therein.

8.1.3 The Date of Substantial Completion of the Work or designated portion thereof is the Date certified by the Architect when construction is sufficiently complete, in accordance with the Contract Documents, so the Owner can occupy or utilize the Work or designated portion thereof for the use for which it is intended.

8.1.4 The term day as used in the Contract Documents shall mean calendar day unless otherwise specifically designated.

8.2 PROGRESS AND COMPLETION

8.2.1 All time limits stated in the Contract Documents are of the essence of the Contract.

8.2.2 The Contractor shall begin the Work on the date of commencement as defined in Subparagraph 8.1.2. He shall carry the Work forward expeditiously with adequate forces and shall achieve Substantial Completion within the Contract Time.

8.3 DELAYS AND EXTENSIONS OF TIME

8.3.1 If the Contractor is delayed at any time in the progress of the Work by any act or neglect of the Owner or the Architect, or by any employee of either, or by any separate contractor employed by the Owner, or by changes ordered in the Work, or by labor disputes, fire, unusual delay in transportation, adverse weather conditions not reasonably anticipatable, unavoidable casualties, or any causes beyond the Contractor's control, or by delay authorized by the Owner pending arbitration, or by any other cause which the Architect determines may justify the delay, then the Contract Time shall be extended by Change Order for such reasonable time as the Architect may determine.

8.3.2 Any claim for extension of time shall be made in writing to the Architect not more than twenty days after the commencement of the delay; otherwise it shall be waived. In the case of a continuing delay only one claim is necessary. The Contractor shall provide an estimate of the probable effect of such delay on the progress of the Work.

8.3.3 If no agreement is made stating the dates upon which interpretations as provided in Subparagraph 2.2.8 shall be furnished, then no claim for delay shall be allowed on account of failure to furnish such interpretations until fifteen days after written request is made for them, and not then unless such claim is reasonable.

8.3.4 This Paragraph 8.3 does not exclude the recovery of damages for delay by either party under other provisions of the Contract Documents.

ARTICLE 9

PAYMENTS AND COMPLETION

9.1 CONTRACT SUM

9.1.1 The Contract Sum is stated in the Owner-Contractor Agreement and, including authorized adjustments thereto, is the total amount payable by the Owner to the Contractor for the performance of the Work under the Contract Documents.

9.2 SCHEDULE OF VALUES

9.2.1 Before the first Application for Payment, the Contractor shall submit to the Architect a schedule of values allocated to the various portions of the Work, prepared in such form and supported by such data to substantiate its accuracy as the Architect may require. This schedule, unless objected to by the Architect, shall be used only as a basis for the Contractor's Applications for Payment.

9.3 APPLICATIONS FOR PAYMENT

9.3.1 At least ten days before the date for each progress payment established in the Owner-Contractor Agreement, the Contractor shall submit to the Architect an itemized Application for Payment, notarized if required, supported by such data substantiating the Contractor's right to payment as the Owner or the Architect may require, and reflecting retainage, if any, as provided elsewhere in the Contract Documents.

9.3.2 Unless otherwise provided in the Contract Documents, payments will be made on account of materials or equipment not incorporated in the Work but delivered and suitably stored at the site and, if approved in advance by the Owner, payments may similarly be made for materials or equipment suitably stored at some other location agreed upon in writing. Payments for materials or equipment stored on or off the site shall be conditioned upon submission by the Contractor of bills of sale or such other procedures satisfactory to the Owner to establish the Owner's title to such materials or equipment or otherwise protect the Owner's interest, including applicable insurance and transportation to the site for those materials and equipment stored off the site.

9.3.3 The Contractor warrants that title to all Work, materials and equipment covered by an Application for Payment will pass to the Owner either by incorporation in the construction or upon the receipt of payment by the Contractor, whichever occurs first, free and clear of all liens, claims, security interests or encumbrances, hereinafter referred to in this Article 9 as "liens"; and that no Work, materials or equipment covered by an Application for Payment will have been acquired by the Contractor, or by any other person performing Work at the site or furnishing materials and equipment for the Project, subject to an agreement under which an interest therein or an encumbrance thereon is retained by the seller or otherwise imposed by the Contractor or such other person.

9.4 CERTIFICATES FOR PAYMENT

9.4.1 The Architect will, within seven days after the receipt of the Contractor's Application for Payment, either issue a Certificate for Payment to the Owner, with a copy to the Contractor, for such amount as the Architect determines is properly due, or notify the Contractor in writing his reasons for withholding a Certificate as provided in Subparagraph 9.6.1.

9.4.2 The issuance of a Certificate for Payment will constitute a representation by the Architect to the Owner, based on his observations at the site as provided in Subparagraph 2.2.3 and the data comprising the Application for Payment, that the Work has progressed to the point indicated; that, to the best of his knowledge, information and belief, the quality of the Work is in accordance with the Contract Documents (subject to an evaluation of the Work for conformance with the Contract Documents upon Substantial Completion, to the results of any subsequent tests required by or performed under the Contract Documents, to minor deviations from the Contract Documents correctable prior to completion, and to any specific qualifications stated in his Certificate); and that the Contractor is entitled to payment in the amount certified. However, by issuing a Certificate for Payment, the Architect shall not thereby be deemed to represent that he has made exhaustive or continuous on-site inspections to check the quality or quantity of the Work or that he has reviewed the construction means, methods, techniques, sequences or procedures, or that he has made any examination to ascertain how or for what purpose the Contractor has used the moneys previously paid on account of the Contract Sum.

9.5 PROGRESS PAYMENTS

9.5.1 After the Architect has issued a Certificate for Payment, the Owner shall make payment in the manner and within the time provided in the Contract Documents.

9.5.2 The Contractor shall promptly pay each Subcontractor, upon receipt of payment from the Owner, out of the amount paid to the Contractor on account of such Subcontractor's Work, the amount to which said Subcontractor is entitled, reflecting the percentage actually retained, if any, from payments to the Contractor on account of such Subcontractor's Work. The Contractor shall, by an appropriate agreement with each Subcontractor, require each Subcontractor to make payments to his Subsubcontractors in similar manner.

9.5.3 The Architect may, on request and at his discretion, furnish to any Subcontractor, if practicable, information regarding the percentages of completion or the amounts applied for by the Contractor and the action taken thereon by the Architect on account of Work done by such Subcontractor.

9.5.4 Neither the Owner nor the Architect shall have any obligation to pay or to see to the payment of any moneys to any Subcontractor except as may otherwise be required by law.

9.5.5 No Certificate for a progress payment, nor any progress payment, nor any partial or entire use or occupancy of the Project by the Owner, shall constitute an acceptance of any Work not in accordance with the Contract Documents.

9.6 PAYMENTS WITHHELD

9.6.1 The Architect may decline to certify payment and may withhold his Certificate in whole or in part, to the extent necessary reasonably to protect the Owner, if in his opinion he is unable to make representations to the Owner as provided in Subparagraph 9.4.2. If the Architect is unable to make representations to the Owner as provided in Subparagraph 9.4.2 and to certify payment in the amount of the Application, he will notify the Contractor as provided in Subparagraph 9.4.1. If the Contractor and the Architect cannot agree on a revised amount, the Architect will promptly issue a Certificate for Payment for the amount for which he is able to make such representations to the Owner. The Architect may also decline to certify payment or, because of subsequently discovered evidence or subsequent observations, he may nullify the whole or any part of any Certificate for Payment previously issued, to such extent as may be necessary in his opinion to protect the Owner from loss because of:

.1 defective work not remedied,

.2 third party claims filed or reasonable evidence indicating probable filing of such claims,

.3 failure of the Contractor to make payments properly to Subcontractors or for labor, materials or equipment,

.4 reasonable evidence that the Work cannot be completed for the unpaid balance of the Contract Sum,

.5 damage to the Owner or another contractor,

.6 reasonable evidence that the Work will not be completed within the Contract Time, or

.7 persistent failure to carry out the Work in accordance with the Contract Documents.

9.6.2 When the above grounds in Subparagraph 9.6.1 are removed, payment shall be made for amounts withheld because of them.

9.7 FAILURE OF PAYMENT

9.7.1 If the Architect does not issue a Certificate for Payment, through no fault of the Contractor, within seven days after receipt of the Contractor's Application for Payment, or if the Owner does not pay the Contractor within seven days after the date established in the Contract Documents any amount certified by the Architect or awarded by arbitration, then the Contractor may, upon seven additional days' written notice to the Owner and the Architect, stop the Work until payment of the amount owing has been received. The Contract Sum shall be increased by the amount of the Contractor's reasonable costs of shut-down, delay and start-up, which shall be effected by appropriate Change Order in accordance with Paragraph 12.3.

9.8 SUBSTANTIAL COMPLETION

9.8.1 When the Contractor considers that the Work, or a designated portion thereof which is acceptable to the Owner, is substantially complete as defined in Subparagraph 8.1.3, the Contractor shall prepare for submission to the Architect a list of items to be completed or corrected. The failure to include any items on such list does not alter the responsibility of the Contractor to complete all Work in accordance with the Contract Documents. When the Architect on the basis of an inspection determines that the Work or designated portion thereof is substantially complete, he will then prepare a Certificate of Substantial Completion which shall establish the Date of Substantial Completion, shall state the responsibilities of the Owner and the Contractor for security, maintenance, heat, utilities, damage to the Work, and insurance, and shall fix the time within which the Contractor shall complete the items listed therein. Warranties required by the Contract Documents shall commence on the Date of Substantial Completion of the Work or designated portion thereof unless otherwise provided in the Certificate of Substantial Completion. The Certificate of Substantial Completion shall be submitted to the Owner and the Contractor for their written acceptance of the responsibilities assigned to them in such Certificate.

9.8.2 Upon Substantial Completion of the Work or designated portion thereof and upon application by the Contractor and certification by the Architect, the Owner shall make payment, reflecting adjustment in retainage, if any, for such Work or portion thereof, as provided in the Contract Documents.

9.9 FINAL COMPLETION AND FINAL PAYMENT

9.9.1 Upon receipt of written notice that the Work is ready for final inspection and acceptance and upon receipt of a final Application for Payment, the Architect will promptly make such inspection and, when he finds the Work acceptable under the Contract Documents and the Contract fully performed, he will promptly issue a final Certificate for Payment stating that to the best of his knowledge, information and belief, and on the basis of his observations and inspections, the Work has been completed in accordance with the terms and conditions of the Contract Documents and that the entire balance found to be due the Contractor, and noted in said final Certificate, is due and payable. The Architect's final Certificate for Payment will constitute a further representation that the conditions precedent to the Contractor's being entitled to final payment as set forth in Subparagraph 9.9.2 have been fulfilled.

9.9.2 Neither the final payment nor the remaining retained percentage shall become due until the Contractor submits to the Architect (1) an affidavit that all payrolls, bills for materials and equipment, and other indebtedness connected with the Work for which the Owner or his property might in any way be responsible, have been paid or otherwise satisfied, (2) consent of surety, if any, to final payment and (3), if required by the Owner, other data establishing payment or satisfaction of all such obligations, such as receipts, releases and waivers of liens arising out of the Contract, to the extent and in such form as may be designated by the Owner. If any Subcontractor refuses to furnish a release or waiver required by the Owner, the Contractor may furnish a bond satisfactory to the Owner to indemnify him against any such lien. If any such lien remains unsatisfied after all payments are made, the Contractor shall refund to the Owner all moneys that the latter may be compelled to pay in discharging such lien, including all costs and reasonable attorneys' fees.

9.9.3 If, after Substantial Completion of the Work, final completion thereof is materially delayed through no fault of the Contractor or by the issuance of Change Orders affecting final completion, and the Architect so confirms, the Owner shall, upon application by the Contractor and certification by the Architect, and without terminating the Contract, make payment of the balance due for that portion of the Work fully completed and accepted. If the remaining balance for Work not fully completed or corrected is less than the retainage stipulated in the Contract Documents, and if bonds have been furnished as provided in Paragraph 7.5, the written consent of the surety to the payment of the balance due for that portion of the Work fully completed and accepted shall be submitted by the Contractor to the Architect prior to certification of such payment. Such payment shall be made under the terms and conditions governing final payment, except that it shall not constitute a waiver of claims.

9.9.4 The making of final payment shall constitute a waiver of all claims by the Owner except those arising from:
    .1 unsettled liens,
    .2 faulty or defective Work appearing after Substantial Completion,
    .3 failure of the Work to comply with the requirements of the Contract Documents, or
    .4 terms of any special warranties required by the Contract Documents.

9.9.5 The acceptance of final payment shall constitute a waiver of all claims by the Contractor except those previously made in writing and identified by the Contractor as unsettled at the time of the final Application for Payment.

ARTICLE 10

PROTECTION OF PERSONS AND PROPERTY

10.1 SAFETY PRECAUTIONS AND PROGRAMS

10.1.1 The Contractor shall be responsible for initiating, maintaining and supervising all safety precautions and programs in connection with the Work.

10.2 SAFETY OF PERSONS AND PROPERTY

10.2.1 The Contractor shall take all reasonable precautions for the safety of, and shall provide all reasonable protection to prevent damage, injury or loss to:
    .1 all employees on the Work and all other persons who may be affected thereby;
    .2 all the Work and all materials and equipment to be incorporated therein, whether in storage on or off the site, under the care, custody or control of the Contractor or any of his Subcontractors or Sub-subcontractors; and
    .3 other property at the site or adjacent thereto, including trees, shrubs, lawns, walks, pavements, roadways, structures and utilities not designated for removal, relocation or replacement in the course of construction.

10.2.2 The Contractor shall give all notices and comply with all applicable laws, ordinances, rules, regulations and lawful orders of any public authority bearing on the safety of persons or property or their protection from damage, injury or loss.

10.2.3 The Contractor shall erect and maintain, as required by existing conditions and progress of the Work, all reasonable safeguards for safety and protection, including posting danger signs and other warnings against hazards, promulgating safety regulations and notifying owners and users of adjacent utilities.

10.2.4 When the use or storage of explosives or other hazardous materials or equipment is necessary for the execution of the Work, the Contractor shall exercise the utmost care and shall carry on such activities under the supervision of properly qualified personnel.

10.2.5 The Contractor shall promptly remedy all damage or loss (other than damage or loss insured under Paragraph 11.3) to any property referred to in Clauses 10.2.1.2 and 10.2.1.3 caused in whole or in part by the Contractor, any Subcontractor, any Sub-subcontractor, or anyone directly or indirectly employed by any of them, or by anyone for whose acts any of them may be liable and for which the Contractor is responsible under Clauses 10.2.1.2 and 10.2.1.3, except damage or loss attributable to the acts or omissions of the Owner or Architect or anyone directly or indirectly employed by either of them, or by anyone for whose acts either of them may be liable, and not attributable to the fault or negligence of the Contractor. The foregoing obligations of the Contractor are in addition to his obligations under Paragraph 4.18.

10.2.6 The Contractor shall designate a responsible member of his organization at the site whose duty shall be the prevention of accidents. This person shall be the Contractor's superintendent unless otherwise designated by the Contractor in writing to the Owner and the Architect.

10.2.7 The Contractor shall not load or permit any part of the Work to be loaded so as to endanger its safety.

10.3 EMERGENCIES

10.3.1 In any emergency affecting the safety of persons or property, the Contractor shall act, at his discretion, to prevent threatened damage, injury or loss. Any additional compensation or extension of time claimed by the Contractor on account of emergency work shall be determined as provided in Article 12 for Changes in the Work.

ARTICLE 11

INSURANCE

11.1 CONTRACTOR'S LIABILITY INSURANCE

11.1.1 The Contractor shall purchase and maintain such insurance as will protect him from claims set forth below which may arise out of or result from the Contractor's operations under the Contract, whether such operations be by himself or by any Subcontractor or by anyone directly or indirectly employed by any of them, or by anyone for whose acts any of them may be liable:

.1 claims under workers' or workmen's compensation, disability benefit and other similar employee benefit acts;

.2 claims for damages because of bodily injury, occupational sickness or disease, or death of his employees;

.3 claims for damages because of bodily injury, sickness or disease, or death of any person other than his employees;

.4 claims for damages insured by usual personal injury liability coverage which are sustained (1) by any person as a result of an offense directly or indirectly related to the employment of such person by the Contractor, or (2) by any other person;

.5 claims for damages, other than to the Work itself, because of injury to or destruction of tangible property, including loss of use resulting therefrom; and

.6 claims for damages because of bodily injury or death of any person or property damage arising out of the ownership, maintenance or use of any motor vehicle.

11.1.2 The insurance required by Subparagraph 11.1.1 shall be written for not less than any limits of liability specified in the Contract Documents, or required by law, whichever is greater.

11.1.3 The insurance required by Subparagraph 11.1.1 shall include contractual liability insurance applicable to the Contractor's obligations under Paragraph 4.18.

11.1.4 Certificates of Insurance acceptable to the Owner shall be filed with the Owner prior to commencement of the Work. These Certificates shall contain a provision that coverages afforded under the policies will not be cancelled until at least thirty days' prior written notice has been given to the Owner.

11.2 OWNER'S LIABILITY INSURANCE

11.2.1 The Owner shall be responsible for purchasing and maintaining his own liability insurance and, at his option, may purchase and maintain such insurance as will protect him against claims which may arise from operations under the Contract.

11.3 PROPERTY INSURANCE

11.3.1 Unless otherwise provided, the Owner shall purchase and maintain property insurance upon the entire Work at the site to the full insurable value thereof. This insurance shall include the interests of the Owner, the Contractor, Subcontractors and Sub-subcontractors in the Work and shall insure against the perils of fire and extended coverage and shall include "all risk" insurance for physical loss or damage including, without duplication of coverage, theft, vandalism and malicious mischief. If the Owner does not intend to purchase such insurance for the full insurable value of the entire Work, he shall inform the Contractor in writing prior to commencement of the Work. The Contractor may then effect insurance which will protect the interests of himself, his Subcontractors and the Sub-subcontractors in the Work, and by appropriate Change Order the cost thereof shall be charged to the Owner. If the Contractor is damaged by failure of the Owner to purchase or maintain such insurance and to so notify the Contractor, then the Owner shall bear all reasonable costs properly attributable thereto. If not covered under the all risk insurance or otherwise provided in the Contract Documents, the Contractor shall effect and maintain similar property insurance on portions of the Work stored off the site or in transit when such portions of the Work are to be included in an Application for Payment under Subparagraph 9.3.2.

11.3.2 The Owner shall purchase and maintain such boiler and machinery insurance as may be required by the Contract Documents or by law. This insurance shall include the interests of the Owner, the Contractor, Subcontractors and Sub-subcontractors in the Work.

11.3.3 Any loss insured under Subparagraph 11.3.1 is to be adjusted with the Owner and made payable to the Owner as trustee for the insureds, as their interests may appear, subject to the requirements of any applicable mortgagee clause and of Subparagraph 11.3.8. The Contractor shall pay each Subcontractor a just share of any insurance moneys received by the Contractor, and by appropriate agreement, written where legally required for validity, shall require each Subcontractor to make payments to his Sub-subcontractors in similar manner.

11.3.4 The Owner shall file a copy of all policies with the Contractor before an exposure to loss may occur.

11.3.5 If the Contractor requests in writing that insurance for risks other than those described in Subparagraphs 11.3.1 and 11.3.2 or other special hazards be included in the property insurance policy, the Owner shall, if possible, include such insurance, and the cost thereof shall be charged to the Contractor by appropriate Change Order.

11.3.6 The Owner and Contractor waive all rights against (1) each other and the Subcontractors, Sub-subcontractors, agents and employees each of the other, and (2) the Architect and separate contractors, if any, and their subcontractors, sub-subcontractors, agents and employees, for damages caused by fire or other perils to the extent covered by insurance obtained pursuant to this Paragraph 11.3 or any other property insurance applicable to the Work, except such rights as they may have to the proceeds of such insurance held by the Owner as trustee. The foregoing waiver afforded the Architect, his agents and employees shall not extend to the liability imposed by Subparagraph 4.18.3. The Owner or the Contractor, as appropriate, shall require of the Architect, separate contractors, Subcontractors and Sub-subcontractors by appropriate agreements, written where legally required for validity, similar waivers each in favor of all other parties enumerated in this Subparagraph 11.3.6.

11.3.7 If required in writing by any party in interest, the Owner as trustee shall, upon the occurrence of an insured loss, give bond for the proper performance of his duties. He shall deposit in a separate account any money so received, and he shall distribute it in accordance with such agreement as the parties in interest may reach, or in accordance with an award by arbitration in which case the procedure shall be as provided in Paragraph 7.9. If after such loss no other special agreement is made, replacement of damaged work shall be covered by an appropriate Change Order.

11.3.8 The Owner as trustee shall have power to adjust and settle any loss with the insurers unless one of the parties in interest shall object in writing within five days after the occurrence of loss to the Owner's exercise of this power, and if such objection be made, arbitrators shall be chosen as provided in Paragraph 7.9. The Owner as trustee shall, in that case, make settlement with the insurers in accordance with the directions of such arbitrators. If distribution of the insurance proceeds by arbitration is required, the arbitrators will direct such distribution.

11.3.9 If the Owner finds it necessary to occupy or use a portion or portions of the Work prior to Substantial Completion thereof, such occupancy or use shall not commence prior to a time mutually agreed to by the Owner and Contractor and to which the insurance company or companies providing the property insurance have consented by endorsement to the policy or policies. This insurance shall not be cancelled or lapsed on account of such partial occupancy or use. Consent of the Contractor and of the insurance company or companies to such occupancy or use shall not be unreasonably withheld.

11.4 LOSS OF USE INSURANCE
11.4.1 The Owner, at his option, may purchase and maintain such insurance as will insure him against loss of use of his property due to fire or other hazards, however caused. The Owner waives all rights of action against the Contractor for loss of use of his property, including consequential losses due to fire or other hazards however caused, to the extent covered by insurance under this Paragraph 11.4.

ARTICLE 12

CHANGES IN THE WORK
12.1 CHANGE ORDERS

12.1.1 A Change Order is a written order to the Contractor signed by the Owner and the Architect, issued after execution of the Contract, authorizing a change in the Work or an adjustment in the Contract Sum or the Contract Time. The Contract Sum and the Contract Time may be changed only by Change Order. A Change Order signed by the Contractor indicates his agreement therewith, including the adjustment in the Contract Sum or the Contract Time.

12.1.2 The Owner, without invalidating the Contract, may order changes in the Work within the general scope of the Contract consisting of additions, deletions or other revisions, the Contract Sum and the Contract Time being adjusted accordingly. All such changes in the Work shall be authorized by Change Order, and shall be performed under the applicable conditions of the Contract Documents.

12.1.3 The cost or credit to the Owner resulting from a change in the Work shall be determined in one or more of the following ways:
.1 by mutual acceptance of a lump sum properly itemized and supported by sufficient substantiating data to permit evaluation;
.2 by unit prices stated in the Contract Documents or subsequently agreed upon;
.3 by cost to be determined in a manner agreed upon by the parties and a mutually acceptable fixed or percentage fee; or
.4 by the method provided in Subparagraph 12.1.4.

12.1.4 If none of the methods set forth in Clauses 12.1.3.1, 12.1.3.2 or 12.1.3.3 is agreed upon, the Contractor, provided he receives a written order signed by the Owner, shall promptly proceed with the Work involved. The cost of such Work shall then be determined by the Architect on the basis of the reasonable expenditures and savings of those performing the Work attributable to the change, including, in the case of an increase in the Contract Sum, a reasonable allowance for overhead and profit. In such case, and also under Clauses 12.1.3.3 and 12.1.3.4 above, the Contractor shall keep and present, in such form as the Architect may prescribe, an itemized accounting together with appropriate supporting data for inclusion in a Change Order. Unless otherwise provided in the Contract Documents, cost shall be limited to the following: cost of materials, including sales tax and cost of delivery; cost of labor, including social security, old age and unemployment insurance, and fringe benefits required by agreement or custom; workers' or workmen's compensation insurance; bond premiums; rental value of equipment and machinery; and the additional costs of supervision and field office personnel directly attributable to the change. Pending final determination of cost to the Owner, payments on account shall be made on the Architect's Certificate for Payment. The amount of credit to be allowed by the Contractor to the Owner for any deletion or change which results in a net decrease in the Contract Sum will be the amount of the actual net cost as confirmed by the Architect. When both additions and credits covering related Work or substitutions are involved in any one change, the allowance for overhead and profit shall be figured on the basis of the net increase, if any, with respect to that change.

12.1.5 If unit prices are stated in the Contract Documents or subsequently agreed upon, and if the quantities originally contemplated are so changed in a proposed Change Order that application of the agreed unit prices to the quantities of Work proposed will cause substantial inequity to the Owner or the Contractor, the applicable unit prices shall be equitably adjusted.

12.2 CONCEALED CONDITIONS

12.2.1 Should concealed conditions encountered in the performance of the Work below the surface of the ground or should concealed or unknown conditions in an existing structure be at variance with the conditions indicated by the Contract Documents, or should unknown physical conditions below the surface of the ground or should concealed or unknown conditions in an existing structure of an unusual nature, differing materially from those ordinarily encountered and generally recognized as inherent in work of the character provided for in this Contract, be encountered, the Contract Sum shall be equitably adjusted by Change Order upon claim by either party made within twenty days after the first observance of the conditions.

12.3 CLAIMS FOR ADDITIONAL COST

12.3.1 If the Contractor wishes to make a claim for an increase in the Contract Sum, he shall give the Architect written notice thereof within twenty days after the occurrence of the event giving rise to such claim. This notice shall be given by the Contractor before proceeding to execute the Work, except in an emergency endangering life or property in which case the Contractor shall proceed in accordance with Paragraph 10.3. No such claim shall be valid unless so made. If the Owner and the Contractor cannot agree on the amount of the adjustment in the Contract Sum, it shall be determined by the Architect. Any change in the Contract Sum resulting from such claim shall be authorized by Change Order.

12.3.2 If the Contractor claims that additional cost is involved because of, but not limited to, (1) any written interpretation pursuant to Subparagraph 2.2.8, (2) any order by the Owner to stop the Work pursuant to Paragraph 3.3 where the Contractor was not at fault, (3) any written order for a minor change in the Work issued pursuant to Paragraph 12.4, or (4) failure of payment by the Owner pursuant to Paragraph 9.7, the Contractor shall make such claim as provided in Subparagraph 12.3.1.

12.4 MINOR CHANGES IN THE WORK

12.4.1 The Architect will have authority to order minor changes in the Work not involving an adjustment in the Contract Sum or an extension of the Contract Time and not inconsistent with the intent of the Contract Documents. Such changes shall be effected by written order, and shall be binding on the Owner and the Contractor.

The Contractor shall carry out such written orders promptly.

ARTICLE 13

UNCOVERING AND CORRECTION OF WORK

13.1 UNCOVERING OF WORK

13.1.1 If any portion of the Work should be covered contrary to the request of the Architect or to requirements specifically expressed in the Contract Documents, it must, if required in writing by the Architect, be uncovered for his observation and shall be replaced at the Contractor's expense.

13.1.2 If any other portion of the Work has been covered which the Architect has not specifically requested to observe prior to being covered, the Architect may request to see such Work and it shall be uncovered by the Contractor. If such Work be found in accordance with the Contract Documents, the cost of uncovering and replacement shall, by appropriate Change Order, be charged to the Owner. If such Work be found not in accordance with the Contract Documents, the Contractor shall pay such costs unless it be found that this condition was caused by the Owner or a separate contractor as provided in Article 6, in which event the Owner shall be responsible for the payment of such costs.

13.2 CORRECTION OF WORK

13.2.1 The Contractor shall promptly correct all Work rejected by the Architect as defective or as failing to conform to the Contract Documents whether observed before or after Substantial Completion and whether or not fabricated, installed or completed. The Contractor shall bear all costs of correcting such rejected Work, including compensation for the Architect's additional services made necessary thereby.

13.2.2 If, within one year after the Date of Substantial Completion of the Work or designated portion thereof or within one year after acceptance by the Owner of designated equipment or within such longer period of time as may be prescribed by law or by the terms of any applicable special warranty required by the Contract Documents, any of the Work is found to be defective or not in accordance with the Contract Documents, the Contractor shall correct it promptly after receipt of a written notice from the Owner to do so unless the Owner has previously given the Contractor a written acceptance of such condition. This obligation shall survive termination of the Contract. The Owner shall give such notice promptly after discovery of the condition.

13.2.3 The Contractor shall remove from the site all portions of the Work which are defective or non-conforming and which have not been corrected under Subparagraphs 4.5.1, 13.2.1 and 13.2.2, unless removal is waived by the Owner.

13.2.4 If the Contractor fails to correct defective or non-conforming Work as provided in Subparagraphs 4.5.1, 13.2.1 and 13.2.2, the Owner may correct it in accordance with Paragraph 3.4.

13.2.5 If the Contractor does not proceed with the correction of such defective or non-conforming Work within a reasonable time fixed by written notice from the Architect, the Owner may remove it and may store the materials or equipment at the expense of the Contractor. If the Contractor does not pay the cost of such removal and storage within ten days thereafter, the Owner may upon ten additional days' written notice sell such Work at auction or at private sale and shall account for the net proceeds thereof, after deducting all the costs that should have been borne by the Contractor, including compensation for the Architect's additional services made necessary thereby. If such proceeds of sale do not cover all costs which the Contractor should have borne, the difference shall be charged to the Contractor and an appropriate Change Order shall be issued. If the payments then or thereafter due the Contractor are not sufficient to cover such amount, the Contractor shall pay the difference to the Owner.

13.2.6 The Contractor shall bear the cost of making good all work of the Owner or separate contractors destroyed or damaged by such correction or removal.

13.2.7 Nothing contained in this Paragraph 13.2 shall be construed to establish a period of limitation with respect to any other obligation which the Contractor might have under the Contract Documents, including Paragraph 4.5 hereof. The establishment of the time period of one year after the Date of Substantial Completion or such longer period of time as may be prescribed by law or by the terms of any warranty required by the Contract Documents relates only to the specific obligation of the Contractor to correct the Work, and has no relationship to the time within which his obligation to comply with the Contract Documents may be sought to be enforced, nor to the time within which proceedings may be commenced to establish the Contractor's liability with respect to his obligations other than specifically to correct the Work.

13.3 ACCEPTANCE OF DEFECTIVE OR NON-CONFORMING WORK

13.3.1 If the Owner prefers to accept defective or non-conforming Work, he may do so instead of requiring its removal and correction, in which case a Change Order will be issued to reflect a reduction in the Contract Sum where appropriate and equitable. Such adjustment shall be effected whether or not final payment has been made.

ARTICLE 14

TERMINATION OF THE CONTRACT

14.1 TERMINATION BY THE CONTRACTOR

14.1.1 If the Work is stopped for a period of thirty days under an order of any court or other public authority having jurisdiction, or as a result of an act of government, such as a declaration of a national emergency making materials unavailable, through no act or fault of the Contractor or a Subcontractor or their agents or employees or any other persons performing any of the Work under a contract with the Contractor, or if the Work should be stopped for a period of thirty days by the Contractor because the Architect has not issued a Certificate for Payment as provided in Paragraph 9.7 or because the Owner has not made payment thereon as provided in Paragraph 9.7, then the Contractor may, upon seven additional days' written notice to the Owner and the Architect, terminate the Contract and recover from the Owner payment for all Work executed and for any proven loss sustained upon any materials, equipment, tools, construction equipment and machinery, including reasonable profit and damages.

14.2 TERMINATION BY THE OWNER

14.2.1 If the Contractor is adjudged a bankrupt, or if he makes a general assignment for the benefit of his creditors, or if a receiver is appointed on account of his insolvency, or if he persistently or repeatedly refuses or fails, except in cases for which extension of time is provided, to supply enough properly skilled workmen or proper materials, or if he fails to make prompt payment to Subcontractors or for materials or labor, or persistently disregards laws, ordinances, rules, regulations or orders of any public authority having jurisdiction, or otherwise is guilty of a substantial violation of a provision of the Contract Documents, then the Owner, upon certification by the Architect that sufficient cause exists to justify such action, may, without prejudice to any right or remedy and after giving the Contractor and his surety, if any, seven days' written notice, terminate the employment of the Contractor and take possession of the site and of all materials, equipment, tools, construction equipment and machinery thereon owned by the Contractor and may finish the Work by whatever method he may deem expedient. In such case the Contractor shall not be entitled to receive any further payment until the Work is finished.

14.2.2 If the unpaid balance of the Contract Sum exceeds the costs of finishing the Work, including compensation for the Architect's additional services made necessary thereby, such excess shall be paid to the Contractor. If such costs exceed the unpaid balance, the Contractor shall pay the difference to the Owner. The amount to be paid to the Contractor or to the Owner, as the case may be, shall be certified by the Architect, upon application, in the manner provided in Paragraph 9.4, and this obligation for payment shall survive the termination of the Contract.

THE AMERICAN INSTITUTE OF ARCHITECTS

AIA Document A101

Standard Form of Agreement Between Owner and Contractor

where the basis of payment is a

STIPULATED SUM

1977 EDITION

*THIS DOCUMENT HAS IMPORTANT LEGAL CONSEQUENCES; CONSULTATION WITH
AN ATTORNEY IS ENCOURAGED WITH RESPECT TO ITS COMPLETION OR MODIFICATION*

Use only with the 1976 Edition of AIA Document A201, General Conditions of the Contract for Construction.

This document has been approved and endorsed by The Associated General Contractors of America.

AGREEMENT

made as of the   day of   in the year of Nineteen
Hundred and

BETWEEN the Owner:

and the Contractor:

The Project:

The Architect:

The Owner and the Contractor agree as set forth below.

ARTICLE 1

THE CONTRACT DOCUMENTS

The Contract Documents consist of this Agreement, the Conditions of the Contract (General, Supplementary and other Conditions), the Drawings, the Specifications, all Addenda issued prior to and all Modifications issued after execution of this Agreement. These form the Contract, and all are as fully a part of the Contract as if attached to this Agreement or repeated herein. An enumeration of the Contract Documents appears in Article 7.

ARTICLE 2

THE WORK

The Contractor shall perform all the Work required by the Contract Documents for
(Here insert the caption descriptive of the Work as used on other Contract Documents.)

ARTICLE 3

TIME OF COMMENCEMENT AND SUBSTANTIAL COMPLETION

The Work to be performed under this Contract shall be commenced

and, subject to authorized adjustments, Substantial Completion shall be achieved not later than

(Here insert any special provisions for liquidated damages relating to failure to complete on time.)

ARTICLE 4

CONTRACT SUM

The Owner shall pay the Contractor in current funds for the performance of the Work, subject to additions and deductions by Change Order as provided in the Contract Documents, the Contract Sum of

The Contract Sum is determined as follows:
(State here the base bid or other lump sum amount, accepted alternates, and unit prices, as applicable.)

ARTICLE 5

PROGRESS PAYMENTS

Based upon Applications for Payment submitted to the Architect by the Contractor and Certificates for Payment issued by the Architect, the Owner shall make progress payments on account of the Contract Sum to the Contractor as provided in the Contract Documents for the period ending the   day of the month as follows:

Not later than   days following the end of the period covered by the Application for Payment   percent (%) of the portion of the Contract Sum properly allocable to labor, materials and equipment incorporated in the Work and   percent (%) of the portion of the Contract Sum properly allocable to materials and equipment suitably stored at the site or at some other location agreed upon in writing, for the period covered by the Application for Payment, less the aggregate of previous payments made by the Owner; and upon Substantial Completion of the entire Work, a sum sufficient to increase the total payments to   percent (%) of the Contract Sum, less such amounts as the Architect shall determine for all incomplete Work and unsettled claims as provided in the Contract Documents.

(If not covered elsewhere in the Contract Documents, here insert any provision for limiting or reducing the amount retained after the Work reaches a certain stage of completion.)

Payments due and unpaid under the Contract Documents shall bear interest from the date payment is due at the rate entered below, or in the absence thereof, at the legal rate prevailing at the place of the Project.
(Here insert any rate of interest agreed upon.)

Usury laws and requirements under the Federal Truth in Lending Act, similar state and local consumer credit laws and other regulations at the Owner's and Contractor's principal places of business, the location of the Project and elsewhere may affect the validity of this provision. Specific legal advice should be obtained with respect to deletion, modification, or other requirements such as written disclosures or waivers.)

ARTICLE 6

FINAL PAYMENT

Final payment, constituting the entire unpaid balance of the Contract Sum, shall be paid by the Owner to the Contractor when the Work has been completed, the Contract fully performed, and a final Certificate for Payment has been issued by the Architect.

ARTICLE 7

MISCELLANEOUS PROVISIONS

7.1 Terms used in this Agreement which are defined in the Conditions of the Contract shall have the meanings designated in those Conditions.

7.2 The Contract Documents, which constitute the entire agreement between the Owner and the Contractor, are listed in Article 1 and, except for Modifications issued after execution of this Agreement, are enumerated as follows:

(List below the Agreement, the Conditions of the Contract (General, Supplementary, and other Conditions), the Drawings, the Specifications, and any Addenda and accepted alternates, showing page or sheet numbers in all cases and dates where applicable.)

This Agreement entered into as of the day and year first written above.

OWNER   CONTRACTOR

_____   _____

_____   _____

BY   BY

_____   _____

APPENDIX *B*

FORM OF CONTRACT FOR ENGINEERING CONSTRUCTION PROJECTS 1966 EDITION

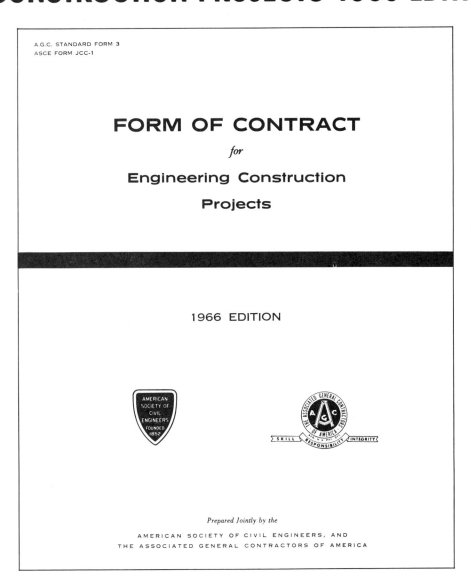

A.G.C. STANDARD FORM 3
ASCE FORM JCC-1

FORM OF CONTRACT

for

Engineering Construction

Projects

1966 EDITION

Prepared Jointly by the
AMERICAN SOCIETY OF CIVIL ENGINEERS, AND
THE ASSOCIATED GENERAL CONTRACTORS OF AMERICA

Courtesy of The Associated General Contractors of America, 1957 E Street, N.W., Washington, D.C. 20006, AGC Standard Form No. 3.; and American Society of Civil Engineers, Form JCC-1.

FOREWORD

These Agreement Forms and the Standard General Conditions of Contract which accompany them pertain to Engineering Construction Contracts between Private or Public Owners and Contractors. They are intended as a guide in the preparation of Contract Documents for such construction and are subject to change or modification to suit particular conditions.

Ordinarily, a complete set of Contract Documents consists of the following:

(a) Advertisement, or Notice to Contractors

(b) Instructions to Bidders

(c) Form of Bid or Proposal

(d) Contract (Agreement)

(e) General and Special Conditions of Contract

(f) Specifications

    1. General

    2. Special

(g) Drawings

The first three items are essential to all contracts that are bid competitively. However, the last four are those essential to the actual Contract. The Forms and Standards contained herein pertain to Items (d) and (e).

The Contract Forms cover three alternative bases of payment to the Contractor:

1. A Lump Sum Basis

2. A Unit Price Basis

3. A Cost-Plus Basis

The Contracting Parties will select the desired form.

The Contract Form for a Cost-Plus Basis of Payment is based upon a payment consisting of Cost Plus a Percentage Fee. Two other types are commonly used: Cost Plus a Fixed Fee and Cost Plus a Percentage or Fixed Fee with a guaranteed cost limit. The text of Articles III and IV can easily be adapted to either of the other two types of Cost-Plus bases of payment. Sometimes, in a Cost-Plus Contract with a guaranteed cost limit, provision is made for dividing equally between the Owner and Contractor any savings below the cost limit that may be achieved in the execution of the work.

FORM OF AGREEMENT FOR ENGINEERING CONSTRUCTION
UNIT PRICE BASIS

THIS AGREEMENT, made on the _____ day of _____ , 19 ____ ,

by and between _____

party of the first part, hereinafter called the OWNER, and _____ _____

_____

_____

party of the second part, hereinafter called the CONTRACTOR.

It is understood ENGINEER representing Owner shall be _____

_____

_____

WITNESSETH, That the Contractor and the Owner, for the considerations hereinafter named, agree as follows:

ARTICLE I — Scope of the Work

The Contractor hereby agrees to furnish all of the materials and all of the equipment and labor necessary, and to perform all of the work shown on the drawings and described in

the specifications for the project entitled _____

_____

_____

all in accordance with the requirements and provisions of the following Documents which are hereby made a part of this Agreement:

(*a*) Drawings prepared for same by _____

numbered _____

and dated _____ , 19 ____ .

(*b*) Specifications consisting of:

1. "Standard General Specifications" issued by _____

_____ , _____ Edition

Unit Price—1

2. "Special Conditions" as prepared by _____

   dated _____ .

3. The "General Conditions of Contract for Engineering Construction"—1966 Edition.

4. Addenda

   No. _____ Date _____

ARTICLE II — Time of Completion

(*a*) The work to be completed under this Contract shall be commenced within _____ calendar days after receipt of notice to proceed.

(*b*) The work shall be completed within _____ calendar days after receipt of notice to proceed.

(*c*) Failure to complete the work within the number of calendar days stated in this Article, including extension granted thereto as determined by Section 19 of the General Conditions, shall entitle the Owner to deduct from the moneys due to the Contractor as "Liquidated Damages" an amount equal to $_____ for each calendar day of delay in the completion of work.

(*d*) If the Contractor completes the work earlier than the date determined in accordance with Paragraph (*b*), and the Engineer shall so certify in writing, the Owner shall pay the Contractor an additional amount equal to $_____ for each calendar day by which the time of completion so determined has been reduced.

ARTICLE III — The Contract Sum

(*a*) The Owner shall pay to the Contractor for the performance of the work the amounts determined for the total number of each of the following units of work completed at the unit price stated thereafter. The number of units contained in this schedule is approximate only, and the final payment shall be made for the actual number of units that are incorporated in or made necessary by the work covered by the contract.

Item No.	Classification	Estimated No. of Units	Unit	Unit Price Bid	Total for Item

(*b*) Should the number of units of completed work of any individual item of the above schedule vary by more than _____% from the number of units stated in such schedule of units, either the Owner or the Contractor may request a revision of the unit price for the item so affected, and both parties agree that under such conditions an equitable revision of the price shall be made.

(*c*) Changes in the work made under Section 18 of the General Conditions, and not included in Article I, that cannot be classified as coming under any of the Contract units may be done at mutually agreed-upon unit prices, or on a lump sum basis, or under the provisions of Article V "Extra Work."

Unit Price—2

ARTICLE IV – Progress Payments

The Owner shall make payments on account of the Contract as follows:

(a) On not later than the fifth day of every month the Contractor shall present to the Engineer an invoice covering the total quantities under each item of work that has been completed from the start of the job up to and including the last day of the preceding month, and the value of the work so completed determined in accordance with the schedule of unit prices for such items together with such supporting evidence as may be required by the Engineer. This invoice shall also include an allowance for the cost of such material required in the permanent work as has been delivered to the site but not as yet incorporated in the work. Measurements of units for payment shall be made in accordance with the Special Conditions of the Contract.*

(b) On not later than the 15th of the month, the Owner shall pay to the Contractor 90 per cent of the amount of the invoice—less previous payments made. The 10 per cent retained percentage may be held by the Owner until the value of the work completed at the end of any month equals 50 per cent of the total amount of the Contract after which, if the Engineer finds that satisfactory progress is being made, he shall recommend that all of the remaining monthly payments be paid in full. Payments for work, under Subcontracts of the General Contractor, shall be subject to the above conditions applying to the general Contract after the work under a Subcontract has been 50 per cent completed.

(c) Final payment of all moneys due on the contract shall be made within 30 days of completion and acceptance of the work.

(d) If the owner fails to make payment as herein provided, or as provided in Article V(d), in addition to those remedies available to the Contractor under Section 25 of the General Conditions, there shall be added to each such payment daily interest at the rate of 6 per cent per annum commencing on the first day after said payment is due and continuing until the payment is delivered or mailed to the Contractor.

ARTICLE V – Extra Work

If the Engineer orders, in writing, the performance of any work not covered by the Drawings or included in the Specifications, and for which no item in the Contract is provided, and for which no unit price or lump sum basis can be agreed upon, then such extra work shall be done on a Cost-Plus-Percentage basis of payment as follows:

(a) The Contractor shall be reimbursed for all costs incurred in doing the work, and shall receive an additional payment of _____ % of all such cost to cover his indirect overhead costs, plus _____ % of all cost, including indirect overhead, as his fee.

---

* In addition to advance payment for materials delivered to site, wording should indicate that, where applicable, advance payment may be in order for materials in storage away from the site, for field plant and equipment, access roads, etc.—details to be spelled out in the Special Conditions.

Unit Price—3

(b) The "Cost of the Work" shall be determined as the net sum of the following items:

1. Job Office and all necessary temporary facilities such as buildings, use of land not furnished by the Owner, access roads and utilities. The costs of these items include construction, furnishings and equipment, maintenance during the period that they are needed, demolition and removal. Salvage values agreed on or received by the Contractor shall be credited to the Owner.

2. All materials used on the work whether for temporary or permanent construction.

3. All small tools and supplies; all fuel, lubricants, power, light, water and telephone service.

4. All plant and equipment at specified rental rates and terms of use. If the rental rates do not include an allowance for running repairs and repair parts needed for ordinary maintenance of the plant and equipment, then such items of cost are to be included in the Cost of the Work.

5. All transportation costs on equipment, materials and men.

6. All labor for the project and including the salaries of superintendents, foremen, engineers, inspectors, clerks and other employees while engaged on the work but excluding salaries of general supervisory employees or officers, who do not devote their full time to the work.

7. All payroll charges such as Social Security payments, unemployment insurance, workmen's compensation insurance premiums, pension and retirement allowances, and social insurance premiums, vacation and sick-leave allowances applicable to wages or salaries paid to employees for work done in connection with the contract.

8. All premiums on fire, public liability, property damage or other insurance coverage authorized or required by the Engineer or the Owner, or regularly paid by the Contractor in the conduct of his business.

9. All sales, use, excise, privilege, business, occupation, gross receipt and all other taxes paid by the Contractor in connection with the work, but excluding state income taxes based solely on net income derived from this contract and Federal income taxes.

10. All travel or other related expense of general supervisory employees for necessary visits to the job excluding expenses of such employees incurred at the Home Office of the Contractor.

11. All Subcontracts approved by the Engineer or Owner.

12. (Insert other costs proper for inclusion in this Contract.)

    a. _____

    b. _____

    c. _____

13. Any other cost incurred by the Contractor as a direct result of executing the Order, subject to approval by the Engineer.

Unit Price—4

356

14. Credit to the Owner for the following items:

    a. Such discounts on invoices as may be obtainable provided that the Owner advances sufficient funds to pay the invoices within the discount period.

    b. The mutually agreed salvage value of materials, tools or equipment charged to the Owner and taken over by the Contractor for his use or sale at the completion of the work.

    c. Any rebates, refunds, returned deposits or other allowances properly credited to the Cost of the Work.

(*c*) The cost of the work done each day shall be submitted to the Engineer in a satisfactory form on the succeeding day, and shall be approved by him or adjusted at once.

(*d*) Monthly payments of all charges for Extra Work in any one month shall be made in full on or before the 15th day of the succeeding month. Those payments shall include the full amount of fee earned on the cost of the work done.

    IN WITNESS WHEREOF the parties hereto have executed this Agreement, the day and year first above written.

_____ OWNER

WITNESS:

_____By: _____
                                                       Title

_____ CONTRACTOR

WITNESS:

_____By: _____
                                                       Title

Unit Price—5

GENERAL CONDITIONS OF CONTRACT FOR ENGINEERING CONSTRUCTION
INDEX

SEC. 1—Definitions

(*a*) The Contract Documents shall consist of Advertisement for Bids or Notice to Contractors, Instructions to Bidders, Form of Bid or Proposal, the signed Agreement, the General and Special Conditions of Contract, the Drawings, and the Specifications, including all modifications thereof incorporated in any of the documents before the execution of Agreement.

(*b*) The Owner, the Contractor and the Engineer are those named as such in the Agreement. They are treated throughout the Contract Documents as if each were of singular number and masculine gender.

(*c*) Wherever in this Contract the word "Engineer" is used it shall be understood as referring to the Engineer of the Owner, acting personally or through assistants duly authorized in writing by the Engineer.

(*d*) Written notice shall be deemed to have been duly served if delivered in person to the individual or to a member of the firm or to an officer of the corporation for whom it is intended, or to an authorized representative of such individual, firm, or corporation, or if delivered at or sent by registered mail to the last business address known to him who gives the notice, with a copy sent to the central office of the Contractor.

(*e*) The term "Subcontractor" shall mean anyone (other than the Contractor) who furnishes at the site, under an Agreement with the Contractor, labor, or labor and materials, or labor and equip-

ment, but shall not include any person who furnishes services of a personal nature.

(*f*) Work shall mean the furnishing of all labor, materials, equipment, and other incidentals necessary or convenient to the successful completion of the Contract and the carrying out of all the duties and obligations imposed by the Contract.

(*g*) Extra work shall mean such additional labor, materials, equipment, and other incidentals as are required to complete the Contract for the purpose for which it was intended but was not shown on the Drawings or called for in the Specifications, or is desired by the Owner in addition to that work called for in the Drawings and Specifications.

(*h*) Dispute shall mean lack of agreement between any parties that have any obligations, duties, or responsibilities under the terms of the Contract, Drawings, or Specifications.

SEC. 2—Execution and Correlation of Documents

The Contract Documents shall be signed in duplicate by the Owner and the Contractor.

The Contract Documents are complementary and what is called for by any one shall be as binding as if called for by all. In case of conflict between Drawings and Specifications, the Specifications shall govern. Materials or work described in words which so applied have a well-known technical or trade meaning shall be held to refer to such recognized standards.

15

SEC. 3.—Design, Drawings and Instructions

It is agreed that the Owner will be responsible for the adequacy of design and sufficiency of the Drawings and Specifications. The Owner, through the Engineer, or the Engineer as the Owner's representative, shall furnish Drawings and Specifications which adequately represent the requirements of the work to be performed under the Contract. All such Drawings and instructions shall be consistent with the Contract Documents and shall be true developments thereof. In the case of lump-sum Contracts, Drawings and Specifications which adequately represent the work to be done shall be furnished prior to the time of entering into the Contract. The Engineer may, during the life of the Contract, and in accordance with Section 18, issue additional instructions by means of Drawings or other media necessary to illustrate changes in the work.

SEC. 4—Copies of Drawings Furnished

Unless otherwise provided in the Contract Documents, the Engineer will furnish to the Contractor, free of charge, all copies of Drawings and Specifications reasonably necessary for the execution of the work.

SEC. 5—Order of Completion

The Contractor shall submit, at such times as may be reasonably requested by the Engineer, schedules which shall show the order in which the Contractor proposes to carry on the work, with dates at which the Contractor will start the several parts of the work, and estimated dates of completion of the several parts.

SEC. 6—Ownership of Drawings

All Drawings, Specifications and copies thereof furnished by the Engineer shall not be reused on other work, and, with the exception of the signed Contract, sets are to be returned to him on request, at the completion of the work.

SEC. 7—Familiarity with Work

The Owner shall make known to all prospective bidders, prior to the receipt of bids, all information that he may have as to subsurface conditions in the vicinity of the work, topographical maps, or other information that might assist the bidder in properly evaluating the amount and character of the work that might be required. Such information is given,

however, as being the best factual information available to the Owner. The Contractor, by careful examination, shall satisfy himself as to the nature and location of the work, the character of equipment and facilities needed preliminary to and during the prosecution of the work, the general and local conditions, and all other matters which can in any way affect the work under this Contract.

SEC. 8—Changed Conditions

The Contractor shall promptly, and before such conditions are disturbed, notify the Owner in writing of: (1) Subsurface or latent physical conditions at the site differing materially from those indicated in this Contract; or (2) previously unknown physical or other conditions at the site, of an unusual nature, differing materially from those ordinarily encountered and generally recognized as inherent in work of the character provided for in this Contract. The Engineer shall promptly investigate the conditions, and if he finds that such conditions do so materially differ and cause an increase or decrease in the cost of, or the time required for, performance of this Contract, an equitable adjustment shall be made and the Contract modified in writing accordingly. Any claim of the Contractor for adjustment hereunder shall not be allowed unless he has given notice as above required; provided that the Engineer may, if he determines the facts so justify, consider and adjust any such claims asserted before the date of final settlement of the Contract. If the parties fail to agree upon the adjustment to be made, the dispute shall be determined as provided in Section 39 hereof.

SEC. 9—Materials and Appliances

Unless otherwise stipulated, the Contractor shall provide and pay for all materials, labor, water, tools, equipment, light, power, transportation and other facilities necessary for the execution and completion of the work. Unless otherwise specified, all materials incorporated in the permanent work shall be new and both workmanship and materials shall be of good quality. The Contractor shall, if required, furnish satisfactory evidence as to the kind and quality of materials.

SEC. 10—Employees

The Contractor shall at all times enforce strict discipline and good order among his employees, and shall seek to avoid employing on the work any unfit person or anyone not skilled in the work assigned to him.

16

Adequate sanitary facilities shall be provided by the Contractor.

SEC. 11—Royalties and Patents

The Contractor shall pay all royalties and license fees. He shall defend all suits or claims for infringement of any patent rights and shall save the Owner harmless from loss on account thereof except that the Owner shall be responsible for all such loss when a particular process or the product of a particular manufacturer or manufacturers is specified, unless the Owner has notified the Contractor prior to the signing of the Contract that the particular process or product is patented or is believed to be patented.

SEC. 12—Surveys

Unless otherwise specified, the Owner shall furnish all land surveys and establish all base lines for locating the principal component parts of the work together with a suitable number of bench marks adjacent to the work. From the information provided by the Owner, the Contractor shall develop and make all detail surveys needed for construction such as slope stakes, batter boards, stakes for pile locations and other working points, lines and elevations.

The Contractor shall carefully preserve bench marks, reference points and stakes and, in case of willful or careless destruction, he shall be charged with the resulting expense and shall be responsible for any mistakes that may be caused by their unnecessary loss or disturbance.

SEC. 13—Permits, Licenses and Regulations

Permits and licenses of a temporary nature necessary for the prosecution of the work shall be secured and paid for by the Contractor. Permits, licenses and easements for permanent structures or permanent changes in existing facilities shall be secured and paid for by the Owner, unless otherwise specified. The Contractor shall give all notices and comply with all laws, ordinances, rules and regulations bearing on the conduct of the work as drawn and specified. If the Contractor observes that the Drawings and Specifications are at variance therewith, he shall promptly notify the Engineer in writing, and any necessary changes shall be adjusted as provided in the Contract for changes in the work.

SEC. 14—Protection of the Public and of Work and Property

The Contractor shall provide and maintain all necessary watchmen, barricades, warning lights and signs and take all necessary precautions for the protection and safety of the public. He shall continuously maintain adequate protection of all work from damage, and shall take all reasonable precautions to protect the Owner's property from injury or loss arising in connection with this Contract. He shall make good any damage, injury or loss to his work and to the property of the Owner resulting from lack of reasonable protective precautions, except such as may be due to errors in the Contract Documents, or caused by agents or employees of the Owner. He shall adequately protect adjacent private and public property, as provided by Law and the Contract Documents.

In an emergency affecting the safety of life, of the work, or of adjoining property, the Contractor is, without special instructions or authorization from the Engineer, hereby permitted to act at his discretion to prevent such threatened loss or injury. He shall also so act, without appeal, if so authorized or instructed by the Engineer.

Any compensation claimed by the Contractor on account of emergency work, shall be determined by agreement or by arbitration.

SEC. 15—Inspection of Work

The Owner shall provide sufficient competent personnel, working under the supervision of a qualified engineer, for the inspection of the work while such work is in progress to ascertain that the completed work will comply in all respects with the standards and requirements set forth in the Specifications. Notwithstanding such inspection, the Contractor will be held responsible for the acceptability of the finished work.

The Engineer and his representatives shall at all times have access to the work whenever it is in preparation or progress, and the Contractor shall provide proper facilities for such access, and for inspection.

If the Specifications, the Engineer's instructions, laws, ordinances, or any public authority require any work to be specially tested or approved, the Contractor shall give the Engineer timely notice of its readiness for inspection, and if the inspection is by an authority other than the Engineer, of the date fixed for such inspection. Inspections by the Engineer shall be made promptly, and where practicable at the source of supply. If any work should be covered up without approval or consent of the Engineer, it must, if required by the Engineer, be uncovered for examination and properly restored at the Contractor's expense, unless the Engineer has unreasonably delayed inspection.

Re-examination of any work may be ordered by the Engineer, and, if so ordered, the work must be uncovered by the Contractor. If such work is found to be in accordance with the Contract Documents, the Owner shall pay the cost of re-examination and replacement. If such work is not in accordance with the Contract Documents, the Contractor shall pay such cost.

SEC. 16—Superintendence

The Contractor shall keep on his work, during its progress, a competent superintendent and any necessary assistants. The superintendent shall represent the Contractor, and all directions given to him shall be binding as if given to the Contractor. Important directions shall immediately be confirmed in writing to the Contractor. Other directions shall be so confirmed on written request in each case. The Contractor shall give efficient superintendence to the work, using his best skill and attention.

SEC. 17—Discrepancies

If the Contractor, in the course of the work, finds any discrepancy between the Drawings and the physical conditions of the locality, or any errors or omissions in Drawings or in the layout as given by survey points and instructions, he shall immediately inform the Engineer, in writing, and the Engineer shall promptly verify the same. Any work done after such discovery, until authorized, will be done at the Contractor's risk.

SEC. 18—Changes in the Work

The Owner may make changes in the Drawings and Specifications or scheduling of the Contract within the general scope at any time by a written order. If such changes add to or deduct from the Contractor's cost of the work, the Contract shall be adjusted accordingly. All such work shall be executed under the conditions of the original Contract except that any claim for extension of time caused thereby shall be adjusted at the time of ordering such change.

In giving instructions, the Engineer shall have authority to make minor changes in the work not involving extra cost, and not inconsistent with the purposes of the work, but otherwise, except in an emergency endangering life or property, no extra work or change shall be made unless in pursuance of a written order by the Engineer, and no claim for an addition to the Contract Sum shall be valid unless the additional work was so ordered.

The Contractor shall proceed with the work as changed and the value of any such extra work or change shall be determined as provided in the Agreement.

SEC. 19—Extension of Time

Extension of time stipulated in the Contract for completion of the work will be made when changes in the work occur, as provided in Section 18; when the work is suspended as provided in Section 23; and when the work of the Contractor is delayed on account of conditions which could not have been foreseen, or which were beyond the control of the Contractor, his Subcontractors or suppliers, and which were not the result of their fault or negligence. Extension of time for completion shall also be allowed for any delays in the progress of the work caused by any act (except as provided elsewhere in these General Conditions) or neglect of the Owner or of his employees or by other contractors employed by the Owner, or by any delay in the furnishing of Drawings and necessary information by the Engineer, or by any other cause which in the opinion of the Engineer entitled the Contractor to an extension of time, including but not restricted to, acts of the public enemy, acts of any government in either its sovereign or any applicable contractual capacity, acts of another contractor in the performance of a contract with the Owner, fires, floods, epidemics, quarantine restrictions, freight embargoes, unusually severe weather, or labor disputes.

The Contractor shall notify the Engineer promptly of any occurrence or conditions which in the Contractor's opinion entitle him to an extension of time. Such notice shall be in writing and shall be submitted in ample time to permit full investigation and evaluation of the Contractor's claim. The Engineer shall acknowledge receipt of the Contractor's notice within 5 days of its receipt. Failure to provide such notice shall constitute a waiver by the Contractor of any claim.

SEC. 20—Claims

If the Contractor claims that any instructions by Drawings or other media issued after the date of the

18

Contract involve extra cost under this Contract, he shall give the Engineer written notice thereof within _____ days after the receipt of such instructions, and in any event before proceeding to execute the work, except in emergency endangering life or property, and the procedure shall then be as provided for changes in the work. No such claim shall be valid unless so made.

SEC. 21—Deductions for Uncorrected Work

If the Engineer deems it inexpedient to correct work that has been damaged or that was not done in accordance with the Contract, an equitable deduction from the Contract price shall be made therefor, unless the Contractor elects to correct the work.

SEC. 22—Correction of Work Before Final Payment

The Contractor shall promptly remove from the premises all materials and work condemned by the Engineer as failing to meet Contract requirements, whether incorporated in the work or not. The Contractor shall promptly replace and re-execute his own work in accordance with the Contract and without expense to the Owner and shall bear the expense of making good all work of other contractors destroyed or damaged by such removal or replacement.

If the Contractor does not take action to remove such condemned materials and work within 10 days after written notice, the Owner may remove them and may store the material at the expense of the Contractor. If the Contractor does not pay the expense of such removal and storage within ten days' time thereafter, the Owner may, upon ten days' written notice, sell such materials at auction or at private sale and shall pay to the Contractor any net proceeds thereof, after deducting all the costs and expenses that should have been borne by the Contractor.

SEC. 23—Suspension of Work

The Owner may at any time suspend the work, or any part thereof, by giving _____ days' notice to the Contractor in writing. The work shall be resumed by the Contractor within ten (10) days after the date fixed in the written notice from the Owner to the Contractor so to do. The Owner shall reimburse the Contractor for expense incurred by the Contractor in connection with the work under this Contract as a result of such suspension.

If the work, or any part thereof, shall be stopped by notice in writing aforesaid, and if the Owner does not give notice in writing to the Contractor to resume work at a date within _____ days of the date fixed in the written notice to suspend, then the Contractor may abandon that portion of the work so suspended and he will be entitled to the estimates and payments for all work done on the portions so abandoned, if any, plus _____% of the value of the work so abandoned, to compensate for loss of overhead, plant expense, and anticipated profit.

SEC. 24—The Owner's Right to Terminate Contract

If the Contractor should be adjudged a bankrupt, or if he should make a general assignment for the benefit of his creditors, or if a receiver should be appointed as a result of his insolvency, or if he should be guilty of a substantial violation of the Contract, then the Owner, upon the certificate of the Engineer that sufficient cause exists to justify such action, may, without prejudice to any other right or remedy and after giving the Contractor and his Surety seven days' written notice, terminate the employment of the Contractor and take possession of the premises and of all materials, tools, equipment and other facilities installed on the work and paid for by the Owner, and finish the work by whatever method he may deem expedient. In such case the Contractor shall not be entitled to receive any further payment until the work is finished. If the unpaid balance of the Contract price shall exceed the expense of finishing the work, including compensation for additional managerial and administrative services, such excess shall be paid to the Contractor. If such expense shall exceed such unpaid balance, the Contractor shall pay the difference to the Owner. The expense incurred by the Owner as herein provided, and the damage incurred through the Contractor's default, shall be certified by the Engineer.

SEC. 25—Contractor's Right to Stop Work or Terminate Contract

If the work should be stopped under an order of any court, or other public authority, for a period of more than three months, through no act or fault of the Contractor or of anyone employed by him, or if the Engineer should fail to issue any estimate for payment within seven days after it is due, or if the Owner should fail to pay the Contractor within seven days of its maturity and presentation any sum certified by the Engineer or awarded by arbitrators, then the Contractor may, upon seven days' written

19

notice to the Owner and the Engineer, stop work or terminate this Contract and recover from the Owner payment for all work executed, plus any loss sustained upon any plant or materials plus reasonable profit and damages.

SEC. 26—Removal of Equipment

In the case of termination of this Contract before completion from any cause whatever, the Contractor, if notified to do so by the Owner, shall promptly remove any part or all of his equipment and supplies from the property of the Owner, failing which the Owner shall have the right to remove such equipment and supplies at the expense of the Contractor.

SEC. 27—Responsibility for Work

The Contractor assumes full responsibility for the work. Until its final acceptance, the Contractor shall be responsible for damage to or destruction of the work (except for any part covered by partial acceptance as set forth in Sec. 28). He agrees to make no claims against the Owner for damages to the work from any cause except negligence or willful acts of the Owner, acts of an Enemy, acts of war or as provided in Sec. 32.

SEC. 28—Partial Completion and Acceptance

If at any time prior to the issuance of the final certificate referred to in Section 42 hereinafter, any portion of the permanent construction has been satisfactorily completed, and if the Engineer determines that such portion of the permanent construction is not required for the operations of the Contractor but is needed by the Owner, the Engineer shall issue to the Contractor a certificate of partial completion, and thereupon or at any time thereafter the Owner may take over and use the portion of the permanent construction described in such certificate, and may exclude the Contractor therefrom.

The issuance of a certificate of partial completion shall not be construed to constitute an extension of the Contractor's time to complete the portion of the permanent construction to which it relates if he has failed to complete it in accordance with the terms of this Contract. The issuance of such a certificate shall not operate to release the Contractor or his sureties from any obligations under this Contract or the performance bond.

If such prior use increases the cost of or delays the work, the Contractor shall be entitled to extra

compensation, or extension of time, or both, as the Engineer may determine, unless otherwise provided.

SEC. 29—Payments Withheld Prior to Final Acceptance of Work

The Owner, as a result of subsequently discovered evidence, may withhold or nullify the whole or part of any payment certificate to such extent as may be necessary to protect himself from loss caused by:

(a) Defective work not remedied.

(b) Claims filed or reasonable evidence indicating probable filing of claims by other parties against the Contractor.

(c) Failure of the Contractor to make payments properly to Subcontractors or for material or labor.

(d) Damage to another contractor.

When the above grounds are removed or the Contractor provides a Surety Bond satisfactory to the Owner which will protect the Owner in the amount withheld, payment shall be made for amounts withheld, because of them.

No moneys may be withheld under (b) and (c) above if a payment bond is included in the Contract.

SEC. 30—Contractor's Insurance

The Contractor shall secure and maintain such insurance policies as will protect himself, his Subcontractors, and unless otherwise specified, the Owner, from claims for bodily injuries, death or property damage which may arise from operations under this Contract whether such operations be by himself or by any Subcontractor or anyone employed by them directly or indirectly. The following insurance policies are required:

(a) Statutory Workmen's Compensation.

(b) Contractor's Public Liability and Property damage—

    Bodily Injury:

        each person _____ $_____

        each accident _____ $_____

    Property Damage:

        each accident _____ $_____

        aggregate _____ $_____

(c) Automobile Public Liability and Property Damage—

    Bodily Injury:

        each person _____ $_____

        each accident _____ $_____

    Property Damage:

        each accident _____ $_____

20

All policies shall be for not less than the amounts set forth above or as stated in the Special Conditions. Other forms of insurance shall also be provided if called for by the Special Conditions.

Certificates and/or copies of policy of such insurance shall be filed with the Engineer, and shall be subject to his approval as to adequacy of protection, within the requirements of the Specifications. Said certificates of insurance shall contain a 10 days' written notice of cancellation in favor of the Owner.

SEC. 31—Surety Bonds

The Owner shall have the right, prior to the signing of the Contract, to require the Contractor to furnish bond covering the faithful performance of the Contract and the payment of all obligations arising thereunder, in such form as the Owner may prescribe in the bidding documents and executed by one or more financially responsible sureties. If such bond is required prior to the receipt of bids, the premium shall be paid by the Contractor; if subsequent thereto, it shall be paid by the Owner. The Owner may require additional bond if the Contract is increased appreciably.

SEC. 32—Owner's Insurance

The Owner shall secure and maintain insurance to 100% of the insurable value thereof against fire, earthquake, flood, and such other perils as he may deem necessary and shall name the Contractor and Subcontractors as additional insured. Such insurance shall be upon the entire work in the Contract and any structures attached or adjacent thereto. He shall also secure and maintain such insurance as will protect him and his officers, agents, servants, and employees from liability to others for damages due to death, bodily injury, or property damage resulting from the performance of the work. The limits of such insurance shall be equal to the amounts stated in subparagraphs (b) and (c), of Section 30.

SEC. 33—Assignment

Neither party to the Contract shall assign the Contract or sublet it as a whole without the written consent of the other, nor shall the Contractor assign any moneys due to him or to become due to him hereunder, except to a bank or financial institution acceptable to the Owner.

SEC. 34—Rights of Various Interests

Whenever work being done by the Owner's or by other contractor's forces is contiguous to work covered by this Contract, the respective rights of the various interests involved shall be established by the Engineer, to secure the completion of the various portions of the work in general harmony.

SEC. 35—Separate Contracts

The Owner reserves the right to let other contracts in connection with this project. The Contractor shall afford other contractors reasonable opportunity for the introduction and storage of their materials and the execution of their work, and shall properly connect and coordinate his work with theirs.

If the proper execution or results of any part of the Contractor's work depends upon the work of any other contractor, the Contractor shall inspect and promptly report to the Engineer any defects in such work that render it unsuitable for such proper execution and results.

SEC. 36—Subcontracts

The Contractor shall, as soon as practicable after signing of the Contract, notify the Engineer in writing of the names of Subcontractors proposed for the work.

The Contractor agrees that he is as fully responsible to the Owner for the acts and omissions of his Subcontractors and of persons either directly or indirectly employed by them, as he is for the acts and omissions of persons directly employed by him.

Nothing contained in the Contract Documents shall create any contractual relation between any Subcontractor and the Owner.

SEC. 37—Engineer's Status

The Engineer shall perform technical inspection of the work. He has authority to stop the work whenever such stoppage may be necessary to insure the proper execution of the Contract. He shall also have authority to reject all work and materials which do not conform to the Contract and to decide questions which arise in the execution of the work.

21

SEC. 38—Engineer's Decisions

The Engineer shall, within a reasonable time after their presentation to him, make decisions in writing on all claims of the Owner or the Contractor and on all other matters relating to the execution and progress of the work or the interpretation of the Contract Documents.

SEC. 39—Arbitration

Any controversy or claim arising out of or relating to this Contract, or the breach thereof which cannot be resolved by mutual agreement, shall be settled by arbitration in accordance with the Rules of the American Arbitration Association, and judgment upon the award rendered by the Arbitrator(s) may be entered in any Court having jurisdiction thereof.

SEC. 40—Lands for Work

The Owner shall provide as indicated on Drawing No. _____ and not later than the date when needed by the Contractor the lands upon which the work under this Contract is to be done, rights-of-way for access to same, and such other lands which are designated on the Drawings for the use of the Contractor. Such lands and rights-of-way shall be adequate for the performance of the Contract. Any delay in the furnishing of these lands by the Owner shall be deemed proper cause for an equitable adjustment in both Contract price and time of completion.

The Contractor shall provide at his own expense and without liability to the Owner any additional land and access thereto that may be required for temporary construction facilities, or for storage of materials.

SEC. 41—Cleaning Up

The Contractor shall remove at his own expense from the Owner's property and from all public and private property all temporary structures, rubbish and waste materials resulting from his operations. This requirement shall not apply to property used for permanent disposal of rubbish or waste materials in accordance with permission of such disposal granted to the Contractor by the Owner thereof.

SEC. 42—Acceptance and Final Payment

(a) Upon receipt of written notice that the work is substantially completed or ready for final inspection and acceptance, the Engineer will promptly make such inspection, and when he finds the work acceptable under the Contract and the Contract fully performed or substantially completed he shall promptly issue a certificate, over his own signature, stating that the work required by this Contract has been completed or substantially completed and is accepted by him under the terms and conditions thereof, and the entire balance found to be due the Contractor, including the retained percentage, less a retention based on the Engineer's estimate of the fair value of the claims against the Contractor and the cost of completing the incomplete or unsatisfactory items of work with specified amounts for each incomplete or defective item of work, is due and payable. The date of substantial completion of a project or specified area of a project is the date when the construction is sufficiently completed in accordance with the Contract Documents as modified by any change orders agreed to by the parties so that the Owner can occupy the project or specified area of the project for the use for which it was intended.

(b) Before issuance of final payment, the Contractor, if required in the Special Conditions, shall certify in writing to the Engineer that all payrolls, material bills, and other indebtedness connected with the work have been paid, or otherwise satisfied, except that in case of disputed indebtedness or liens, if the Contract does not include a payment bond, the Contractor may submit in lieu of certification of payment a surety bond in the amount of the disputed indebtedness or liens, guaranteeing payment of all such disputed amounts, including all related costs and interest in connection with said disputed indebtedness or liens which the Owner may be compelled to pay upon adjudication.

(c) The making and acceptance of the final payment shall constitute a waiver of all claims by the Owner, other than those arising from unsettled liens, from faulty work appearing within the guarantee period provided in the Special Conditions, from the requirements of the Drawings and Specifications, or from manufacturer's guarantees. It shall also constitute a waiver of all claims by the Contractor, except those previously made and still unsettled.

(d) If after the work has been substantially completed, full completion thereof is materially delayed through no fault of the Contractor, and the Engineer so certifies, the Owner shall, upon certificate of the Engineer, and without terminating the Contract, make payment of the balance due for that portion of the work fully completed and accepted. Such payment shall be made under the terms and conditions governing final payment, except that it shall not constitute a waiver of claims.

(e) If the Owner fails to make payment as herein provided, there shall be added to each such payment daily interest at the rate of 6 per cent per annum commencing on the first day after said payment is due and continuing until the payment is delivered or mailed to the Contractor.

22

APPENDIX C

CONSTRUCTION CONTRACT— U.S. GOVERNMENT STANDARD FORM 23 WITH GENERAL PROVISIONS

Courtesy of Corps of Engineers, U.S. Army.

CONSTRUCTION CONTRACT
(See instructions on reverse)

CONTRACT NO.

DATE OF CONTRACT

NAME AND ADDRESS OF CONTRACTOR

CHECK APPROPRIATE BOX

☐ Individual

☐ Partnership

☐ Joint Venture

☐ Corporation, incorporated in the

State of _____.

DEPARTMENT OR AGENCY

CONTRACT FOR (*Work to be performed*)

PLACE

CONTRACT PRICE (*Express in words and figures*)

ADMINISTRATIVE DATA (*Optional*)

The United States of America (hereinafter called the Government), represented by the Contracting Officer executing this contract, and the individual, partnership, joint venture, or corporation named above (hereinafter called the Contractor), mutually agree to perform this contract in strict accordance with the General Provisions (Standard Form 23–A), Labor Standards Provisions Applicable to Contracts in Excess of $2,000 (Standard Form 19–A), and the following designated specifications, schedules, drawings, and conditions:

WORK SHALL BE STARTED

WORK SHALL BE COMPLETED

Alterations. The following alterations were made in this contract before it was signed by the parties hereto:

In witness whereof, the parties hereto have executed this contract as of the date entered on the first page hereof.

THE UNITED STATES OF AMERICA   CONTRACTOR

By _____

_____
(Name of Contractor)

_____   By _____
(Official title)   *(Signature)*

_____
(Title)

INSTRUCTIONS

1. The full name and business address of the Contractor must be inserted in the space provided on the face of the form. The Contractor shall sign in the space provided above with his usual signature and typewrite or print his name under the signature.

2. An officer of a corporation, a member of a partnership, or an agent signing for the Contractor shall place his signature and title after the word "By" under the name of the Contractor. A contract executed by an attorney or agent on behalf of the Contractor shall be accompanied by two authenticated copies of his power of attorney or other evidence of his authority to act on behalf of the Contractor.

INDEX OF GENERAL PROVISIONS
(Construction Contract)

(31 Mar 72)

(Const. Gen. Prov.)

GENERAL PROVISIONS
(Construction Contract)
Issued By: Department of the Army, Corps of Engineers

(General Provisions 1 through 23 and 24 through 31 are those prescribed by the General Services Administration in Standard Form 23-A, Oct 1969 edition, and Standard Form 19-A, Apr 1965 edition, respectively, as amended pursuant to the latest revisions of the Armed Services Procurement Regulation and Engineer Contract Instructions, ER 1180-1-1.)

1.1 DEFINITIONS

(The following clause is applicable if the procurement instrument identification number is prefixed by the letters "DACW")

(a) The term "head of the agency" or "Secretary" as used herein means the Secretary of the Army; and the term "his duly authorized representative" means the Chief of Engineers, Department of the Army, or an individual or board designated by him.

(b) The term "Contracting Officer" as used herein means the person executing this contract on behalf of the Government and includes a duly appointed successor or authorized representative. (ASPR 7-602.1 and ECI 7-070)

1.2 DEFINITIONS (1964 JUN)

(The following clause is applicable if the procurement instrument identification number is prefixed by the letters "DACA")

(a) The term "head of the agency" or "Secretary" as used herein means the Secretary, the Under Secretary, any Assistant Secretary, or any other head or assistant head of the executive or military department or other Federal agency; and the term "his duly authorized representative" means any person or persons or board (other than the Contracting Officer) authorized to act for the head of the agency or the Secretary.

(b) The term "Contracting Officer" as used herein means the person executing this contract on behalf of the Government and includes a duly appointed successor or authorized representative. (ASPR 7-602.1)

2. SPECIFICATIONS AND DRAWINGS (1964 JUN)

The Contractor shall keep on the work a copy of the drawings and specifications and shall at all times give the Contracting Officer access thereto. Anything mentioned in the specifications and not shown on the drawings, or shown on the drawings and not mentioned in the specifications, shall be of like effect as if shown or mentioned in both. In case of difference between drawings and specifications, the specifications shall govern. In case of discrepancy either in the figures, in the drawings, or in the specifications, the matter shall be promptly submitted to the Contracting Officer, who shall promptly make a determination in writing. Any adjustment by the Contractor without such a determination shall be at his own risk and expense. The Contracting Officer shall furnish from time to time such detail drawings and other information as he may consider necessary, unless otherwise provided. (ASPR 7-602.2)

3. CHANGES (1968 FEB)

(a) The Contracting Officer may, at any time, without notice to the sureties, by written order designated or indicated to be a change order, make any change in the work within the general scope of the contract, including but not limited to changes:

    (i)   in the specifications (including drawings and designs);

    (ii)   in the method or manner of performance of the work;

    (iii) in the Government-furnished facilities, equipment, materials, services, or site; or

    (iv) directing acceleration in the performance of the work.

(b) Any other written order or an oral order (which terms as used in this paragraph (b) shall include direction, instruction, interpretation or determination) from the Contracting Officer, which causes any such change, shall be treated as a change order under this clause, *provided* that the Contractor gives the Contracting Officer written notice stating the date, circumstances, and source of the order and that the Contractor regards the order as a change order.

(c) Except as herein provided, no order, statement, or conduct of the Contracting Officer shall be treated as a change under this clause or entitle the Contractor to an equitable adjustment hereunder.

(d) If any change under this clause causes an increase or decrease in the Contractor's cost of, or the time required for, the performance of any part of the work under this contract, whether or not changed by any order, an equitable adjustment shall be made and the contract modified in writing accordingly: *Provided, however,* That except for claims based on defective specifications, no claim for any change under (b) above shall be allowed for any costs incurred more than 20 days before the Contractor gives written notice as therein required: *And provided further,* That in the case of defective specifications for which the Government is responsible, the equitable adjustment shall include any increased cost reasonably incurred by the Contractor in attempting to comply with such defective specifications.

(e) If the Contractor intends to assert a claim for an equitable adjustment under this clause, he must, within 30 days after receipt of a written change order under (a) above or the furnishing of a written notice under (b) above, submit to the Contracting Officer a written statement setting forth the general nature and monetary extent of such claim, unless this period is extended by the Government. The statement of claim hereunder may be included in the notice under (b) above.

(f) No claim by the Contractor for an equitable adjustment hereunder shall be allowed if asserted after final payment under this contract. (ASPR 7-602.3)

4. DIFFERING SITE CONDITIONS (1968 FEB)

(a) The Contractor shall promptly, and before such conditions are disturbed, notify the Contracting Officer in writing of: (1) subsurface or latent physical conditions at the site differing materially from those indicated in this contract, or (2) unknown physical conditions at the site, of an unusual nature, differing materially from those ordinarily encountered and generally recognized as inhering in work of the character provided for in this contract. The Contracting Officer shall promptly investigate the conditions, and if he finds that such conditions do materially so differ and cause an increase or decrease in the Contractor's cost of, or the time required for, performance of any part of the work under this contract, whether or not changed as a result of such conditions, an equitable adjustment shall be made and the contract modified in writing accordingly.

(b) No claim of the Contractor under this clause shall be allowed unless the Contractor has given the notice required in (a) above; *provided,* however, the time prescribed therefor may be extended by the Government.

(c) No claim by the Contractor for an equitable adjustment hereunder shall be allowed if asserted after final payment under this contract. (ASPR 7-602.4)

5. TERMINATION FOR DEFAULT - DAMAGES FOR DELAY - TIME EXTENSIONS (1969 AUG)

(a) If the Contractor refuses or fails to prosecute the work, or any separable part thereof, with such diligence as will insure its completion within the time specified in this contract, or any extension thereof, or fails to complete said work within such time, the Government may, by written notice to the Contractor, terminate his right to proceed with the work or such part of the work as to which there has been delay. In such event the Government may take over the work and prosecute the same to completion, by contract or otherwise, and may take possession of and utilize in completing the work such materials, appliances, and plant as may be on the site of the work and necessary therefor. Whether or not the Contractor's right to proceed with the work is terminated, he and his sureties shall be liable for any damage to the Government resulting from his refusal or failure to complete the work within the specified time.

(b) If fixed and agreed liquidated damages are provided in the contract and if the Government so terminates the Contractor's right to proceed, the resulting damage will consist of such liquidated damages until such reasonable time as may be required for final completion of the work together with any increased costs occasioned the Government in completing the work.

(c) If fixed and agreed liquidated damages are provided in the contract and if the Government does not so terminate the Contractor's right to proceed, the resulting damage will consist of such liquidated damages until the work is completed or accepted.

(d) The Contractor's right to proceed shall not be so terminated nor the Contractor charged with resulting damage if:

(1) The delay in the completion of the work arises from causes other than normal weather beyond the control and without the fault or negligence of the Contractor, including but not restricted to, acts of God, acts of the public enemy, acts of the Government in either its sovereign or contractual capacity, acts of another contractor in the performance of a contract with the Government, fires, floods, epidemics, quarantine restrictions, strikes, freight embargoes, unusually severe weather, or delays of subcontractors or suppliers arising from causes other than normal weather beyond the control and without the fault or negligence of both the Contractor and such subcontractors or suppliers; and

(2) The Contractor, within 10 days from the beginning of any such delay (unless the Contracting Officer grants a further period of time before the date of final payment under the contract), notifies the Contracting Officer in writing of the causes of delay. The Contracting Officer shall ascertain the facts and the extent of the delay and extend the time for completing the work when, in his judgment, the findings of fact justify such an extension, and his findings of fact shall be final and conclusive on the parties, subject only to appeal as provided in the "Disputes" clause of this contract.

(e) If, after notice of termination of the Contractor's right to proceed under the provisions of this clause, it is determined for any reason that the Contractor was not in default under the provisions of this clause, or that the delay was excusable under the provisions of this clause, the rights and obligations of the parties shall, if the contract contains a clause providing for termination for convenience of the Government, be the same as if the notice of termination had been issued pursuant to such clause. If, in the foregoing circumstances, this contract does not contain a clause providing for termination for convenience of the Government, the contract shall be equitably adjusted to compensate for such termination and the contract modified accordingly; failure to agree to any such adjustment shall be a dispute concerning a question of fact within the meaning of the clause of this contract entitled "Disputes".

(f) The rights and remedies of the Government provided in this clause are in addition to any other rights and remedies provided by law or under this contract.

(g) As used in paragraph (d)(1) of this clause, the term "subcontractors or suppliers" means subcontractors or suppliers at any tier. (ASPR 7-602.5 and 8-709(b))

6. DISPUTES (1964 JUN)

(a) Except as otherwise provided in this contract, any dispute concerning a question of fact arising under this contract which is not disposed of by agreement shall be decided by the Contracting Officer, who shall reduce his decision to writing and mail or otherwise furnish a copy thereof to the Contractor. The decision of the Contracting Officer shall be final and conclusive unless, within 30 days from the date of receipt of such copy, the Contractor mails or otherwise furnishes to the Contracting Officer a written appeal addressed to the head of the agency involved. The decision of the head of the agency or his duly authorized representative for the determination of such appeals shall be final and conclusive. This provision shall not be pleaded in any suit involving a question of fact arising under this contract as limiting judicial review of any such decision to cases where fraud by such official or his representative or board is alleged: *Provided, however,* that any such decision shall be final and conclusive unless the same is fraudulent or capricious or arbitrary or so grossly erroneous as necessarily to imply bad faith or is not supported by substantial evidence. In connection with any appeal proceeding under this clause, the Contractor shall be afforded an opportunity to be heard and to offer evidence in support of his appeal. Pending final decision of a dispute hereunder, the Contractor shall proceed diligently with the performance of the contract and in accordance with the Contracting Officer's decision.

(b) This "Disputes" clause does not preclude consideration of questions of law in connection with decisions provided for in paragraph (a) above. Nothing in this contract, however, shall be construed as making final the decision of any administrative official, representative, or board on a question of law. (ASPR 7-602.6(a))

7. PAYMENTS TO CONTRACTOR (1964 JUN)

(The last two sentences of paragraph (c) of the following clause are applicable only where the contract amount exceeds $1,000,000 and the time of performance exceeds one year)

(a) The Government will pay the contract price as hereinafter provided.

(b) The Government will make progress payments monthly as the work proceeds, or at more frequent intervals as determined by the Contracting Officer, on estimates approved by the Contracting Officer. If requested by the Contracting Officer, the Contractor shall furnish a breakdown of the total contract price showing the amount included therein for each principal category of the work, in such detail as requested, to provide a basis for determining progress payments. In the preparation of estimates the Contracting Officer, at his discretion, may authorize material delivered on the site and preparatory work done to be taken into consideration. Material delivered to the Contractor at locations other than the site may also be taken into consideration (1) if such consideration is specifically authorized by the contract and (2) if the Contractor furnishes satisfactory evidence that he has acquired title to such material and that it will be utilized on the work covered by this contract.

(c) In making such progress payments, there shall be retained 10 percent of the estimated amount until final completion and acceptance of the contract work. However, if the Contracting Officer, at any time after 50 percent of the work has been completed, finds that satisfactory progress is being made, he may authorize any of the remaining progress payments to be made in full. Also, whenever the work is substantially complete, the Contracting Officer, if he considers the amount retained to be in excess of the amount adequate for the protection of the Government, at his discretion, may release to the Contractor all or a portion of such excess amount. Furthermore, on completion and acceptance of each separate building, public work, or other division of the contract, on which the price is stated separately in the contract, payment may be made therefor without retention of a percentage. Where the time originally specified for completion of this contract exceeds one year, the Contracting Officer, at any time after 50 percent of the work has been completed, if he finds that satisfactory progress is being made, may reduce the total amount retained from progress payments to an amount not less than 10 percent of the estimated value of the work remaining to be done under the contract or 1-1/2 percent of the total contract amount, whichever is the higher. In computing the total contract amount, for the purposes of the preceding sentence, the contract amount for any separate building, public work, or other division of the contract on which the price is stated separately in the contract and on which payment has been made in full, including retained percentage thereon under this clause shall be excluded.

(d) All material and work covered by progress payments made shall thereupon become the sole property of the Government, but this provision shall not be construed as relieving the Contractor from the sole responsibility for all material and work upon which payments have been made or the restoration of any damaged work, or as waiving the right of the Government to require the fulfillment of all of the terms of the contract.

(e) Upon completion and acceptance of all work, the amount due the Contractor under this contract shall be paid upon the presentation of a properly executed voucher and after the Contractor shall have furnished the Government with a release, if required, of all claims against the Government arising by virtue of this contract, other than claims in stated amounts as may be specifically excepted by the Contractor from the operation of the release. If the Contractor's claim to amounts payable under the contract has been assigned under the Assignment of Claims Act of 1940, as amended (31 U.S.C. 203, 41 U.S.C. 15), a release may also be required of the assignee. (ASPR 7-602.7(a) and (b))

8. ASSIGNMENT OF CLAIMS (1964 JUN)

(a) Pursuant to the provisions of the Assignment of Claims Act of 1940, as amended (31 U.S.C. 203, 41 U.S.C. 15), if this contract provides for payments aggregating $1,000 or more, claims for moneys due or to become due the Contractor from the Government under this contract may be assigned to a bank, trust company, or other financing institution, including any Federal lending agency, and may thereafter be further assigned and reassigned to any such institution. Any such assignment or reassignment shall cover all amounts payable under this contract and not already paid, and shall not be made to more than one party, except that any such assignment or reassignment may be made to one party as agent or trustee for two or more parties participating in such financing. Unless otherwise provided in this contract, payments to assignee of any moneys due or to become due under this contract shall not, to the extent provided in said Act, as amended, be subject to reduction or setoff. (The preceding sentence applies only if this contract is made in time of war or national emergency as defined in said Act and is with the Department of Defense, the General Services Administration, the Atomic Energy Commission, the National Aeronautics and Space Administration, the Federal Aviation Agency, or any other department or agency of the United States designated by the President pursuant to Clause 4 of the proviso of section 1 of the Assignment of Claims Act of 1940, as amended by the Act of May 15, 1951, 65 Stat. 41.)

(b) In no event shall copies of this contract or of any plans, specifications, or other similar documents relating to work under this contract, if marked "Top Secret," "Secret," or "Confidential," be furnished to any assignee of any claim arising under this contract or to any other person not entitled to receive the same. However, a copy of any part or all of this contract so marked may be furnished, or any information contained therein may be disclosed, to such assignee upon the prior written authorization of the Contracting Officer. (ASPR 7-602.8)

9. MATERIAL AND WORKMANSHIP (1964 JUN)

(a) Unless otherwise specifically provided in this contract, all equipment, material, and articles incorporated in the work covered by this contract are to be new and of the most suitable grade for the purpose intended. Unless otherwise specifically provided in this contract, reference to any equipment, material, article, or patented process, by trade name, make, or catalog number, shall be regarded as establishing a standard of quality and shall not be construed as limiting competition, and the Contractor may, at his option, use any equipment, material, article, or process which, in the judgment of the Contracting Officer, is equal to that named. The Contractor shall furnish to the Contracting Officer for his approval the name of the manufacturer, the model number, and other identifying data and information respecting the performance, capacity, nature, and rating of the machinery and mechanical and other equipment which the Contractor contemplates incorporating in the work. When required by this contract or when called for by the Contracting Officer, the Contractor shall furnish the Contracting Officer for approval full information concerning the material or articles which he contemplates incorporating in the work. When so directed, samples shall be submitted for approval at the Contractor's expense, with all shipping charges prepaid. Machinery, equipment, material, and articles installed or used without required approval shall be at the risk of subsequent rejection.

(b) All work under this contract shall be performed in a skillful and workmanlike manner. The Contracting Officer may, in writing, require the Contractor to remove from the work any employee the Contracting Officer deems incompetent, careless, or otherwise objectionable. (ASPR 7-602.9)

10. INSPECTION AND ACCEPTANCE (1964 JUN)

(a) Except as otherwise provided in this contract, inspection and test by the Government of material and workmanship required by this contract shall be made at reasonable times and at the site of the work, unless the Contracting Officer determines that such inspection or test of material which is to be incorporated in the work shall be made at the place of production, manufacture, or shipment of such material. To the extent specified by the Contracting Officer at the time of determining to make off-site inspection or test, such inspection or test shall be conclusive as to whether the material involved conforms to the contract requirements. Such off-site inspection or test shall not relieve the Contractor of responsibility for damage to or loss of the material prior to acceptance, nor in any way affect the continuing rights of the Government after acceptance of the completed work under the terms of paragraph (f) of this clause, except as hereinabove provided.

(b) The Contractor shall, without charge, replace any material or correct any workmanship found by the Government not to conform to the contract requirements, unless in the public interest the Government consents to accept such material or workmanship with an appropriate adjustment in contract price. The Contractor shall promptly segregate and remove rejected material from the premises.

(c) If the Contractor does not promptly replace rejected material or correct rejected workmanship, the Government (1) may, by contract or otherwise, replace such material or correct such workmanship and charge the cost thereof to the Contractor, or (2) may terminate the Contractor's right to proceed in accordance with the "Termination for Default - Damages for Delay - Time Extensions" clause of this contract.

(d) The Contractor shall furnish promptly, without additional charge, all facilities, labor, and material reasonably needed for performing such safe and convenient inspection and test as may be required by the Contracting Officer. All inspection and test by the Government shall be performed in such manner as not unnecessarily to delay the work. Special, full size, and performance tests shall be performed as described in this contract. The Contractor shall be charged with any additional cost of inspection when material and workmanship are not ready at the time specified by the Contractor for its inspection.

(e) Should it be considered necessary or advisable by the Government at any time before acceptance of the entire work to make an examination of work already completed, by removing or tearing out same, the Contractor shall, on request, promptly furnish all necessary facilities, labor and material. If such work is found to be defective or nonconforming in any material respect, due to the fault of the Contractor or his subcontractors, he shall defray all the expenses of such examination and of satisfactory reconstruction. If, however, such work is found to meet the requirements of the contract, an equitable adjustment shall be made in the contract price to compensate the Contractor for the additional services involved in such examination and reconstruction and, if completion of the work has been delayed thereby, he shall, in addition, be granted a suitable extension of time.

(f) Unless otherwise provided in this contract, acceptance by the Government shall be made as promptly as practicable after completion and inspection of all work required by this contract. Acceptance shall be final and conclusive except as regards latent defects, fraud, or such gross mistakes as may amount to fraud or as regards the Government's rights under any warranty or guarantee. (ASPR 7-602.11)

11. SUPERINTENDENCE BY CONTRACTOR (1964 JUN)

The Contractor shall give his personal superintendence to the work or have a competent foreman or superintendent, satisfactory to the Contracting Officer, on the work at all times during progress, with authority to act for him. (ASPR 7-602.12)

12. PERMITS AND RESPONSIBILITIES (1964 JUN)

The Contractor shall, without additional expense to the Government, be responsible for obtaining any necessary licenses and permits, and for complying with any applicable Federal, State, and municipal laws, codes, and regulations, in connection with the prosecution of the work. He shall be similarly responsible for all damages to persons or property that occur as a result of his fault or negligence. He shall take proper safety and health precautions to protect the work, the workers, the public, and the property of others. He shall also be responsible for all materials delivered and work performed until completion and acceptance of the entire construction work, except for any completed unit of construction thereof which theretofore may have been accepted. (ASPR 7-602.13)

13. CONDITIONS AFFECTING THE WORK (1964 JUN)

The Contractor shall be responsible for having taken steps reasonably necessary to ascertain the nature and location of the work, and the general and local conditions which can affect the work or the cost thereof. Any failure by the Contractor to do so will not relieve him from responsibility for successfully performing the work without additional expense to the Government. The Government assumes no responsibility for any understanding or representations concerning conditions made by any of its officers or agents prior to the execution of this contract, unless such understanding or representations by the Government are expressly stated in the contract. (ASPR 7-602.14)

14. OTHER CONTRACTS (1964 JUN)

The Government may undertake or award other contracts for additional work, and the Contractor shall fully cooperate with such other contractors and Government employees and carefully fit his own work to such additional work as may be directed by the Contracting Officer. The Contractor shall not commit or permit any act which will interfere with the performance of work by any other contractor or by Government employees. (ASPR 7-602.15)

15. PATENT INDEMNITY (1964 JUN)

Except as otherwise provided, the Contractor agrees to indemnify the Government and its officers, agents, and employees against liability, including costs and expenses, for infringement upon any Letters Patent of the United States (except Letters Patent issued upon an application which is now or may hereafter be, for reasons of national security, ordered by the Government to be kept secret or otherwise withheld from issue) arising out of the performance of this contract or out of the use or disposal by or for the account of the Government of supplies furnished or construction work performed hereunder. (ASPR 7-602.16)

16. ADDITIONAL BOND SECURITY (1949 JUL)

If any surety upon any bond furnished in connection with this contract becomes unacceptable to the Government, or if any such surety fails to furnish reports as to his financial condition from time to time as requested by the Government, the Contractor shall promptly furnish such additional security as may be required from time to time to protect the interests of the Government and of persons supplying labor or materials in the prosecution of the work contemplated by this contract. (ASPR 7-103.9)

17. COVENANT AGAINST CONTINGENT FEES (1958 JAN)

The contractor warrants that no person or selling agency has been employed or retained to solicit or secure this contract upon an agreement or understanding for a commission, percentage, brokerage, or contingent fee, excepting bona fide employees or bona fide established commercial or selling agencies maintained by the Contractor for the purpose of securing business. For breach or violation of this warranty the Government shall have the right to annul this contract without liability or in its discretion, to deduct from the contract price or consideration, or otherwise recover, the full amount of such commission, percentage, brokerage or contingent fee. (ASPR 7-103.20)

18. OFFICIALS NOT TO BENEFIT (1964 JUN)

No Member of Congress or resident Commissioner shall be admitted to any share or part of this contract, or to any benefit that may arise therefrom; but this provision shall not be construed to extend to this contract if made with a corporation for its general benefit. (ASPR 7-602.19)

19. BUY AMERICAN ACT (1966 OCT)

(a) *Agreement.* In accordance with the Buy American Act (41 U.S.C. 10a-10d), the Contractor agrees that only domestic construction material will be used (by the Contractor, subcontractors, materialmen, and suppliers) in the performance of this contract, except for nondomestic construction material listed in the "Nondomestic Construction Materials" clause, if any, of this contract.

(b) *Domestic construction material.* "Construction material" means any article, material, or supply brought to the construction site for incorporation in the building or work. An unmanufactured construction material is a "domestic construction material" if it has been mined or produced in the United States. A manufactured construction material is a "domestic construction material" if it has been manufactured in the United States and if the cost of its components which have been mined, produced, or manufactured in the United States exceeds 50 percent of the cost of all its components. "Component" means any article, material, or supply directly incorporated in a construction material.

(c) *Domestic component.* A component shall be considered to have been "mined, produced, or manufactured in the United States" (regardless of its source in fact) if the article, material, or supply in which it is incorporated was manufactured in the United States and the component is of a class or kind determined by the Government to be not mined, produced, or manufactured in the United States in sufficient and reasonably available commercial quantities and of a satisfactory quality. (ASPR 7-602.20)

20. CONVICT LABOR (1949 MAR)

In connection with the performance of work under this contract, the Contractor agrees not to employ any person undergoing sentence of imprisonment at hard labor. (ASPR 7-104.17)

21. EQUAL OPPORTUNITY (1971 APR)

(The following clause is applicable unless this contract is exempt under the rules, regulations and relevant orders of the Secretary of Labor (41 CFR, Chapter 60). Exemptions include contracts and subcontracts (i) not exceeding $10,000, and (ii) under which work is performed outside the United States and no recruitment of workers within the United States is involved.)

During the performance of this contract, the contractor agrees as follows:

(a) The Contractor will not discriminate against any employee or applicant for employment because of race, color, religion, sex, or national origin. The Contractor will take affirmative action to ensure that applicants are employed, and that employees are treated during employment, without regard to their race, color, religion, sex, or national origin. Such action shall include, but not be limited to, the following: Employment, upgrading, demotion, or transfer; recruitment or recruitment advertising; layoff or termination; rates of pay or other forms of compensation; and selection for training, including apprenticeship. The Contractor agrees to post in conspicuous places, available to employees and applicants for employment, notices to be provided by the Contracting Officer setting forth the provisions of this Equal Opportunity clause.

(b) The Contractor will, in all solicitations or advertisements for employees placed by or on behalf of the Contractor, state that all qualified applicants will receive consideration for employment without regard to race, color, religion, sex, or national origin.

(c) The Contractor will send to each labor union or representative of workers with which he has a collective bargaining agreement or other contract or understanding, a notice, to be provided by the agency Contracting Officer, advising the labor union or workers' representative of the contractor's commitments under this Equal Opportunity clause, and shall post copies of the notice in conspicuous places available to employees and applicants for employment.

(d) The Contractor will comply with all provisions of Executive Order No. 11246 of September 24, 1965, and of the rules, regulations, and relevant orders of the Secretary of Labor.

(e) The Contractor will furnish all information and reports required by Executive Order No. 11246 of September 24, 1965, and by the rules, regulations, and orders of the Secretary of Labor, or pursuant thereto, and will permit access to his books, records, and accounts by the contracting agency and the Secretary of Labor for purposes of investigation to ascertain compliance with such rules, regulations, and orders.

(f) In the event of the Contractor's noncompliance with the Equal Opportunity clause of this contract or with any of the said rules, regulations, or orders, this contract may be canceled, terminated, or suspended, in whole or in part, and the Contractor may be declared ineligible for further Government contracts in accordance with procedures authorized in Executive Order No. 11246 of September 24, 1965, and such other sanctions may be imposed and remedies invoked as provided in Executive Order No. 11246 of September 24, 1965, or by rule, regulation, or order of the Secretary of Labor, or as otherwise provided by law.

(g) The Contractor will include the provisions of paragraphs (a) through (g) in every subcontract or purchase order unless exempted by rules, regulations, or orders of the Secretary of Labor issued pursuant to Section 204 of Executive Order No. 11246 of September 24, 1965, so that such provisions will be binding upon each subcontractor or vendor. The Contractor will take such action with respect to any subcontract or purchase order as the contracting agency may direct as a means of enforcing such provisions, including sanctions for noncompliance: *Provided, however,* That in the event the Contractor becomes involved in, or is threatened with, litigation with a subcontractor or vendor as a result of such direction by the contracting agency, the Contractor may request the United States to enter into such litigation to protect the interests of the United States. (ASPR 7-103.18(a))

22. UTILIZATION OF SMALL BUSINESS CONCERNS (1958 JAN)

(The following clause is applicable if this contract is in excess of $5,000)

(a) It is the policy of the Government as declared by the Congress that a fair proportion of the purchases and contracts for supplies and services for the Government be placed with small business concerns.

(b) The Contractor agrees to accomplish the maximum amount of subcontracting to small business concerns that the Contractor finds to be consistent with the efficient performance of this contract. (ASPR 7-104.14(a))

23. SUSPENSION OF WORK (1968 FEB)

(a) The Contracting Officer may order the Contractor in writing to suspend, delay, or interrupt all or any part of the work for such period of time as he may determine to be appropriate for the convenience of the Government.

(b) If the performance of all or any part of the work is, for an unreasonable period of time, suspended, delayed, or interrupted by an act of the Contracting Officer in the administration of this contract, or by his failure to act within the time specified in this contract (or if no time is specified, within a reasonable time), an adjustment shall be made for any increase in the cost of performance of this contract (excluding profit) necessarily caused by such unreasonable suspension, delay, or interruption and the contract modified in writing accordingly. However, no adjustment shall be made under this clause for any suspension, delay, or interruption to the extent (1) that performance would have been so suspended, delayed, or interrupted by any other cause, including the fault or negligence of the Contractor or (2) for which an equitable adjustment is provided for or excluded under any other provision of this contract.

(c) No claim under this clause shall be allowed (1) for any costs incurred more than 20 days before the Contractor shall have notified the Contracting Officer in writing of the act or failure to act involved (but this requirement shall not apply as to a claim resulting from a suspension order), and (2) unless the claim, in an amount stated, is asserted in writing as soon as practicable after the termination of such suspension, delay, or interruption, but not later than the date of final payment under the contract. (ASPR 7-602.46)

24. DAVIS–BACON ACT (40 U.S.C. 276a to a–7) (1972 FEB)

(a) All mechanics and laborers employed or working directly upon the site of the work shall be paid unconditionally and not less often than once a week, and

without subsequent deduction or rebate on any account (except such payroll deductions as are permitted by the Copeland Regulations (29 CFR, Part 3)), the full amounts due at time of payment computed at wage rates not less than the aggregate of the basic hourly rates and the rates of payments, contributions, or costs for any fringe benefits contained in the wage determination decision of the Secretary of Labor which is attached hereto and made a part hereof, regardless of any contractual relationship which may be alleged to exist between the Contractor or subcontractor and such laborers and mechanics. A copy of such wage determination decision shall be kept posted by the Contractor at the site of the work in a prominent place where it can be easily seen by the workers.

(b) The Contractor may discharge his obligation under this clause to workers in any classification for which the wage determination decision contains:

(1) Only a basic hourly rate of pay, by making payment at not less than such basic hourly rate, except as otherwise provided in the Copeland Regulation (29 CFR, Part 3); or

(2) Both a basic hourly rate of pay and fringe benefits payments, by making payment in cash, by irrevocably making contributions pursuant to a fund, plan, or program for, and/or by assuming an enforceable commitment to bear the cost of, bona fide fringe benefits contemplated by the Davis–Bacon Act, or by any combination thereof. Contributions made, or costs assumed, on other than a weekly basis shall be considered as having been constructively made or assumed, during a weekly period to the extent that they apply to such period. Where a fringe benefit is expressed in a wage determination in any manner other than as an hourly rate and the Contractor pays a cash equivalent or provides an alternative fringe benefit, he shall furnish information with his payrolls showing how he determined that the cost incurred to make the cash payment or to provide the alternative fringe benefit is equal to the cost of the wage determination fringe benefit. In any case where the Contractor provides a fringe benefit different from any contained in the wage determination, he shall similarly show how he arrived at the hourly rate shown therefor. In the event of disagreement between or among the interested parties as to an equivalent of any fringe benefit, the Contracting Officer shall submit the question, together with his recommendation, to the Secretary of Labor for final determination.

(c) The assumption of an enforceable commitment to bear the cost of fringe benefits, or the provision of any fringe benefits not expressly listed in section 1(b)(2) of the Davis–Bacon Act or in the wage determination decision forming a part of the contract, may be considered as payment of wages only with the approval of the Secretary of Labor pursuant to a written request by the Contractor. The Secretary of Labor may require the Contractor to set aside assets, in a separate account, to meet his obligations under any unfunded plan or program.

(d) The Contracting Officer shall require that any class of laborers or mechanics, including apprentices and trainees, which is not listed in the wage determination decision and which is to be employed under the contract shall be classified or reclassified conformably to the wage determination decision and shall report the action taken to the Secretary of Labor. If the interested parties cannot agree on the proper classification or reclassification of a particular class of laborers or mechanics, including apprentices and trainees, to be used, the Contracting Officer shall submit the question, together with his recommendation, to the Secretary of Labor for final determination.

(e) In the event it is found by the Contracting

Officer that any laborer or mechanic employed by the Contractor or any subcontractor directly on the site of the work covered by this contract has been or is being paid at a rate of wages less than the rate of wages required by paragraph (a) of this clause, the Contracting Officer may (i) by written notice to the Government Prime Contractor terminate his right to proceed with the work, or such part of the work as to which there has been a failure to pay said required wages, and (ii) prosecute the work to completion by contract or otherwise, whereupon such Contractor and his sureties shall be liable to the Government for any excess costs occasioned the Government thereby.

(f) Paragraphs (a) through (e) of the clause shall apply to this contract to the extent that it is (i) a prime contract with the Government subject to the Davis–Bacon Act or (ii) a subcontract also subject to the Davis–Bacon Act under such prime contract. (ASPR 7–602.23(a)(i))

25. CONTRACT WORK HOURS AND SAFETY STANDARDS ACT – OVERTIME COMPENSATION (40 U.S.C. 327–333) (1972 FEB)

This contract is subject to the Contract Work Hours and Safety Standards Act and to the applicable rules, regulations, and interpretations of the Secretary of Labor.

(a) The Contractor shall not require or permit any laborer or mechanic in any workweek in which he is employed on any work under this contract to work in excess of eight (8) hours in any calendar day or in excess of forty (40) hours in such workweek on work subject to the provisions of the Contract Work Hours and Safety Standards Act unless such laborer or mechanic receives compensation at a rate not less than one and one–half times his basic rate of pay for all such hours worked in excess of eight (8) hours in any calendar day or in excess of forty (40) hours in such workweek, whichever is the greater number of overtime hours. The "basic rate of pay," as used in this clause, shall be the amount paid per hour, exclusive of the Contractor's contribution or cost for fringe benefits and any cash payment made in lieu of providing fringe benefits, or the basic hourly rate contained in the wage determination, whichever is greater.

(b) In the event of any violation of the provisions of paragraph (a), the Contractor shall be liable to any affected employee for any amounts due, and to the United States for liquidated damages. Such liquidated damages shall be computed with respect to each individual laborer or mechanic employed in violation of the provisions of paragraph (a) in the sum of $10 for each calendar day on which such employee was required or permitted to be employed on such work in excess of eight (8) hours or in excess of the standard workweek of forty (40) hours without payment of the overtime wages required by paragraph (a). (ASPR 7–602.23(a)(ii))

26. APPRENTICES AND TRAINEES (1972 FEB)

(a) Apprentices shall be permitted to work as such only when they are registered, individually, under a bona fide apprenticeship program registered with a State apprenticeship agency which is recognized by the Bureau of Apprenticeship and Training, United States Department of Labor; or, if no such recognized agency exists in a State, under a program registered with the aforesaid Bureau of Apprenticeship and Training. The allowable ratio of apprentices to journeymen in any craft classification shall not be greater than the ratio permitted to the Contractor as to his entire work force under the registered program. Any employee listed on a payroll at an apprentice wage rate,

who is not a trainee as defined in subparagraph (b) of this clause, who is not registered as above, shall be paid the wage rate determined by the Secretary of Labor for the classification of work he actually performed. The Contractor shall furnish written evidence of the registration of his program and apprentices as well as of the ratios allowed and the wage rates required to be paid thereunder for the area of construction, prior to using any apprentices in the contract work. "Apprentice" means a person employed and individually registered in a bona fide apprenticeship program registered with the United States Department of Labor, Bureau of Apprenticeship and Training, or with a State apprenticeship agency recognized by the Bureau, or a person in his first 90 days of probationary employment as an apprentice in such an apprenticeship program, who is not individually registered in the program, but who has been certified by the Bureau of Apprenticeship and Training or a State Apprenticeship Council to be eligible for probationary employment as an apprentice.

(b) Trainees shall be permitted to work as such when they are bona fide trainees employed pursuant to a program approved by the U.S. Department of Labor, Manpower Administration, Bureau of Apprenticeship and Training. "Trainee" means a person receiving on–the–job training in a construction occupation under a program which is approved (but not necessarily sponsored) by the U.S. Department of Labor, Manpower Administration, Bureau of Apprenticeship and Training, and which is reviewed from time to time by the Manpower Administration to insure that the training meets adequate standards.

(c) The Contractor shall make a diligent effort to hire for performance of work under this contract a number of apprentices or trainees, or both, in each occupation, which bears to the average number of the journeymen in that occupation to be employed in the performance of the contract the applicable ratio as set forth in paragraph (c)(6) of this clause.

  (1) The Contractor shall assure that twenty–five per cent (25%) of such apprentices or trainees in each occupation are in their first year of training, where feasible. Feasibility here involves a consideration of
    (i) the availability of training opportunities for first year apprentices,
    (ii) the hazardous nature of the work for beginning workers and
    (iii) excessive unemployment of apprentices in their second and subsequent years of training.

  (2) The Contractor shall, during the performance of the contract, to the greatest extent possible, employ the number of apprentices or trainees necessary to meet currently the requirements of paragraphs (c) and (c)(1) of this clause.

  (3) The Contractor shall maintain records of employment on this contract by trade of the number of apprentices and trainees, apprentices and trainees in first year of training, and of journeymen, and the wages paid and hours of work of such apprentices, trainees and journeymen. In addition, the Contractor who claims compliance based on the criterion set forth in paragraph (4)(ii) of this clause

shall maintain such records of employment on all his construction work in the same labor market area, both public and private, during the performance of this contract.

(4) The Contractor will be deemed to have made a "diligent effort" as required by paragraph (c) if during the performance of this contract, he accomplishes at least one of the following three objectives:

(i) the Contractor employs under this contract a number of apprentices and trainees by craft, at least equal to the ratios established in accordance with paragraph (6) of this clause, or

(ii) the Contractor employs, on all his construction work, both public and private, in the same labor market area, an average number of apprentices and trainees by craft at least equal to the ratios established in accordance with paragraph (6) of this clause, or

(iii) the Contractor

(A) if covered by a collective bargaining agreement, before commencement of any work on the project, has given written notice to all joint apprenticeship committees, the local U.S. Employment Security Office, local chapter of the Urban League, Workers Defense League, or other local organizations concerned with minority employment, and the Bureau of Apprenticeship and Training Representatives, U.S. Department of Labor for the locality of the work;

(B) if not covered by a collective bargaining agreement, has given written notice to all of the groups stated above, except joint apprenticeship committees, and will in addition notify all non–joint apprenticeship sponsors in the labor market area;

(C) has employed all qualified applicants referred to him through normal channels (such as the Employment Service, the Joint Apprenticeship Committees, and where applicable, minority organizations and apprentice outreach programs who have been delegated this function) at least up to the number of such apprentices and trainees required by paragraph (6) of this clause;

(D) notice, as referred to herein, will include at least the Contractor's name and address, job site address, value of the contract, expected starting and completion dates, the estimated average number of employees in each occupation to be employed over the duration of the contract work, and a statement of his willingness to employ a number of apprentices and trainees at least equal to the ratios established in accordance with paragraph (6) of this clause. A copy of this notice shall be furnished to the Contracting Officer upon request.

(5) The Contractor shall supply, to the Contracting Officer, and to the Secretary of Labor, a report at three month intervals during performance of the contract and after completion of contract performance a statement describing steps taken toward making a diligent effort and containing a breakdown by craft, of hours worked and wages paid for first year apprentices and trainees, other apprentices and trainees, and journeymen.

(6) The applicable ratios of apprentices and trainees to journeymen in any occupation for the purpose of this clause shall be as follows:

(i) In any occupation the applicable ratio of apprentices and trainees to journeymen shall be equal to the predominant ratio for the occupation in the area where the construction is being undertaken, set forth in collective bargaining agreements, or other employment agreements, and available through the Bureau of Apprenticeship and Training Representative, U.S. Department of Labor for the locality of the work.

(ii) For any occupation for which no ratio is found, the ratio of apprentices and trainees to journeymen shall be determined by the contractor in accordance with the recommendations set forth in the Standards of the National Joint Apprentice Committee for the occupation, which are on file at offices of the U.S. Department of Labor's Bureau of Apprenticeship and Training.

(iii) For any occupation for which no such recommendations are found, the ratio of apprentices and trainees to journeymen shall be at least one apprentice or trainee for every five journeymen.

NOTE: Paragraphs (a) and (b) of this clause apply to contracts in excess of $2,000; in addition, paragraph (c) applies to contracts in excess of $10,000. (ASPR 7–602.23(a)(iii))

27. PAYROLLS AND BASIC RECORDS (1969 JUN)

(a) The Contractor shall maintain payrolls and basic records relating thereto during the course of the work and shall preserve them for a period of three (3) years thereafter for all laborers and mechanics working at the site of the work. Such records shall contain the name and address of each such employee, his correct classification, rate of pay (including rates of contributions for, or costs assumed to provide, fringe benefits), daily and weekly number of hours worked, deductions made and actual wages paid. Whenever the Contractor has obtained approval from the Secretary of Labor as provided in paragraph (c) of the clause entitled "Davis–Bacon Act," he shall maintain records which show the commitment, its approval, written communication of the plan or program to the laborers or mechanics affected, and the costs anticipated or incurred under the plan or program.

(b) The Contractor shall submit weekly a copy of all payrolls to the Contracting Officer. The Government Prime Contractor shall be responsible for the submission of copies of payrolls of all subcontractors. The copy shall be accompanied by a statement signed by the Contractor indicating that the payrolls are correct and complete, that the wage rates contained therein are not less than those determined by the Secretary of Labor, and that the classifications set forth for each laborer or mechanic conform with the work he performed. Weekly submission of the "Statement of Compliance" required under this contract and the Copeland Regulations of the Secretary of Labor (29 CFR, Part 3) shall satisfy the requirement for submission of the above statement. The Contractor shall submit also a copy of any approval by the Secretary of Labor with respect to fringe benefits which is required by paragraph (c) of the clause entitled "Davis–Bacon Act."

(c) The Contractor shall make the records required under this clause available for inspection by authorized representatives of the Contracting Officer and the Department of Labor, and shall permit such representatives to interview employees during working hours on the job. (ASPR 7–602.23(a)(iv))

28. COMPLIANCE WITH COPELAND REGULATIONS (1964 JUN)

The Contractor shall comply with the Copeland Regulations of the Secretary of Labor (29 CFR, Part 3) which are incorporated herein by reference. (ASPR 7–602.23(a)(v))

29. WITHHOLDING OF FUNDS (1972 FEB)

(a) The Contracting Officer may withhold or cause to be withheld from the Government Prime Contractor so much of the accrued payments or advances as may be considered necessary (i) to pay laborers and mechanics, including apprentices and trainees, employed by the Contractor or any subcontractor on the work the full amount of wages required by the contract, and (ii) to satisfy any liability of any Contractor for liquidated damages under the clause hereof entitled "Contract Work Hours and Safety Standards Act – Overtime Compensation."

(b) If any Contractor fails to pay any laborer or mechanic including any apprentice or trainee employed or working on the site of the work, all or part of the wages required by the contract, the Contracting Officer may, after written notice to the Government Prime Contractor, take such action as may be necessary to cause suspension of any further payments or advances until such violations have ceased. (ASPR 7–602.23(a)(vi))

30. SUBCONTRACTS (1972 FEB)

The Contractor agrees to insert the clauses hereof entitled "Davis–Bacon Act," "Contract Work Hours and Safety Standards Act – Overtime Compensation," "Apprentices and Trainees," "Payrolls and Basic Records," "Compliance with Copeland Regulations," "Withholding of Funds," "Subcontracts," and "Contract Termination – Debarment" in all subcontracts. The term "Contractor" as used in such clauses in any subcontract shall be deemed to refer to the subcontractor except in the phrase "Government Prime Contractor." (ASPR 7–602.23(a)(vii))

31. CONTRACT TERMINATION – DEBARMENT (1972 APR)

A breach of the clauses hereof entitled "Davis–Bacon Act," "Contract Work Hours and Safety Standards Act – Overtime Compensation," "Apprentices and Trainees," "Payrolls and Basic Records," "Compliance with Copeland Regulations," "Withholding of Funds," and "Subcontracts" may be grounds for termination of the contract, and for debarment as provided in 29 CFR 5.6. (ASPR 7–602.23(a)(viii))

32. CONTRACTOR INSPECTION SYSTEM (1964 NOV)

The Contractor shall (i) maintain an adequate inspection system and perform such inspections as will assure that the work performed under the contract conforms to contract requirements, and (ii) maintain and make available to the Government adequate records of such inspections. (ASPR 7-602.10(a))

33. GRATUITIES (1952 MAR)

(a) The Government may, by written notice to the Contractor, terminate the right of the Contractor to proceed under this contract if it is found, after notice and hearing, by the Secretary or his duly authorized representative, that gratuities (in the form of entertainment, gifts, or otherwise) were offered or given by the Contractor, or any agent or representative of the Contractor, to any officer or employee of the Government with a view toward securing a contract or securing favorable treatment with respect to the awarding or amending, or the making of any determinations with respect to the performing of such contract; *provided,* that the existence of the facts upon which the Secretary or his duly authorized representative makes such findings shall be in issue and may be reviewed in any competent court.

(b) In the event this contract is terminated as provided in paragraph (a) hereof, the Government shall be entitled (i) to pursue the same remedies against the Contractor as it could pursue in the event of a breach of the contract by the Contractor, and (ii) as a penalty in addition to any other damages to which it may be entitled by law, to exemplary damages in an amount (as determined by the Secretary or his duly authorized representative) which shall be not less than three nor more than ten times the costs incurred by the Contractor in providing any such gratuities to any such officer or employee.

(c) The rights and remedies of the Government provided in this clause shall not be exclusive and are in addition to any other rights and remedies provided by law or under this contract. (ASPR 7-104.16)

34. SMALL BUSINESS SUBCONTRACTING PROGRAM (MAINTENANCE, REPAIR AND CONSTRUCTION) (1967 JUN)

(The following clause is applicable if this contract is in excess of $500,000)

(a) The Contractor agrees to establish and conduct a small business subcontracting program which will enable small business concerns to be considered fairly as subcontractors, including suppliers, under this contract. In this connection, the Contractor shall designate an individual to (i) maintain liaison with the Government on small business matters, and (ii) administer the Contractor's Small Business Subcontracting Program.

(b) Notwithstanding the instructions on DD Form 1140-1, prior to completion of the contract and as soon as the final information is available, the Contractor shall submit a one-time completed DD Form 1140-1 to the Government addressees prescribed thereon. The DD Form 1140-1 shall show the prime contract number in lieu of identifying a quarterly report period. This subparagraph (b) is not applicable if the Contractor is a small business concern.

(c) The Contractor further agrees (i) to insert the "Utilization of Small Business Concerns" clause in subcontracts which offer substantial subcontracting opportunities, and (ii) to insert in each such subcontract exceeding $500,000 a clause conforming substantially to the language of this clause except that subcontractors shall submit DD Form 1140-1 direct to the Government addressees prescribed on the Form. The Contractor will notify the Contracting Officer of the name and address of each subcontractor that will be required to submit a report on DD Form 1140-1. (ASPR 7-602.26(b))

35. FEDERAL, STATE, AND LOCAL TAXES (1971 NOV)

(a) Except as may be otherwise provided in this contract, the contract price includes all applicable Federal, State, and local taxes and duties.

(b) Nevertheless, with respect to any Federal excise tax or duty on the transactions or property covered by this contract, if a statute, court decision, written ruling, or regulation takes effect after the contract date, and --

(1) results in the Contractor being required to pay or bear the burden of any such Federal excise tax or duty or increase in the rate thereof which would not otherwise have been payable on such transactions or property, the contract price shall be increased by the amount of such tax or duty or rate increase, *provided* the Contractor warrants in writing that no amount for such newly imposed Federal excise tax or duty or rate increase was included in the contract price as a contingency reserve or otherwise; or

(2) results in the Contractor not being required to pay or bear the burden of, or in his obtaining a refund or drawback of, any such Federal excise tax or duty which would otherwise have been payable on such transactions or property or which was the basis of an increase in the contract price, the contract price shall be decreased by the amount of the relief, refund, or drawback, or that amount shall be paid to the Government, as directed by the Contracting Officer. The contract price shall be similarly decreased if the Contractor, through his fault or negligence or his failure to follow instructions of the Contracting Officer, is required to pay or bear the burden of, or does not obtain a refund or drawback of, any such Federal excise tax or duty.

(c) Paragraph (b) above shall not be applicable to social security taxes or to any other employment tax.

(d) No adjustment of less than $100 shall be made in the contract price pursuant to paragraph (b) above.

(e) As used in paragraph (b) above, the term "contract date" means the date set for bid opening, or if this is a negotiated contract, the contract date. As to additional supplies or services procured by modification to this contract, the term "contract date" means the date of such modification.

(f) Unless there does not exist any reasonable basis to sustain an exemption, the Government upon the request of the Contractor shall, without further liability, furnish evidence appropriate to establish exemption from any Federal, State, or local tax; *provided* that, evidence appropriate to establish exemption from any Federal excise tax or duty which may give rise to either an increase or decrease in the contract price will be furnished only at the discretion of the Government.

(g) The Contractor shall promptly notify the Contracting Officer of matters which will result in either an increase or decrease in the contract price, and shall take action with respect thereto as directed by the Contracting Officer. (ASPR 7-103.10(a))

36. RENEGOTIATION (1959 OCT)

(a) To the extent required by law, this contract is subject to the Renegotiation Act of 1951 (50 U.S.C. App. 1211, et seq.), as amended, and to any subsequent act of Congress providing for the renegotation of contracts. Nothing contained in this clause shall impose any renegotiation obligation with respect to this contract or any subcontract hereunder which is not imposed by an act of Congress heretofore or hereafter enacted. Subject to the foregoing this contract shall be deemed to contain all the provisions required by section 104 of the Renegotiation Act of 1951, and by any such other act, without subsequent contract amendment specifically incorporating such provisions.

(b) The Contractor agrees to insert the provisions of this clause, including this paragraph (b), in all subcontracts, as that term is defined in section 103g of the Renegotiation Act of 1951, as amended. (ASPR 7-103.13(a))

37. TERMINATION FOR CONVENIENCE OF THE GOVERNMENT (1971 NOV)

(a) The performance of work under this contract may be terminated by the Government in accordance with this clause in whole, or from time to time in part, whenever the Contracting Officer shall determine that such termination is in the best interest of the Government. Any such termination shall be effected by delivery to the Contractor of a Notice of Termination specifying the extent to which performance of work under the contract is terminated, and the date upon which such termination becomes effective.

(b) After receipt of a Notice of Termination, and except as otherwise directed by the Contracting Officer, the Contractor shall,

(i) stop work under the contract on the date and to the extent specified in the Notice of Termination;

(ii) place no further orders or subcontracts for materials, services or facilities, except as may be necessary for completion of such portion of the work under the contract as is not terminated;

(iii) terminate all orders and subcontracts to the extent that they relate to the performance of work terminated by the Notice of Termination;

(iv) assign to the Government, in the manner, at the times, and to the extent directed by the Contracting Officer, all of the right, title, and interest of the Contractor under the orders and subcontracts so terminated, in which case the Government shall have the right, in its discretion, to settle or pay any or all claims arising out of the termination of such orders and subcontracts;

(v) settle all outstanding liabilities and all claims arising out of such termination of orders and subcontracts, with the approval or ratification of the Contracting Officer, to the extent he may require, which approval or ratification shall be final for all the purposes of this clause;

(vi) transfer title and deliver to the Government, in the manner, at the times, and to the extent, if any, directed by the Contracting Officer, (A) the fabricated or unfabricated parts, work in process, completed work, supplies, and other material produced as a part of, or acquired in connection with the performance of, the work terminated by the Notice of Termination, and (B) the completed or partially completed plans, drawings, information, and other property which, if the contract had been completed, would have been required to be furnished to the Government;

(vii) use his best efforts to sell, in the manner, at the times, to the extent, and at the price or prices directed or authorized by the Contracting Officer, any property of the types referred to in (vi) above; provided, however, that the Contractor (A) shall not be required to extend credit to any purchaser, and (B) may acquire any such property under the conditions prescribed by and at a price or prices approved by the Contracting Officer; and provided further that the proceeds of any such transfer or disposition shall be applied in reduction of any payments to be made by the Government to the Contractor under this contract or shall otherwise be credited to the price or cost of the work covered by this contract or paid in such other manner as the Contracting Officer may direct;

(viii) complete performance of such part of the work as shall not have been terminated by the Notice of Termination; and

(ix) take such action as may be necessary, or as the Contracting Officer may direct, for the protection and preservation of the property related to this contract which is in the possession of the Contractor and in which the Government has or may acquire an interest.

At any time after expiration of the plant clearance period, as defined in Section VIII, Armed Services Procurement Regulation, as it may be amended from time to time, the Contractor may submit to the Contracting Officer a list, certified as to quantity and quality, of any or all items of termination inventory not previously disposed of, exclusive of items the disposition of which has been directed or authorized by the Contracting Officer, and may request the Government to remove such items or enter into a storage agreement covering them. Not later than fifteen (15) days thereafter, the Government will accept title to such items

and remove them or enter into a storage agreement covering the same; provided, that the list submitted shall be subject to verification by the Contracting Officer upon removal of the items, or if the items are stored, within forty—five (45) days from the date of submission of the list, and any necessary adjustment to correct the list as submitted shall be made prior to final settlement.

(c) After receipt of a Notice of Termination, the Contractor shall submit to the Contracting Officer his termination claim, in the form and with certification prescribed by the Contracting Officer. Such claim shall be submitted promptly but in no event later than one year from the effective date of termination, unless one or more extensions in writing are granted by the Contracting Officer, upon request of the Contractor made in writing within such one year period or authorized extension thereof. However, if the Contracting Officer determines that the facts justify such action, he may receive and act upon any such termination claim at any time after such one year period or any extension thereof. Upon failure of the Contractor to submit his termination claim within the time allowed, the Contracting Officer may, subject to any Settlement Review Board approvals required by Section VIII of the Armed Services Procurement Regulation in effect as of the date of execution of this contract, determine, on the basis of information available to him, the amount, if any, due to the Contractor by reason of the termination and shall thereupon pay to the Contractor the amount so determined.

(d) Subject to the provisions of paragraph (c), and subject to any Settlement Review Board approvals required by Section VIII of the Armed Services Procurement Regulation in effect as of the date of execution of this contract, the Contractor and the Contracting Officer may agree upon the whole or any part of the amount or amounts to be paid to the Contractor by reason of the total or partial termination of work pursuant to this clause, which amount or amounts may include a reasonable allowance for profit on work done; provided, that such agreed amount or amounts, exclusive of settlement costs, shall not exceed the total contract price as reduced by the amount of payments otherwise made and as further reduced by the contract price of work not terminated. The contract shall be amended accordingly, and the Contractor shall be paid the agreed amount. Nothing in paragraph (e) of this clause, prescribing the amount to be paid to the Contractor in the event of failure of the Contractor and the Contracting Officer to agree upon the whole amount to be paid to the Contractor by reason of the termination of work pursuant to this clause, shall be deemed to limit, restrict, or otherwise determine or affect the amount or amounts which may be agreed upon to be paid to the Contractor pursuant to this paragraph (d).

(e) In the event of the failure of the Contractor and the Contracting Officer to agree, as provided in paragraph (d), upon the whole amount to be paid to the Contractor by reason of the termination of work pursuant to this clause, the Contracting Officer shall, subject to any Settlement Review Board approvals required by Section VIII of the Armed Services Procurement Regulation in effect as of the date of execution of this contract, pay to the Contractor the amounts determined by the Contracting Officer as follows, but without duplication of any amounts agreed upon in accordance with paragraph (d):

(i) with respect to all contract work performed prior to the effective date of the Notice of Termination, the total (without duplication of any items) of — —

(A) the cost of such work;

(B) the cost of settling and paying claims arising out of the termination of work under subcontracts or orders as provided in paragraph (b)(v) above, exclusive of the amounts paid or payable on account of supplies or materials delivered or services furnished by the subcontractor prior to the effective date of the Notice of Termination of Work under this contract, which amounts shall be included in the cost on account of which payment is made under (A) above; and

(C) a sum, as profit on (A) above, determined by the Contracting Officer pursuant to 8-303 of the Armed Services Procurement Regulation, in effect as of the date of execution of this contract, to be fair and reasonable; *provided,* however, that if it appears that the Contractor would have sustained a loss on the entire contract had it been completed, no profit shall be included or allowed under this subdivision (C) and an appropriate adjustment shall be made reducing the amount of the settlement to reflect the indicated rate of loss; and

(ii) the reasonable cost of the preservation and protection of property incurred pursuant to paragraph (b)(ix); and any other reasonable cost incidental to termination of work under this contract, including expense incidental to the determination of the amount due to the Contractor as the result of the termination of work under this contract.

The total sum to be paid to the Contractor under (i) above shall not exceed the total contract price as reduced by the amount of payments otherwise made and as further reduced by the contract price of work not terminated. Except for normal spoilage, and except to the extent that the Government shall have otherwise expressly assumed the risk of loss, there shall be excluded from the amounts payable to the Contractor under (i) above, the fair value, as determined by the Contracting Officer, of property which is destroyed, lost, stolen, or damaged so as to become undeliverable to the Government, or to a buyer pursuant to paragraph (b)(vii).

(f) Costs claimed, agreed to, or determined pursuant to (c), (d), and (e) hereof shall be in accordance with Section XV of the Armed Services Procurement Regulation as in effect on the date of this contract.

(g) The Contractor shall have the right of appeal, under the clause of this contract entitled "Disputes," from any determination made by the Contracting Officer under paragraph (c) or (e) above, except that if the Contractor has failed to submit his claim within the time provided in paragraph (c) above and has failed to request extension of

such time, he shall have no such right of appeal. In any case where the Contracting Officer has made a determination of the amount due under paragraph (c) or (e) above, the Government shall pay to the Contractor the following: (i) if there is no right of appeal hereunder or if no timely appeal has been taken, the amount so determined by the Contracting Officer, or (ii) if an appeal has been taken, the amount finally determined on such appeal.

(h) In arriving at the amount due the Contractor under this clause there shall be deducted (i) all unliquidated advance or other payments on account theretofore made to the Contractor, applicable to the terminated portion of this contract, (ii) any claim which the Government may have against the Contractor in connection with this contract, and (iii) the agreed price for, or the proceeds of sale of, any materials, supplies, or other things acquired by the Contractor or sold, pursuant to the provisions of this clause, and not otherwise recovered by or credited to the Government.

(i) If the termination hereunder be partial, prior to the settlement of the terminated portion of this contract, the Contractor may file with the Contracting Officer a request in writing for an equitable adjustment of the price or prices specified in the contract relating to the continued portion of the contract (the portion not terminated by the Notice of Termination), and such equitable adjustment as may be agreed upon shall be made in such price or prices.

(j) The Government may from time to time, under such terms and conditions as it may prescribe, make partial payments and payments on account against costs incurred by the Contractor in connection with the terminated portion of this contract whenever in the opinion of the Contracting Officer the aggregate of such payments shall be within the amount to which the Contractor will be entitled hereunder. If the total of such payments is in excess of the amount finally agreed or determined to be due under this clause, such excess shall be payable by the Contractor to the Government upon demand, together with interest computed at the rate of 6 percent per annum, for the period from the date such excess payment is received by the Contractor to the date on which such excess is repaid to the Government; *provided,* however, that no interest shall be charged with respect to any such excess payment attributable to a reduction in the Contractor's claim by reason of retention or other disposition of termination inventory until ten days after the date of such retention or disposition, or such later date as determined by the Contracting Officer by reason of the circumstances.

(k) Unless otherwise provided for in this contract, or by applicable statute, the Contractor shall — from the effective date of termination until the expiration of three years after final settlement under this contract — preserve and make available to the Government at all reasonable times at the office of the Contractor but without direct charge to the Government, all his books, records, documents and other evidence bearing on the costs and expenses of the Contractor under this contract and relating to the work terminated hereunder, or, to the extent approved by the Contracting Officer, photographs, microphotographs, or other authentic reproductions thereof. (ASPR 7-103.21(b) & 7-602.29(a))

38. NOTICE AND ASSISTANCE REGARDING PATENT AND COPYRIGHT INFRINGEMENT (1965 JAN)

(The provisions of this clause shall be applicable only if the amount of this contract exceeds $10,000.)

(a) The Contractor shall report to the Contracting Officer, promptly and in reasonable written detail, each notice or claim of patent or copyright infringement based on the performance of this contract of which the Contractor has knowledge.

(b) In the event of any claim or suit against the Government on account of any alleged patent or copyright infringement arising out of the performance of this contract or out of the use of any supplies furnished or work or services performed hereunder, the Contractor shall furnish to the Government, when requested by the Contracting Officer, all evidence and information in possession of the Contractor pertaining to such suit or claim. Such evidence and information shall be furnished at the expense of the Government except where the Contractor has agreed to indemnify the Government.

(c) This clause shall be included in all subcontracts. (ASPR 7-103.23)

39. AUTHORIZATION AND CONSENT (1964 MAR)

The Government hereby gives its authorization and consent (without prejudice to any rights of indemnification) for all use and manufacture, in the performance of this contract or any part hereof or any amendment hereto or any subcontract hereunder (including any lower-tier subcontract), of any invention described in and covered by a patent of the United States (i) embodied in the structure or composition of any article the delivery of which is accepted by the Government under this contract, or (ii) utilized in the machinery, tools, or methods the use of which necessarily results from compliance by the Contractor or the using subcontractor with (a) specifications or written provisions now or hereafter forming a part of this contract, or (b) specific written instructions given by the Contracting Officer directing the manner of performance. The entire liability to the Government for infringement of a patent of the United States shall be determined solely by the provisions of the indemnity clauses, if any, included in this contract or any subcontract hereunder (including any lower-tier subcontract), and the Government assumes liability for all other infringement to the extent of the authorization and consent hereinabove granted. (ASPR 7-103.22)

40. COMPOSITION OF CONTRACTOR (1965 JAN)

If the Contractor hereunder is comprised of more than one legal entity, each such entity shall be jointly and severally liable hereunder. (ASPR 7-602.32)

41. SITE INVESTIGATION (1965 JAN)

The Contractor acknowledges that he has investigated and satisfied himself as to the conditions affecting the work, including but not restricted to those bearing upon transportation, disposal, handling and storage of materials, availability of labor, water, electric power, roads and uncertainties of weather, river stages, tides or similar physical conditions at the site, the conformation and conditions of the ground, the character of equipment and facilities needed preliminary to and during prosecution of the work. The Contractor further acknowledges that he has satisfied himself as to the character, quality and quantity of surface and subsurface materials or obstacles to be encountered insofar as this information is reasonably ascertainable from an inspection of the site, including all exploratory work done by the Government, as well as from information presented by the drawings and specifications made a part of this contract. Any failure by the Contractor to acquaint himself with the available information will not relieve him from responsibility for estimating properly the difficulty or cost of successfully performing the work. The

Government assumes no responsibility for any conclusions or interpretations made by the Contractor on the basis of the information made available by the Government. (ASPR 7-602.33)

42. PROTECTION OF EXISTING VEGETATION, STRUCTURES, UTILITIES, AND IMPROVEMENTS (1965 JAN)

(a) The Contractor will preserve and protect all existing vegetation such as trees, shrubs, and grass on or adjacent to the site of work which is not to be removed and which does not unreasonably interfere with the construction work. Care will be taken in removing trees authorized for removal to avoid damage to vegetation to remain in place. Any limbs or branches of trees broken during such operations or by the careless operation of equipment, or by workmen, shall be trimmed with a clean cut and painted with an approved tree pruning compound as directed by the Contracting Officer.

(b) The Contractor will protect from damage all existing improvements or utilities at or near the site of the work, the location of which is made known to him, and will repair or restore any damage to such facilities resulting from failure to comply with the requirements of this contract or the failure to exercise reasonable care in the performance of the work. If the Contractor fails or refuses to repair any such damage promptly, the Contracting Officer may have the necessary work performed and charge the cost thereof to the Contractor. (ASPR 7-602.34)

43. OPERATIONS AND STORAGE AREAS (1965 JAN)

(a) All operations of the Contractor (including storage of materials) upon Government premises shall be confined to areas authorized or approved by the Contracting Officer. The Contractor shall hold and save the Government, its officers and agents, free and harmless from liability of any nature occasioned by his operations.

(b) Temporary buildings (storage sheds, shops, offices, etc.) may be erected by the Contractor only with the approval of the Contracting Officer, and shall be built with labor and materials furnished by the Contractor without expense to the Government. Such temporary buildings and utilities shall remain the property of the Contractor and shall be removed by him at his expense upon the completion of the work. With the written consent of the Contracting Officer, such buildings and utilities may be abandoned and need not be removed.

(c) The Contractor shall, under regulations prescribed by the Contracting Officer, use only established roadways or construct and use such temporary roadways as may be authorized by the Contracting Officer. Where materials are transported in the prosecution of the work, vehicles shall not be loaded beyond the loading capacity recommended by the manufacturer of the vehicle or prescribed by any Federal, State or local law or regulation. When it is necessary to cross curbings or sidewalks, protection against damage shall be provided by the Contractor and any damaged roads, curbings, or sidewalks shall be repaired by, or at the expense of the Contractor. (ASPR 7-602.35)

44. MODIFICATION PROPOSALS — PRICE BREAKDOWN (1968 APR)

The Contractor, in connection with any proposal he makes for a contract modification, shall furnish a price breakdown, itemized as required by the Contracting Officer. Unless otherwise directed, the breakdown shall be in sufficient detail to permit an analysis of all material, labor, equipment, subcontract, and overhead costs, as well as profit, and shall cover all work involved in the modification, whether such work was deleted, added or changed. Any amount claimed for subcontracts shall be supported by a similar price breakdown. In addition, if the proposal includes a time extension, a justification therefor shall also be furnished. The proposal, together with the price breakdown and time extension justification, shall be furnished by the date specified by the Contracting Officer. (ASPR 7-602.36)

45. SUBCONTRACTORS (1972 FEB)

Within seven days after the award of any subcontract either by himself or a subcontractor, the Contractor shall deliver to the Contracting Officer a statement setting forth the name and address of the subcontractor and a summary description of the work subcontracted. The Contractor shall at the same time furnish a statement signed by the subcontractor acknowledging the inclusion in his subcontract of the clauses of this contract entitled "Equal Opportunity," "Davis–Bacon Act," "Contract Work Hours and Safety Standards Act – Overtime Compensation," "Apprentices and Trainees," "Payrolls and Basic Records," "Compliance with Copeland Regulations," "Withholding of Funds," "Subcontracts" and "Contract Termination – Debarment." Nothing contained in this contract shall create any contractual relation between the subcontractor and the Government. (ASPR 7–602.37)

46. USE AND POSSESSION PRIOR TO COMPLETION (1965 JAN)

The Government shall have the right to take possession of or use any completed or partially completed part of the work. Such possession or use shall not be deemed an acceptance of any work not completed in accordance with the contract. While the Government is in such possession, the Contractor, notwithstanding the provisions of the clause of this contract entitled "Permits and Responsiblities," shall be relieved of the responsibility for loss or damage to the work other than that resulting from the Contractor's fault or negligence. If such prior possession or use by the Government delays the progress of the work or causes additional expense to the Contractor, an equitable adjustment in the contract price or the time of completion will be made and the contract shall be modified in writing accordingly. (ASPR 7-602.39)

47. CLEANING UP (1965 JAN)

The Contractor shall at all times keep the construction area, including storage areas used by him, free from accumulations of waste material or rubbish and prior to completion of the work remove any rubbish from the premises and all tools, scaffolding, equipment, and materials not the property of the Government. Upon completion of the construction the Contractor shall leave the work and premises in a clean, neat and workmanlike condition satisfactory to the Contracting Officer. (ASPR 7-602.40)

48. ADDITIONAL DEFINITIONS (1965 JAN)

(a) Wherever in the specifications or upon the drawings the words "directed," "required," "ordered," "designated," "prescribed," or words of like import are used, it shall be understood that the "direction," "requirement," "ordered," "designation," or "prescription," of the Contracting Officer is intended and similarly the words "approved," "acceptable," "satisfactory" or words of like import shall mean "approved by" or "acceptable to," or "satisfactory to" the Contracting Officer, unless otherwise expressly stated.

(b) Where "as shown," "as indicated," "as detailed," or words of similar import are used, it shall be

understood that the reference is made to the drawings accompanying this contract unless stated otherwise. The word "provided" as used herein shall be understood to mean "provided complete in place", that is "furnished and installed." (ASPR 7-602.41)

49. ACCIDENT PREVENTION (1967 JUN)

(a) In order to provide safety controls for protection to the life and health of employees and other persons; for prevention of damage to property, materials, supplies, and equipment; and for avoidance of work interruptions in the performance of this contract, the Contractor shall comply with all pertinent provisions of Corps of Engineers Manual, EM 385-1-1, dated 1 March 1967, entitled "General Safety Requirements", as amended, and will also take or cause to be taken such additional measures as the Contracting Officer may determine to be reasonably necessary for the purpose.

(b) The Contractor will maintain an accurate record of, and will report to the Contracting Officer in the manner and on the forms prescribed by the Contracting Officer, exposure data and all accidents resulting in death, traumatic injury, occupational disease, and damage to property, materials, supplies and equipment incident to work performed under this contract.

(c) The Contracting Officer will notify the Contractor of any noncompliance with the foregoing provisions and the action to be taken. The Contractor shall, after receipt of such notice, immediately take corrective action. Such notice, when delivered to the Contractor or his representative at the site of the work, shall be deemed sufficient for the purpose. If the Contractor fails or refuses to comply promptly, the Contracting Officer may issue an order stopping all or part of the work until satisfactory corrective action has been taken. No part of the time lost due to any such stop orders shall be made the subject of claim for extension of time or for excess costs or damages by the Contractor.

(d) Compliance with the provisions of this article by subcontractors will be the responsibility of the Contractor.

(e) Prior to commencement of the work the Contractor will:

(1) submit in writing his proposals for effectuating this provision for accident prevention;

(2) meet in conference with representatives of the Contracting Officer to discuss and develop mutual understandings relative to administration of the over-all safety program. (ASPR 7-602.42(a) & (b))

50. GOVERNMENT INSPECTORS (1965 JAN)

The work will be conducted under the general direction of the Contracting Officer and is subject to inspection by his appointed inspectors to insure strict compliance with the terms of the contract. No inspector is authorized to change any provision of the specifications without written authorization of the Contracting Officer, nor shall the presence or absence of an inspector relieve the Contractor from any requirements of the contract. (ASPR 7-602.43)

51. RIGHTS IN SHOP DRAWINGS (1966 APR)

(Applicable to all contracts calling for the delivery of shop drawings)

(a) Shop drawings for construction means drawings, submitted to the Government by the Construction Contractor, subcontractor or any lower tier subcontractor pursuant to a construction contract, showing in detail (i) the proposed fabrication and assembly of structural elements and (ii) the installation (i.e., form, fit, and attachment details) of materials or equipment. The Government may duplicate, use, and disclose in any manner and for any purpose shop drawings delivered under this contract.

(b) This clause, including this paragraph (b), shall be included in all subcontracts hereunder at any tier. (ASPR 7-602.47)

52. NOTICE TO THE GOVERNMENT OF LABOR DISPUTES (1958 SEP)

(a) Whenever the Contractor has knowledge that any actual or potential labor dispute is delaying or threatens to delay the timely performance of this contract, the Contractor shall immediately give notice thereof, including all relevant information with respect thereto, to the Contracting Officer.

(b) The Contractor agrees to insert the substance of this clause, including this paragraph (b), in any subcontract hereunder as to which a labor dispute may delay the timely performance of this contract; except that each such subcontract shall provide that in the event its timely performance is delayed or threatened by delay by any actual or potential labor dispute, the subcontractor shall immediately notify his next higher tier subcontractor, or the prime contractor, as the case may be, of all relevant information with respect to such dispute. (ASPR 7-104.4)

53. CONTRACT PRICES - BIDDING SCHEDULE (1968 APR)

(The following clause is applicable to contracts containing unit prices)

Payment for the various items listed in the Bidding Schedule shall constitute full compensation for furnishing all plant, labor, equipment, appliances, and materials, and for performing all operations required to complete the work in conformity with the drawings and specifications. All costs for work not specifically mentioned in the Bidding Schedule shall be included in the contract prices for the items listed. (ASPR 7-603.5)

54. EXAMINATION OF RECORDS BY COMPTROLLER GENERAL (1971 MAR)

(a) This clause is applicable if the amount of this contract exceeds $2,500 and was entered into by means of negotiation, including small business restricted advertising, but is not applicable if this contract was entered into by means of formal advertising.

(b) The Contractor agrees that the Comptroller General of the United States or any of his duly authorized representatives shall, until the expiration of three years after final payment under this contract or such lesser time specified in either Appendix M of the Armed Services Procurement Regulation or the Federal Procurement Regulations Part 1-20, as appropriate, have access to and the right to examine any directly pertinent books, documents, papers, and records of the Contractor involving transactions related to this contract.

(c) The Contractor further agrees to include in all his subcontracts hereunder a provision to the effect that the subcontractor agrees that the Comptroller General of the United States or any of his duly authorized representatives shall, until the expiration of three years after final payment under the subcontract or such lesser time specified in either Appendix M of the Armed Services Procurement Regulation or the Federal Procurement Regulations Part 1-20, as appropriate, have access to and the right to examine any directly pertinent books, documents, papers, and records of such subcontractor,

involving transactions related to the subcontract. The term "subcontract" as used in this clause excludes (i) purchase orders not exceeding $2,500 (ii) subcontracts or purchase orders for public utility services at rates established for uniform applicability to the general public.

(d) The periods of access and examination described in (b) and (c) above for records which relate to (i) appeals under the "Disputes" clause of this contract, (ii) litigation or the settlement of claims arising out of the performance of this contract, or (iii) costs and expenses of this contract as to which exception has been taken by the Comptroller General or any of his duly authorized representatives, shall continue until such appeals, litigation, claims or exceptions have been disposed of. (ASPR 7-104.15)

55. PRIORITIES, ALLOCATIONS, AND ALLOTMENTS (1971 APR)

(The following clause is applicable to rateable contracts)

The Contractor shall follow the provisions of DMS Reg. 1 and all other applicable regulations and orders of the Bureau of Domestic Commerce in obtaining controlled materials and other products and materials needed to fill this order. (ASPR 7-104.18)

56. PRICE REDUCTION FOR DEFECTIVE COST OR PRICING DATA – PRICE ADJUSTMENTS (1970 JAN)

(The following clause is applicable if this contract is in excess of $100,000)

(a) This clause shall become operative only with respect to any modification of this contract which involves aggregate increases and/or decreases in costs plus applicable profits in excess of $100,000 unless the modification is priced on the basis of adequate competition, established catalog or market prices of commercial items sold in substantial quantities to the general public, or prices set by law or regulation. The right to price reduction under this clause is limited to defects in data relating to such modification.

(b) If any price, including profit, or fee, negotiated in connection with any price adjustment under this contract was increased by any significant sums because:

(i)   the Contractor furnished cost or pricing data which was not complete, accurate and current as certified in the Contractor's Certificate of Current Cost or Pricing Data;

(ii) a subcontractor, pursuant to the clause of this contract entitled "Subcontractor Cost or Pricing Data" or "Subcontractor Cost or Pricing Data - Price Adjustments" or any subcontract clause therein required, furnished cost or pricing data which was not complete, accurate and current as certified in the subcontractor's Certificate of Current Cost or Pricing Data;

(iii) a subcontractor or prospective subcontractor furnished cost or pricing data which was required to be complete, accurate and current and to be submitted to support a subcontract cost estimate furnished by the Contractor but which was not complete, accurate and current as of the date certified in the Contractor's Certificate of Current Cost or Pricing Data; or

(iv) the Contractor or a subcontractor or prospective subcontractor furnished any data, not within (i), (ii) or (iii) above, which was not accurate, as submitted;

the price shall be reduced accordingly and the contract

shall be modified in writing as may be necessary to reflect such reduction. However, any reduction in the contract price due to defective subcontract data of a prospective subcontractor, when the subcontract was not subsequently awarded to such subcontractor, will be limited to the amount (plus applicable overhead and profit markup) by which the actual subcontract, or actual cost to the Contractor if there was no subcontract, was less than the prospective subcontract cost estimate submitted by the Contractor, *provided* the actual subcontract price was not affected by defective cost or pricing data.

Note: Since the contract is subject to reduction under this clause by reason of defective cost or pricing data submitted in connection with certain subcontracts, it is expected that the contractor may wish to include a clause in each such subcontract requiring the subcontractor to appropriately indemnify the contractor. However, the inclusion of such a clause and the terms thereof are matters for negotiation and agreement between the contractor and the subcontractor, *provided* that they are consistent with ASPR 23–203 relating to Disputes provisions in subcontracts. It is also expected that any subcontractor subject to such indemnification will generally require substantially similar indemnification for defective cost or pricing data required to be submitted by his lower tier subcontractors. (ASPR 7–104.29(b))

57. INTEREST (1972 MAY)

Notwithstanding any other provision of this contract, unless paid within thirty (30) days, all amounts that become payable by the Contractor to the Government under this contract (net of any applicable tax credit under the Internal Revenue Code) shall bear interest from the date due until paid and shall be subject to adjustments as provided by Part 6 of Appendix E of the Armed Services Procurement Regulation, as in effect on the date of this contract. The interest rate per annum shall be the interest rate in effect which has been established by the Secretary of the Treasury pursuant to Public Law 92–41; 85 STAT 97 for the Renegotiation Board, as of the date the amount becomes due as herein provided. Amounts shall be due upon the earliest one of (i) the date fixed pursuant to this contract; (ii) the date of the first written demand for payment, consistent with this contract, including demand consequent upon default termination; (iii) the date of transmittal by the Government to the Contractor of a proposed supplemental agreement to confirm completed negotiations fixing the amount; or (iv) if this contract provides for revision of prices, the date of written notice to the Contractor stating the amount of refund payable in connection with a pricing proposal or in connection with a negotiated pricing agreement not confirmed by contract supplement. (ASPR 7–104.39)

58. AUDIT BY DEPARTMENT OF DEFENSE (1971 APR)

(The following clause is applicable unless this contract was entered into by formal advertising and is not in excess of $100,000)

(a) *General.* The Contracting Officer or his representatives shall have the audit and inspection rights described in the applicable paragraphs (b), (c) and (d) below.

(b) *Examination of Costs.* If this is a cost reimbursement type, incentive, time and materials, labor hour, or price redeterminable contract, or any combination thereof, the Contractor shall maintain, and the Contracting Officer or his representatives shall have the right to examine books, records, documents, and other evidence and accounting procedures and practices, sufficient to reflect properly all direct and indirect costs of whatever nature claimed to have been incurred and anticipated to be incurred for the performance of this contract. Such right of examination shall include inspection at all reasonable times of the Contractor's plants, or such parts thereof, as may be engaged in the performance of this contract.

(c) *Cost or Pricing Data.* If the Contractor submitted cost or pricing data in connection with the pricing of this contract or any change or modification thereto, unless such pricing was based on adequate price competition, established catalog or market prices of commercial items sold in substantial quantities to the general public, or prices set by law or regulation, the Contracting Officer or his representatives who are employees of the United States Government shall have the right to examine all books, records, documents and other data of the Contractor related to the negotiation, pricing or performance of such contract, change or modification, for the purpose of evaluating the accuracy, completeness and currency of the cost or pricing data submitted. Additionally, in the case of pricing any change or modification exceeding $100,000 to formally advertised contracts, the Comptroller General of the United States or his representatives who are employees of the United States Government shall have such rights. The right of examination shall extend to all documents necessary to permit adequate evaluation of the cost or pricing data submitted, along with the computations and projections used therein.

(d) *Reports.* If the Contractor is required to furnish Cost Information Reports (CIR) or Contract Fund Status Reports (CFSR), the Contracting Officer or his representatives shall have the right to examine books, records, documents, and supporting materials, for the purpose of evaluating (i) the effectiveness of the Contractor's policies and procedures to produce data compatible with the objectives of these reports, and (ii) the data reported.

(e) *Availability.* The materials described in (b), (c) and (d) above shall be made available at the office of the Contractor, at all reasonable times, for inspection, audit, or reproduction, until the expiration of three years from the date of final payment under this contract or such lesser time specified in Appendix M of the Armed Services Procurement Regulation, and for such longer period, if any, as is required by applicable statute, or by other clauses of this contract, or by (1) and (2) below:

(1) If this contract is completely or partially terminated, the records relating to the work terminated shall be made available for a period of three years from the date of any resulting final settlement.

(2) Records which relate to appeals under the "Disputes" clause of this contract, or litigation or the settlement of claims arising out of the performance of this contract, shall be made available until such appeals, litigation, or claims have been disposed of.

(f) The Contractor shall insert a clause containing all the provisions of this clause, including this paragraph (f), in all subcontracts hereunder, except altered as necessary for proper identification of the contracting parties and the Contracting Officer under the Government prime contract. (ASPR 7–104.41(a))

59. SUBCONTRACTOR COST OR PRICING DATA – PRICE ADJUSTMENTS (1970 JAN)

(The following clause is applicable if this contract is in excess of $100,000)

(a) Paragraphs (b) and (c) of this clause shall become operative only with respect to any modification made pursuant to one or more provisions of this contract which involves aggregate increases and/or decreases in costs plus applicable profits expected to exceed $100,000. The requirements of this clause shall be limited to such modifications.

(b) The Contractor shall require subcontractors hereunder to submit cost or pricing data under the following circumstances: (i) prior to the award of any subcontract the amount of which is expected to exceed $100,000 when entered into; (ii) prior to the pricing of any subcontract modification which involves aggregate increases and/or decreases in costs plus applicable profits expected to exceed $100,000, except where the price is based on adequate price competition, established catalog or market prices of commercial items sold in substantial quantities to the general public, or prices set by law or regulation.

(c) The Contractor shall require subcontractors to certify that to the best of their knowledge and belief the cost and pricing data submitted under (b) above is accurate, complete, and current as of the date of agreement on the negotiated price of the subcontract or subcontract change or modification.

(d) The Contractor shall insert the substance of this clause including this paragraph (d) in each subcontract which exceeds $100,000. (ASPR 7-104.42(b))

60.1 G O V E R N M E N T – F U R N I S H E D PROPERTY (SHORT FORM) (1964 NOV)

(The following clause is applicable when Government Property having an acquisition cost of $25,000 or less is furnished to or acquired by the Contractor)

(a) The Government shall deliver to the Contractor, for use only in connection with this contract, the property described in the schedule or specifications (hereinafter referred to as "Government-furnished property"), at the times and locations stated therein. If the Government-furnished property, suitable for its intended use, is not so delivered to the Contractor, the Contracting Officer shall, upon timely written request made by the Contractor, and if the facts warrant such action, equitably adjust any affected provision of this contract pursuant to the procedures of the "Changes" clause hereof.

(b) Title to Government-furnished property shall remain in the Government. The Contractor shall maintain adequate property control records of Government-furnished property in accordance with sound industrial practice.

(c) Unless otherwise provided in this contract, the Contractor, upon delivery to him of any Government-furnished property, assumes the risk of, and shall be responsible for, any loss thereof or damage thereto except for reasonable wear and tear, and except to the extent that such property is consumed in the performance of this contract.

(d) The Contractor shall, upon completion of this contract, prepare for shipment, deliver f.o.b. origin, or dispose of all Government-furnished property not consumed in the performance of this contract or not theretofore delivered to the Government, as may be directed or authorized by the Contracting Officer. The net proceeds of any such disposal shall be credited to the contract price or paid in such other manner as the Contracting Officer may direct. (ASPR 7-104.24(f))

60.2 GOVERNMENT PROPERTY (FIXED PRICE) (1968 SEP)

(The following clause is applicable when Government Property having an acquisition cost in excess of $25,000 is furnished to or acquired by the Contractor)

(a) *Government-Furnished Property.* The Government shall deliver to the Contractor, for use in connection with and under the terms of this contract, the property described as Government-furnished property in the Schedule or specifications, together with such related data and information as the Contractor may request and as may reasonably be required for the intended use of such property (hereinafter referred to as "Government-furnished property"). The delivery or performance dates for the supplies or services to be furnished by the Contractor under this contract are based upon the expectation that Government-furnished property suitable for use (except for such property furnished "as is") will be delivered to the Contractor at the times stated in the Schedule or, if not so stated, in sufficient time to enable the Contractor to meet such delivery or performance dates. In the event that Government-furnished property is not delivered to the Contractor by such time or times, the Contracting Officer shall, upon timely written request made by the Contractor, make a determination of the delay, if any, occasioned the Contractor thereby, and shall equitably adjust the delivery or performance dates or the contract price, or both, and any other contractual provision affected by any such delay, in accordance with the procedures provided for in the clause of this contract entitled "Changes." Except for Government-furnished property furnished "as is," in the event the Government-furnished property is received by the Contractor in a condition not suitable for the intended use the Contractor shall, upon receipt thereof, notify the Contracting Officer of such fact and, as directed by the Contracting Officer, either (i) return such property at the Government's expense or otherwise dispose of the property, or (ii) effect repairs or modifications. Upon the completion of (i) or (ii) above, the Contracting Officer upon written request of the Contractor shall equitably adjust the delivery or performance dates or the contract price, or both, and any other contractual provision affected by the rejection or disposition, or the repair or modification, in accordance with the procedures provided for in the clause of this contract entitled "Changes." The foregoing provisions for adjustment are exclusive and the Government shall not be liable to suit for breach of contract by reason of any delay in delivery of Government-furnished property or delivery of such property in a condition not suitable for its intended use.

(b) *Changes in Government-furnished Property.*

    (1) By notice in writing, the Contracting Officer may (i) decrease the property provided or to be provided by the Government under this contract, or (ii) substitute other Government-owned property for property to be provided by the Government, or to be acquired by the Contractor for the Government, under this contract. The Contractor shall promptly take such action as the Contracting Officer may direct with respect to the removal and shipping of property covered by such notice.

    (2) In the event of any decrease in or substitution of property pursuant to subparagraph (1) above, or any withdrawal of authority to use property provided under any other contract or lease, which property the Government had agreed in the Schedule to make available for the performance of this contract, the Contracting Officer, upon the written request of the Contractor (or, if the substitution of property causes a decrease in the cost of performance, on his own initiative), shall equitably adjust such contractual provisions as may be affected by the decrease, substitution, or withdrawal, in accordance with the procedures provided

for in the "Changes" clause of this contract.

(c) *Title.* Title to all property furnished by the Government shall remain in the Government. In order to define the obligations of the parties under this clause, title to each item of facilities, special test equipment, and special tooling (other than that subject to a "Special Tooling" clause) acquired by the Contractor for the Government pursuant to this contract shall pass to and vest in the Government when its use in the performance of this contract commences, or upon payment therefor by the Government, whichever is earlier, whether or not title previously vested. All Government-furnished property, together with all property acquired by the Contractor title to which vests in the Government under this paragraph, is subject to the provisions of this clause and is hereinafter collectively referred to as "Government property." Title to Government property shall not be affected by the incorporation or attachment thereof to any property not owned by the Government, nor shall such Government property, or any part thereof, be or become a fixture or lose its identity as personalty by reason of affixation to any realty.

(d) *Property Administration.* The Contractor shall comply with the provisions of Appendix B, Armed Services Procurement Regulation, as in effect on the date of the contract, which is hereby incorporated by reference and made a part of this contract. Material to be furnished by the Government shall be ordered or returned by the Contractor, when required, in accordance with the "Manual for Military Standard Requisitioning and Issue Procedure (MILSTRIP) for Defense Contractors" (Appendix H, Armed Services Procurement Regulation) as in effect on the date of this contract, which Manual is hereby incorporated by reference and made a part of this contract.

(e) *Use of Government Property.* The Government property shall, unless otherwise provided herein or approved by the Contracting Officer, be used only for the performance of this contract.

(f) *Utilization, Maintenance and Repair of Government Property.* The Contractor shall maintain and administer, in accordance with sound industrial practice, and in accordance with applicable Provisions of Appendix B, a program for the utilization, maintenance, repair, protection and preservation of Government property, until disposed of by the Contractor in accordance with this clause. In the event that any damage occurs to Government property the risk of which has been assumed by the Government under this contract, the Government shall replace such items or the Contractor shall make such repair of the property as the Government directs; *provided,* however, that if the Contractor cannot effect such repair within the time required, the Contractor shall dispose of such property in the manner directed by the Contracting Officer. The contract price includes no compensation to the Contractor for the performance of any repair or replacement for which the Government is responsible, and an equitable adjustment will be made in any contractual provisions affected by such repair or replacement of Government property made at the direction of the Government, in accordance with the procedures provided for in the "Changes" clause of this contract. Any repair or replacement for which the Contractor is responsible under the provisions of this contract shall be accomplished by the Contractor at his own expense.

(g) *Risk of Loss.* Unless otherwise provided in this contract, the Contractor assumes the risk of, and shall be responsible for, any loss of or damage to Government property provided under this contract upon its delivery to him or upon passage of title thereto to the Government as provided in paragraph (c) hereof, except for reasonable wear and tear and except to the extent that such property is consumed in the performance of this contract.

(h) *Access.* The Government, and any persons designated by it, shall at all reasonable times have access to the premises wherein any Government property is located, for the purpose of inspecting the Government property.

(i) *Final Accounting and Disposition of Government Property.* Upon the completion of this contract, or at such earlier dates as may be fixed by the Contracting Officer, the Contractor shall submit, in a form acceptable to the Contracting Officer, inventory schedules covering all items of Government property not consumed in the performance of this contract (including any resulting scrap) or not theretofore delivered to the Government, and shall prepare for shipment, deliver f.o.b. origin, or dispose of the Government property, as may be directed or authorized by the Contracting Officer. The net proceeds of any such disposal shall be credited to the contract price or shall be paid in such other manner as the Contracting Officer may direct.

(j) *Restoration of Contractor's Premises and Abandonment.* Unless otherwise provided herein, the Government:

(i) may abandon any Government property in place, and thereupon all obligations of the Government regarding such abandoned property shall cease; and

(ii) has no obligation to the Contractor with regard to restoration or rehabilitation of the Contractor's premises, neither in case of abandonment (paragraph (j) (i) above), disposition on completion of need or of the contract (paragraph (i) above), nor otherwise, except for restoration or rehabilitation costs which are properly included in an equitable adjustment under paragraph (b) above.

(k) *Communications.* All communications issued pursuant to this clause shall be in writing or in accordance with the "Manual for Military Standard Requisitioning and Issue Procedure (MILSTRIP) for Defense Contractors" (Appendix H, Armed Services Procurement Regulation). (ASPR 7-104.24(a))

61. DISPUTES CONCERNING LABOR STANDARDS (1965 JAN)

Disputes arising out of the labor standards provisions of this contract shall be subject to the Disputes clause except to the extent such disputes involve the meaning of classifications or wage rates contained in the wage determination decision of the Secretary of Labor or the applicability of the labor provisions of the contract which questions shall be referred to the Secretary of Labor in accordance with the procedures of the Department of Labor. (ASPR 7-603.26)

62. VARIATIONS IN ESTIMATED QUANTITIES (1968 APR)

Where the quantity of a pay item in this contract is an estimated quantity and where the actual quantity of such pay item varies more than fifteen percent (15%) above or below the estimated quantity stated in this contract, an equitable adjustment in the contract price shall be made upon demand of either party. The equitable adjustment shall be based upon any increase or decrease in costs due solely to the variation above one hundred fifteen percent (115%) or below eighty-five percent (85%) of the estimated quantity. If the quantity variation is such as to cause an increase in the time necessary for completion, the

Contracting Officer shall, upon receipt of a written request for an extension of time within ten (10) days from the beginning of such delay, or within such further period of time which may be granted by the Contracting Officer prior to the date of final settlement of the contract, ascertain the facts and make such adjustment for extending the completion date as in his judgment the findings justify. (ASPR 7-603.27)

63. PROGRESS CHARTS AND REQUIREMENTS FOR OVERTIME WORK (1965 JAN)

(a) The Contractor shall within 5 days or within such time as determined by the Contracting Officer, after date of commencement of work, prepare and submit to the Contracting Officer for approval a practicable schedule, showing the order in which the Contractor proposes to carry on the work, the date on which he will start the several salient features (including procurement of materials, plant and equipment) and the contemplated dates for completing the same. The schedule shall be in the form of a progress chart of suitable scale to indicate appropriately the percentage of work scheduled for completion at any time. The Contractor shall enter on the chart the actual progress at such intervals as directed by the Contracting Officer, and shall immediately deliver to the Contracting Officer three copies thereof. If the Contractor fails to submit a progress schedule within the time herein prescribed, the Contracting Officer may withhold approval of progress payment estimates until such time as the Contractor submits the required progress schedule.

(b) If, in the opinion of the Contracting Officer, the Contractor falls behind the progress schedule, the Contractor shall take such steps as may be necessary to improve his progress and the Contracting Officer may require him to increase the number of shifts, or overtime operations, days of work, or the amount of construction plant, or all of them, and to submit for approval such supplementary schedule or schedules in chart form as may be deemed necessary to demonstrate the manner in which the agreed rate of progress will be regained, all without additional cost to the Government.

(c) Failure of the Contractor to comply with the requirements of the Contracting Officer under this provision shall be grounds for determination by the Contracting Officer that the Contractor is not prosecuting the work with such diligence as will insure completion within the time specified. Upon such determination the Contracting Officer may terminate the Contractor's right to proceed with the work, or any separable part thereof, in accordance with the clause of the contract entitled "Termination for Default - - Damages for Delay - - Time Extensions." (ASPR 7-603.48)

64. VALUE ENGINEERING INCENTIVE (1971 MAY)

(The following clause is applicable if this contract is in excess of $100,000)

(a) (1) This clause applies to those cost reduction proposals initiated and developed by the Contractor for changing the drawings, designs, specifications, or other requirements of this contract. This clause does not, however, apply to any such proposal unless it is identified by the Contractor, at the time of its submission to the Contracting Officer, as a proposal submitted pursuant to this clause. Furthermore, if this contract also contains a "Value Engineering Program Requirement" clause, this clause applies to any given value engineering change proposal only to the extent the Contracting Officer affirmatively determines that it resulted from value engineering efforts clearly outside the scope of the program requirement; to the extent the Contracting Officer does not affirmatively so determine, the proposal shall be considered for all purposes as having been submitted pursuant to the Value Engineering Program Requirement clause, even if it was purportedly submitted pursuant to this clause.

(2) The cost reduction proposals contemplated are those that:

(i) would require, in order to be applied to this contract, a change to this contract; and

(ii) would result in savings to the Government by providing a decrease in the cost of performance of this contract, without impairing any of the items' essential functions and characteristics such as service life, reliability, economy of operation, ease of maintenance, and necessary standardized features.

(b) As a minimum, the following information shall be submitted by the Contractor with each proposal:

(i) a description of the difference between the existing contract requirement and the proposed change, and the comparative advantages and disadvantages of each;

(ii) an itemization of the requirements of the contract which must be changed if the proposal is adopted, and a recommendation as to how to make each such change (e.g., a suggested revision);

(iii) an estimate of the reduction in performance costs, if any, that will result from adoption of the proposal, taking into account the costs of development and implementation by the Contractor (including any amount attributable to subcontracts in accordance with paragraph (e) below) and the basis for the estimate;

(iv) a prediction of any effects the proposed change would have on collateral costs to the Government such as Government-furnished property costs, costs of related items, and costs of maintenance and operation;

(v) a statement of the time by which a change order adopting the proposal must be issued so as to obtain the maximum cost reduction during the remainder of this contract, noting any effect on the contract completion time or delivery schedule; and

(vi) the dates of any previous submissions of the proposal, the numbers of the Government contracts under which submitted, and the previous actions by the Government, if known.

(c) (1) Cost reduction proposals shall be submitted to the Procuring Contracting Officer (PCO). When the contract is administered by other than the procuring activity, a copy of the proposal shall also be submitted to the Administrative Contracting Officer (ACO). Cost reduction proposals shall be processed expeditiously; however, the Government shall not be liable for any delay in acting upon any proposal submitted pursuant to this clause. The Contractor does have the right to withdraw, in whole or in part, any value engineering change proposal not accepted by the Government within the period specified in the proposal. The decision of the Contracting Officer as to the acceptance of any such proposal under this contract (including the decision as to which clause is applicable to the proposal if this contract contains both a "Value Engineering Incentive" and a "Value Engineering Program Requirement" clause) shall be final and shall not be subject to the "Disputes" clause of this contract.

(2) The Contracting Officer may accept, in whole or in part, either before or within a reasonable time after performance has been completed under this contract, any cost reduction proposal submitted pursuant to this clause by giving the Contractor written notice thereof reciting acceptance under this clause. Where performance under this contract has not yet been completed, this written notice may be given by issuance of a change order to this contract. Unless and until a change order applies a value engineering change proposal to this contract, the Contractor shall remain obligated to perform in accordance with the terms of the existing contract. If a proposal is accepted after performance under this contract has been completed, the adjustment required shall be effected by contract modification in accordance with this clause.

(3) If a cost reduction proposal submitted pursuant to this clause is accepted by the Government, the Contractor is entitled to share in instant contract savings, collateral savings, and future acquisition savings not as alternatives, but rather to the full extent provided for in this clause.

(4) Contract modification made as a result of this clause will state that they are made pursuant to it.

(d) If a cost reduction proposal submitted pursuant to this clause is accepted and applied to this contract, an equitable adjustment in the contract price and in any other affected provisions of this contract shall be made in accordance with this clause and the "Termination for Convenience," "Changes," or other applicable clause of this contract. The equitable adjustment shall be established by determining the effect of the proposal on the Contractor's cost of performance, taking into account the Contractor's cost of developing the proposal, insofar as such is properly a direct charge not otherwise reimbursed under this contract, and the Contractor's cost of implementing the change (including any amount attributable to subcontracts in accordance with paragraph (e) below). When the cost of performance of this contract is decreased as a result of the change, the contract price shall be reduced by the following amount: the total estimated decrease in the Contractor's cost of performance less * percent (* %) of the difference between the amount of such total estimated decrease and any net increase in ascertainable collateral costs to the Government which must reasonably be incurred as a result of application of the cost reduction proposal to this contract. When the cost of performance of this contract is increased as a result of the change, the equitable adjustment increasing the contract price shall be in accordance with the

---

*Fifty percent (50%) for the first two approved proposals, fifty-five percent (55%) for the next two approved proposals, and sixty percent (60%) for all other approved proposals.

"Changes" clause rather than under this clause, but the resulting contract modification shall state that it is made pursuant to this clause. (1967 JUN)

(e) The Contractor will use his best efforts to include appropriate value engineering arrangements in any subcontract which, in the judgment of the Contractor, is of such a size and nature as to offer reasonable likelihood of value engineering cost reductions. For the purpose of computing any equitable adjustment in the contract price under paragraph (d) above, the Contractor's cost of development and implementation of a cost reduction proposal which is accepted under this contract shall be deemed to include any development and implementation costs of a subcontractor and any value engineering incentive payments to a subcontractor, or cost reduction shares accruing to a subcontractor, which clearly pertain to such proposal and which are incurred, paid, or accrued in the performance of a subcontract under this contract.

(f) Omitted pursuant to ASPR 7-104.44(f).

(g) (1) A cost reduction proposal identical to one submitted under any other contract with the Contractor or another contractor may also be submitted under this contract.

(2) If the Contractor submits under this clause a proposal which is identical to one previously received by the Contracting Officer under a different contract with the Contractor or another contractor for substantially the same items and both proposals are accepted by the Government, the Contractor shall share instant contract savings realized under this contract, pursuant to paragraph (d) of this clause, but he shall not share collateral or future savings pursuant to paragraphs (f) and (j) (if included) of this clause.

(h) The Contractor may restrict the Government's right to use any sheet of a value engineering proposal or of the supporting data, submitted pursuant to this clause, in accordance with the terms of the following legend if it is marked on such sheet:

This data furnished pursuant to the Value Engineering clause of contract _____

shall not be disclosed outside the Government, or duplicated, used, or disclosed, in whole or in part, for any purpose other than to evaluate a value engineering proposal submitted under said clause. This restriction does not limit the Government's right to use information contained in this data if it is or has been obtained, or is otherwise available, from the Contractor or from another source, without limitations. If such a proposal is accepted by the Government under said contract after the use of this data in such an evaluation, the Government shall have the right to duplicate, use, and disclose any data reasonably necessary to the full utilization of such proposal as accepted, in any manner and for any purpose whatsoever, and have others so do.

In the event of acceptance of a value engineering proposal, the Contractor hereby grants to the Government all rights to use, duplicate or disclose, in whole or in part, in any manner and for any purpose whatsoever, and to have or permit others to do so, any data reasonably necessary to fully utilize such proposal.

(i) (1) For purposes of sharing under paragraph (d) above, the term "instant contract" shall not include any supplemental agreements to or other modifications of the instant contract, executed subsequent to acceptance of the particular value engineering change proposal, by which the Government increases the quantity of any item or adds any item, nor shall it include any extension of the instant contract through exercise of an option (if any) provided under this contract after acceptance of the proposal. Such supplemental agreements, modifications, and extensions shall be considered "future contracts" within paragraph (j) (if included) of this clause.

(2) If this contract is an estimated requirements or other indefinite quantity type contract, the term "instant contract" for purposes of sharing under paragraph (d) above shall include only those orders actually placed by the Government up to the time the particular value engineering change proposal is accepted. All orders placed subsequent to the acceptance of the particular change proposal shall be considered "future contracts" within paragraph (j) (if included) of this clause.

(3) If this clause is included in a basic ordering agreement, the "instant contract" for purposes of sharing under paragraph (d) above shall be the order under which the particular value engineering change proposal is submitted. Other orders under the same agreement shall be considered either "existing contracts" (if awarded prior to acceptance of the proposal) or "future contracts" (if awarded after acceptance of the proposal), within paragraph (j) (if included) of this clause.

(4) If this contract is a multi-year contract, the "instant contract" shall be the entire contract for the total multi-year quantity.

(j) Omitted pursuant to ASPR 1-1703.3(a)(1). (ASPR 7-104.44(a), (c), (e)(i) and (f))

65. PRICING OF ADJUSTMENTS (1970 JUL)

When costs are a factor in any determination of a contract price adjustment pursuant to the "Changes" clause or any other provision of this contract, such costs shall be in accordance with Section XV of the Armed Services Procurement Regulation as in effect on the date of this contract. (ASPR 7-103.26)

66. LISTING OF EMPLOYMENT OPENINGS FOR VETERANS (1971 NOV)

(This clause is applicable pursuant to 41 CFR 50-250 if this contract is for $10,000 or more and will generate 400 or more man-days of employment.)

(1) The Contractor agrees that all employment openings of the Contractor which exist at the time of the execution of this contract and those which occur during the performance of this contract, including those not generated by this contract and including those occurring at an establishment of the Contractor other than the one wherein the contract is being performed but excluding those of independently operated corporate affiliates, shall, to the maximum extent feasible, be offered for listing at an appropriate local office of the State employment service system wherein the opening occurs and to provide such periodic reports to such local office regarding employment openings and hires as may be required.

(2) Listing of employment openings with the employment service system pursuant to this clause shall be made at least concurrently with the use of any other recruitment source of effort and shall involve only the normal obligations which attach to the placing of a bona fide job order but does not require the hiring of any job applicant referred by the employment service system.

(3) The periodic reports required by paragraph (1) above shall be filed at least quarterly with the appropriate local office or, where the Contractor has more than one establishment in a State, with the central office of that State employment service. Such reports shall indicate for each establishment the number of individuals who were hired during the reporting period and the number of hires

who were veterans who served in the Armed Forces on or after August 5, 1964, and who received other than a dishonorable discharge. The Contractor shall maintain copies of the reports submitted until the expiration of one year after final payment under the contract, during which time they shall be made available, upon request, for examination by any authorized representatives of the Contracting Officer or of the Secretary of Labor.

(4) Whenever the Contractor becomes contractually bound to the listing provisions of this clause, it shall advise the employment service system in each State wherein it has establishments of the name and location of each such establishment in the State. As long as the Contractor is contractually bound to these provisions and has so advised the State employment service system, there is no need to advise the State system of subsequent contracts. The Contractor may advise the State systems when it is no longer bound by this contract clause.

(5) This clause does not apply (i) to the listing of employment openings which occur outside of the 50 States, the District of Columbia, Guam, Puerto Rico, and the Virgin Islands; and (ii) contracts with state and local governments.

(6) This clause does not apply to openings which the Contractor proposes to fill from within his own organization or to fill pursuant to a customary and traditional employer–union hiring arrangement. This exclusion does not apply to a particular opening once an employer decides to consider applicants outside of his own organization or employer–union arrangement for that opening.

(7) As used in this clause:

(i) "All employment openings" include, but are not limited to, openings which occur in the following job categories: production and nonproduction; plant and office; laborers and mechanics; supervisory and nonsupervisory; technical; and executive, administrative, and professional openings which are compensated on a salary basis of less than $18,000 per year. This term includes full–time employment, temporary employment of more than three (3) days' duration, and part–time employment.

(ii) "Appropriate office of the State employment service system" means the local office of the Federal–State national system of public employment offices with assigned responsibility for serving the area of the establishment where the employment opening is to be filled, including the District of Columbia, Guam, Puerto Rico, and the Virgin Islands.

(iii) "Openings which the Contractor proposes to fill from within his own organization or to fill pursuant to a customary and traditional employer–union hiring arrangement," means employment openings for which no consideration will be given to persons outside the Contractor's organization (including any affiliates, subsidiaries, and parent companies) or outside of a special hiring arrangement which is part of the customary and traditional employment relationship which exists between the Contractor and representative of its employees and includes any openings which the Contractor proposes to fill from regularly established "recall" or "rehire" lists or from union hiring halls.

(iv) "Man–day of employment" means any day during which an employee performs more than one hour of work.

(8) The Contractor agrees to place this clause (excluding this paragraph (8)) in any subcontract directly under this contract *provided,* such subcontract is for $10,000 or more and will generate 400 or more man–days of employment. (ASPR 7-103.27)

67. UTILIZATION OF MINORITY BUSINESS ENTERPRISES (1971 NOV)

(a) It is the policy of the Government that Minority Business Enterprises shall have the maximum practicable opportunity to participate in the performance of Government contracts.

(b) The Contractor agrees to use his best efforts to carry out this policy in the award of his subcontracts to the fullest extent consistent with the efficient performance of this contract. As used in this contract, the term "minority business enterprise" means a business, at least 50 percent of which is owned by minority group members or, in case of publicly—owned businesses, at least 51 percent of the stock of which is owned by minority group members. For the purposes of this definition, minority group members are Negroes, Spanish—speaking American persons, American—Orientals, American—Indians, American Eskimos, and American Aleuts. Contractors may rely on written representations by subcontractors regarding their status as minority business enterprises in lieu of an independent investigation. (ASPR 7—104.36(a))

68. PAYMENT OF INTEREST ON CONTRACTOR'S CLAIMS (1972 MAY)

(a) If an appeal is filed by the Contractor from a final decision of the Contracting Officer under the "Disputes" clause of this contract, denying a claim arising under the contract, simple interest on the amount of the claim finally determined owed by the Government shall be payable to the Contractor. Such interest shall be at the rate established by the Secretary of the Treasury pursuant to Public Law 92—41; 85 STAT 97 for the Renegotiation Board, from the date the Contractor furnishes to the Contracting Officer his written appeal pursuant to the "Disputes" clause of this contract, to the date of (i) a final judgment by a court of competent jurisdiction, or (ii) mailing to the Contractor of a supplemental agreement for execution either confirming completed negotiations between the parties or carrying out a decision of a Board of Contract Appeals.

(b) Notwithstanding (a) above, (i) interest shall be applied only from the date payment was due, if such date is later than the filing of appeal; and (ii) interest shall not be paid for any period of time that the Contracting Officer determines the Contractor has unduly delayed in pursuing his remedies before a Board of Contract Appeals or a court of competent jurisdiction. (ASPR 7—104.82)

NOTICE OF REQUIREMENT FOR SUBMISSION OF AFFIRMATIVE ACTION PLAN
TO ENSURE EQUAL EMPLOYMENT OPPORTUNITY

TO BE ELIGIBLE FOR AWARD OF A CONTRACT EXCEEDING $10,000, EACH BIDDER MUST FULLY COMPLY WITH THE REQUIREMENTS, TERMS, AND CONDITIONS OF THIS NOTICE.

The bidder shall submit with his bid an Affirmative Action Plan that complies with one of the following alternatives:

(a) he shall certify that he is signatory to the Hometown Plan (as hereinafter defined), either individually or through an association, and that he will comply with the requirements of the contract clause entitled "LOCAL AFFIRMATIVE ACTION PLAN" in the performance of the contract that may be awarded pursuant to this solicitation; or

(b) he shall submit percentage goals for minority manpower utilization, within the ranges set forth below, for all his construction work in the covered area (whether done under contract with the Federal Government or otherwise) during the term of the contract that may be awarded pursuant to this solicitation, and he shall agree to pursue these goals in accordance with, and to comply with, the "LOCAL AFFIRMATIVE ACTION PLAN" clause of the contract. Each goal is to be expressed as a percentage, representing the ratio of manhours of work or training of minority persons to the total manhours to be worked on all of the Bidder's construction work in the covered area, including the project to result from this solicitation. The acceptable ranges of minority manpower utilization, expressed in percentage terms, are as follows:

Goals need to be submitted only for construction trades that the Bidder contemplates using in the performance of the contract, and only for years during which the Bidder contemplates performing any work or engaging in any activity under the contract.

In the event that a construction trade for which no goal has been submitted is used in the performance of the contract, or that any work or activity under the contract takes place in a year for which no goals have been submitted, the Bidder will be considered to have submitted as goals, to the extent that goals may be necessary, the minimum percentage from the acceptable range for the particular trade for the appropriate year.

In the event that any work or activity under the contract takes place in a year later than the latest year for which an acceptable range of minority manpower utilization has been provided in this solicitation, the goals for said latest year shall apply.

If the Bidder fails or refuses to complete and submit an Affirmative Action Plan with his bid, or if any percentage goals fall below the ranges of minority manpower utilization set forth above, the bid or proposal shall be considered nonresponsive and will be rejected.

The Bidder's attention is called to the "EQUAL OPPORTUNITY" and "LOCAL AFFIRMATIVE ACTION PLAN" clauses of the contract.

The Bidder shall, within 5 days after a request therefor by the Contracting Officer or his duly authorized representative, submit the following information:

    (a)   A list of the construction trades he intends to use, either directly or through subcontractors at any tier, in the performance of the work covered by this solicitation;

    (b)   A list of the labor organizations with which he has collective bargaining agreements and which are signatories to the Hometown Plan with respect to trades for which specific commitments to goals of minority manpower utilization are set forth in Hometown Plan;

    (c)   A list of the labor organizations with which he has collective bargaining agreements and which are not signatories to the Hometown Plan or which are signatories thereto but with respect to trades for which no specific commitments to goals of minority manpower utilization are set forth in the Hometown Plan; and

    (d)   A list of all current construction work or contracts to which he is a party in any capacity in the covered area.

As used in this NOTICE, the Affirmative Action Plan, and the contract to result from this solicitation:

    (a)   "Hometown Plan" or "Plan" means the   Plan consisting of

    (b)   "the covered area" means

    (c)   "Director, OFCC" means the Director, Office of Federal Contract Compliance, United States Department of Labor, or any person to whom he delegates authority; and

    (d)   "minority" means Negro, Spanish—surnamed American, Oriental, and American Indian.

AFFIRMATIVE ACTION PLAN

THE BIDDER MUST SUBMIT THIS FORM WITH HIS BID, AND INDICATE HEREON THAT EITHER (a) OR (b) BELOW IS APPLICABLE. IF (b) IS APPLICABLE, THE BIDDER MUST ALSO INSERT HIS MINORITY MANPOWER UTILIZATION GOALS.

(a)   [   ] The Bidder hereby certifies that he is signatory, either individually or through an association to the Hometown Plan, as defined in this solicitation, and that he will comply with the requirements of the contract clause entitled "LOCAL AFFIRMATIVE ACTION PLAN" in the performance of the contract that may be awarded pursuant to this solicitation.

(b)   [   ] The Bidder hereby submits the amounts set forth below as his minority manpower utilization goals for all his construction work in the covered area during the term of the contract that may be awarded pursuant to this solicitation, and he agrees to pursue these goals in accordance with, and to comply with, the "LOCAL AFFIRMATIVE ACTION PLAN" clause of the contract. These goals are as follows:

BIDDER'S MINORITY MANPOWER UTILIZATION GOALS
(Percentage of Total Manhours)

TRADE

LOCAL AFFIRMATIVE ACTION PLAN

(a)   As used in this clause:

(1)   "Hometown Plan" or "Plan" means the formal agreement among contractors, unions, and minority representatives described in the solicitation from which this contract resulted;

(2)   "the covered area" means the geographical area described in the solicitation from which this contract resulted;

(3)   "Director, OFCC" means the Director, Office of Federal Contract Compliance, United States Department of Labor, or any person to whom he delegates authority; and

(4)   "minority" means Negro, Spanish–surnamed American, Oriental, and American Indian.

(b)   Whenever the Contractor, or any subcontractor at any tier, subcontracts a portion of the work in any construction trade, he shall include in such subcontract the provisions of this clause and any applicable minority manpower utilization goals under this contract, which shall be adopted by his subcontractor, who shall with regard to his own employees and subcontractors be bound thereby to the full extent as if he were the Contractor.

(c)   If the Contractor is signatory to the Plan, either individually or through an association, and so certified with his bid, his affirmative action program shall be in accordance with the Plan. In such case, paragraphs (a), (b), (c), (m), (n) and (o) of this clause shall apply. However, to the extent that the Contractor, or any subcontractor at any tier, employs construction trades that are not covered by the Plan, all of the provisions of this clause, and minority manpower utilization goals as determined in accordance with (k) and (l) below, shall apply. For the purposes of this paragraph (c), construction trades shall be considered not covered by the Plan: (i) if the Contractor violates a substantial requirement of the Plan or ceases to be signatory thereto; (ii) if the Director, OFCC, determines that the Plan is no longer an acceptable affirmative action plan; (iii) if the employing subcontractor is not or ceases to be signatory to the Plan, either individually or through an association; (iv) if the construction trade is not one of the trades participating in the Plan; or (v) if the construction trade, though participating, is not subject to a specific goal for minority manpower utilization and has not been exempted from such a goal by the Director, OFCC.

(d)   If the Contractor is not signatory to the Plan, or, even if signatory, did not so certify with his bid, his affirmative action program shall be in accordance with this clause and his minority manpower utilization goals submitted with his bid. However, if at any time after the opening of his bid, the Contractor certifies that he is signatory to the Plan, his affirmative action program from the time of such certification shall be in accordance with the Plan, subject to the provisions of paragraph (c) above.

(e)   The Contractor's minority manpower utilization goals submitted with his bid express his commitment of the manhours of employment and training of minority workers he will undertake in each construction trade as a percentage of the total manhours to be worked in that construction trade on all of the Contractor's construction work in the covered area during the term of this contract. The percentage of manhours for minority employment and training shall be substantially uniform throughout the term of this contract for each construction trade and for all projects. Minority employees or trainees shall not be transferred from employer–to–employer or from project–to–project for the sole purpose of meeting

the Contractor's minority manpower utilization goals.

(f)   The Contractor's minority manpower utilization goals shall be satisfied, whenever possible, by employment of qualified minority journeymen. However, minority trainees in pre–apprenticeship, apprenticeship, and journeyman training or similar programs may be used when necessary.

(g)   In order for the nonworking training hours of trainees to be counted in meeting the Contractor's minority manpower utilization goals, such trainees must be employed by the Contractor during the training period, the Contractor must have made a commitment to employ the trainees at the completion of their training, and the trainees must be trained pursuant to established training programs which must be the equivalent, with respect to the nature, extent and duration of training offered, of the training programs provided for in the Plan.

(h)   The Contractor shall take affirmative action to increase his minority manpower utilization, including but not limited to the following steps:

(i)   Notify community organizations that the contractor has employment opportunities available and maintain records of the organizations' response.

(ii)   Maintain a file of the names and addresses of each minority worker referred to him and what action was taken with respect to each such referred worker, and if the worker was not employed, the reasons therefor. If such worker was not sent to the union hiring hall for referral, the contractor's file should document this fact and the reasons therefor.

(iii)   Promptly notify the Director, OFCC, when any union with whom the contractor has a collective bargaining agreement has not referred to the contractor a minority worker sent by the contractor, or the contractor has other information that the union referral process has impeded him in his efforts to meet his goal.

(iv)   Participate in training programs in the area, especially those funded by the Department of Labor.

(v)   Disseminate his EEO policy within his own organization by including it in any policy manual; by publicizing it in company newspapers, annual report, etc.; by conducting staff, employee and union representatives' meetings to explain and discuss the policy; by posting of the policy; and by specific review of the policy with minority employees.

(vi)   Disseminate his EEO policy externally by informing, and discussing it with, all recruitment sources; by advertising in news media, specifically including minority news media; and by notifying, and discussing it with, all subcontractors and suppliers.

(vii)   Make specific and constant personal (both written and oral) recruitment efforts directed at all minority organizations, schools with minority students, minority recruitment organizations and minority training organizations, within the contractor's recruitment area.

APPENDIX D

SUBCONTRACT FORMS

THE AMERICAN INSTITUTE OF ARCHITECTS

AIA Document A401

SUBCONTRACT

Standard Form of Agreement Between Contractor and Subcontractor

1978 EDITION

Use with the latest edition of the appropriate AIA Documents as follows:

A101, Owner-Contractor Agreement — Stipulated Sum
A107, Abbreviated Owner-Contractor Agreement with General Conditions
A111, Owner-Contractor Agreement — Cost plus Fee
A201, General Conditions of the Contract for Construction.

THIS DOCUMENT HAS IMPORTANT LEGAL CONSEQUENCES; CONSULTATION WITH AN ATTORNEY IS ENCOURAGED WITH RESPECT TO ITS COMPLETION OR MODIFICATION

This document has been approved and endorsed by the American Subcontractors Association and the Associated Specialty Contractors, Inc.

AGREEMENT

made as of the   day of   in the year Nineteen
Hundred and

BETWEEN the Contractor:

and the Subcontractor:

The Project:

The Owner:

The Architect:

The Contractor and Subcontractor agree as set forth below.

ARTICLE 1

THE CONTRACT DOCUMENTS

1.1 The Contract Documents for this Subcontract consist of this Agreement and any Exhibits attached hereto, the Agreement between the Owner and Contractor dated as of   , the Conditions of the Contract between the Owner and Contractor (General, Supplementary and other Conditions), the Drawings, the Specifications, all Addenda issued prior to and all Modifications issued after execution of the Agreement between the Owner and Contractor and agreed upon by the parties to this Subcontract. These form the Subcontract, and are as fully a part of the Subcontract as if attached to this Agreement or repeated herein.

1.2 Copies of the above documents which are applicable to the Work under this Subcontract shall be furnished to the Subcontractor upon his request. An enumeration of the applicable Contract Documents appears in Article 15.

ARTICLE 2

THE WORK

2.1 The Subcontractor shall perform all the Work required by the Contract Documents for

(Here insert a precise description of the Work covered by this Subcontract and refer to numbers of Drawings and pages of Specifications including Addenda, Modifications and accepted Alternates.)

ARTICLE 3

TIME OF COMMENCEMENT AND SUBSTANTIAL COMPLETION

3.1 The Work to be performed under this Subcontract shall be commenced
and, subject to authorized adjustments, shall be substantially completed not later than

(Here insert the specific provisions that are applicable to this Subcontract including any information pertaining to notice to proceed or other method of modification for commencement of Work, starting and completion dates, or duration, and any provisions for liquidated damages relating to failure to complete on time.)

3.2 Time is of the essence of this Subcontract.

3.3 No extension of time will be valid without the Contractor's written consent after claim made by the Subcontractor in accordance with Paragraph 11.10.

ARTICLE 4

THE CONTRACT SUM

4.1 The Contractor shall pay the Subcontractor in current funds for the performance of the Work, subject to additions and deductions authorized pursuant to Paragraph 11.9, the Contract Sum of
dollars ($).

The Contract Sum is determined as follows:

(State here the base bid or other lump sum amount, accepted alternates, and unit prices, as applicable.)

ARTICLE 5
PROGRESS PAYMENTS

5.1   The Contractor shall pay the Subcontractor monthly progress payments in accordance with Paragraph 12.4 of this Subcontract.

5.2   Applications for monthly progress payments shall be in writing and in accordance with Paragraph 11.8, shall state the estimated percentage of the Work in this Subcontract that has been satisfactorily completed and shall be submitted to the Conractor on or before the   day of each month.

(Here insert details on (1) payment procedures and date of monthly applications, or other procedure if on other than a monthly basis, (2) the basis on which payment will be made on account of materials and equipment suitably stored at the site or other location agreed upon in writing, and (3) any provisions consistent with the Contract Documents for limiting or reducing the amount retained after the Work reaches a certain stage of completion.)

5.3   When the Subcontractor's Work or a designated portion thereof is substantially complete and in accordance with the Contract Documents, the Contractor shall, upon application by the Subcontractor, make prompt application for payment of such Work. Within thirty days following issuance by the Architect of the Certificate for Payment covering such substantially completed Work, the Contractor shall, to the full extent provided in the Contract Documents, make payment to the Subcontractor of the entire unpaid balance of the Contract Sum or of that portion of the Contract Sum attributable to the substantially completed Work, less any portion of the funds for the Subcontractor's Work withheld in accordance with the Certificate to cover costs of items to be completed or corrected by the Subcontractor.

(Delete the above Paragraph if the Contract Documents do not provide for, and the Subcontractor agrees to forego, release of retainage for the Subcontractor's Work prior to completion of the entire Project.)

5.4   Progress payments or final payment due and unpaid under this Subcontract shall bear interest from the date payment is due at the rate entered below or, in the absence thereof, at the legal rate prevailing at the place of the Project.

(Here insert any rate of interest agreed upon.)

(Usury laws and requirements under the Federal Truth in Lending Act, similar state and local consumer credit laws and other regulations at the Owner's, Contractor's and Subcontractor's principal places of business, the location of the Project and elsewhere may affect the validity of this provision. Specific legal advice should be obtained with respect to deletion, modification, or other requirements such as written disclosures or waivers.)

ARTICLE 6
FINAL PAYMENT

6.1   Final payment, constituting the entire unpaid balance of the Contract Sum, shall be due when the Work described in this Subcontract is fully completed and performed in accordance with the Contract Documents and is satisfactory to the Architect, and shall be payable as follows, in accordance with Article 5 and with Paragraph 12.4 of this Subcontract:

(Here insert the relevant conditions under which or time in which final payment will become payable.)

6.2   Before issuance of the final payment, the Subcontractor, if required, shall submit evidence satisfactory to the Contractor that all payrolls, bills for materials and equipment, and all known indebtedness connected with the Subcontractor's Work have been satisfied.

ARTICLE 7
PERFORMANCE BOND AND LABOR AND MATERIAL PAYMENT BOND

(Here insert any requirement for the furnishing of bonds by the Subcontractor.)

ARTICLE 8
TEMPORARY FACILITIES AND SERVICES

8.1 Unless otherwise provided in this Subcontract, the Contractor shall furnish and make available at no cost to the Subcontractor the following temporary facilities and services:

ARTICLE 9
INSURANCE

9.1 Prior to starting work, the Subcontractor shall obtain the required insurance from a responsible insurer, and shall furnish satisfactory evidence to the Contractor that the Subcontractor has complied with the requirements of this Article 9. Similarly, the Contractor shall furnish to the Subcontractor satisfactory evidence of insurance required of the Contractor by the Contract Documents.

9.2 The Contractor and Subcontractor waive all rights against each other and against the Owner, the Architect, separate contractors and all other subcontractors for damages caused by fire or other perils to the extent covered by property insurance provided under the General Conditions, except such rights as they may have to the proceeds of such insurance.

(Here insert any insurance requirements and Subcontractor's responsibility for obtaining, maintaining and paying for necessary insurance with limits equalling or exceeding those specified in the Contract Documents and inserted below, or required by law. If applicable, this shall include fire insurance and extended coverage, public liability, property damage, employer's liability, and workers' or workmen's compensation insurance for the Subcontractor and his employees. The insertion should cover provisions for notice of cancellation, allocation of insurance proceeds, and other aspects of insurance.)

ARTICLE 10
WORKING CONDITIONS

(Here insert any applicable arrangements concerning working conditions and labor matters for the Project.)

ARTICLE 11
SUBCONTRACTOR

11.1 RIGHTS AND RESPONSIBILITIES

11.1.1 The Subcontractor shall be bound to the Contractor by the terms of this Agreement and, to the extent that provisions of the Contract Documents between the Owner and Contractor apply to the Work of the Subcontractor as defined in this Agreement, the Subcontractor shall assume toward the Contractor all the obligations and responsibilities which the Contractor, by those Documents, assumes toward the Owner and the Architect, and shall have the benefit of all rights, remedies and redress against the Contractor which the Contractor, by those Documents, has against the Owner, insofar as applicable to this Subcontract, provided that where any provision of the Contract Documents between the Owner and Contractor is inconsistent with any provision of this Agreement, this Agreement shall govern.

11.1.2 The Subcontractor shall not assign this subcontract without the written consent of the Contractor, nor subcontract the whole of this Subcontract without the written consent of the Contractor, nor further subcontract portions of this Subcontract without written notification to the Contractor when such notification is requested by the Contractor. The Subcontractor shall not assign any amounts due or to become due under this Subcontract without written notice to the Contractor.

11.2 EXECUTION AND PROGRESS OF THE WORK

11.2.1 The Subcontractor agrees that the Contractor's equipment will be available to the Subcontractor only at the Contractor's discretion and on mutually satisfactory terms.

11.2.2 The Subcontractor shall cooperate with the Contractor in scheduling and performing his Work to avoid conflict or interference with the work of others.

11.2.3 The Subcontractor shall promptly submit shop drawings and samples required in order to perform his Work efficiently, expeditiously and in a manner that will not cause delay in the progress of the Work of the Contractor or other subcontractors.

11.2.4 The Subcontractor shall furnish periodic progress reports on the Work as mutually agreed, including information on the status of materials and equipment under this Subcontract which may be in the course of preparation or manufacture.

11.2.5 The Subcontractor agrees that all Work shall be done subject to the final approval of the Architect. The Architect's decisions in matters relating to artistic effect shall be final if consistent with the intent of the Contract Documents.

11.2.6 The Subcontractor shall pay for all materials, equipment and labor used in, or in connection with, the performance of this Subcontract through the period covered by previous payments received from the Contractor, and shall furnish satisfactory evidence, when requested by the Contractor, to verify compliance with the above requirements.

11.3 LAWS, PERMITS, FEES AND NOTICES

11.3.1 The Subcontractor shall give all notices and comply with all laws, ordinances, rules, regulations and orders of any public authority bearing on the performance of the Work under this Subcontract. The Subcontractor shall secure and pay for all permits and governmental fees, licenses and inspections necessary for the proper execution and completion of the Subcontractor's Work, the furnishing of which is required of the Contractor by the Contract Documents.

11.3.2 The Subcontractor shall comply with Federal, State and local tax laws, social security acts, unemployment compensation acts and workers' or workmen's compensation acts insofar as applicable to the performance of this Subcontract.

11.4 WORK OF OTHERS

11.4.1 In carrying out his Work, the Subcontractor shall take necessary precautions to protect properly the finished work of other trades from damage caused by his operations.

11.4.2 The Subcontractor shall cooperate with the Contractor and other subcontractors whose work might interfere with the Subcontractor's Work, and shall participate in the preparation of coordinated drawings in areas of congestion as required by the Contract Documents, specifically noting and advising the Contractor of any such interference.

11.5 SAFETY PRECAUTIONS AND PROCEDURES

11.5.1 The Subcontractor shall take all reasonable safety precautions with respect to his Work, shall comply with all safety measures initiated by the Contractor and with all applicable laws, ordinances, rules, regulations and orders of any public authority for the safety of persons or property in accordance with the requirements of the Contract Documents. The Subcontractor shall report within three days to the Contractor any injury to any of the Subcontractor's employees at the site.

11.6 CLEANING UP

11.6.1 The Subcontractor shall at all times keep the premises free from accumulation of waste materials or rubbish arising out of the operations of this Subcontract. Unless otherwise provided, the Subcontractor shall not be held responsible for unclean conditions caused by other contractors or subcontractors.

11.7 WARRANTY

11.7.1 The Subcontractor warrants to the Owner, the Architect and the Contractor that all materials and equipment furnished shall be new unless otherwise specified, and that all Work under this Subcontract shall be of good quality, free from faults and defects and in conformance with the Contract Documents. All Work not conforming to these requirements, including substitutions not properly approved and authorized, may be considered defecive. The warranty provided in this Paragraph 11.7 shall be in addition to and not in limitation of any other warranty or remedy required by law or by the Contract Documents.

11.8 APPLICATIONS FOR PAYMENT

11.8.1 The Subcontractor shall submit to the Contractor applications for payment at such times as stipulated in Article 5 to enable the Contractor to apply for payment.

11.8.2 If payments are made on the valuation of Work done, the Subcontractor shall, before the first application, submit to the Contractor a schedule of values of the various parts of the Work aggregating the total sum of this Subcontract, made out in such detail as the Subcontractor and Contractor may agree upon or as required by the Owner, and supported by such evidence as to its correctness as the Contractor may direct. This schedule, when approved by the Contractor, shall be used only as a basis for Applications for Payment, unless it be found to be in error. In applying for payment, the Subcontractor shall submit a statement based upon this schedule.

11.8.3 If payments are made on account of materials or equipment not incorporated in the Work but delivered and suitably stored at the site or at some other location agreed upon in writing, such payments shall be in accordance with the terms and conditions of the Contract Documents.

11.9 CHANGES IN THE WORK

11.9.1 The Subcontractor may be ordered in writing by the Contractor, without invalidating this Subcontract, to make changes in the Work within the general scope of this Subcontract consisting of additions, deletions or other revisions, the Contract Sum and the Contract Time being adjusted accordingly. The Subcontractor, prior to the commencement of such changed or revised Work, shall submit promptly to the Contractor written copies of any claim for adjustment to the Contract Sum and Contract Time for such revised Work in a manner consistent with the Contract Documents.

11.10 CLAIMS OF THE SUBCONTRACTOR

11.10.1 The Subcontractor shall make all claims promptly to the Contractor for additional cost, extensions of time, and damages for delays or other causes in accordance with the Contract Documents. Any such claim which will affect or become part of a claim which the Contractor is required to make under the Contract Documents within a specified time period or in a specified manner shall be made in sufficient time to permit the Contractor to satisfy the requirements of the Contract Documents. Such claims shall be received by the Contractor not less than two working days preceding the time by which the Contractor's claim must be made. Failure of the Subcontractor to make such a timely claim shall bind the Subcontractor to the same consequences as those to which the Contractor is bound.

11.11 INDEMNIFICATION

11.11.1 To the fullest extent permitted by law, the Subcontractor shall indemnify and hold harmless the Owner the Architect and the Contractor and all of their agents and employees from and against all claims, damages, losses and expenses, including but not limited to attor-

ney's fees, arising out of or resulting from the performance of the Subcontractor's Work under this Subcontract, provided that any such claim, damage, loss, or expense is attributable to bodily injury, sickness, disease, or death, or to injury to or destruction of tangible property (other than the Work itself) including the loss of use resulting therefrom, to the extent caused in whole or in part by any negligent act or omission of the Subcontractor or anyone directly or indirectly employed by him or anyone for whose acts he may be liable, regardless of whether it is caused in part by a party indemnified hereunder. Such obligation shall not be construed to negate, or abridge, or otherwise reduce any other right or obligation of indemnity which would otherwise exist as to any party or person described in this Paragraph 11.11.

11.11.2 In any and all claims against the Owner, the Architect, or the Contractor or any of their agents or employees by any employee of the Subcontractor, anyone directly or indirectly employed by him or anyone for whose acts he may be liable, the indemnification obligation under this Paragraph 11.11 shall not be limited in any way by any limitation on the amount or type of damages, compensation or benefits payable by or for the Subcontractor under workers' or workmen's compensation acts, disability benefit acts or other employee benefit acts.

11.11.3 The obligations of the Subcontractor under this Paragraph 11.11 shall not extend to the liability of the Architect, his agents or employees arising out of (1) the preparation or approval of maps, drawings, opinions, reports, surveys, Change Orders, designs or specifications, or (2) the giving of or the failure to give directions or instructions by the Architect, his agents or employees provided such giving or failure to give is the primary cause of the injury or damage.

11.12 SUBCONTRACTOR'S REMEDIES

11.12.1 If the Contractor does not pay the Subcontractor through no fault of the Subcontractor, within seven days from the time payment should be made as provided in Paragraph 12.4, the Subcontractor may, without prejudice to any other remedy he may have, upon seven additional days' written notice to the Contractor, stop his Work until payment of the amount owing has been received. The Contract Sum shall, by appropriate adjustment, be increased by the amount of the Subcontractor's reasonable costs of shutdown, delay and start-up.

ARTICLE 12
CONTRACTOR

12.1 RIGHTS AND RESPONSIBILITIES

12.1.1 The Contractor shall be bound to the Subcontractor by the terms of this Agreement, and to the extent that provisions of the Contract Documents between the Owner and the Contractor apply to the Work of the Subcontractor as defined in this Agreement, the Contractor shall assume toward the Subcontractor all the obligations and responsibilities that the Owner, by those Documents, assumes toward the Contractor, and shall have the benefit of all rights, remedies and redress against the Subcontractor which the Owner, by those Documents, has against the Contractor. Where any provision of the

Contract Documents between the Owner and the Contractor is inconsistent with any provisions of this Agreement, this Agreement shall govern.

12.2 SERVICES PROVIDED BY THE CONTRACTOR

12.2.1 The Contractor shall cooperate with the Subcontractor in scheduling and performing his Work to avoid conflicts or interference in the Subcontractor's Work, and shall expedite written responses to submittals made by the Subcontractor in accordance with Paragraphs 11.2, 11.9 and 11.10. As soon as practicable after execution of this Agreement, the Contractor shall provide the Subcontractor a copy of the estimated progress schedule of the Contractor's entire Work which the Contractor has prepared and submitted for the Owner's and the Architect's information, together with such additional scheduling details as will enable the Subcontractor to plan and perform his Work properly. The Subcontractor shall be notified promptly of any subsequent changes in the progress schedule and the additional scheduling details.

12.2.2 The Contractor shall provide suitable areas for storage of the Subcontractor's materials and equipment during the course of the Work. Any additional costs to the Subcontractor resulting from the relocation of such facilities at the direction of the Contractor shall be reimbursed by the Contractor.

12.3 COMMUNICATIONS

12.3.1 The Contractor shall promptly notify the Subcontractor of all modifications to the Contract between the Owner and the Contractor which affect this Subcontract and which were issued or entered into subsequent to the execution of this Subcontract.

12.3.2 The Contractor shall not give instructions or orders directly to employees or workmen of the Subcontractor except to persons designated as authorized representatives of the Subcontractor.

12.4 PAYMENTS TO THE SUBCONTRACTOR

12.4.1 Unless otherwise provided in the Contract Documents, the Contractor shall pay the Subcontractor each progress payment and the final payment under this Subcontract within three working days after he receives payment from the Owner, except as provided in Subparagraph 12.4.3. The amount of each progress payment to the Subcontractor shall be the amount to which the Subcontractor is entitled, reflecting the percentage of completion allowed to the Contractor for the Work of this Subcontractor applied to the Contract Sum of this Subcontract, and the percentage actually retained, if any, from payments to the Contractor on account of such Subcontractor's Work, plus, to the extent permitted by the Contract Documents, the amount allowed for materials and equipment suitably stored by the Subcontractor, less the aggregate of previous payments to the Subcontractor.

12.4.2 The Contractor shall permit the Subcontractor to request directly from the Architect information regarding the percentages of completion or the amount certified on account of Work done by the Subcontractor.

12.4.3 If the Architect does not issue a Certificate for Payment or the Contractor does not receive payment for any cause which is not the fault of the Subcontractor, the Contractor shall pay the Subcontractor, on demand, a progress payment computed as provided in Subparagraph 12.4.1 or the final payment as provided in Article 6.

12.5 CLAIMS BY THE CONTRACTOR

12.5.1 The Contractor shall make no demand for liquidated damages for delay in any sum in excess of such amount as may be specifically named in this Subcontract, and liquidated damages shall be assessed against this Subcontractor only for his negligent acts and his failure to act in accordance with the terms of this Agreement, and in no case for delays or causes arising outside the scope of this Subcontract, or for which other subcontractors are responsible.

12.5.2 Except as may be indicated in this Agreement, the Contractor agrees that no claim for payment for services rendered or materials and equipment furnished by the Contractor to the Subcontractor shall be valid without prior notice to the Subcontractor and unless written notice thereof is given by the Contractor to the Subcontractor not later than the tenth day of the calendar month following that in which the claim originated.

12.6 CONTRACTORS' REMEDIES

12.6.1 If the Subcontractor defaults or neglects to carry out the Work in accordance with this Agreement and fails within three working days after receipt of written notice from the Contractor to commence and continue correction of such default or neglect with diligence and promptness, the Contractor may, after three days following receipt by the Subcontractor of an additional written notice, and without prejudice to any other remedy he may have, make good such deficiencies and may deduct the cost thereof from the payments then or thereafter due the Subcontractor, provided, however, that if such action is based upon faulty workmanship or materials and equipment, the Architect shall first have determined that the workmanship or materials and equipment are not in accordance with the Contract Documents.

ARTICLE 13

ARBITRATION

13.1 All claims, disputes and other matters in question arising out of, or relating to, this Subcontract, or the breach thereof, shall be decided by arbitration, which shall be conducted in the same manner and under the same procedure as provided in the Contract Documents with respect to disputes between the Owner and the Contractor, except that a decision by the Architect shall not be a condition precedent to arbitration. If the Contract Documents do not provide for arbitration or fail to specify the manner and procedure for arbitration, it shall be conducted in accordance with the Construction Industry Arbitration Rules of the American Arbitration Association then obtaining unless the parties mutually agree otherwise.

13.2 Except by written consent of the person or entity sought to be joined, no arbitration arising out of or relating to the Contract Documents shall include, by consolidation, joinder or in any other manner, any person or entity not a party to the Agreement under which such arbitration arises, unless it is shown at the time the demand for arbitration is filed that (1) such person or entity is substantially involved in a common question of fact or law,

(2) the presence of such person or entity is required if complete relief is to be accorded in the arbitration, (3) the interest or responsibility of such person or entity in the matter is not insubstantial, and (4) such person or entity is not the Architect, his employee or his consultant. This agreement to arbitrate and any other written agreement to arbitrate with an additional person or persons referred to herein shall be specifically enforceable under the prevailing arbitration law.

13.3 The Contractor shall permit the Subcontractor to be present and to submit evidence in any arbitration proceeding involving his rights.

13.4 The Contractor shall permit the Subcontractor to exercise whatever rights the Contractor may have under the Contract Documents in the choice of arbitrators in any dispute, if the sole cause of the dispute is the Work, materials, equipment, rights or responsibilities of the Subcontractor; or if the dispute involves the Subcontractor and any other subcontractor or subcontractors jointly, the Contractor shall permit them to exercise such rights jointly.

13.5 The award rendered by the arbitrators shall be final, and judgment may be entered upon it in accordance with applicable law in any court having jurisdiction thereof.

13.6 This Article shall not be deemed a limitation of any rights or remedies which the Subcontractor may have under any Federal or State mechanics' lien laws or under any applicable labor and material payment bonds unless such rights or remedies are expressly waived by him.

ARTICLE 14
TERMINATION

14.1 TERMINATION BY THE SUBCONTRACTOR

14.1.1 If the Work is stopped for a period of thirty days through no fault of the Subcontractor because the Contractor has not made payments thereon as provided in this Agreement, then the Subcontractor may without prejudice to any other remedy he may have, upon seven additional days' written notice to the Contractor, terminate this Subcontract and recover from the Contractor payment for all Work executed and for any proven loss resulting from the stoppage of the Work, including reasonable overhead, profit and damages.

14.2 TERMINATION BY THE CONTRACTOR

14.2.1 If the Subcontractor persistently or repeatedly fails or neglects to carry out the Work in accordance with the Contract Documents or otherwise to perform in accordance with this Agreement and fails within seven days after receipt of written notice to commence and continue correction of such default or neglect with diligence and promptness, the Contractor may, after seven days following receipt by the Subcontractor of an additional written notice and without prejudice to any other remedy he may have, terminate the Subcontract and finish the Work by whatever method he may deem expedient. If the unpaid balance of the Contract Sum exceeds the expense of finishing the Work, such excess shall be paid to the Subcontractor, but if such expense exceeds such unpaid balance, the Subcontractor shall pay the difference to the Contractor.

ARTICLE 15
MISCELLANEOUS PROVISIONS

15.1 Terms used in this Agreement which are defined in the Conditions of the Contract shall have the meanings designated in those Conditions.

15.2 The Contract Documents, which constitute the entire Agreement between the Owner and the Contractor, are listed in Article 1, and the documents which are applicable to this Subcontract, except for Addenda and Modifications issued after execution of this Subcontract, are enumerated as follows:

(List below the Agreement, the Conditions of the Contract [General, Supplementary, and other Conditions], the Drawings, the Specifications, and any Addenda and accepted Alternates, showing page or sheet numbers in all cases and dates where applicable. Continue on succeeding pages as required.)

This Agreement entered into as of the day and year first written above.

CONTRACTOR

_____

_____

_____

SUBCONTRACTOR

_____

_____

_____

AIA DOCUMENT A401 • CONTRACTOR-SUBCONTRACTOR AGREEMENT • ELEVENTH EDITION • APRIL 1978 • AIA®
©1978 • THE AMERICAN INSTITUTE OF ARCHITECTS, 1735 NEW YORK AVE., N.W., WASHINGTON, D.C. 20006

ASC

STANDARD SUBCONTRACT
AGREEMENT

THIS AGREEMENT made this   day of   in the year Nineteen

Hundred and   by and between

hereinafter

called the Subcontractor and

hereinafter called the Contractor.

WITNESSETH, That the Subcontractor and Contractor for the consideration hereinafter named agree as follows:

(Developed as a guide by The Associated General Contractors of America, The National Electrical Contractors Association, The Mechanical Contractors Association of America. The Sheet Metal and Air Conditioning Contractors National Association and the National Association of Plumbing - Heating - Cooling Contractors °1966 by the Associated General Contractors of America and the Associated Specialty Contractors, Inc.)

1

ARTICLE I

The Subcontractor agrees to furnish all material and perform all work as described in Article II hereof
for

(Here name the project.)
or

(Here name the Contractor.)
at

(Here insert the location of the work and name of Owner.)

in accordance with this Agreement, the Agreement between the Owner and Contractor, and in accordance with the General Conditions of the Contract, Supplementary General Conditions, the Drawings and Specifications and addenda prepared by

hereinafter called the Architect or Owner's authorized agent, all of which documents, signed by the parties thereto or identified by the Architect or Owner's authorized agent, form a part of a Contract

between the Contractor and the Owner dated   , 19   , and hereby become a part of this contract, and herein referred to as the Contract Documents, and shall be made available to the Subcontractor upon his request prior to and at anytime subsequent to signing this Subcontract.

ARTICLE II

The Subcontractor and the Contractor agree that the materials and equipment to be furnished and work to be done by the Subcontractor are:

(Here insert a precise description of the work, preferably by reference to the numbers of the drawings and the pages of the specifications including addenda and accepted alternates.)

2

ARTICLE III

Time is of the essence and the Subcontractor agrees to commence and to complete the work as described in Article II as follows:

(Here insert any information pertaining to the method of notification for commencement of work, starting and completion dates, or duration, and any liquidated damage requirements.)

(a) No extension of time of this contract will be recognized without the written consent of the Contractor which consent shall not be withheld unreasonably consistent with Article X-4 of this Contract, subject to the arbitration provisions herein provided.

ARTICLE IV

The Contractor agrees to pay the Subcontractor for the performance of this work the

sum of   ($
in current funds, subject to additions and deductions for changes as may be agreed upon in writing and to make monthly payments on account thereof in accordance with Article X, Sections 20-23 inclusive.

(Here insert additional details—unit prices, etc., payment procedure including date of monthly applications for payment, payment procedure if other than on a monthly basis, consideration of materials safely and suitably stored at the site or at some other location agreed upon in writing by the parties—and any provisions made for limiting or reducing the amount retained after the work reaches a certain stage of completion which should be consistent with the Contract Documents.)

ARTICLE V

Final payment shall be due when the work described in this contract is fully completed and performed in accordance with the Contract Documents, and payment to be consistent with Article IV and Article X, Sections 18, 20-23 inclusive of this contract.

Before issuance of the final payment the Subcontractor if required shall submit evidence satisfactory to the Contractor that all payrolls, material bills, and all known indebtedness connected with the Subcontractor's work have been satisfied.

ARTICLE VI
Performance and Payment Bonds

(Here insert any requirement for the furnishing of performance and payment bonds.)

4

ARTICLE VII

Temporary Site Facilities

(Here insert any requirements and terms concerning temporary site facilities, i.e., storage, sheds, water, heat, light, power, toilets, hoists, elevators, scaffolding, cold weather protection, ventilating, pumps, watchman service, etc.)

ARTICLE VIII

Insurance

Unless otherwise provided herein, the Subcontractor shall have a direct liability for the acts of his employees and agents for which he is legally responsible, and the Subcontractor shall not be required to assume the liability for the acts of any others.

Prior to starting work the insurance required to be furnished shall be obtained from a responsible company or companies to provide proper and adequate coverage and satisfactory evidence will be furnished to the Contractor that the Subcontractor has complied with the requirements as stated in this Section.

(Here insert any insurance requirements and Subcontractor's responsibility for obtaining, maintaining and paying for necessary insurance, not less than limits as may be specified in the Contract Documents or required by laws. This to include fire insurance and extended coverage, consideration of public liability, property damage, employer's liability, and workmen's compensation insurance for the Subcontractor and his employees. The insertion should provide the agreement of the Contractor and the Subcontractor on subrogation waivers, provision for notice of cancellation, allocation of insurance proceeds, and other aspects of insurance.)

(It is recommended that the AGC Insurance and Bonds Checklist (AGC Form No. 29) be referred to as a guide for other insurance coverages.)

5

ARTICLE IX
Job Conditions

Here insert any applicable arrangements and necessary cooperation concerning labor matters for the project.)

ARTICLE X

In addition to the foregoing provisions the parties also agree:
That the Subcontractor shall:

(1) Be bound to the Contractor by the terms of the Contractor Documents and this Agreement, and assume toward the Contractor all the obligations and responsibilities that the Contractor, by those documents, assumes toward the Owner, as applicable to this Subcontract. (a) Not discriminate against any employee or applicant for employment because of race, creed, color, or national origin.

(2) Submit to the Contractor applications for payment at such times as stipulated in Article IV so as to enable the Contractor to apply for payment.

If payments are made on valuations of work done, the Subcontractor shall, before the first application, submit to the Contractor a schedule of values of the various parts of the work, aggregating the total sum of the Contract, made out in such detail as the Subcontractor and Contractor may agree upon, or as required by the Owner, and, if required, supported by such evidence as to its correctness as the Contractor may direct. This schedule, when approved by the Contractor, shall be used as a basis for Certificates for Payment, unless it be found to be in error. In applying for payment, the Subcontractor shall submit a statement based upon this schedule.

If payments are made on account of materials not incorporated in the work but delivered and suitably stored at the site, or at some other location agreed upon in writing, such payments shall be in accordance with the terms and conditions of the Contract Documents.

(3) Pay for all materials and labor used in, or in connection with, the performance of this contract, through the period covered by previous payments received from the Contractor, and furnish satisfactory evidence when requested by the Contractor, to verify compliance with the above requirements.

(4) Make all claims for extras, for extensions of time and for damage for delays or otherwise, promptly to the Contractor consistent with the Contract Documents.

(5) Take necessary precaution to properly protect the finished work of other trades.

(6) Keep the building and premises clean at all times of debris arising out of the operation of this subcontract. The Subcontractor shall not be held responsible for unclean conditions caused by other contractors or subcontractors, unless otherwise provided for.

(7) Comply with all statutory and/or contractual safety requirements applying to his work and/or initiated by the Contractor, and shall report within 3 days to the Contractor any injury to the Subcontractor's employees at the site of the project.

6

(8) (a) Not assign this subcontract or any amounts due or to become due thereunder without the written consent of the contractor. (b) Nor subcontract the whole of this subcontract without the written consent of the contractor. (c) Nor further subcontract portions of this subcontract without written notification to the contractor when such notification is requested by the contractor.

(9) Guarantee his work against all defects of materials and/or workmanship as called for in the plans, specifications and addenda, or if no guarantee is called for, then for a period of one year from the dates of partial or total acceptance of the Subcontractor's work by the Owner.

(10) And does hereby agree that if the Subcontractor should neglect to prosecute the work diligently and properly or fail to perform any provision of this contract, the Contractor, after three days written notice to the Subcontractor, may, without prejudice to any other remedy he may have, make good such deficiencies and may deduct the cost thereof from the payment then or thereafter due the Subcontractor, provided, however, that if such action is based upon faulty workmanship the Architect or Owner's authorized agent, shall first have determined that the workmanship and/or materials is defective.

(11) And does hereby agree that the Contractor's equipment will be available to the Subcontractor only at the Contractor's discretion and on mutually satisfactory terms.

(12) Furnish periodic progress reports of the work as mutually agreed including the progress of materials or equipment under this Agreement that may be in the course of preparation or manufacture.

(13) Make any and all changes or deviations from the original plans and specifications without nullifying the original contract when specifically ordered to do so in writing by the Contractor. The Subcontractor prior to the commencement of this revised work, shall submit promptly to the Contractor written copies of the cost or credit proposal for such revised work in a manner consistent with the Contract Documents.

(14) Cooperate with the Contractor and other Subcontractors whose work might interfere with the Subcontractor's work and to participate in the preparation of coordinated drawings in areas of congestion as required by the Contract Documents, specifically noting and advising the Contractor of any such interference.

(15) Cooperate with the Contractor in scheduling his work so as not conflict or interfere with the work of others. To promptly submit shop drawings, drawings, and samples, as required in order to carry on said work efficiently and at speed that will not cause delay in the progress of the Contractor's work or other branches of the work carried on by other Subcontractors.

(16) Comply with all Federal, State and local laws and ordinances applying to the building or structure and to comply and give adequate notices relating to the work to proper authorities and to secure and pay for all necessary licenses or permits to carry on the work as described in the Contract Documents as applicable to this Subcontract.

(17) Comply with Federal, State and local tax laws, Social Security laws and Unemployment Compensation laws and Workmen's Compensation Laws insofar as applicable to the performance of this subcontract.

(18) And does hereby agree that all work shall be done subject to the final approval of the Architect or Owner's authorized agent, and his decision in matters relating to artistic effect shall be final, if within the terms of the Contract Documents.

That the Contractor shall—

(19) Be bound to the Subcontractor by all the obligations that the Owner assumes to the Contractor under the Contract Documents and by all the provisions thereof affording remedies and redress to the Contractor from the Owner insofar as applicable to this Subcontract.

(20) Pay the Subcontractor within seven days, unless otherwise provided in the Contract Documents, upon the payment of certificates issued under the Contractor's schedule of values, or as described in Article IV herein. The amount of the payment shall be equal to the percentage of completion certified by the Owner or his authorized agent for the work of this Subcontractor applied to the amount set forth under Article IV and allowed to the Contractor on account of the Subcontractor's work to the extent of the Subcontractor's interest therein.

(21) Permit the Subcontractor to obtain direct from the Architect or Owner's authorized agent, evidence of percentages of completion certified on his account.

(22) Pay the Subcontractor on demand for his work and/or materials as far as executed and fixed in place, less the retained percentage, at the time the payment should be made to the Subcontractor if the Architect or Owner's authorized agent fails to issue the certificate for any fault of the Contractor and not the fault of the Subcontractor or as otherwise provided herein.

(23) And does hereby agree that the failure to make payments to the Subcontractor as herein provided for any cause not the fault of the Subcontractor, within 7 days from the Contractor's receipt of payment or from time payment should be made as

7

provided in Article X, Section 22, or maturity, then the Subcontractor may upon 7 days written notice to the Contractor stop work without prejudice to any other remedy he may have.

24) Not issue or give any instructions, order or directions directly to employees or workmen of the Subcontractor other than to the persons designated as the authorized representative(s) of the Subcontractor.

25) Make no demand for liquidated damages in any sum in excess of such amount as may be specifically named in the subcontract, provided, however, no liquidated damages shall be assessed for delays or causes attributable to other Subcontractors or arising outside the scope of this Subcontract.

26) And does hereby agree that no claim for services rendered or materials furnished by the Contractor to the Subcontractor shall be valid unless written notice thereof is given by the Contractor to the Subcontractor during the first ten days of the calendar month following that in which the claim originated.

27) Give the Subcontractor an opportunity to be present and to submit evidence in any arbitration involving his rights.

28) Name as arbitor under arbitration proceedings as provided in the General Conditions the person nominated by the Subcontractor, if the sole cause of dispute is the work, materials, rights or responsibilities of the Subcontractor; or if, of the Subcontractor and any other Subcontractor jointly, to name as such arbitrator the person upon whom they agree.

That the Contractor and the Subcontractor agree—

(29) That in the matter of arbitration, their rights and obligations and all procedure shall be analogous to those set forth in the Contract Documents provided, however, that a decision by the Architect or Owner's authorized agent, shall not be a condition precedent to arbitration.

(30) This subcontract is solely for the benefit of the signatories hereto.

ARTICLE XI

IN WITNESS WHEREOF the parties hereto have executed this Agreement under seal, the day and year first above written.

Attest: Subcontractor
_____   _____
        (Seal)   By   (Title)

Attest: Contractor
_____   _____
        (Seal)   By   (Title)

8

LENOX CONTRACTING & ENGINEERING CO. INC
3408 Lenox Rd. N. E.
Atlanta, Ga., 30327

CONTRACT

STATE OF GEORGIA
COUNTY OF FULTON

THIS AGREEMENT, entered into this *7* day of *January, 1980,* by and between LENOX CON-TRACTING AND ENGINEERING COMPANY, INC., a corporation of the State of Deleware, hereinafter referred to as "Contractor", and *Buford Plumbing Co., a corporation, and James C. Buford, individually and severally,* hereinafter referred to as "Subcontractor" witnesseth:

In consideration of the mutual and reciprocal obligations hereinafter stipulated, Contractor and Subcontractor do contract and agree as follows:

Article I. Subcontractor hereby agrees to furnish and pay for all of the materials and equipment, and perform all labor necessary to complete the following work in Lenox Square Shopping Center, Atlanta, Georgia, in all respects, namely: *All work as described in Sections 5 and 6 of the specifications, and all plumbing work required by the plans together with all work incidental thereto, in accordance with the plans and specifications prepared by Toombs, Amisano and Wells, including addenda through No. 9 and revisions through January 1, 1978,* and also in accordance with the following additional drawings:

None

Article II. Subcontractor hereby certifies and agrees that he has examined all the plans and read all of the plans and specifications for the entire work, of which the work covered by this contract is a part, and that he and his subcontractors will be and are bound by any and all parts of said plans and specifications insofar as they relate in any part or in any way to the work undertaken herein, and shall be further bound to this contract and the General Conditions. It is specifically agreed that Subcontractor shall not sublet, assign or transfer this contract, or any part thereof, without the written consent of Contractor. All employees of Subcontractor and his subcontractors shall carry on the project a designation sufficiently clear to identify their employer. All sub-subcontracts shall carry the designation of the project sufficiently clear to identify the sub-subcontract with the probject.

Article III. It is understood and agreed by and between the parties hereto that the work included in this contract is to be performed under the direction of the Project Manager, hereinafter defined, and that his decisions as to all matters relating to this contract shall be final (subject only to arbitration as hereinafter set forth). It is also understood and agreed by and between the parties hereto that such additional drawings and explanations as may be necessary to detail and illustrate the work to be done, are to be furnished by the Project Manager, and the parties hereto agree to conform to, and abide by, the same insofar as they may be consistent with the purpose and intent of the original drawings and specifications hereinabove referred to. No claim for extra compensation shall be made or allowed for such detail unless Subcontractor can show that he could have provided a different detail which would comply with the original drawings and specifications, have been good construction practice, and which would have been included in the contract price; and that Subcontractor proposed in writing to provide such a detail before proceeding with any work for which extra payment is claimed.

Article IV. Where "Owner" is referred to in the general and special conditions of the specifications, it shall refer to the Contractor in this contract; and where "Contractor" is used in the General and Special Conditions, it shall refer to "Subcontractor" in this contract.

Changes to the work may be directed at any time by the contractor. If such changes reduce the cost of the work, the contract amount shall be reduced by the net cost of such savings to the Subcontractor. If a change increases the cost of the work, the increase in amount shall be as mutually agreed, or at actual cost not to include more than 25% of labor, material, and equipment operation cost as an allowance for profit

Courtesy of Ed Noble.

and costs not directly associated with the work. A sub-subcontract is a cost in this respect only to the extent it represents job costs for labor, material, and equipment.

The subcontractor may not delay any change because of disagreement, so long as he receives a clear written order for the work to be done; and compliance with an order stating "at no extra cost" or words of similar intent, after protest, does not prevent the subcontractor from later claiming extra payment for such work.

Article V. Subcontractor shall provide sufficient, safe and proper facilities at all times for the inspection of the work by Contractor, or its Project Manager, and shall within twenty-four (24) hours after receiving written notice from Contractor to that effect, proceed to take down all portions of the work and remove from the grounds and buildings all material, whether worked or unworked, which the Project Manager shall condemn as unsound or improper, or as in any way failing to conform to the drawings and specifications, and shall make good all work so condemned, and all other work damaged or destroyed in making good such condemned work. In the event Subcontractor considers such work acceptable, he shall remove it and claim extra payment for replacement as previously prescribed for extra claims.

Article VI. Subcontractor hereby agrees that the work under this contract is to be begun, carried on, and completed to the satisfaction of Contractor, as directed by the Project Manager, and specifically as follows: *With a work force of 15 men by Mar. 1, 1980, maintaining this minimum force so long as there is work to be done under this contract.*

When other trades are ahead of this Subcontractor and minimum work force has not been maintained, this shall be considered proof that he is unable to maintain specified progress and Contractor may take such measures as are outlined under Article VII. It is agreed that work will be carried on as required by Contractor, promptly and efficiently, and without delaying other branches of the work. Since it is necessary to have the various stores completed in accordance with Tenant's requirements, Contractor may direct that certain stores be completed in preference to others, even though this may not be most convenient for Subcontractor. In general, larger stores will be completed earlier than adjacent smaller stores.

In order to secure the execution of this work at, and within, the time specified, it is hereby distinctly agreed that damages arising from the nonfulfillment of this contract as regards time shall be deducted from the contract price.

If performance and payment bond is not required of the Subcontractor, the cost of same shall be allowed the Contractor as a deduction from the amount of this contract.

Time is of the essence of this contract.

Article VII. Should Subcontractor at any time refuse, neglect, or be unable to supply a sufficient number of properly skilled workmen, or a sufficient quantity of materials of proper quality, or fail in any respect to prosecute the work covered by this contract with promptness and diligence, or fail in the performance of any of the agreements herein contained, Contractor may, at its option, after forty-eight (48) hours written notice to Subcontractor, provide any such labor and materials and deduct the costs thereof from any money then due, or thereafter to become due, the Subcontractor under this contract , or Contractor may, at its option, terminate the employment of Subcontractor for the said work, and shall have the right to enter upon the premises and take possession, for the purpose of completing the work included under this contract, of all the materials, tools and appliances thereon, and may employ any other person or persons to finish the work and provide the materials therefor; and in case of such discontinuance of the employment by Contractor, Subcontractor shall not be entitled to receive any further payment under this contract until the said work shall be wholly finished; at which time, if the unpaid balance of the amount to be paid under this contract exceeds the expenses incurred by Contractor in finishing the work, such excess shall be paid by Contractor to Subcontractor; but if such expense shall exceed such unpaid balance, then Subcontractor shall pay the difference to Contractor. The expense incurred by Contractor, as herein provided, either for furnishing materials or for finishing the work, and any damages incurred by such default, shall be chargeable to, and paid by, Subcontractor, and Contractor shall have a lien upon all materials, tools and appliances taken possession of, as aforesaid, to secure the payment thereof.

Should Contractor, under the provisions of this article, provide labor and materials for any part of the work, it shall not act to terminate the employment of Subcontractor for any other work, and Subcontractor shall remain obligated to complete the balance of the work. The deduction from the contract amount for such work shall be as stated in the breakdown furnished by Subcontractor before the contract is signed. Contractor shall have the option of using such materials as Subcontractor has purchased for that portion of the work performed by Contractor under the provisions of this article, in which event Subcontractor shall receive his actual net cost for such materials.

Article VIII. Subcontractor hereby agrees to save and indemnify and keep harmless Contractor against all liability claims, judgments or demands or damages arising from accidents to persons or property (including, but not limited to, Tenant's fixtures and merchandise) occasioned by Subcontractor, his agents or employees, and against all claims or demands for damages arising from accidents to Subcontractor, his agents or employees, if occasioned by any cause or condition, by Subcontractor or his employees, and Subcontractor will defend any and all suits that may be brought against Contractor on account of any such accidents and will make good to, and reimburse, Contractor for any expenditures that Contractor may make by reason of such accidents.

In order to ensure the fulfillment of the foregoing, Subcontractor hereby agrees to carry the following insurance coverages:

(a) Comprehensive General Liability with bodily injury limits of $100,000 for any one person, and $300,000 for any one occurrence; property damage limits of $200,000 for any one occurrence and $500,000 for any one accident; such coverage to further include the following: Independent Contractor, Contractual, Completed Operations, and Owners and Contractors Protective, plus waiver of subrogation against *Lenox Square, Inc.,* Contractor or any controlled subsidiary;

(b) Comprehensive Automobile Liability Policy with bodily limits of $200,000 for any one occurrence and $500,000 for any one accident and property damage limit of $100,000, such policy shall include "Hired Car" and "Non-Ownership" coverage;

(c) Workmen's Compensation coverage complying with the statutory requirements of the State of Georgia and Employer's Liability with limits of $100,000, as well as any other municipal, State and Federal insurance required by law; all of such policies shall contain a waiver of subrogation clause or an indorsement indemnifying Contractor, or any controlled subsidiary of Contractor, against any claim brought by an employee on the basis of a master and servant relationship;

(d) Performance and payment bond in the amount of this contract, with Contractor being named as the obligee, when and if requested by Contractor.

Subcontractor shall furnish Contractor suitable evidence that such insurance and all special conditions relating thereto have been complied with and is in effect before any work is commenced hereunder. Should Subcontractor sublet any work covered by this contract to a third party or parties, Subcontractor agrees to require such third party or parties, to secure and have in effect insurance coverage as set forth prior to commencing work, and in such amounts as may be required by Contractor. Subcontractor shall furnish, or cause to be furnished Contractor, suitable evidence of such coverage.

All insurance policies herein required shall contain a stipulation that such policies may be canceled only after ten (10) days' notice to Contractor.

Article IX. Contractor hereby agrees to pay to Subcontractor for such labor and material herein undertaken to be done and furnished the sum of $75,000 which sum shall be for full payment for all work to be performed hereunder, subject to additions and deductions as hereinbefore provided, and such sum shall be paid by Contractor to Subcontractor as the work progresses in installments based upon estimates as approved by Project Manager, less 10%.

Final payment shall be made within 30 days after the completion of all work in this project, and written accpetance by Contractor after satisfactory proof has been furnished that all bills have been paid for labor and material furnished hereunder. Subcontractor shall give to Contractor a sworn statement that all bills for labor and materials used in the performance of the contract have been paid at the time of final payment.

Article X. Subcontractor hereby agrees to turn said work over to Contractor in good condition and free and clear from all claims, encumbrances and liens for taxes, labor or material and to protect and save harmless Contractor from all claims, encumbrances and liens growing out of the performance of this contract, and in the event of the failure of Subcontractor during the progress of the said work, or at any time thereafter, to pay for all materials and labor used in the prosecution of said work, Contractor may at its option, and without notice to Subcontractor prior thereto, pay all such claims for labor and materials and charge the amounts to Subcontractor. In case action or filing to establish a lien is brought by any person, firm or corporation employed by, or furnishing material to, Subcontractor, under this contract, Subcontractor will upon written demand by Contractor, post such bond as may be necessary to legally dissolve said lien or liens, and at his own cost and expense (including attorney's fees) initiate or defend any action to effect a cancellation of such lien promptly and without delay and pay any such lien established in court.

Subcontractor shall, as often as requested in writing by Contractor, make out and give a sworn statement of persons furnishing labor or materials to Subcontractor, giving their name, and how much, if any, is due, or will be due each. Such statement shall be furnished by the Subcontractor to Contractor upon special request therefor, each time Subcontractor receives payments from Contractor under this contract and shall accompany requisition for same. Like statement shall be required from Subcontractors of Subcontractor on demand of Contractor.

Article XI. It is further mutually agreed between the parties hereto, that no payments made under this contract shall be conclusive evidence of the performance of this contract, either in whole or in part, and that no payment shall be construed to be an acceptance of defective work or improper materials.

Article XII. Contractor shall cause there to be effected and maintained Installation Floater & Builder's Risk insurance upon the entire structure on which the work of this contract is to be done to one hundred percent of the insurable value thereof, including items of building equipment and materials connected therewith whether in or adjacent to the structure insured, materials in place or to be used as part of the permanent construction including surplus materials, shanties, protective fences, bridges, or temporary structures, miscellaneous materials and supplies incident to the work, and such scaffoldings, stagings, towers, forms, and equipment as are not owned or rented by Subcontractor, the cost of which is included in the cost of the work. This insurance does not cover any tools owned by mechanics, any tools, equipment, scaffolding, staging, towers, and forms owned or rented by Subcontractor, the capital value of which is not included in the cost of the work, or any cook shanties, bunkhouses or other structures erected for housing for workmen. The loss, if any, is to be made adjustable with and payable to Lenox Square, Inc., as Trustee for the insureds as their interests may appear, except in such cases as may require payment of all or a portion of said insurance to be made to any mortgagee as his interests may appear.

Subcontractors shall be named or designated in such capacity as insured jointly with Contractor in all policies, all of which shall be open to Subcontractor's inspection. Certificates of such insurance shall be filed with Subcontractor if he so requires. If Contractor fails to cause there to be effected or maintained insurance as above and so notifies Subcontractor, Subcontractor may insure his own interest and that of the subcontractors and charge the cost thereof to Contractor. If Subcontractor is damaged by failure of Contractor to cause there to be maintained such insurance or to so notify Subcontractor, he may recover as stipulated in the contract for recovery of damages. If extended coverage or other special insurance not herein provided for is required by Subcontractor, Contractor shall, if possible, effect such insurance at Subcontractor's expense by appropriate riders to the Installation Floater and Builder's Risk insurance policy.

Lenox Square, Inc. shall deposit any money received from insurance in an account separate from all its other funds and it shall distribute it in accordance with such agreement as the parties in interest may reach, or under an award of arbitrators in accordance with the Construction Industry Arbitration Rules of the American Arbitration Association. If, after loss, no special agreement is made, replacement of injured work shall be ordered and executed as provided for changes in the work.

The Trustee shall have power to adjust and settle any loss with the insurers unless one of the contractors interested shall object in writing within three working days of the occurrence of loss, and thereupon arbitrators shall be chosen as above. The Trustee shall in that case make settlement with the insurers in accordance with the directions of such arbitrators, who shall also, if distribution by arbitration is required, direct such distribution.

Article XIII. Subcontractor shall immediately remove from the premises, as often as directed, by Contractor, all rubbish, debris, and surplus material which may accumulate from the prosecution of the work covered by this contract, and should Subcontractor fail to do so upon such notice, Contractor may, at its option, after twenty-four (24) hours, cause the same to be removed and charge the expense of such removal to Subcontractor. All materials placed on site shall be at the risk of Subcontractor. All materials placed on site, if placed as approved by the Project Manager, shall be moved as necessary for other work upon three (3) days' notice. Materials placed on the job without any such approval shall be moved immediately where they interfere with other work. Upon failure to move such materials, Contractor may move them at Subcontractor's expense.

Article XIV. None but labor compatible with other labor on the job, shall be employed on work covered by this contract.

Article XV. Claims under this contract or arising out of the breach thereof shall be subject to arbitration in accordance with the Construction Industry Arbitration Rules of the American Arbitration Association.

It is expressly agreed that the arbitration herein provided shall be a condition precedent to any legal action between the parties hereto, and that judgment may be entered in accordance with such findings by any Court having jurisdiction on application therefor by either party.

Article XVI. Contractor shall not be obligated to pay for any work of this trade made necessary by Subcontractor's failure to check all drawings available to him of the work of other trades; except when specific directions have been given him by the Project Manager.

Article XVII. When "Project Manager" is used in this contract, it refers to the Contractor's Project Manager, who may delegate portions of his authority. This contract is administered and interpreted by Contractor's Project Manager.

Article XVIII. Invoices for payment under this contract shall be to Contractor at 3408 Lenox Road, N. E., Atlanta, Georgia 30327.

Article XIX. Acceptance of final payment of this contract, by Subcontractor, shall constitute a release of all other claims of Subcontractor arising out of this contract.

Article XX. Property liens being a source of embarrassment and interference, Subcontractor agrees that no such liens shall be filed prior to completion of the entire project.

Article XXI. Subcontractor at his own risk may make use of installed electrical, gas, and water lines of Contractor without charge; but Contractor assumes no obligation to furnish any of these services or to furnish distribution lines for same, except where specifically provided elsewhere in this contract; provided, however, Subcontractor shall be responsible for and reimburse Contractor for the cost of any wasteful or unnecessary use.

Article XXII. Should any of the work to be performed hereunder be designated "Tenant's Work," Subcontractor agrees as follows:

1. Tenant, in whose behalf such work is done, or its representatives, shall have the right to inspect Subcontractor's books and records pertaining to such work;

2. Subcontractor understands that, in the event of such arbitration, Tenant, in whose behalf such work is done, may conduct Contractor's case in such arbitration.

Article XXIII. Should any provision of this contract be held invalid by any Court, the remaining provisions hereof shall not be affected thereby.

IN WITNESS WHEREOF, the parties hereto have hereunto set their hands and seals, the day and year first above written.

Witness: Joe Butler Date: 1/7/80 LENOX CONTRACTING & ENGINEERING CO., INC.

Joe Butler By _King Royer_
                      King Royer, Secy.

Witness: Jane Smith Date: 1/7/80 _James C. Buford_

Jane Smith   Buford Plumbing Co.

                by James C. Buford, and James C. Buford individually.

INDEX